Foundations of Nanomechanics

Advanced Texts in Physics

This program of advanced texts covers a broad spectrum of topics which are of current and emerging interest in physics. Each book provides a comprehensive and yet accessible introduction to a field at the forefront of modern research. As such, these texts are intended for senior undergraduate and graduate students at the MS and PhD level; however, research scientists seeking an introduction to particular areas of physics will also benefit from the titles in this collection.

Springer
Berlin
Heidelberg
New York
Hong Kong
London
Milan
Paris
Tokyo

Andrew N. Cleland

Foundations of Nanomechanics

From Solid-State Theory to Device Applications

With 215 Figures
and 112 Worked Examples and Exercises

Springer

Prof. Andrew N. Cleland
Department of Physics
University of California
Santa Barbara, CA 93106-9530
USA
e-mail: Cleland@iquest.ucsb.edu

Library of Congress Cataloging-in-Publication Data.

Cleland, Andrew N., 1961–
Foundations of nanomechanics: from solid-state theory to device applications
Andrew N. Cleland
p.cm. ISBN 3-540-43661-8 (alk. paper)
1. Nanotechnology. 2. Solid-state physics. I. Title.
T174.7.C554 2003 620'.5–dc21 2002070543

ISSN 1439-2674

ISBN 978-3-642-07821-7

Springer-Verlag Berlin Heidelberg New York
a member of BertelsmannSpringer Science+Business Media GmbH

http://www.springer.de

Softcover reprint of the hardcover 1st edition 2003

Final page layout: EDV-Beratung Frank Herweg, Leutershausen
Cover design: *design & production* GmbH, Heidelberg

Printed on acid-free paper

Preface

Why write another book on mechanics? There are, after all, a number of excellent texts that describe in great detail the way classical solids behave when acted upon by static and time-varying forces; many of these are cited in this monograph. These texts treat solids as continuous objects, and quantum mechanics does not enter the discussion. Furthermore, the atomic nature of the solid is implicit, but does not enter in a central role. At the other end of the spectrum, texts on condensed matter physics focus on the quantum mechanical nature of the solid; these contain quite clear descriptions of acoustic waves in solids, describing their dynamic and thermal properties, and how they interact with electrons in the solid, but contain little information regarding bulk deformations.

This text, focussing on the mechanics of very small objects, attempts to provide a link between these two approaches; in addition to describing the theories of both the classical and the quantum mechanical solid, I attempt to outline where the classical description breaks down, and quantum mechanics must be applied, to understand the behavior of a nanoscale object. I have tried to merge the continuum description of the solid with the atomic one, and to show how and where quantum mechanics plays a role, especially as the size scale of the system is reduced, making the quantized energy scale *larger* and the role of thermal vibrations more important.

This text is designed to be an introduction for physicists and engineers to the basic foundations of solid mechanics, treating both the static and dynamic theories. We begin with a simplified atomistic description of solids, starting in Chap. 1 with the problem of two atoms in a bound state and then extending the discussion to three and then N atoms in a linear chain. In Chap. 2 we cover the microscopic description of the mechanics of three dimensional insulating solids, from the same basic condensed matter viewpoint, and in Chap. 3 touch on the thermal and transport properties for phonons, the quantized mechanical excitations of a solid. In Chaps. 4–7 we connect the microscopic description to the conventional description of continuum dynamics, introducing the concepts of strain and stress, their interplay and their control of the mechanical function of deformable solids, and then dealing with both static and dynamic problems in continuum mechanics. In Chap. 8 we touch on the topic of dissipation and noise in mechanical systems.

In the last three chapters, Chaps. 9–11, we describe a number of experimental implementations of nanomechanical devices, and also give an outline of the techniques involved in patterning and fabricating nanomechanical structures.

This text is written at an advanced undergraduate or beginning graduate student level. It should also prove useful for the practicing engineer or scientist. The reader is expected to have a good grounding in classical, rigid-body mechanics, covered in most first-year courses. An understanding of vector calculus and linear algebra is assumed, so that the reader should be familiar with divergence and gradient operators, as well as how to invert and diagonalize a matrix, and take a determinant. No background in solid mechanics is assumed, nor is any knowledge of semiconductor processing, the main tool for the fabrication of nanometer-scale devices. Exercises are provided at the end of each chapter, and range in difficulty as the exercise number goes up.

Some notes on symbols: I have chosen to use bold, italic roman symbols for vectors such as $\boldsymbol{u}$, and bold, sans-serif symbols for tensors such as R. Scalars are written in italic roman type as in G, and components of vector or tensors written as italic roman with roman subscripts such as R_{ij}. A list of commonly used symbols, and their units and conversion in the Systéme Internationale (SI) and centimeter-gram-second (CGS) systems appears at the end of the book.

I would like to express my appreciation to those who read and commented on sections of the unfinished manuscript, including Dr. Robert Knobel, Derek Barge, and Kang-kuen Ni. I would also like to express my thanks to my wife Ning and my children Agnetta and Nicholas for their patience while I was working on this book.

Santa Barbara and Los Angeles
July, 2002

Andrew N. Cleland

Contents

1. Introduction: Linear Atomic Chains

This book is aimed at developing a coherent description of what we term "nanomechanics", the mechanical behavior of nanometer-scale objects, that is, objects for which at least one dimension is significantly less than 1 μm. We will not concern ourselves with rigid body dynamics, those related to the motion of an object's center of mass, and the rotation of the object about its center of mass; these are quite well described in a large number of texts, for both the classical and quantum mechanical limits. Instead we will focus on the static and dynamic deformations of solid objects, both with and without external forces. The self-resonant vibrational modes determine some of the thermodynamic properties of solids, such as the heat capacity and thermal conductivity; we touch on this connection as well. In the smallest structures, quantum mechanics must be applied to the description of the motion.

In the first part of this text, we provide the formalism required to understand these aspects in a unified manner, and we present a number of simple examples showing how the small size scale can impact and alter the bulk properties. In the second part of the text, we describe some approaches to fabricating and measuring the properties of nanoscale objects, providing some examples of actual devices.

We begin this chapter by discussing a very simple mechanical problem, that of a molecule of two atoms bound together by their mutual interaction. We restrict the motion of the atoms to one dimension, along the line connecting the atoms, and consider the response to external forces, and then work out the molecule's natural vibrational resonance frequency. Next we move to a similar problem with three atoms, and then to the more general problem of the N atom chain. Following this heuristic discussion, we enter into a discussion of the quantum mechanical description of these systems, and conclude with the theory for the thermodynamic properties of the N atom chain.

1.1 A Model Binary Molecule: Two Atoms

Consider a molecule consisting of just two atoms. We restrict the motion of the two atoms to one dimension, along the line connecting them, so the atoms can only move directly towards or away from one another. We assume that there is a net attractive force between the atoms; this may be due to an electrostatic attraction, if the atoms have opposite electric charges, to a chemical covalent bonding, where the outermost electrons are shared between the atoms, or to an attraction known as the *van der Waals* force, generated dynamically by the mutually induced dipole moments in each atom. At any rate, we assume that there is a force $f(r)$ between the two atoms, that depends only on the distance r between them. If this force is purely attractive, the atoms will accelerate and merge with one another. This however does not occur, because when the atoms get too close to one another, the electrons surrounding each atomic nucleus repel one another through their electrostatic interaction, and are furthermore limited by a fundamental law of quantum mechanics from occupying the same volume of space. The attractive force therefore becomes repulsive when the atoms get very close to one another.

As we shall see, it is often more useful to deal with the interaction potential energy $\phi(r)$ rather than the force $f(r)$, which is defined through the relation

$$f(r) \equiv -\frac{d\phi}{dr}. \tag{1.1}$$

Note that this differential equation defines the potential energy only up to an additive constant, whose value is arbitrary and has no physical meaning. The potential energy can also be described as the negative of the work done by the force for a displacement $r - r_0$ from the point of zero potential energy r_0, or

$$\phi(r) = -W = -\int_{r_0}^{r} f(r)\, dr, \tag{1.2}$$

which is equivalent to (1.1).

As a specific example, we will consider the model known as the *Lennard–Jones* interaction, which applies to atoms interacting through the van der Waals interaction. The interaction potential energy for the Lennard–Jones model has the algebraic form

$$\phi(r) = -\frac{A}{r^6} + \frac{B}{r^{12}}, \tag{1.3}$$

with the parameter A determining the strength of the attractive interaction, and B the repulsive interaction. The attractive $1/r^6$ dependence is characteristic of the van der Waals interaction, while the repulsive $1/r^{12}$ dependence is somewhat phenomenological. The repulsive interaction is strongest for small r, with the atoms close together, and decreases more rapidly with r than the attractive interaction, which therefore dominates for large r. The zero for the potential energy $\phi(r)$ is chosen so that the energy is zero when the atoms are infinitely far apart. The force corresponding to (1.3) is

$$f(r) = 6\,\frac{A}{r^7} - 12\,\frac{B}{r^{13}}. \tag{1.4}$$

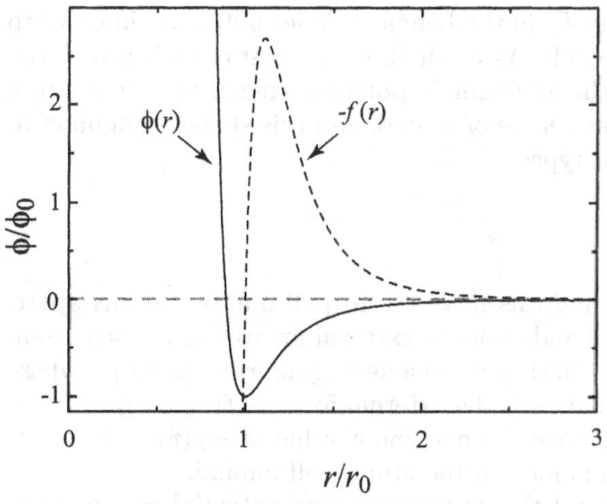

Fig. 1.1. Lennard–Jones model interaction force $-f(r)$ and potential energy $\phi(r)$, as a function of distance r/r_0. Note we have plotted the negative of the force, so that a positive value corresponds to an attractive force. The vertical axis is for the potential, in units of the potential at the minimum, $\phi_0 = -\phi(r_0)$.

In Fig. 1.1 we have plotted the Lennard–Jones potential $\phi(r)$ and the force $-f(r)$, as a function of distance r, in units of the equilibrium distance $r_0 = (2B/A)^{1/6}$. At $r = r_0$, the force is zero, and equivalently, the potential energy has a minimum. The *binding energy* E_b, the difference between the potential energy minimum and that when the atoms are infinitely far apart, is given by $E_b = \phi(\infty) - \phi(r_0) = A^2/4B$. This amount of energy must be transferred to the atoms in order to break the bond between them.

The Lennard–Jones potential gives an excellent, quantitative description of the interactions between noble gas atoms, such as in argon, krypton and xenon. In argon, for example, the equilibrium spacing in the Ar_2 molecule is found to be $r_0 = 3.8$ Å, and the binding energy is $E_b = 10.4$ meV $= 1.7 \times 10^{-21}$ J [1]. Equivalently, we can write down the constants $A = 2r_0^6 E_b = 63$ Å^6-eV and $B = r_0^{12} E_b = 9.4 \times 10^4$ Å^{12}-eV.

The binding energy for argon is less than the thermal energy scale at room temperature, $k_B T = 26$ meV. Solid Ar, formed through multiple bonds of the kind we have just described, therefore only forms at quite low temperatures, below 100 K. The van der Waals interaction, with typical binding energies in the 10-100 meV range, is particularly weak; while the equilibrium distance r_0 is typical for all atomic interactions, the binding energy for the much stronger ionic (electrostatic), metallic and covalent interactions in typical solids is in the range of one to several tens of electron volts ($\sim 10^{-19} - 10^{-18}$ J), rather than a few meV. These types of bonds are however not as simply parameterized as the Lennard–Jones interaction, but require a more sophisticated approach. However, meaningful results can be obtained by simply increasing the binding energy E_b in the Lennard–Jones potential, in order to mimic these other types of bonds: As we shall see, in most cases it is only the shape of the potential near the minimum in potential energy that determines the important mechanical properties of a solid, and this shape is common to almost all the different bond types.

1.1.1 External Forces

We can now imagine what happens if we try to pull our two atoms apart. Let's say that we apply equal and opposite external forces f_{ext} to each atom. The atoms will move apart until they reach a new equilibrium point r_0', where their attractive interaction balances the external force, $-f(r_0') = f_{\text{ext}}$. If the external force is too large, above the maximum value of $-f(r)$ in Fig. 1.1, there will be no equilibrium point and the atoms will unbind.

Another way to understand this is to define the potential energy associated with the external force, $\phi_{\text{ext}}(r) = -f_{\text{ext}}r$. Note that the zero for the external potential is chosen at $r = 0$. Note there is no factor of two in this expression: Both atoms are acted upon by the external force, but each atom is only displaced from $r = 0$ by $r/2$, so the work done by the force, which is the negative of the potential energy, is $W = 2 \times f_{\text{ext}}r/2$.

The total potential energy is then $U_{\text{tot}} = \phi(r) + \phi_{\text{ext}}(r)$. For $f_{\text{ext}} = 0$, the total potential energy is the same as the interaction potential. For small f_{ext}, the minimum for the total potential U_{tot} will shift to the new equilibrium point r_0'; for f_{ext} too large, no minimum occurs. In Fig. 1.2, we show a family of potential energy curves for different external forces, showing how the minimum energy point moves away from r_0 until it disappears at large enough f_{ext} (see Exercise 1.2).

Note that as soon as we apply the external force, the energy minimum at r_0' becomes *metastable*; the atoms can achieve a lower total energy if they can cross over the potential barrier and escape to infinity. This provides interesting questions in the case where we allow the atoms to have non-zero temperature, so that there is a certain probability that they can be thermally activated over the potential barrier; the same question occurs when we consider quantum mechanical *tunnelling*, through the barrier, also allowing the atoms to escape.

In addition to the question of binding, it is often useful to know how a solid, or in our case, the two atoms, respond to very weak forces, such that the atoms only displace a very small amount from their equilibrium positions, from r_0 to r_0'. We can use our model Lennard–Jones interaction to see how this works. For a very weak force f_{ext}, the very small shift in the equilibrium point allows us to approximate the interaction potential by using a *Taylor series* expansion of the potential:

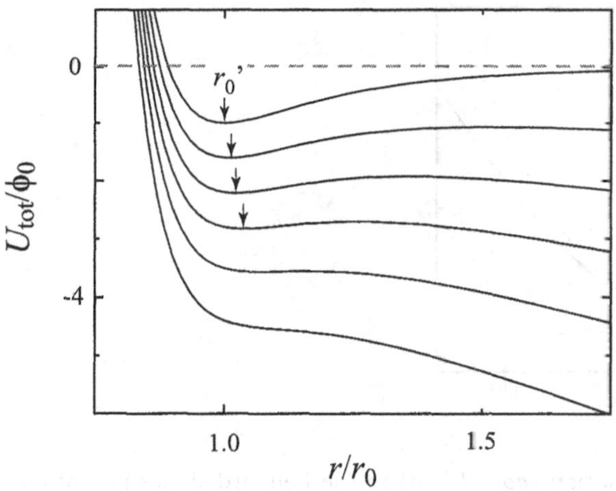

Fig. 1.2. Total potential energy U_{tot} for the Lennard–Jones potential in the presence of a constant external force; the family of curves is for external forces ranging from zero (*top*) to a force larger than the maximum Lennard–Jones binding force (*bottom*). The arrows indicate the new equilibrium point r_0' for each value of the external force. Vertical axis is in units of the interaction potential at the minimum point.

$$\begin{aligned}\phi(r) &= \phi(r_0) + \left.\frac{\mathrm{d}\phi}{\mathrm{d}r}\right|_{r_0}(r-r_0) + \frac{1}{2!}\left.\frac{\mathrm{d}^2\phi}{\mathrm{d}r^2}\right|_{r_0}(r-r_0)^2 \\ &\quad + \frac{1}{3!}\left.\frac{\mathrm{d}^3\phi}{\mathrm{d}r^3}\right|_{r_0}(r-r_0)^3 + \ldots \\ &\approx \phi(r_0) + \frac{1}{2}\left.\frac{\mathrm{d}^2\phi}{\mathrm{d}r^2}\right|_{r_0}(r-r_0)^2, \end{aligned} \tag{1.5}$$

where in the second, approximate equality, we have used the fact that $\mathrm{d}\phi/\mathrm{d}r(r_0) = 0$, and we have dropped the higher order terms in the Taylor expansion. We are thus left with a *harmonic potential* approximation for the interaction, that depends quadratically on the square of the displacement $u = r - r_0$ from equilibrium.

For the Lennard–Jones potential, the curvature is given in terms of the equilibrium spacing and binding energy by

$$\left.\frac{\mathrm{d}^2\phi}{\mathrm{d}r^2}\right|_{r_0} = 72\,\frac{E_b}{r_0^2}. \tag{1.6}$$

In Fig. 1.3 we show the harmonic approximation to the Lennard–Jones potential; the approximation is seen to work well for very small displacements from equilibrium, but rapidly fails as one moves further away.

In the presence of a weak external force, the equilibrium point shifts to where $\mathrm{d}U_{tot}/\mathrm{d}r = 0$; using the expansion (1.5) for the interaction potential, this is when

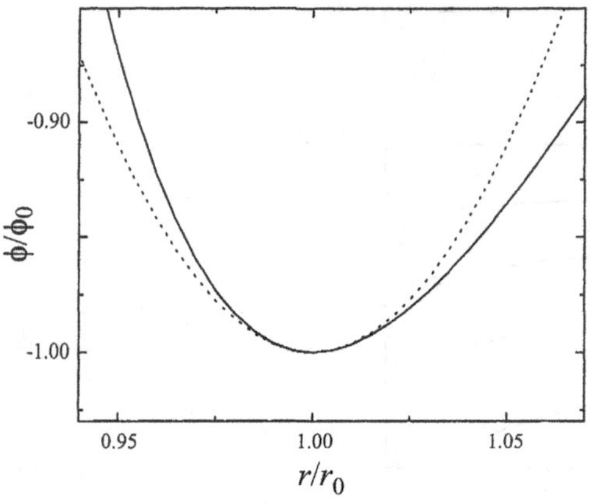

Fig. 1.3. Harmonic approximation (*dotted line*) to the Lennard–Jones potential.

$$-f_{\mathrm{ext}} + \left.\frac{\mathrm{d}^2\phi}{\mathrm{d}r^2}\right|_{r_0}(r - r_0) = 0, \tag{1.7}$$

or

$$u \equiv r - r_0 = \frac{1}{\mathrm{d}^2\phi/\mathrm{d}r^2}\, f_{\mathrm{ext}} = \frac{1}{k}\, f_{\mathrm{ext}}. \tag{1.8}$$

We thus find that the displacement u from equilibrium, for small forces f_{ext}, is linear in the external force, with a response in inverse proportion to the curvature of the interaction potential $\phi(r)$. The interatomic potential is therefore equivalent to a spring with spring constant k, with restoring force proportional to the displacement. Applied to argon, the curvature, or equivalently the spring constant, is given by $\mathrm{d}^2\phi/\mathrm{d}r^2(r_0) = 52$ meV/Å^2, with an equivalent spring constant of $k = 0.83$ N/m.

The linear response for small displacements u is a generic property of almost all solids, and holds for complex single-crystal materials as well as for amorphous solids made of plastics and proteins.

1.1.2 Dynamic Motion

We now consider the dynamical behavior of the atoms in our model interaction potential. What this means is that we will allow the atoms to move, so that they have a kinetic energy T in addition to the potential energy $U = \phi(r)$.

We assume that the *center of mass* of our system remains at rest. With the atoms at distances r_1 and r_2 from the origin, their separation is $r = r_2 - r_1$. If the atoms have masses M_1 and M_2, the location of the center of mass r_{cm} is the weighted sum of their locations, or

$$r_{\mathrm{cm}} \equiv \frac{M_1 r_1 + M_2 r_2}{M_1 + M_2}. \tag{1.9}$$

The atom positions can be written in terms of $r_{\rm cm}$ and r as

$$\left.\begin{aligned} r_1 &= r_{\rm cm} - \frac{M_2}{M_1+M_2} r, \\ r_2 &= r_{\rm cm} + \frac{M_1}{M_1+M_2} r. \end{aligned}\right\} \tag{1.10}$$

If the center of mass is at rest, so $\dot{r}_{\rm cm} = 0$ (where we use the notation $\dot{r} = {\rm d}r/{\rm d}t$), then the atom velocities must satisfy

$$M_1 \dot{r}_1 = -M_2 \dot{r}_2. \tag{1.11}$$

The kinetic energy can then be written

$$\begin{aligned} T &= \frac{1}{2} M_1 \dot{r}_1^2 + \frac{1}{2} M_2 \dot{r}_2^2 \\ &= \frac{1}{2} \mu \dot{r}^2, \end{aligned} \tag{1.12}$$

using the *reduced mass* $1/\mu = 1/M_1 + 1/M_2$. With the momentum $p = \mu \dot{r}$, the kinetic energy is

$$T = \frac{p^2}{2\mu}. \tag{1.13}$$

The Hamiltonian for the system, $H = T + U$, is then

$$H = \frac{1}{2\mu} p^2 + \phi(r), \tag{1.14}$$

and Hamilton's equations of motion (see e.g. Goldstein [2]) then yield

$$\mu \ddot{r} = -\frac{{\rm d}\phi}{{\rm d}r}(r) = f(r). \tag{1.15}$$

Let's again consider only very small displacements $u = r - r_0$ from the equilibrium spacing r_0. Using the Taylor expansion (1.5) for the interaction potential, the equation of motion for $u(t)$ is easily shown to be

$$\mu \ddot{u} = -\left.\frac{{\rm d}^2\phi}{{\rm d}r^2}\right|_{r_0} u. \tag{1.16}$$

This is the equation of motion of a *simple harmonic oscillator*, and has the usual harmonic solution of the form

$$u(t) = u_0 \cos(\omega_0 t + \varphi), \tag{1.17}$$

where the resonance frequency ω_0 is given by

$$\omega_0 = \sqrt{\frac{1}{\mu} \frac{{\rm d}^2\phi}{{\rm d}r^2}}, \tag{1.18}$$

and the amplitude u_0 and phase φ are determined by the initial conditions. A convenient shorthand for writing harmonic solutions of the form (1.17) is to use complex exponential notation, of the form

$$u(t) = u_0 e^{-i\omega_0 t}, \tag{1.19}$$

where the amplitude u_0 can be complex, to allow for the phase factor φ, and the actual solution is obtained by taking the real part of (1.19).

We see that the natural resonance frequency ω_0 of our system is determined by the reduced mass μ and by the curvature of the interaction potential $d^2\phi/dr^2$, just as the displacement due to a static external force, (1.8), is inversely proportional to the same curvature. This close relation is due to our use of the harmonic approximation for the potential; another way to represent the interaction in this approximation is to think of the atoms as being linked by a simple linear spring, with spring constant k given by

$$k = \left. \frac{d^2\phi}{dr^2} \right|_{r_0}. \tag{1.20}$$

The equilibrium length of the spring is r_0, and the potential energy for separation r is thus $U = k(r - r_0)^2/2 = ku^2/2$, the same as for the harmonic approximation, with the trivial change that our zero of energy is now at the equilibrium point r_0 rather than at infinite separation. Both the response to weak external forces, and the natural dynamic resonant response (1.15), are captured by this simple model, shown schematically in Fig. 1.4.

For the Lennard–Jones potential, and two argon atoms, with masses $M_1 = M_2 = 2\mu = 6.6 \times 10^{-23}$ g, we have already calculated a spring constant $k = 52$ meV/Å^2 $= 0.83$ N/m; we now find a natural resonance frequency $\omega_0/2\pi = 0.8$ THz. This is somewhat low for a mechanical atomic resonance frequency; the shallow van der Waals interaction has a gentle curvature, reducing the frequency from that for covalently or ionically bonded atoms, which typically have resonance frequencies of order 10 THz $= 10^{13}$ Hz. The spring constants for such bonds are correspondingly larger, with k in the range of 10-100 N/m.

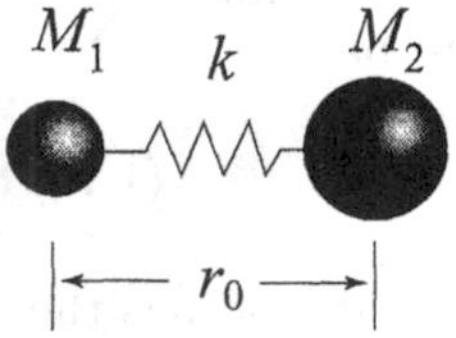

Fig. 1.4. Spring model for the interaction potential between two atoms.

1.2 The Three Atom Chain

We now add a third atom to our system, and for simplicity assume all the atoms are identical, with mass M. We arrange the atoms in a line, and again restrict motion to one dimension, along the line of atoms. We further assume that the force which holds the three atoms together is a *two-particle* interaction, depending only on the positions r_1 and r_2 of, say, atoms 1 and 2. Hence we write the interaction potential energy as $\phi(r_1, r_2)$. Furthermore, we assume the potential is a function only of the distance between the atoms, so that $\phi(r_1, r_2) = \phi(r_2 - r_1)$. For three identical atoms, the total potential energy U in the absence of any external forces is then given by

$$U(r_1, r_2, r_3) = \phi(r_2 - r_1) + \phi(r_3 - r_2) + \phi(r_1 - r_3). \tag{1.21}$$

We will further assume that the potential interaction is very short-range, so that we only need include the interactions between atoms adjacent to one another; for a chain with the atoms in the sequence $(1, 2, 3)$, as shown in Fig. 1.5, this means we can drop the third term in (1.21).

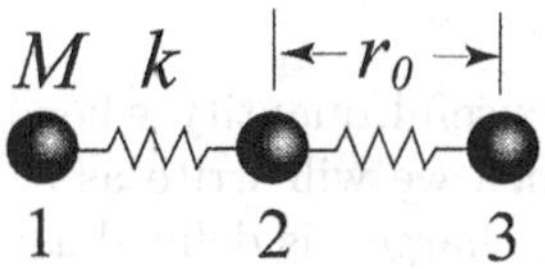

Fig. 1.5. Spring model for the three atom chain.

Our final simplification will be to replace each potential interaction by its harmonic approximation, so that for atoms 2 and 3,

$$\phi(r = r_3 - r_2) = \frac{1}{2} \left.\frac{\mathrm{d}^2\phi}{\mathrm{d}r^2}\right|_{r_0} (r - r_0)^2, \tag{1.22}$$

choosing the zero of energy at the equilibrium separation $r_3 - r_2 = r_0$.

As discussed above, the response of the two-atom potential to both weak external forces and for small dynamic motion is entirely equivalent to that when the atoms are connected by a linear spring with spring constant k, whose value is given by the curvature of $\phi(r)$. We therefore find that our simplified, one-dimensional three atom system now consists of three masses M connected by linear springs k, as shown in Fig. 1.5.

1.2.1 External Force on a Trimer

We first consider what happens when we pull on either end of the chain with equal and opposite external forces f_{ext}, as shown in Fig. 1.6. By symmetry, atom 2, in the middle of the chain, does not move. The two end atoms 1 and 3 will be displaced until the restoring force provided by their interaction with atom 2 matches that of the external force. The displacement u_1 of atom 1 from its rest point is therefore identical to that obtained when pulling on the two atom chain,

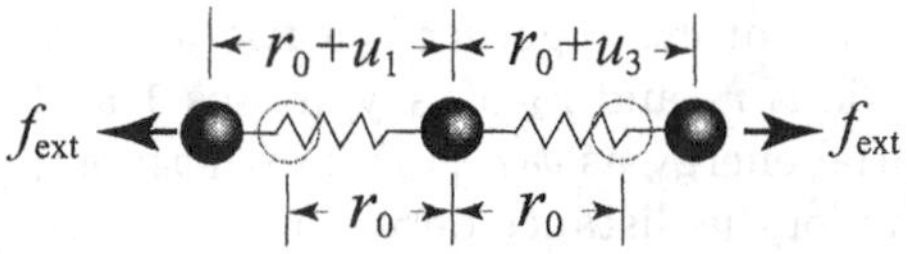

Fig. 1.6. Three atom chain stretched by an external force f_{ext}.

$$\begin{aligned} u_1 &= r_2 - r_1 - r_0 = f_{\text{ext}}/k \\ &= \frac{1}{\mathrm{d}^2\phi/\mathrm{d}r^2} f_{\text{ext}}, \end{aligned} \tag{1.23}$$

with an identical expression for $u_3 = r_3 - r_2 - r_0$.

1.2.2 Strain in the Trimer

We are now in a position to define the *strain*, a very useful quantity when describing distortions of solid objects. The strain, which we will write as e (we will be careful to distinguish this from the electron charge), is defined as the fractional displacement of a point in the solid from its rest point. For our three-atom chain, the strain e between atoms 1 and 2 is then

$$e = \frac{u_1}{r_0} = \frac{f_{\text{ext}}}{kr_0}. \tag{1.24}$$

The strain between atoms 3 and 2 is the same, $e = u_3/r_0 = f_{\text{ext}}/kr_0$. The strain is a dimensionless quantity, and for a uniform external force applied to a uniform object, will be constant through the object.

1.2.3 Dynamic Motion

We now briefly explore the dynamic behavior of our three atom chain; we will assume, as we did with the two-atom system, that the center of mass of the system is at rest. The dynamic motion can be described by the relative displacements u_1, u_2 and u_3 of the atoms from their rest points; the center of mass condition is then, for equal mass atoms,

$$\dot{u}_1 + \dot{u}_2 + \dot{u}_3 = 0. \tag{1.25}$$

The momentum associated with each atom is $p_n = M\dot{u}_n$.

The kinetic energy of the three atoms, and their (nearest-neighbor) spring interactions, yield the Hamiltonian

$$H = T + U = \frac{1}{2M}\left(p_1^2 + p_2^2 + p_3^2\right) + \frac{k}{2}(u_1 - u_2)^2 + \frac{k}{2}(u_2 - u_3)^2. \tag{1.26}$$

The corresponding equations of motion are

$$\left.\begin{aligned} M\ddot{u}_1 &= k(u_2 - u_1), \\ M\ddot{u}_2 &= k(u_1 - 2u_2 + u_3), \\ M\ddot{u}_3 &= k(u_2 - u_3). \end{aligned}\right\} \tag{1.27}$$

To solve these coupled equations, we look for the *normal modes* of the system, the solutions where all the degrees of freedom (u_1, $\dot{u}_2$ and u_3) have the same, harmonic, time dependence. Using the convenient exponential notation, this means we look for solutions of the form

$$u_n = A_n e^{-i\omega t}, \qquad (n = 1 \text{ to } 3) \tag{1.28}$$

where the common frequency ω and the (complex) amplitudes A_n are still to be determined. Inserting these solutions in (1.27), we find the linear system of equations

$$\left.\begin{aligned} -M\omega^2 A_1 &= k(A_2 - A_1), \\ -M\omega^2 A_2 &= k(A_1 - 2A_2 + A_3), \\ -M\omega^2 A_3 &= k(A_2 - A_3), \end{aligned}\right\} \tag{1.29}$$

where we have cancelled out the common time dependence. Defining the frequency $\omega_0 = (k/M)^{1/2}$, the system of equations (1.29) can be written as an *eigenvalue-eigenvector equation,*

$$\left(\frac{\omega}{\omega_0}\right)^2 \begin{bmatrix} A_1 \\ A_2 \\ A_3 \end{bmatrix} = \begin{bmatrix} 1 & -1 & 0 \\ -1 & 2 & -1 \\ 0 & -1 & 1 \end{bmatrix} \begin{bmatrix} A_1 \\ A_2 \\ A_3 \end{bmatrix}. \tag{1.30}$$

The eigenvalues are the three values of ω/ω_0 that yield non-trivial solutions to this equation, and the eigenvectors the set of amplitudes (A_1, A_2, A_3) that correspond to each eigenvalue; we refer the reader unfamiliar with this approach to a text on linear algebra, or a book on mathematical physics, such as that of Arfken [3] or Morse and Feshbach [4].

With $z = (\omega/\omega_0)^2$, we subtract the term on the left from both sides, leaving the equation

$$\begin{bmatrix} 1-z & -1 & 0 \\ -1 & 2-z & -1 \\ 0 & -1 & 1-z \end{bmatrix} \begin{bmatrix} A_1 \\ A_2 \\ A_3 \end{bmatrix} = 0. \tag{1.31}$$

This has a non-trivial solution only if the determinant of the matrix is zero:

$$\det \begin{vmatrix} 1-z & -1 & 0 \\ -1 & 2-z & -1 \\ 0 & -1 & 1-z \end{vmatrix} = -z^3 + 4z^2 - 3z = 0. \tag{1.32}$$

This is known as the *characteristic equation.* The solutions to this cubic equation are the set $z = (\omega/\omega_0)^2 = 0$, 1 and 3.

Let us deal first with the non-zero solutions. If we take the solution $z = 1$, i.e. $\omega = \omega_0$, the corresponding eigenvector (A_1, A_2, A_2) is obtained from

(1.31). We find that the amplitudes must satisfy $A_1 = -A_3$ and $A_2 = 0$. The overall amplitude of the eigenvector is arbitrary; the eigenvector equation (1.31) only tells us the *relative* amplitudes of motion for the three atoms. It is convenient to make the length of the eigenvector (A_1, A_2, A_3) equal to unity, that is, to *normalize* the solution, so we find the solution pair

$$\omega = \omega_0 \;\leftrightarrow\; \begin{bmatrix} A_1 \\ A_2 \\ A_3 \end{bmatrix} = \begin{bmatrix} 1/\sqrt{2} \\ 0 \\ -1/\sqrt{2} \end{bmatrix} . \tag{1.33}$$

The second non-zero frequency is $z = 3$ or $\omega = \sqrt{3}\,\omega_0$; we find the corresponding eigenvector must satisfy $A_2 = -2A_1 = -2A_3$, so the normalized solution is

$$\omega = \sqrt{3}\omega_0 \;\leftrightarrow\; \begin{bmatrix} A_1 \\ A_2 \\ A_3 \end{bmatrix} = \begin{bmatrix} -1/\sqrt{6} \\ 2/\sqrt{6} \\ -1/\sqrt{6} \end{bmatrix} . \tag{1.34}$$

The final frequency is $\omega = 0$. Inserting this solution in (1.31), we find the corresponding eigenvector $(A_1, A_2, A_3) = (1/\sqrt{3}, 1/\sqrt{3}, 1/\sqrt{3})$. This last solution, with equal values for all the displacement amplitudes A_n, is actually somewhat special: The zero frequency solution corresponds to an overall motion of the center of mass, where the actual time dependence is not an exponential $\mathrm{e}^{-\mathrm{i}\omega t}$, but instead has the form $u_n = at + b$, corresponding to the uniform translation of the chain at constant velocity. This can be seen by inserting this solution into the original system of equations (1.27); however, it violates our condition on the center of mass, (1.25), unless the velocity a is zero. We therefore ignore this now trivial solution; we have, after all, found

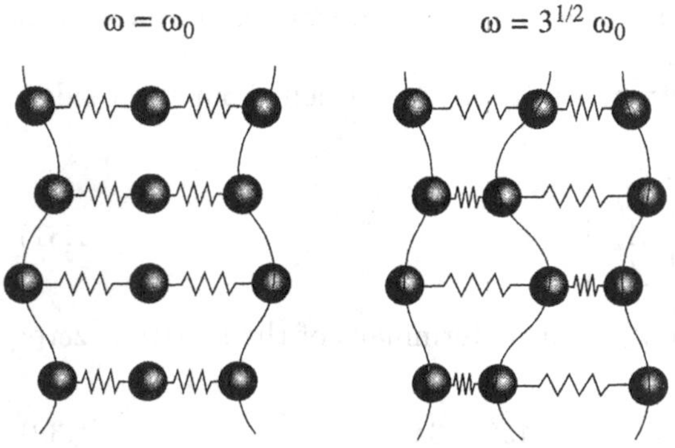

Fig. 1.7. The two normal modes for the three atom chain. The time axis runs vertically; note that if the time axis were proportional, the oscillations for the higher frequency mode on the right would be three times faster than for that on the left. The relative amplitudes and phase of motion for the three atoms are however correct.

two independent solutions to a system with two independent degrees of freedom, the third degree of freedom having been removed by our condition on the center of mass motion.

We note that if we had used the center of mass equation to eliminate one degree of freedom, say u_2, in terms of the other two, and then written the corresponding Hamiltonian and two equations of motion, we would have obtained the same two (nontrivial) eigenfrequencies and eigenvectors found above.

In Fig. 1.7 we illustrate the two normal modes, (1.33) and (1.34).

1.3 An N-mer Molecule: The N Atom Linear Chain

We now turn to an N-atom one-dimensional chain, where the atoms are connected by linear springs k with equilibrium spacing r_0; the atoms all have mass M (see Fig. 1.8). The nth atom will have relative displacement u_n from its equilibrium point, so the potential energy in the spring connecting it to its $(n+1)$th neighbor is $\phi = k(u_n - u_{n+1})^2/2$. Motion is restricted, as before, to along the line of atoms.

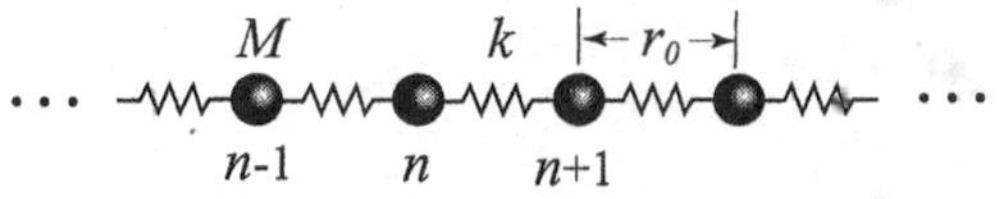

Fig. 1.8. Model for the N atom chain, with masses M connected by springs k. Equilibrium spacing is r_0.

If we apply a force f_{ext} to the ends of the chain, we find the usual result that each atom displaces by an amount

$$u_n = f_{\text{ext}}/k. \tag{1.35}$$

The strain is therefore again uniform, with value $e = f_{\text{ext}}/kr_0$.

We now look at the dynamic behavior of the chain, and try to find the normal mode solutions. The potential energy U of the whole chain of N atoms, summing over atoms n, is given by

$$U = \frac{1}{2}\sum_{n=1}^{N-1} k(u_{n+1} - u_n)^2, \tag{1.36}$$

and the total kinetic energy T, in terms of the momenta $p_n = M\dot{u}_n$, by the sum

$$T = \frac{1}{2M}\sum_{n=1}^{N} p_n^2. \tag{1.37}$$

Hamilton's equations of motion are given by

$$\left.\begin{array}{rcl} M\ddot{u}_1 & = & k(u_2 - u_1), \\ M\ddot{u}_n & = & k(u_{n+1} - 2u_n + u_{n-1}) \quad (n = 2 \text{ to } N-1), \\ M\ddot{u}_N & = & k(u_{N-1} - u_N). \end{array}\right\} \tag{1.38}$$

We notice from (1.38) that the atoms $n = 1$ and $n = N$ need special treatment, as they only have one neighboring atom. This presents an annoying aspect of the problem, and dealing with it subtracts somewhat from the message we are trying to convey. We therefore take this problem and transform it somewhat: We connect the atom $n = 1$ to the atom $n = N$ by a spring k, so that effectively the chain of atoms is now a ring, shown in Fig. 1.9. Performing this sleight of hand is known as applying *periodic boundary conditions.*

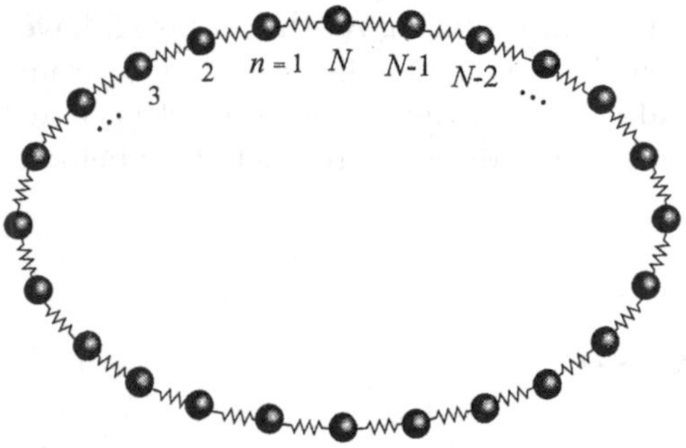

Fig. 1.9. Ring of N atoms, with atom 1 connected to atom N.

Our new equations of motion, with this modification, are

$$\left.\begin{array}{rcl} M\ddot{u}_1 & = & k(u_2 - 2u_1 + u_N), \\ M\ddot{u}_n & = & k(u_{n+1} - 2u_n + u_{n-1}) \quad (n = 2 \text{ to } N-1), \\ M\ddot{u}_N & = & k(u_{N-1} - 2u_N + u_1). \end{array}\right\} \tag{1.39}$$

The eigenmodes of the chain are those where all the atoms move at the same frequency,

$$u_n = \frac{A_n}{\sqrt{N}} \mathrm{e}^{-\mathrm{i}\omega t}, \tag{1.40}$$

where A_n is the amplitude associated with the nth atom, and has units of length; the factor $1/\sqrt{N}$ will prove convenient later.

Inserting this in the system of equations (1.38), we find the eigenvector-eigenvalue equation

$$\left(\frac{\omega}{\omega_0}\right)^2 \begin{bmatrix} A_1 \\ A_2 \\ A_3 \\ A_4 \\ \vdots \end{bmatrix} = \begin{bmatrix} 2 & -1 & 0 & 0 & \dots \\ -1 & 2 & -1 & 0 & \dots \\ 0 & -1 & 2 & -1 & \dots \\ 0 & 0 & -1 & 2 & \dots \\ \vdots & \vdots & \vdots & & \end{bmatrix} \begin{bmatrix} A_1 \\ A_2 \\ A_3 \\ A_4 \\ \vdots \end{bmatrix}. \tag{1.41}$$

Using the assumption of normal modes has transformed the coupled differential system of equations (1.38) into an N-dimensional linear system (1.41). This is a rather horrendous system to solve. We can however use the fact that the chain of atoms is extremely regular, with each atom equivalent to its neighbors (this holds for the atoms 1 and N as well, due to our joining them together). This *translational symmetry* allows us to apply a version of a theorem known as *Bloch's theorem*, about which we will say more later. This theorem says that we can look for normal modes that have the form

$$u_n = \frac{A}{\sqrt{N}}\, \mathrm{e}^{\mathrm{i}qr_n}\, \mathrm{e}^{-\mathrm{i}\omega t}, \tag{1.42}$$

where A is the overall amplitude (with dimensions of distance) and r_n the equilibrium position of the nth atom, $r_n = nr_0$. The spatial dependence of the mode, that is, the dependence on the index n, is sinusoidal, with *wavevector* q, or wavelength $\lambda = 2\pi/q$. This is equivalent to saying that the amplitudes A_n in the normal modes (1.40) depend on the index n through $A_n = A\mathrm{e}^{\mathrm{i}qr_n}$. As we shall see, the equations of motion will force a relationship between the frequency ω and the wavevector q, although the amplitude A is arbitrary, as it is for the three-atom chain.

Inserting the form (1.42) into the equations of motion (1.39), we find a set of three equations,

$$\left.\begin{aligned} -\omega^2 &= \omega_0^2 \left(\mathrm{e}^{\mathrm{i}qr_0} + \mathrm{e}^{\mathrm{i}(N-1)qr_0} - 2\right), \\ -\omega^2 &= \omega_0^2 \left(2\cos qr_0 - 2\right) \\ -\omega^2 &= \omega_0^2 \left(\mathrm{e}^{-\mathrm{i}qr_0} + \mathrm{e}^{-\mathrm{i}(N-1)qr_0} - 2\right). \end{aligned}\right\} \tag{1.43}$$

This set of three equations for the two unknowns ω and q appears to overdetermine the problem. However, the third equation is simply the complex conjugate of the first, and as the frequency ω is real (not complex), it merely emphasizes that the right side of these equations must also be real. Hence we find that q must be given by

$$q = \frac{2\pi}{r_0}\frac{m}{N}, \tag{1.44}$$

for integer values m; we will write these discrete values as q_m. With this condition, the first and third equations are the same as the middle equation, which gives us the relation between ω_m and q_m:

$$\begin{aligned} \frac{\omega_m}{\omega_0} &= \sqrt{2 - 2\cos q_m r_0} \\ &= 2\left|\sin\frac{q_m r_0}{2}\right|. \end{aligned} \tag{1.45}$$

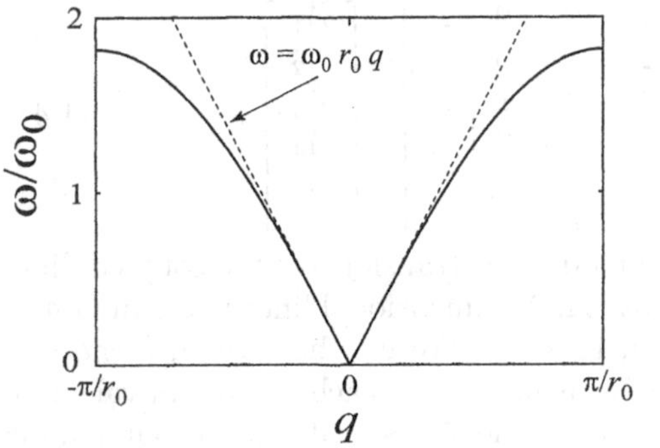

Fig. 1.10. The eigenfrequency solution $\omega(q)$ of the simple linear chain. The dotted line shows the linear solution at small wavevector q Note that the values of q are actually discrete, but for a large number N of atoms, the spacing is very fine, and the discreteness can be ignored.

In Fig. 1.10 we display this solution, along with the linear approximation that applies for small values of the wavevector q. We have only plotted the solution for $-\pi/r_0 < q < \pi/r_0$, and we have ignored the discreteness in the values of q specified by (1.44).

The values of q are discrete, given by the set q_m, because they relate to the motion of a finite number of atoms; there cannot be more than N values of q. The spacing of the values of q is set by (1.44), but this equation does not limit the range of q. However, if we examine the Bloch form (1.42), we see that values of q larger than π/r_0, and smaller than $-\pi/r_0$, do not represent physically different displacements. If we take q larger than π/r_0, the displacements u_n are the same as if we take $q' = q - 2\pi/r_0$:

$$\begin{aligned} u_n &= \frac{A}{\sqrt{N}} \mathrm{e}^{\mathrm{i}qnr_0} \, \mathrm{e}^{-\mathrm{i}\omega t} \\ &= \frac{A}{\sqrt{N}} \mathrm{e}^{\mathrm{i}q'nr_0 + \mathrm{i}2\pi n} \, \mathrm{e}^{-\mathrm{i}\omega t} \\ &= \frac{A}{\sqrt{N}} \mathrm{e}^{\mathrm{i}q'nr_0} \, \mathrm{e}^{-\mathrm{i}\omega t}. \end{aligned} \tag{1.46}$$

A similar argument applies if $q < -\pi/r_0$, where we would take $q' = q + 2\pi/r_0$. It is therefore not physically meaningful to take q outside the range $-\pi/r_0 < q < \pi/r_0$. From the relation (1.44), we therefore see that the index m can range from $-N/2$ to $N/2$, yielding N distinct values for q, as desired.

The solutions (1.42) are written using complex notation; the actual displacements are of course real, and are obtained by taking the real part of (1.42).

The most general form for the displacement for the nth atom, in other words for motion that is not restricted to a single normal mode, is built by superposing all the normal modes with arbitrary amplitude and relative phase,

$$u_n = \frac{1}{\sqrt{N}} \sum_{m=-N/2}^{N/2} |A_m| \cos(q_m r_n - \omega_m t + \varphi_m). \tag{1.47}$$

There is a physical difference between positive and negative values of q (or m). For $q > 0$, the solution (1.47) is a travelling wave moving towards larger n, counterclockwise in Fig. 1.9; the *phase* of the cosine remains constant if, as time t increases, n increases as well. For $q < 0$, the wave travels towards smaller n, or clockwise, in the periodic chain. Our solutions therefore include waves travelling in both directions around the periodic chain. This result is not obtained for the original problem, with the end atoms not connected to one another: In that case, one finds *standing waves*, as we discuss in Sect. 1.3.3.

1.3.1 Normal Mode Coordinates

In the previous section, we determined the relation between the normal mode frequencies and wavevectors. In that section, the relative displacements u_n of each atom provided the coordinates in which the system's motion was described. Another way to look at this system is to treat the amplitude of each normal mode as a coordinate of the system, leaving the amplitudes as time-dependent variables. The idea is to write (1.42) in the form

$$\begin{aligned} u_n(t) &= \frac{1}{\sqrt{N}} \sum_{m=-N/2}^{N/2} \mathcal{U}_m(t)\, \mathrm{e}^{\mathrm{i} q_m r_n} \\ &= \frac{1}{\sqrt{N}} \sum_{m=-N/2}^{N/2} \mathcal{U}_m(t)\, \mathrm{e}^{2\pi \mathrm{i} mn/N}, \end{aligned} \tag{1.48}$$

where the amplitudes $\mathcal{U}_m$ are time-dependent. We can invert (1.48) to find an expression for the mode amplitudes $\mathcal{U}_m$ in terms of the atom displacements u_n:

$$\mathcal{U}_m(t) = \frac{1}{\sqrt{N}} \sum_{n=1}^{N} u_n(t)\, \mathrm{e}^{-2\pi \mathrm{i} mn/N}. \tag{1.49}$$

The proof that this relation is equivalent to (1.48) is left to the exercises (see Exercise 1.11).

The momentum associated with each atom is $p_n = M\dot{u}_n$; we can define a momentum $\mathcal{P}_m$ associated with the normal mode amplitudes $\mathcal{U}_m$ using the definition of the *canonical momentum*,

$$\mathcal{P}_m = \frac{\partial \mathcal{L}}{\partial \dot{\mathcal{U}}_m} = \frac{\partial T}{\partial \dot{\mathcal{U}}_m}, \tag{1.50}$$

in terms of the Lagrangian $\mathcal{L} = T - U$ and then the kinetic energy T of the system; as the potential energy U does not depend on $\dot{\mathcal{U}}_m$, it drops out of (1.50) (see e.g. Goldstein [2]). We write the kinetic energy in terms of the time derivatives $\dot{\mathcal{U}}_m$:

$$\begin{aligned} T &= \frac{M}{2}\sum_{n=1}^{N}\dot{u}_n^2 = \frac{1}{N}\frac{M}{2}\sum_{n=1}^{N}\left(\sum_{m=-N/2}^{N/2}\dot{\mathcal{U}}_m\, \mathrm{e}^{2\pi \mathrm{i} mn/N}\right)^2 \\ &= \frac{M}{2}\sum_{m=-N/2}^{N/2}\dot{\mathcal{U}}_m\dot{\mathcal{U}}_{-m}. \end{aligned} \tag{1.51}$$

(see Exercise 1.12). From this expression we construct the momenta $\mathcal{P}_m$,

$$\mathcal{P}_m = \frac{\partial T}{\partial \dot{\mathcal{U}}_m} = M\dot{\mathcal{U}}_{-m} = \frac{1}{\sqrt{N}}\sum_{n=1}^{N} p_n\, \mathrm{e}^{2\pi \mathrm{i} mn/N}. \tag{1.52}$$

The Hamiltonian for the system is

$$H = T + U = \sum_{n=1}^{N}\frac{p_n^2}{2M} + \sum_{n=1}^{N}\frac{k}{2}\left(u_n - u_{n-1}\right)^2, \tag{1.53}$$

where we use the harmonic approximation for the atomic interaction, with spring constant k, and for notional convenience we identify u_0 with u_N and u_{N+1} with u_1. This can be written in terms of the normal mode coordinates and their momenta as (see Exercise 1.13)

$$H = \sum_{m=-N/2}^{N/2}\frac{\mathcal{P}_m\mathcal{P}_{-m}}{2M} + \sum_{m=-N/2}^{N/2} k\left(1 - \cos 2\pi m/N\right)\mathcal{U}_m\mathcal{U}_{-m}. \tag{1.54}$$

The equations of motion for the normal mode amplitudes and momenta can then be obtained from Hamilton's equations of motion:

$$\left.\begin{aligned} \dot{\mathcal{U}}_m &= \frac{\partial H}{\partial \mathcal{P}_m} = \frac{\mathcal{P}_{-m}}{M}, \\ \dot{\mathcal{P}}_m &= \frac{\partial H}{\partial \mathcal{U}_m} = 2k(1 - \cos 2\pi m/N)\mathcal{U}_{-m} \\ &= 4k\sin^2(\pi m/N)\,\mathcal{U}_{-m}. \end{aligned}\right\} \tag{1.55}$$

From this set of equations we can obtain the dispersion relations; we take a harmonic time dependence $\mathcal{U}_m = A\mathrm{e}^{\mathrm{i}\omega_m t}$ and $\mathcal{P}_m = \mathrm{i}\omega_m MA\mathrm{e}^{\mathrm{i}\omega_m t}$, and immediately find the same relation as before, (1.45).

We now return to (1.54), and write the Hamiltonian in terms of the complex conjugates of the amplitudes and momenta, which are given by $\mathcal{P}_m^* = \mathcal{P}_{-m}$ and $\mathcal{U}_m^* = \mathcal{U}_{-m}$, so that

$$H = \frac{1}{2M}\sum_m \mathcal{P}_m\mathcal{P}_m^* + k\sum_m\left(1 - \cos 2\pi m/N\right)\mathcal{U}_m\mathcal{U}_m^*. \tag{1.56}$$

If we isolate a single mode m in (1.54), we have a Hamiltonian H_m given by

$$H_m = \frac{1}{2M}\mathcal{P}_m\mathcal{P}_m^* + k\left(1 - \cos 2\pi m/N\right)\mathcal{U}_m\mathcal{U}_m^*. \tag{1.57}$$

This expression is quadratic in the momentum $\mathcal{P}_m$ and amplitude $\mathcal{U}_m$, and is identical to that for a simple harmonic oscillator with mass M and effective spring constant $k_{\text{eff}} = k(1-\cos 2\pi m/N)$. Each mode therefore looks formally identical to a harmonic oscillator; this identification will prove very useful when we apply quantum mechanics to the normal modes.

We now define a new set of *normal coordinates* a_m and their complex conjugates a_m^* by

$$\left.\begin{aligned} a_m &= \frac{1}{\sqrt{2M}}\mathcal{P}_m - \mathrm{i}\sqrt{k(1-\cos 2\pi m/N)}\,\mathcal{U}_m^* \\ a_m^* &= \frac{1}{\sqrt{2M}}\mathcal{P}_m^* + \mathrm{i}\sqrt{k(1-\cos 2\pi m/N)}\,\mathcal{U}_m. \end{aligned}\right\} \tag{1.58}$$

Using these the Hamiltonian can be written

$$H = \frac{1}{2}\sum_m (a_m^* a_m + a_m a_m^*) = \sum_m a_m^* a_m\,. \tag{1.59}$$

The last step in (1.59) depends on the coordinates a_m *commuting*, which holds for the classical system we have been considering so far, but does not in quantum mechanical system, as we shall see below. Equation (1.59) is a purely diagonal form, in that the Hamiltonian does not involve any cross-products $a_m a_n$, although it does use complex-valued coordinates. The energy of the entire system is thus just the sum of the square amplitudes of the normal mode coordinates a_m, and (classically) can take on any value.

1.3.2 Phase and Group Velocity

The relation (1.45) relates the frequency ω to the wavevector q of the travelling waves; this relation is known as a *dispersion relation.* In a continuous medium, the dispersion relations are usually linear, of the form $\omega = cq$, where c is known as the *wave velocity.* This is true for light travelling through vacuum, sound through air, and also holds for our N atom chain, if the wavelength q and frequency ω are small enough. For small q, we can approximate the dispersion relation (1.45) as

$$\omega \approx \omega_0 r_0 |q|. \tag{1.60}$$

The wave velocity is therefore $c = \omega_0 r_0$. If we wish to make a connection with the Lennard–Jones model used for the two and three atom problems, we have $\omega_0/2\pi = 0.8$ THz and $r_0 = 3.8$ Å, so the phase velocity is $c = 1900$ m/s. This is in the range for acoustic velocities in solids; more typically, the phase

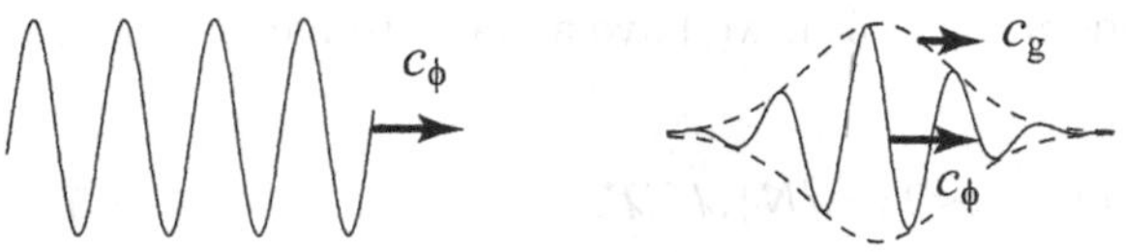

Fig. 1.11. Phase velocity of a single-frequency oscillation (*left*), and the combined phase and group velocity for a wave train (*right*), composed of a central frequency (*solid line*) modulated by an envelope (*dotted line*).

velocity is in the range of 3000-6000 m/s, but the weak interaction potential in the Lennard–Jones model reduces c somewhat.

For values of q large enough that the approximation (1.60) does not hold, the frequency is lower than the linearized expression; the wave velocity is therefore smaller at larger wavevector and frequency. When this type of non-linearity appears in the dispersion relation, the medium through which the wave travels is known as *dispersive*. We then have to make a distinction between the velocity at which a pure sinusoidal wave travels through the medium, known as the *phase velocity* c_ϕ, and the velocity at which a *wave train* containing a range of frequencies travels, which is known as the *group velocity* c_g. In Fig. 1.11 we illustrate this difference, with a pure sine wave moving at c_ϕ, and a wave train, composed of the superposition of a central frequency and a range of frequencies around it, has its envelope (shown by the dotted line) travelling at the group velocity c_g. Interestingly, the center frequency oscillation in the wave train travels at the phase velocity; the phase velocity is typically higher than the group velocity , so in Fig. 1.11, the center frequency will travel faster than the envelope, disappearing at the leading edge of the envelope and appearing at the trailing edge. This phenomenon can be observed with surface waves on water, which is a highly dispersive medium.

The group velocity c_g for a wave train is given by the derivative of the frequency with wavevector, so we have the relations

$$\left.\begin{aligned} c_\phi &= \omega/q, \\ c_g &= \mathrm{d}\omega/\mathrm{d}q. \end{aligned}\right\} \tag{1.61}$$

This is discussed in more detail in Sect. 3.2.

For the N-atom chain, the group velocity is $c_g = \omega_0 r_0 |\cos q r_0/2|$. In Fig. 1.12 we show the phase and group velocity for this system; note that the group velocity is always less than the phase velocity, and as q approaches $\pm\pi/r_0$, the group velocity goes to zero. At small wavevectors, both the phase and group velocities approach the linearized value $\omega_0 r_0$.

1.3.3 Fixed Boundary Conditions

We have treated the problem of the N atom chain using periodic boundary conditions. It is perhaps more appealing to see how this problem is treated

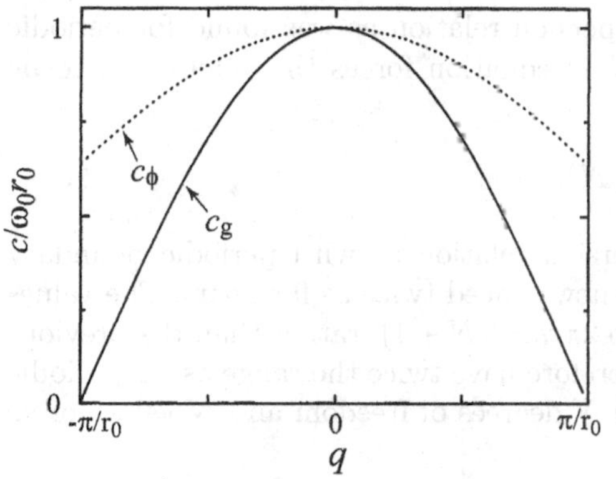

Fig. 1.12. Phase velocity (*dotted line*) and group velocity (*solid line*) for the N atom chain.

using more "realistic" boundary conditions; we will therefore briefly treat the problem of fixed boundary conditions, where we connect atoms 1 and N by springs to fixed, rigid supports, shown in Fig. 1.13.

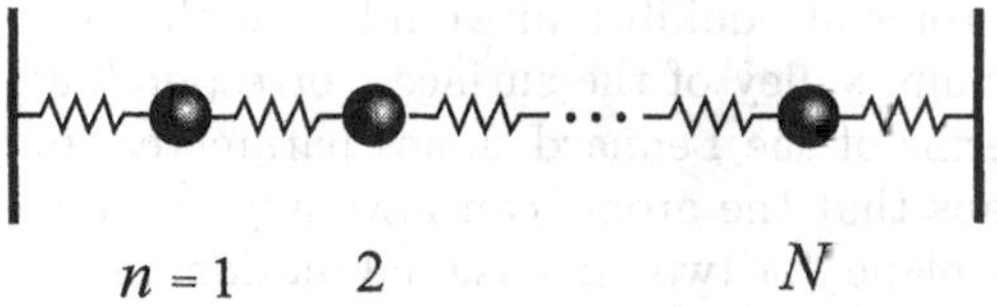

Fig. 1.13. Fixed boundary conditions for a chain with N atoms.

The equations of motion for this system are

$$\left.\begin{aligned} M\ddot{u}_1 &= k(u_2 - 2u_1), \\ M\ddot{u}_n &= k(u_{n+1} - 2u_n + u_{n-1}) \quad (n = 2 \text{ to } N-1), \\ M\ddot{u}_N &= k(u_{N-1} - 2u_N). \end{aligned}\right\} \tag{1.62}$$

The solutions to this system of equations are not travelling waves but standing waves, made up of equal amplitudes of the right- and left-going waves found for periodic boundary conditions. These have the form

$$u_n = \frac{A}{\sqrt{N}} \sin q r_0 n \, \mathrm{e}^{\mathrm{i}\omega t}. \tag{1.63}$$

Inserting this in (1.62), we find after some trigonometric manipulations the pair of equations

$$\left.\begin{aligned} \omega^2 &= 4\omega_0^2 \sin^2 q r_0/2, \\ \tan N q r_0 &= -\tan q r_0 \, . \end{aligned}\right\} \tag{1.64}$$

The first equation is the same dispersion relation as was found for periodic boundary conditions, and the second equation forces the values of q to be from the set

$$q = m\frac{2\pi}{(N+1)r_0} \quad (1 \leq m \leq N). \tag{1.65}$$

Hence we find the same dispersion relation as with periodic boundary conditions, but the values of q are now spaced twice as far apart. The values of q now run from $2\pi r_0/(N+1)$ to $2\pi r_0 N/(N+1)$, rather than the previous range from $-\pi/r_0$ to π/r_0, and therefore have twice the range as for periodic boundary conditions: we still have N degrees of freedom and N values of q.

1.3.4 Transverse Motion

Our discussion has been restricted to one dimension, with our N atoms moving only towards or away from one another. We now discuss how one approaches the problem of transverse motion, where the atoms are free to move perpendicular to one another. The spherically symmetric model potential $\phi(r)$ we have been using can in principle still serve for this type of motion. For two-dimensional motion, for example, the Lennard–Jones potential has the form shown in Fig. 1.14. Two atoms at equilibrium would sit with their separation $r = r_2 - r_1$ in the minimum valley of the surface, corresponding to a distance $r_0 = (2B/A)^{1/6}$ in terms of the Lennard–Jones parameters in (1.3). The spherical symmetry means that the atoms can have any orientation on the circle of radius r_0 in the plane, for two dimensional motion, or on a sphere of the same radius r_0 for three dimensions.

We can approximate the potential function for small displacements from this equilibrium distance, using the harmonic approximation, $\phi(r) \approx k(r - r_0)^2/2$, with effective spring constant k. This works well for in-line motion;

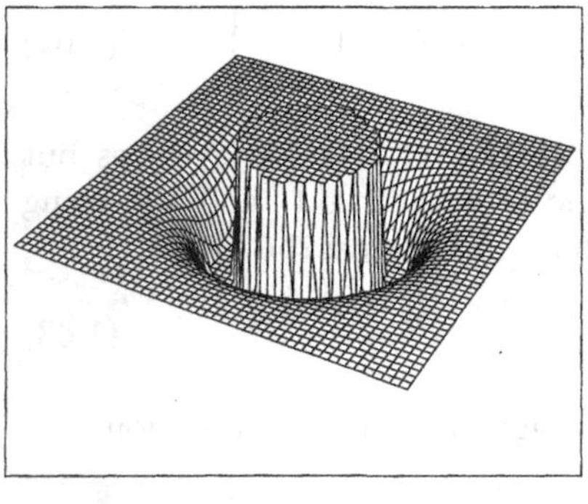

Fig. 1.14. A perspective view of the Lennard–Jones potential for motion in the plane; the vertical height indicates the strength of the potential. Note that for small r the potential diverges to $+\infty$, so the plot is truncated.

if our atoms are arranged along the x axis, with the nth atom at x_n and the $(n-1)$th atom at x_{n-1}, their distance at equilibrium is $r_0 = x_n - x_{n-1}$. In-line motion results in the nth atom moving to $x_n + u_n$ and the $(n-1)$th atom moving to $x_{n-1} + u_{n-1}$, so their separation changes by $\delta r = u_n - u_{n-1} \ll r_0$. Considering only these two atoms, the potential energy changes quadratically in the displacements, with $U \approx \phi(r_0) + k(u_n - u_{n-1})^2/2$.

Consider however the effect of a transverse displacement on the potential energy, when we allow the atoms to move in the (x, y) plane. The nth atom moves from $(x, y) = (x_n, 0)$ to (x_n, u_n), and the $(n-1)$th atom to (x_{n-1}, u_{n-1}). Their separation changes to

$$r = \left(r_0^2 + (u_n - u_{n-1})^2\right)^{1/2}, \tag{1.66}$$

and the potential energy is

$$U \approx \phi(r_0) + \frac{k}{2}\frac{r_0^2}{4}\left(\frac{u_n - u_{n-1}}{r_0}\right)^4. \tag{1.67}$$

Hence we find that the potential energy changes with the fourth power in the displacement $u_n - u_{n-1}$, so the harmonic term vanishes for purely transverse motion. This is not specific to the Lennard–Jones potential: Any spherically symmetric potential will have a quartic dependence for motion transverse to the separation, if the initial point is a local potential minimum. All of our previous discussion thus falls apart: for small displacements u, the potential energy changes as u^4, so the force changes as u^3, and we cannot model the response as a spring, as we did for in-line motion. The spherically symmetric potential is thus not capable of dealing with this type of motion in a simple, perturbative fashion; in fact, linear chains bound by van der Waals forces are not stable because of this behavior. As we shall see later, however, many types of three dimensional structures can be treated using spherically symmetric potentials, and their static and dynamic response can be quite accurately modelled. Their stability is however due to the three-dimensional geometry, which breaks the spherical symmetry of the potential and provides resistance to transverse motion.

To return to our one-dimensional problem, we now have to take an interaction between the atoms that is not spherically symmetric. Many types of atoms bind through non-spherically symmetric interactions. Atoms such as Si and C, in Group IV in the periodic table for example, have their outermost valence electrons in a spherically-symmetric valence s orbital, and three valence p orbitals, one along each of the rectangular axes, p_x, p_y and p_z; these are shown schematically in Fig. 1.15. Atoms with larger numbers of valence electrons also occupy the higher-energy d, f, g orbitals, and so on. In covalently-bonded atoms, the s and p orbitals can *hybridize*, mixing together to form different combinations. An sp hybrid consists of two lobes, similar in appearance to a single p orbital. Two p orbitals can hybridize with the s orbital to form an sp^2 orbital, with triangular symmetry in the plane. In Si

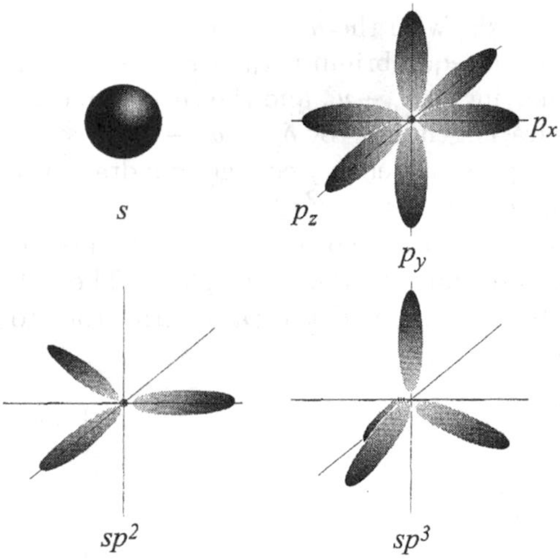

Fig. 1.15. Sketch of the s, p_x, p_y and p_z orbitals, and the hybrid orbitals sp^2 and sp^3.

and C, and other Group IV atoms, the three p and the s orbitals hybridize to form sp^3 orbitals with three-dimensional tetrahedral symmetry.

These orbitals form the basis for highly non-symmetric interactions; the local orbital density is proportional to the probability of finding an electron at that point. Atoms with sp orbitals will position themselves to maximize the overlap of their orbitals, forming linear chains; those with sp^2 orbitals will form triangular lattices in the plane, and those with sp^3 orbitals form tetrahedral geometries in three dimensions. The diamond crystal structures of C and Si are due to this tetrahedral symmetry; covalently bonded materials in general form crystal structures determined by the geometry of their binding orbitals.

To add a little formalism to this argument, we consider two identical atoms, with hybridized sp orbitals, interacting in the $x-y$ plane, as shown schematically in Fig. 1.16. Each atom is free to position and orient itself in the $x-y$ plane; note the asymmetry of the hybridized orbitals means that the relative orientation of each atom is now significant. We define the orientation by the angle θ between the x-axis and the perpendicular to the line through the sp orbital. The atoms will position themselves to minimize the interaction potential energy, which is achieved at relative spacing $|\boldsymbol{r}_1-\boldsymbol{r}_2| = r_0$; they also try to align the axes of their binding orbitals, so that $\theta_1 = \theta_2$. At equilibrium, the energy will have zero first derivative with respect to these two coordinates, or degrees of freedom, and the first non-zero term will be quadratic in the variation:

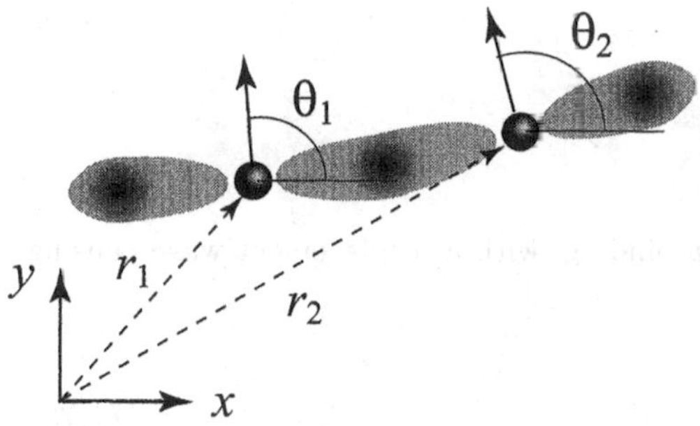

Fig. 1.16. Two atoms binding through their hybridized *sp* orbitals; the positions $\boldsymbol{r}_1$ and $\boldsymbol{r}_2$, and the orientations θ_1 and θ_2 of the atoms with respect to the x axis, form the degrees of freedom for motion in the $x-y$ plane. The angles are measured between the perpendicular to the line through the *sp* orbital and the x-axis.

$$U = U_0 + \frac{k}{2}(|\boldsymbol{r}_2 - \boldsymbol{r}_1| - r_0)^2 + \frac{G}{2}(\theta_1 - \theta_2)^2. \tag{1.68}$$

The first term U_0 represents the (negative) binding energy; the second term involves energy in stretching the bond away from the equilibrium r_0; and the third term the torsional energy associated with different binding axis angles, with torsional spring constant G. More complex models can be built that account for shear energy associated with linear displacements of the atoms perpendicular to the binding axis (without rotating the axes), and so on.

At equilibrium, the atoms are spaced by r_0 with equal angles $\theta_1 = \theta_2$. A linear force F pulling the atoms in different directions along the binding axis causes a displacement $\Delta r = F/k$; a torque N, bending the relative orientation of one atom with respect to the other will cause a rotation $\theta_2 - \theta_1 = N/G$.

Consider now an N atom chain bound together by this type of interaction, with the line of atoms aligned with the x axis, and the atoms free to move in the $x-y$ plane. The atoms are identical, with mass M, and equilibrium spacing r_0. All the calculations for response to static forces and dynamic motion along the x axis, which we worked out for the Lennard–Jones potential, still hold. Here we will treat motion along y, and write the y displacement of the nth atom as $u_n(t)$. We now assume a displacement wave $d(x,t)$, giving the y displacement of an atom initially at x, so that $u_n(t) = d(nr_0, t)$. The wave will be a travelling wave, of the form

$$d(x,t) = A\mathrm{e}^{\mathrm{i}qx - \mathrm{i}\omega t}. \tag{1.69}$$

This represents a harmonic wave with amplitude A, frequency ω, wavevector q, and phase velocity $c_\phi = \omega/q$. We will assume that under the influence of this displacement, the atoms reorient as they are displaced, in an attempt to maintain the alignment of their bonding axes with one another; this is sketched in Fig. 1.17. For long wavelengths, $\lambda \gg r_0$, that is, wavevectors $q \ll 2\pi/r_0$, and small amplitudes A, the difference in bonding angles $\theta_n - \theta_{n-1}$ between the nth and $(n-1)$th atoms in the chain is approximately given by

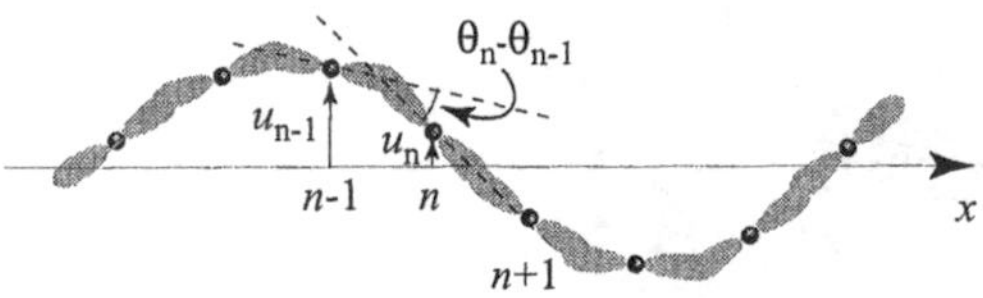

Fig. 1.17. An N atom chain with *sp* binding, with a displacement wave causing torsion of the bonding angles.

$$\theta_n - \theta_{n-1} \approx \left.\frac{\partial d}{\partial x}\right|_{x=x_n} - \left.\frac{\partial d}{\partial x}\right|_{x=x_{n-1}}. \tag{1.70}$$

The total potential energy is approximately given by

$$U = N\phi(r_0) + \sum_{n=1}^{N} \frac{G}{2}(\theta_n - \theta_{n-1})^2. \tag{1.71}$$

Note that we assume the torsional energy is more important than the energy due to the change in interatomic spacing; those changes only give terms proportional to the fourth power of the displacement, as discussed earlier, and therefore do not show up in this second-order expansion.

The kinetic energy associated with the displacements $u_n(t)$ is given by

$$T = \sum_{n=1}^{N} \frac{M}{2}\left(\frac{\partial u_n}{\partial t}\right)^2. \tag{1.72}$$

Hamilton's equations of motion then yield the system

$$M\frac{\partial^2 u_n}{\partial t^2} = G(\theta_{n+1} - 2\theta_n + \theta_{n-1}). \tag{1.73}$$

This system can be shown to yield exactly the same type of dispersion relation $\omega(q)$ as was found for the longitudinal displacement waves discussed earlier, although with a different prefactor (see Exercise 1.14).

An equivalent argument can be created using not the torsional energy of the linear chain, but instead using the transverse displacement energy; again, a similar result is found. In our discussion of two and three dimensional solids, we will introduce the concept of *shear* forces and displacements, which are similar to these.

1.4 Linear Chain with Optical Modes

The last one-dimensional problem we will discuss is what happens if we add a second set of atoms of mass M' to the one dimensional chain, alternating these with the original atoms of mass M; as we shall see, this generates a new set, or *branch*, of frequencies, which are known as *optical modes* to distinguish them from the *acoustic modes* discussed so far.

The masses are linked by identical linear springs k, and the masses have equilibrium spacing r_0, as sketched in Fig. 1.18. We number the masses in pairs, so that the nth atom with mass M has displacement u_n, while nth atom with mass M' has displacement v_n. The equations of motion can be shown to be

$$\left.\begin{aligned} M\ddot{u}_n &= k(v_{n+1} - 2u_n + v_{n-1}), \\ M'\ddot{v}_n &= k(u_{n+1} - 2v_n + u_{n-1}). \end{aligned}\right\} \tag{1.74}$$

We will take periodic boundary conditions, so the $n = 1$ atom of mass M is coupled to the $n = N$ atom of mass M'; this is not reflected in (1.74), but we will enforce this at the end of the calculation, when we determine the set of allowed values of the wavevector.

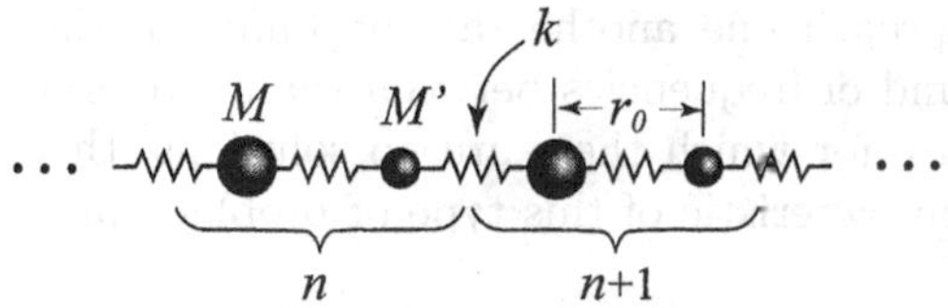

Fig. 1.18. Model for a linear chain with alternating atoms of mass M and M'. The index n counts pairs of atoms.

The displacements are written in Bloch form,

$$\left.\begin{aligned} u_n(t) &= \frac{A}{\sqrt{N}} e^{iqn2r_0 - i\omega t} \\ v_n(t) &= \frac{B}{\sqrt{N}} e^{iqn2r_0 - i\omega t}, \end{aligned}\right\} \tag{1.75}$$

where the amplitudes A and B can be complex. We define two frequencies ω_0 and ω_0',

$$\left.\begin{aligned} \omega_0 &= \sqrt{k/M} \\ \omega_0' &= \sqrt{k/M'}, \end{aligned}\right\} \tag{1.76}$$

and using these and the forms (1.75), the equations of motion become

$$\left.\begin{aligned} \omega^2 A &= 2\omega_0^2 A - 2\omega_0^2 \cos(2qr_0)\, B \\ \omega^2 B &= -2\omega_0'^2 \cos(2qr_0)\, A + 2\omega_0'^2 B. \end{aligned}\right\} \tag{1.77}$$

This is a 2×2 eigenvalue-eigenvector equation, where the values of ω are the eigenvalues and the amplitude pairs (A, B) the eigenvectors. Written in matrix form, the equation is

$$\begin{bmatrix} 2\omega_0^2 - \omega^2 & -2\omega_0^2 \cos(2qr_0) \\ -2\omega_0'^2 \cos(2qr_0) & 2\omega_0'^2 - \omega^2 \end{bmatrix} \begin{bmatrix} A \\ B \end{bmatrix} = 0. \tag{1.78}$$

The characteristic equation, obtained by setting the determinant of the matrix of coefficients to zero, is given by

$$\omega^4 - 2(\omega_0^2 + \omega_0'^2)\omega^2 + 4\omega_0^2\omega_0'^2 \sin^2 2qr_0 = 0. \tag{1.79}$$

The two eigenfrequencies for each value of wavevector q are thus given by

$$\omega_\pm^2(q) = \omega_0^2 + \omega_0'^2 \pm \sqrt{\omega_0^4 + \omega_0'^4 + 2\omega_0^2\omega_0'^2 \cos 4qr_0}. \tag{1.80}$$

There are two solutions, ω_+ and ω_-, for each value of the wavevector q. These two solutions are shown in Fig. 1.19, as a function of q, for a frequency ratio $\omega_0'/\omega_0 = 1.3$ (corresponding to a mass ratio $M/M' = 1.3^2 = 1.69$). The lower set of solutions, $\omega_-(q)$, looks very similar to that for the single mass chain, but the upper set, $\omega_+(q)$, is rather different. For small wavevectors q, $\omega_-(q)$ increases linearly with q, while $\omega_+(q)$ is constant, and then falls gradually. The two frequencies approach one another at the limits of the plot, as $q \to \pm\pi/4r_0$. There is a band of frequencies between the maximum value of ω_- and the minimum of ω_+ for which there are no solutions; this gap in the frequency spectrum is characteristic of this type of problem, and is known as a *bandgap*.

The range of q in Fig. 1.19 is $-\pi/4r_0 < q < \pi/4r_0$. We have not yet determined the spacing of the values of q; this is set by the periodic boundary conditions, where we have u_1 connected to v_N. The equations for these two atoms are

$$\left.\begin{aligned} -\omega^2 u_1 &= \omega_0^2(v_1 - 2u_1 + v_N) \\ -\omega^2 v_N &= \omega_0'^2(u_1 - 2v_N + u_N). \end{aligned}\right\} \tag{1.81}$$

Substituting the forms (1.75) in this pair of equations ultimately yields the condition

$$q_m = \frac{\pi}{2(N-1)r_0} m, \tag{1.82}$$

for integer values m. There are thus N values of q (counting $q = 0$), and as there are two frequencies ω_- and ω_+ for each value of q, we find $2N$ independent values, equal to the number of degrees of freedom.

The two distinct frequencies, and the spectral gap in their dispersion relation, is characteristic of the perturbation in the periodicity of the linear chain, here from the variation in atomic mass. An equivalent result appears if the spring constant alternates between two values, k and k', even if the masses are identical (see Exercise 1.15). The lower set of frequencies $\omega_-(q)$ is known as the *acoustic* band, while the upper set $\omega_+(q)$ is known as the *optical band.* For all the normal modes in the acoustic band, the two atoms in the unit cell move in the same direction, so that the displacements u_n and v_n have the same sign. For a normal mode in the optical band, by contrast, the two atoms move in opposite directions, so u_n and v_n have opposite signs. If the charge on the atoms is different, as for example in an ionic system where neighboring atoms have opposite charge, or in a chain where the bonding electrons are

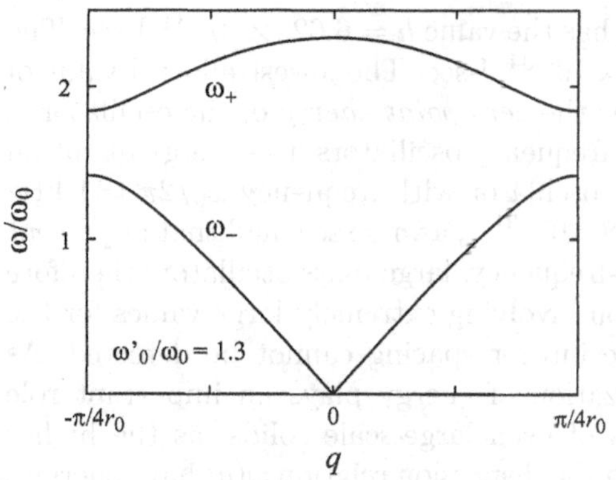

Fig. 1.19. The eigenfrequency solutions $\omega_{\pm}(q)$ for the two mass linear chain. Calculations were made for $\omega_0'/\omega_0 = 1.3$, corresponding to a mass ratio $M'/M = 1.7$.

closer to one atom than its neighbor, the optical band motion will couple to the electric field in an electromagnetic wave, through the induced dipole moment. The upper frequency band is therefore optically active; this is the origin of the terminology.

1.5 Quantum Mechanics and Thermodynamics of the Linear Chain

We now begin a treatment of the quantum mechanical properties of our atomic chain. Developing the formalism of quantum mechanics for mechanical motion will help to explain the thermodynamic and statistical properties of the chain, and will also provide a basis for calculating the interaction of the mechanical vibrations with one another and with electrons in a solid.

In (1.56) and (1.57) we showed that the Hamiltonian for the N atom chain, written in terms of the normal mode amplitudes $\mathcal{U}_m$ and momenta $\mathcal{P}_m$, is formally identical to a series of harmonic oscillators with mass M and frequencies $\omega_m = \omega(q_m)$. The quantum mechanical behavior of a simple harmonic oscillator is often the first problem solved in introductory courses in quantum mechanics; through the use of Schrödinger's equation, one can work out the quantum wavefunctions and quantized energies E_n. The energies for a harmonic oscillator with natural (classical) resonance frequency ω_0 are given by

$$E_n = \left(n + \frac{1}{2}\right) \hbar\omega_0, \tag{1.83}$$

where the index n can take on any non-negative integer value; n is known as the quantum number or *occupation number* of the oscillator. The quantity $\hbar = h/2\pi$ (pronounced "h-bar") is ubiquitous in quantum mechanical systems. h

is known as Planck's constant, and has the value $h = 6.628 \times 10^{-34}$ J-sec. The corresponding value for $\hbar$ is 1.055×10^{-34} J-sec. The lowest allowed value of the energy, $E_0 = \hbar\omega/2$, is known as the *zero-point energy* of the oscillator.

Note that, except for very high frequency oscillators, the energy quantum is very small: A classical harmonic oscillator with frequency $\omega_0/2\pi = 1$ kHz has quantized levels spaced by 6.628×10^{-31} J, a value so small that at present there is no way to measure it. Low-frequency, large mass oscillators therefore behave classically, with their motion involving extremely large values for the quantum number n, such that the integer spacing cannot be detected. As we shall see, however, the quantization of energy plays an important role in describing the thermodynamics of even large-scale solids, as the higher frequency acoustic modes we find in the dispersion relation $\omega(q)$ have energies whose quantization is quite noticeable.

We can apply the principle of the quantization of energy to the normal modes of the N atom chain. The energy eigenstates consist of a set of N integers $|n_1 n_2 \dots n_m \dots n_N\rangle$, where n_m indicates the number of energy quanta in the normal mode with wavevector q_m and frequency ω_m. The set of integers is written in what is known as "bracket" form, $|\dots\rangle$, useful in quantum calculations.

The total energy of a chain with the set of occupation numbers

$$|n_1 n_2 \dots n_m \dots n_N\rangle \tag{1.84}$$

is given by

$$E_{\text{tot}} = \sum_{m=-N/2}^{N/2} \left(n_m + \frac{1}{2}\right) \hbar\omega_m. \tag{1.85}$$

Changes in the energy of the system are accommodated by changing the values of the indices n_m, with a corresponding change in energy in quanta of $\hbar\omega_m$. The zero-point energy of the chain is the sum of the individual zero point energies of the normal modes.

1.6 Effect of Temperature on the Linear Chain

Using the quantum mechanical description of the N atom chain developed in the previous section, we now turn to the last topic in this chapter, the effect of non-zero temperature. We will allow each atom to be in equilibrium with a thermal bath at temperature T; this means that the energy E of each atom will fluctuate, as it exchanges energy with the bath.

Classically, an atom can have any velocity $\dot{u}$ and any displacement u (in one dimension). The probability $P(E)\,\mathrm{d}u\,\mathrm{d}\dot{u}$ that the atom has energy $E = E(u, \dot{u})$, with velocity between $\dot{u}$ and $\dot{u} + \mathrm{d}\dot{u}$ and displacement between u and $u + \mathrm{d}u$, is given by the Gibbs distribution,

$$P(E)\,\mathrm{d}u\,\mathrm{d}\dot{u} = \mathcal{N} \mathrm{e}^{-E/k_{\mathrm{B}}T}\,\mathrm{d}u\,\mathrm{d}\dot{u} \tag{1.86}$$

where k_B is Boltzmann's constant, $k_B = 1.38 \times 10^{-23}$ J/K. The factor $du\, d\dot{u}$ is the volume in *phase space*, which includes all possible combinations of displacement and velocity. $\mathcal{N}$ is the normalization for the probability distribution, set by the requirement that the integral of $P(E)$ over all of phase space be unity. For a more complete discussion on this topic, there are a number of standard texts; see e.g. Landau and Lifshitz [5].

In a classical system, the energy E is a continuous function of the displacement u and velocity $\dot{u}$. For instance, in our two-atom system, with fixed center of mass, we have $E(u, \dot{u}) = T(\dot{u}) + U(u) = \frac{1}{2}\mu\, \dot{u}^2 + \phi(u)$, using the reduced mass μ and the interaction potential $\phi(u)$. We can simplify by taking the harmonic approximation for the potential, equivalent to connecting the atoms with a spring k, so $\phi(u) \approx ku^2/2$. The thermally averaged energy $\langle E\rangle = \langle T\rangle + \langle U\rangle$ can then be calculated from the distribution function:

$$\begin{aligned} \langle E\rangle &= \int\!\!\int E\, P(E)\, du\, d\dot{u} = \mathcal{N} \int\!\!\int E(u,\dot{u}) e^{-\beta E} du\, d\dot{u} \\ &= \mathcal{N}_1 \int_{-\infty}^{\infty} \frac{M}{2} \dot{u}^2 e^{-\beta M \dot{u}^2/2} d\dot{u} + \mathcal{N}_2 \int_{-\infty}^{\infty} \frac{k}{2} u^2 e^{-\beta k u^2/2} du, \end{aligned} \tag{1.87}$$

using the inverse temperature $\beta = 1/k_B T$. The normalizations $\mathcal{N}_1$ and $\mathcal{N}_2$ are related to $\mathcal{N}$ by $\mathcal{N} = \mathcal{N}_1 \mathcal{N}_2$, where

$$\left. \begin{aligned} \frac{1}{\mathcal{N}_1} &= \int_{-\infty}^{\infty} e^{-\beta M \dot{u}^2/2} d\dot{u} = \sqrt{\frac{2\pi k_B T}{M}}, \\ \frac{1}{\mathcal{N}_2} &= \int_{-\infty}^{\infty} e^{-\beta k u^2/2} du = \sqrt{\frac{2\pi k_B T}{k}}, \end{aligned} \right\} \tag{1.88}$$

(see Exercise 1.16). Note that the two integrals in (1.87) correspond to the thermally averaged kinetic and potential energies, and we complete these integrals to find the classical result

$$\langle T\rangle = \langle U\rangle = \frac{1}{2} k_B T. \tag{1.89}$$

This result is an example of the virial theorem. The corresponding average energy is thus $\langle E\rangle = k_B T$.

The classical heat capacity for our system with two degrees of freedom (u and $\dot{u}$) is thus $C = \partial E/\partial T = k_B$. This is a standard result: Classically, the heat capacity is obtained by adding $\frac{1}{2}k_B$ for each degree of freedom. For an N atom system, with fixed center of mass, there are $2N - 2$ degrees of freedom, and the heat capacity is $C = (N-1)k_B$. This is the one-dimensional form of the *Dulong–Petit* law for the heat capacity. In this classical limit, C is independent of temperature.

Here, however, we must deal with a system that is quantum mechanical, so the energies of the system are discrete, as described in the previous section. We look first at the thermal statistics of a single harmonic oscillator, with resonance frequency ω_0; our model will be an oscillator consisting of a pair of argon atoms bound by the van der Waal's interaction, as described in Sect. 1.1. The energy of the oscillator is quantized, with occupation number n corresponding to an energy $E_n = (n + 1/2)\hbar\omega_0$. The probability $P(n)$ that the oscillator, when in thermal equilibrium with a bath a temperature T, will be found in the nth energy state, is given by (1.86), in the form $P(n) = \mathcal{N}e^{-\beta E_n}$. The normalization $\mathcal{N}$ is set by requirement that the sum of all the probabilities $P(n)$ over possible occupation numbers n equal unity,

$$\frac{1}{\mathcal{N}} = \sum_{n=0}^{\infty} e^{-\beta E_n} = \frac{e^{\beta\hbar\omega_0/2}}{e^{\beta\hbar\omega_0} - 1}. \tag{1.90}$$

Hence the normalized probability distribution $P(n)$ of finding our thermally-equilibrated oscillator in the state with occupation number n is

$$P(n) = 2\sinh\frac{\beta\hbar\omega_0}{2}\, e^{-(n+1/2)\beta\hbar\omega_0}. \tag{1.91}$$

The expectation value for the oscillator's thermally-averaged energy $\langle E\rangle$ is then given by

$$\begin{aligned}\langle E\rangle &= \sum_{n=0}^{\infty} E_n P(n)\\ &= \left[\frac{1}{e^{\beta\hbar\omega_0} - 1} + \frac{1}{2}\right]\hbar\omega_0.\end{aligned} \tag{1.92}$$

This can be written in terms of the *Bose–Einstein* distribution $n(E)$,

$$n(E) = \frac{1}{e^{\beta E} - 1}, \tag{1.93}$$

so that

$$\langle E\rangle = \left(n(\hbar\omega_0) + \frac{1}{2}\right)\hbar\omega_0. \tag{1.94}$$

The heat capacity for the harmonic oscillator is then given by

$$C = \frac{\partial\langle E\rangle}{\partial T} = \frac{e^{\beta\hbar\omega_0}}{(e^{\beta\hbar\omega_0} - 1)^2}(\beta\hbar\omega_0)^2\, k_B. \tag{1.95}$$

This is plotted in Fig. 1.20.

We note that the vertical scale in Fig. 1.20 is fixed by Boltzmann's constant, while the horizontal temperature scale is set by the resonance frequency ω_0. For our model system of two argon atoms interacting through a van der Waal's interaction, with $\omega_0/2\pi = 0.8$ THz, the temperature scale is $\hbar\omega_0/k_B = 40$ K.

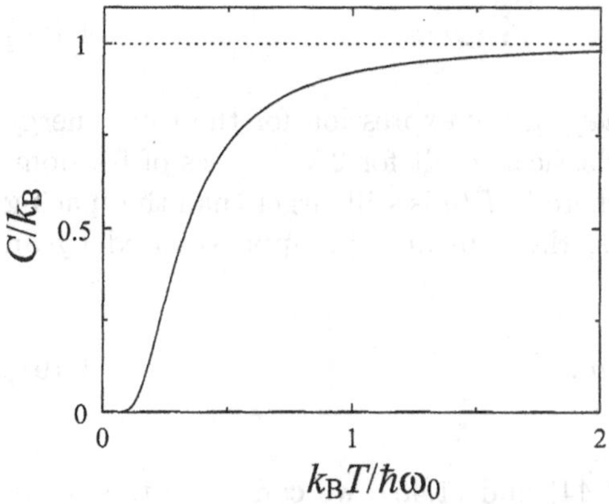

Fig. 1.20. Heat capacity for a harmonic oscillator with resonance frequency ω_0.

1.6.1 Specific Heat for an N Atom Chain

We can now expand from the two-atom system to calculating the energy and *specific heat*, that is, the heat capacity per unit length, of a one-dimensional chain of N atoms. This involves understanding the thermal properties of the normal modes of the chain. We have already shown that the energy E_m associated with the mth normal mode is quantized according to the formula

$$E_m = \left(n_m + \frac{1}{2}\right)\hbar\omega_m, \tag{1.96}$$

where n_m is the number of energy quanta in the mth normal mode; we can also say that n_m is the number of *phonons* in the mode. The total energy of the system is given by (1.85),

$$E_{\mathrm{tot}} = \sum_{m=-N/2}^{N/2} \left(n_m + \frac{1}{2}\right)\hbar\omega_m. \tag{1.97}$$

At non-zero temperature, the probability distribution for the energy E_m of each mode is distributed in the same way as that for a simple harmonic oscillator, given by (1.92). The thermally averaged total energy is thus

$$\langle E_{\mathrm{tot}}\rangle = \sum_m \langle E_m\rangle = \sum_m \left[\frac{1}{\mathrm{e}^{\beta\hbar\omega_m}-1} + \frac{1}{2}\right]\hbar\omega_m. \tag{1.98}$$

Henceforth we will drop the zero-point energy terms $\hbar\omega/2$. Each term in the square brackets is then the Bose–Einstein distribution for the thermally-averaged occupation $n(\hbar\omega_m)$ of the mth mode,

$$n(\hbar\omega_m) = \frac{1}{\mathrm{e}^{\beta\hbar\omega_m}-1}, \tag{1.99}$$

and the zero-point corrected average energy is

$$\langle E_{\mathrm{tot}} \rangle = \sum_m n(\hbar\omega_m)\hbar\omega_m. \tag{1.100}$$

At high temperatures $k_{\mathrm{B}}T \gg \hbar\omega_{N/2}$, the expression for the total energy simplifies to $\langle E_{\mathrm{tot}} \rangle \to Nk_{\mathrm{B}}T$, the classical result for $2N$ degrees of freedom. At moderately low temperatures, where $k_{\mathrm{B}}T/\hbar$ is still larger than the spacing $\Delta\omega = \omega_{m+1} - \omega_m$ between modes, the sum may be approximated by an integral:

$$\langle E_{\mathrm{tot}} \rangle \approx \int_{-N/2}^{N/2} \frac{\hbar\omega(m)}{\mathrm{e}^{\beta\hbar\omega(m)} - 1}\, \mathrm{d}m, \tag{1.101}$$

treating m as a continuous variable.

From the dispersion relation (1.44) and (1.45), we can write this as an integral over frequency, defining the *density of states* $\mathcal{D}(\omega) = (1/Nr_0)\mathrm{d}m/\mathrm{d}\omega$, the number of states per unit frequency per unit length:

$$\langle E_{\mathrm{tot}} \rangle \approx 2Nr_0 \int_0^{2\omega_0} \frac{\hbar\omega}{\mathrm{e}^{\beta\hbar\omega} - 1} \mathcal{D}(\omega)\, \mathrm{d}\omega. \tag{1.102}$$

The density of states is given by

$$\mathcal{D}(\omega) = \frac{1}{Nr_0} \left| \frac{1}{\mathrm{d}\omega/\mathrm{d}m} \right| = \frac{1}{2\pi\omega_0 r_0} \frac{1}{\sqrt{1 - \omega^2/4\omega_0^2}}. \tag{1.103}$$

Defining the dimensionless parameters $z = \beta\hbar\omega$ and $z_0 = 2\beta\hbar\omega_0$, this can be written

$$\langle E_{\mathrm{tot}} \rangle \approx \frac{2N}{\pi\beta} \int_0^{z_0} \frac{z}{\mathrm{e}^z - 1} \frac{1}{\sqrt{z_0^2 - z^2}}\, \mathrm{d}z. \tag{1.104}$$

At very high temperatures, z is always small, and the integral is approximately $\pi/2$, reproducing the classical result $\langle E_{\mathrm{tot}} \rangle \to Nk_{\mathrm{B}}T$. The high-temperature specific heat, that is, the heat capacity per unit length, is then given by

$$c_V = \frac{1}{Nr_0} \frac{\partial \langle E_{\mathrm{tot}} \rangle}{\partial T} = \frac{k_{\mathrm{B}}}{r_0} \qquad (k_{\mathrm{B}}T/\hbar\omega_0 \gg 1). \tag{1.105}$$

At low temperatures, with $k_{\mathrm{B}}T/\hbar\omega_0 \ll 1$, we can take the upper limit of the integral in (1.104) as infinite, so we find

$$\begin{aligned} \langle E_{\mathrm{tot}} \rangle &\to \frac{2N}{\pi\beta z_0} \int_0^{\infty} \frac{z}{\mathrm{e}^z - 1}\, \mathrm{d}z \approx \frac{\pi N}{3\beta z_0} \\ &\approx \frac{\pi}{6} N \frac{1}{\hbar\omega_0} (k_{\mathrm{B}}T)^2 \qquad (k_{\mathrm{B}}T/\hbar\omega_0 \ll 1). \end{aligned} \tag{1.106}$$

The total energy scales as T^2. The low-temperature specific heat is thus proportional to temperature:

$$c_V = \frac{\pi}{3r_0} \frac{k_{\mathrm{B}}^2 T}{\hbar\omega_0} \qquad (k_{\mathrm{B}}T/\hbar\omega_0 \ll 1). \tag{1.107}$$

This is the one-dimensional form of the Debye law for the specific heat of an insulator; in three dimensions, as we shall see, the low-temperature specific heat scales as T^3.

For moderate temperatures, between these two extremes, the integrals can be evaluated numerically. The integral form for the specific heat is

$$\begin{aligned} c_V &= 2k_B \int_0^{2\omega_0} (\beta\hbar\omega)^2 \frac{e^{\beta\hbar\omega}}{(e^{\beta\hbar\omega}-1)^2} \mathcal{D}(\omega)\, d\omega \\ &= \frac{2k_B}{\pi r_0} \int_0^{z_0} \frac{z^2 e^z}{(e^z-1)^2} \frac{1}{\sqrt{z_0^2 - z^2}}\, dz. \end{aligned} \tag{1.108}$$

The numerically calculated integral is plotted in Fig. 1.21, along with the result from the classical Dulong–Petit result and the specific heat for the simple harmonic oscillator.

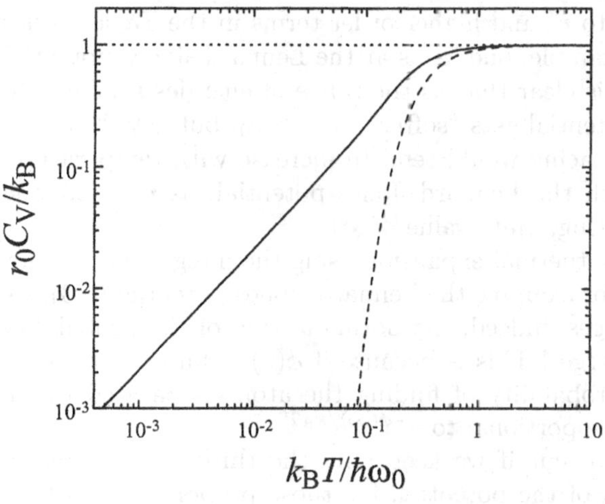

Fig. 1.21. Specific heat for the N-atom chain (*solid line*), as well as that from the Dulong–Petit classical result (*dotted line*) and for the simple harmonic oscillator with resonance frequency ω_0 (*dashed line*).

1.6.2 Thermal Expansion

We now look at the problem of thermal expansion. Increasing the temperature of most solids causes their dimensions to increase, with typical expansion coefficients in the range of 10-100 parts per million per °C, so an increase of 10°C will lengthen a 1 m bar by 0.1-1 mm. This thermally-induced strain in the bar is uniform, so the distance between any two atoms will change by the same proportion as the overall length. Let's look at what our models predict for the average spacing r between two atoms as a function of temperature; we assume the atoms are coupled by a spring, with rest length r_0. The potential energy is $\phi(r) = k(r - r_0)^2/2$, and the mean spacing r as a function of temperature is given by the classical expression

$$\begin{aligned} \langle r \rangle &= \int_{-\infty}^{\infty} r P(r)\, \mathrm{d}r & (1.109) \\ &= \frac{\int_{-\infty}^{\infty} r\, \mathrm{e}^{-k(r-r_0)^2/2k_\mathrm{B}T}\, \mathrm{d}r}{\int_{-\infty}^{\infty} \mathrm{e}^{-k(r-r_0)^2/2k_\mathrm{B}T}\, \mathrm{d}r} & (1.110) \\ &= r_0 & (1.111) \end{aligned}$$

(note the same result holds for the quantized system). The thermally averaged distance remains fixed at r_0, independent of temperature. Hence for a spring, or in other words, the harmonic approximation to a real atomic potential, there is no thermal expansion!

Thermal expansion is actually due to the anharmonic terms in the potential, those proportional to r^3 and higher order terms in the Taylor series expansion of $\phi(r)$. If, for example, one looks at the Lennard–Jones potential (1.3), shown in Fig. 1.1, it is clear that as the range of energies available to the atoms increases, the potential gets "softer" for $r > r_0$, but gets "harder" for $r < r_0$, so the atomic spacing would tend to increase with temperature. In fact, as can be seen with the Lennard–Jones potential, as $r \to \infty$, the potential approaches a limiting, finite value $\phi(\infty)$.

If we try to calculate the thermal expansion using the integral (1.109) for a non-harmonic potential, for example the Lennard–Jones potential (1.3), we find that the integral diverges; indeed, the normalization of the probability distribution $P(r)$ itself diverges! This is because if $\phi(r)$ is finite as $r \to \infty$, then there is a non-zero probability of finding the atoms separated by an arbitrarily large distance, proportional to $\mathrm{e}^{-\phi(\infty)/k_\mathrm{B}T}$.

The same problem can occur if we keep only the third-order terms in the Taylor series expansion of the potential, i.e. those proportional to $(r - r_0)^3$. If the potential gets softer as r increases, the third order term will be negative, and at large r will diverge towards $-\infty$, so again the integrals present problems. However, it is actually possible to find a value for the thermal expansion due to the third-order term in the potential, by using a variational approach coupled with an integration technique known as the *method of steepest descent*. Following Marder [6], we can write the mean position $\langle r \rangle$ as a formal variation

$$\begin{aligned} \langle r \rangle &= \frac{\int r \mathrm{e}^{-\phi(r)/k_\mathrm{B}T} \mathrm{d}r}{\int \mathrm{e}^{-\phi(r)/k_\mathrm{B}T} \mathrm{d}r} \\ &= \frac{\partial}{\partial z}\left(\ln \int_0^{\infty} \mathrm{e}^{zr-\phi(r)/k_\mathrm{B}T} \right)\Bigg|_{z=0}, & (1.112) \end{aligned}$$

where z is a variational parameter, taken equal to zero at the end of the calculation. Writing the potential to third order in r as

$$\phi(r) \approx \frac{1}{2}k(r-r_0)^2 - \frac{1}{6}\gamma(r-r_0)^3, \tag{1.113}$$

where γ is the size of the anharmonicity, then using the method of steepest descent, (1.112) can be shown to be given to first order in T by

$$\langle r \rangle \approx r_0 + \frac{\gamma}{2k^2}k_B T. \tag{1.114}$$

Hence the thermal expansion is proportional to γ (note that the sign of γ is positive if the potential becomes softer than a harmonic potential for large r, as is the case for most real potential functions).

1.7 Quantum Operators for Normal Modes

Here we provide a brief discussion on how to generate the quantum mechanical operators for the normal modes of the N atom chain. We begin, however, with the simple harmonic oscillator.

1.7.1 Quantum Operators for the Simple Harmonic Oscillator

Any system, when described in the language of quantum mechanics, can have its quantum state specified by a *ket*; if the system is in a quantum eigenstate of energy, this is written $|E\rangle = |n\rangle$, for the nth energy eigenstate, while an eigenstate of momentum is written $|p\rangle$. The energy, position and momentum quantum mechanical operators $\hat{H}$, $\hat{r}$, and $\hat{p}$ all operate on the ket (the hat indicates a quantum mechanical operator rather than a classical variable).

If a simple harmonic oscillator is in the nth energy eigenstate, with the state $|n\rangle$, then the energy is $\hat{H}\,|n\rangle = (n+1/2)\hbar\omega_0\,|n\rangle$. A harmonic oscillator in an eigenstate of momentum, with eigenvalue p_0, has $\hat{p}\,|p_0\rangle = p_0\,|p_0\rangle$.

Quantum operators in general do not *commute*, that is, the result of operating first with one operator on a ket, and then with another, is not the same as reversing the order of operation; for example, the position and momentum operators $\hat{r}$ and $\hat{p}$ have the commutation relation

$$[\hat{r},\hat{p}] = \hat{r}\,\hat{p} - \hat{p}\,\hat{r} = i\hbar. \tag{1.115}$$

When two operators do not commute, that is, the expression $[\hat{a},\hat{b}] \neq 0$, the eigenstates of one of the operators cannot be eigenstates of the other operator. If we have a system whose state is an eigenstate of the operator $\hat{a}$, so that $\hat{a}\,|a_1\rangle = a_1\,|a_1\rangle$, then when we operate with $\hat{b}$, the result will be a different state $|x\rangle$, that may not be an eigenstate of either operator. A subsequent operation with $\hat{a}$ will then yield a result that may have no relation to a_1. A more complete discussion of operator algebra, and bra and ket notation, can be found in most introductory books on quantum mechanics, for example that of Cohen–Tannoudji, Diu and Laloé [7].

For the simple harmonic oscillator, the position, momentum and energy operators do not commute. The Hamiltonian operator is written

$$\hat{H} = \frac{\hat{p}^2}{2M} + \frac{k}{2}(\hat{r} - r_0)^2, \tag{1.116}$$

in terms of the momentum and position operators, with the resonance frequency $\omega_0 = \sqrt{k/M}$. The eigenstates of the Hamiltonian are written as $|n\rangle$, with

$$\hat{H}|n\rangle = \left(n + \frac{1}{2}\right)\hbar\omega_0 |n\rangle. \tag{1.117}$$

The commutator of the Hamiltonian with the position and momentum operators is

$$\left.\begin{aligned} \left[\hat{H}, \hat{r}\right] &= \hat{H}\hat{r} - \hat{r}\hat{H} = -\mathrm{i}\hbar\frac{\hat{p}}{M}, \\ \left[\hat{H}, \hat{p}\right] &= \hat{H}\hat{p} - \hat{p}\hat{H} = +\mathrm{i}\hbar k\hat{r}. \end{aligned}\right\} \tag{1.118}$$

From these relations we can build the *raising* operator $\hat{a}^\dagger$ (also known as the *creation* operator), and the *lowering* operator $\hat{a}$ (also known as the *destruction* or *annihilation* operator):

$$\left.\begin{aligned} \hat{a}^\dagger &= \sqrt{\frac{M\omega}{2\hbar}}\hat{r} - i\frac{1}{\sqrt{2\hbar M\omega}}\hat{p}, \\ \hat{a} &= \sqrt{\frac{M\omega}{2\hbar}}\hat{r} + i\frac{1}{\sqrt{2\hbar M\omega}}\hat{p}. \end{aligned}\right\} \tag{1.119}$$

These operators are constructed to have the following effect on the eigenstates $|n\rangle$ of the Hamiltonian:

$$\left.\begin{aligned} \hat{a}^\dagger|n\rangle &= \sqrt{n+1}\,|n+1\rangle, \\ \hat{a}|n\rangle &= \sqrt{n}\,|n-1\rangle. \end{aligned}\right\} \tag{1.120}$$

These relations can be shown by working out the commutator of the raising and lowering operators with the Hamiltonian $\hat{H}$. In other words, the raising operator $\hat{a}^\dagger$ generates an eigenstate of the Hamiltonian with n increased by one, and the lowering operator $\hat{a}$ does the same, but lowers n by one. Furthermore one can show that the Hamiltonian (1.116) may be written in the form

$$\hat{H} = \hbar\omega_0\left(\hat{a}^\dagger\hat{a} + \frac{1}{2}\right). \tag{1.121}$$

We can also define the *number operator*

$$\hat{n} = \hat{a}^\dagger\hat{a}. \tag{1.122}$$

1.7.2 Quantum Operators for the N-Atom Chain

Building on the formalism for the simple harmonic oscillator, we want a similar operator formalism for the normal modes of the N-atom chain. We have already shown that the Hamiltonian for the chain can be simply written in the form (1.56), which we rewrite here using operator formalism:

$$\hat{H} = \frac{1}{2M}\sum_m \hat{\mathcal{P}}_m \hat{\mathcal{P}}_m^\dagger + k\sum_m (1 - \cos 2\pi m/N)\, \hat{\mathcal{U}}_m \hat{\mathcal{U}}_m^\dagger, \tag{1.123}$$

for the amplitude operators $\hat{\mathcal{U}}_m$ and momentum operators $\hat{\mathcal{P}}_m$ of the normal modes m. This form is very similar to that of (1.116). The energy eigenstates of this Hamiltonian, of the form $|n_1\, n_2\, n_3\, \ldots\, n_m\, \ldots\rangle$, with n_1 energy quanta in the first mode, n_2 in the second, and so on, satisfy

$$\hat{H}\,|n_1\, n_2\, \ldots\rangle = \sum_m \left(n_m + \frac{1}{2}\right) \hbar\omega_m\, |n_1\, n_2\, \ldots\rangle\,. \tag{1.124}$$

The commutation relations (1.115) and (1.118) will hold for each position and momentum operator. We can, for example, write

$$\left[\hat{\mathcal{U}}_m, \hat{\mathcal{P}}_n\right] = \hat{\mathcal{U}}_m\, \hat{\mathcal{P}}_n - \hat{\mathcal{P}}_n \hat{\mathcal{U}}_m = \mathrm{i}\hbar\, \delta_{mn}, \tag{1.125}$$

using the Kronecker delta function, $\delta_{mn} = 1$ if $m = n$ and $\delta_{mn} = 0$ if $m \neq n$.

Now we define the set of normal mode raising and lowering operators $\hat{a}_m^\dagger$ and $\hat{a}_m$ by

$$\left.\begin{aligned} \hat{a}_m^\dagger &= \sqrt{\frac{M\omega_m}{2\hbar}}\, \hat{\mathcal{U}}_m^\dagger - \mathrm{i}\frac{1}{\sqrt{2\hbar M\omega_m}}\, \hat{\mathcal{P}}_m, \\ \hat{a}_m &= \sqrt{\frac{M\omega_m}{2\hbar}}\, \hat{\mathcal{U}}_m + \mathrm{i}\frac{1}{\sqrt{2\hbar M\omega_m}}\, \hat{\mathcal{P}}_m^\dagger. \end{aligned}\right\} \tag{1.126}$$

Note the close similarity with the classical normal coordinates given by (1.58). These operators, acting on the eigenstates $|n_1\, n_2\, n_3\, \ldots\rangle$ of the Hamiltonian, yield

$$\left.\begin{aligned} \hat{a}_m^\dagger\, |n_1 \ldots n_m \ldots\rangle &= \sqrt{n_m + 1}\, |n_1 \ldots n_m + 1 \ldots\rangle\,, \\ \hat{a}_m\, |n_1 \ldots n_m \ldots\rangle &= \sqrt{n_m}\, |n_1 \ldots n_m - 1 \ldots\rangle\,, \end{aligned}\right\} \tag{1.127}$$

leaving the quanta in all the other modes unchanged.

The number operator for the mth mode is

$$\hat{n}_m = \hat{a}_m^\dagger\, \hat{a}_m, \tag{1.128}$$

and the Hamiltonian can be written as

$$\hat{H} = \sum_m \hbar\omega_m \left(\hat{a}_m^\dagger\, \hat{a}_m + \frac{1}{2}\right) = \sum_m \hbar\omega_m \left(\hat{n}_m + \frac{1}{2}\right). \tag{1.129}$$

We will find this formalism useful later in the text, when dealing with quantum mechanical interactions between the lattice vibrations and other excitations in the solid.

Exercises

1.1 For the Lennard–Jones potential, in the presence of an external force, find the expression for the new equilibrium point r_0' as a function of f_{ext}, and the maximum force for which an equilibrium point still exists. Draw a plot showing the displacement from equilibrium $\delta = r_0' - r_0$ as a function of f_{ext}, showing the linear region and where the linear approximation breaks down.

1.2 Find an expression for the location r_1 of the maximum force in the Lennard–Jones potential, in terms of the equilibrium spacing r_0, and an expression for the maximum force in terms of the binding energy E_b r_0.

1.3 Develop an interaction model for two ions with charges $\pm e$, the charge on the electron; assume a $1/r^{12}$ repulsive potential, with equilibrium spacing of 3 Å. What is the binding energy? Find the natural resonance frequency if the ions are sodium (mass number 23) and chlorine (mass number 35).

1.4 Work out in detail the calculations for the dispersion relation and allowed values of the wavevector for the N atom chain with fixed boundary conditions.

1.5 Work out the equivalent of the periodic boundary condition problem for the three atom chain, with atom 1 connected to atom 3 by a spring. Find the normal mode frequencies and amplitude eigenvectors.

1.6 Calculate the dispersion relations and allowed values of wavevector for a linear chain of masses, where all the masses are equal but are connected by springs that alternate between k and k', as shown in Fig. 1.22. Assume equilibrium spacing r_0.

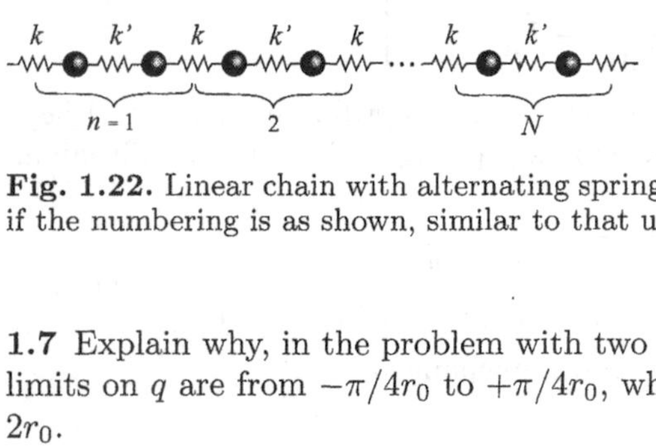

Fig. 1.22. Linear chain with alternating spring constants; the calculation is easiest if the numbering is as shown, similar to that used for alternating mass atoms.

1.7 Explain why, in the problem with two different masses in the chain, the limits on q are from $-\pi/4r_0$ to $+\pi/4r_0$, where the mass pairs are spaced by $2r_0$.

1.8 Derive the dispersion relation $\omega(q)$ for transverse waves whose form is given by (1.69).

1.9 Derive the full equation of motion for the N atom chain under transverse motion, without making any of the approximations regarding angles and wavelengths that were used to obtain (1.73).

1.10 Find the dispersion relation for transverse waves, starting from the equation (1.73).

1.11 Show that the normal coordinates $\mathcal{U}_m$ defined through (1.48) are consistent with the relations (1.49); a key to proving this result is the identity

$$\sum_{m=1}^{N} \mathrm{e}^{\mathrm{i}2\pi(n+n')m/N} = N\delta_{n,-n'} = \begin{cases} 0 & n \neq -n' \\ N & n = -n'. \end{cases} \tag{1.130}$$

1.12 Fill in the steps to show that (1.51) is correct.

1.13 Derive the expression (1.54) from the expressions for the normal mode coordinates and their momenta.

1.14 Find the dispersion relation $\omega(q)$ for the linear chain bound by torsional springs, as given by (1.73).

1.15 Find the dispersion relations $\omega_{\pm}(q)$ for a linear chain of atoms of mass M, linked by alternating springs k and k'; this is analogous to the two-mass problem discussed in Sect. 1.4. Plot the two frequencies as a function of q, for $k/k' = 2$. What is the bandgap width in terms of the parameters in the problem?.

1.16 Fill in the steps in the calculation of the thermally-averaged kinetic and potential energy, from (1.87) through (1.89).

1.17 In the derivation of the expression for the thermal energy, (1.102), and that for the specific heat, (1.108), we made the assumption that the temperature was high enough that the thermal energy $k_{\mathrm{B}}T \gg \hbar\Delta\omega = \hbar(\omega_{m+1} - \omega_m)$, the energy spacing between modes. Find out what the correct expressions are for the thermal energy and specific heat when the temperature is so low that $k_{\mathrm{B}}T \ll \hbar\Delta\omega$.

1.18 Find the thermal expansion coefficient for a one-dimensional chain of argon atoms.

2. Two- and Three-Dimensional Lattices

In this chapter we move from the one-dimensional atomic chain to two- and three-dimensional systems. We introduce the concept of the crystal lattice and the corresponding concept of the reciprocal lattice; we then work out the formalism for calculating the normal modes in a three-dimensional infinite solid, with a brief discussion of how this can also be applied to finite objects. We conclude our discussion by discussing the quantized excitations of the vibrational normal modes of the solid, known as *phonons*. The properties of phonons and of the phonon gas will be worked out in Chap. 3.

2.1 Crystal Lattices

In going to more than one dimension, we add a degree of freedom that we have not yet considered: The geometry in which the atoms are arranged. In one dimension, we had the freedom to choose the atomic spacing, and consider systems with more than one type (mass or spring constant) of atom. In two or three dimensions, we can consider a much wider range of different arrangements for the atoms, adding a range of geometries not available in one dimension. We will restrict our discussion to regular, repeated structures known as *crystals*. These consist of an infinite set of points, arranged in a repeated structure that fills space, a geometry that can be generated by repeated translations of a single starting point. This set of points is known as the *space lattice*. Each point in the space lattice is associated with an arrangement of one or more atoms, known as the *basis*; the basis, repeated in the space lattice, provides a complete description of a crystal. For simplicity, we will usually assume the solid is infinite, treating the boundaries of a real, finite solid as perturbations of the infinite system. Note that some of the formalism that we will use is strictly correct only in infinite solids, but is a good approximation in all but the smallest systems; an example of this is Bloch's theorem, discussed below.

The space lattice is defined by the set of repeated points $\{\boldsymbol{R}\}$ that fill space. In two dimensions, the set $\{\boldsymbol{R}\}$ can be generated from two linearly independent vectors $\boldsymbol{a}_1$ and $\boldsymbol{a}_2$, known as the *generating vectors* for the lattice; the lattice points are formed by the linear combinations

$$R = n_1 a_1 + n_2 a_2, \tag{2.1}$$

with integers n_1 and n_2. In three dimensions it takes three such independent generating vectors, a_1, a_2 and a_3, with lattice points R given by

$$R = n_1 a_1 + n_2 a_2 + n_3 a_3, \tag{2.2}$$

for integers n_1, n_2 and n_3.

The space surrounding the lattice points R can be divided up into a regular, repeated arrangement of *unit cells*. In two dimensions, the unit cell is an area, and in three dimensions the unit cell is a volume. The unit cells, which are all identical, are defined in such a way that they fill space without overlapping. The smallest unit cell that can be formed for a particular lattice is called the *primitive unit cell*; it may contain one or more atoms, depending on the structure of the crystal, and it is not unique. A simple construction exists that generates a particular, uniquely defined primitive unit cell known as the *Wigner–Seitz cell*: You take all the vectors from the origin $R = 0$ to all the other points in the lattice, and create the set of planes that bisect these vectors. You then form the smallest area, or volume in three dimensions, centered on the origin, that is bounded by these planes. This area (volume) is by construction a primitive unit cell, and when translated to each lattice point, fills space without overlapping. The Wigner–Seitz cell is characterized by the fact that all points in a unit cell associated with a lattice point R are closer to that point than any other point in the space lattice.

2.2 Two-Dimensional Crystals

There are an infinite number of possible bases, or atomic arrangements, for a given space lattice. There are however a very limited number of geometrically distinct space lattices, which are sorted into groups based on the symmetries of the set of lattice points. Each group, or symmetry class, is known as a *Bravais lattice.* In two dimensions, it turns out there are only five distinct Bravais lattices; these are shown in Fig. 2.1.

The simplest two-dimensional Bravais lattice is the simple cubic lattice, consisting of all points $R = (n_1 d, n_2 d)$, for integers n_1 and n_2, both positive and negative, with lattice spacing d. The generating vectors for this space lattice are $a_1 = (d, 0)$ and $a_2 = (0, d)$. The Wigner–Seitz cell is a square of side d.

The rectangular Bravais lattice is similar to the simple cubic, but has two different lattice spacings, d_1 and d_2, with generating vectors $a_1 = (d_1, 0)$ and $a_2 = (0, d_2)$. The lattice is then the set of points $R = (n_1 d_1, n_2 d_2)$, for all integers n_1 and n_2. The Wigner–Seitz cell is a $d_1 \times d_2$ rectangle.

A somewhat more complicated two-dimensional lattice is the hexagonal lattice, with generating vectors $a_1 = (d, 0)$ and $a_2 = (d/2, \sqrt{3}d/2)$. The

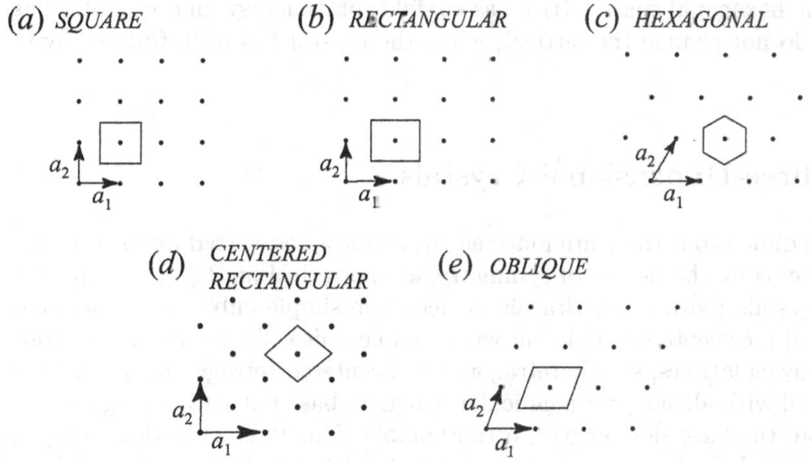

Fig. 2.1a–e. The set of five Bravais lattices in two dimensions. The lattice points, generating vectors, and unit cells are shown for each lattice type.

Wigner–Seitz cell is a hexagon, centered on a lattice point, with vertices $d/2$ from the center.

The simplest crystals have one atom per lattice point, or in other words a one-atom basis. The symmetry of the crystal is then that of the Bravais lattice. As we add more atoms to the basis, the particular arrangement of atoms can reduce the symmetry of the crystal from that of the underlying Bravais lattice; a sufficiently asymmetric basis can reduce a highly symmetric Bravais lattice to one with the lowest degree of symmetry (the oblique lattice). An example is shown in Fig. 2.2, where the Bravais lattice is hexagonal, but the three atoms in the basis reduce the symmetry to that of the oblique lat-

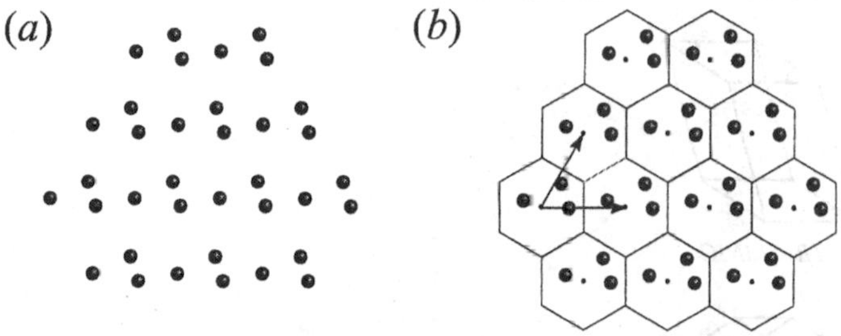

Fig. 2.2. (**a**) A two-dimensional oblique crystal, showing the arrangement of atoms, and (**b**) the space lattice (*shown as solid points*), with the Bravais symmetry of the hexagonal lattice, the unit cell (*the set of space-filling hexagons*), and the three-atom basis. The two generating vectors for the hexagonal Bravais lattice are also shown.

tice: The hexagonal space lattice has six-fold rotational symmetry (rotations by 60° do not change the lattice), while the crystal has no rotational symmetry.

2.3 Three-Dimensional Crystals

In three dimensions there are fourteen Bravais lattices, sorted into seven crystal systems by the degree of symmetry, as shown in Fig. 2.3. These are the cubic crystals (with three Bravais lattices, the simple cubic, body-centered cubic and face-centered cubic, all with a cubic unit cell), tetragonal crystals (two Bravais lattices, simple tetragonal and centered tetragonal, which have a unit cell with all sides perpendicular, a square base but with a height different from the base side length), orthorhombic (four Bravais lattices, simple, base-centered, body-centered, face-centered, with a unit cell with all faces perpendicular but all sides of different lengths), monoclinic (two Bravais lattices, simple and centered, with a unit cell similar to the orthorhombic but base is no longer rectangular), trigonal (one Bravais lattice, formed by three equal-length generating vectors at equal angles to one another), hexagonal (one Bravais lattice, with a regular hexagon as a base and perpendicular height), and finally the triclinic (one Bravais lattice, where all three sides are non-perpendicular and possibly of different lengths; the triclinic can be formed by any three generating vectors, and has the lowest symmetry, that

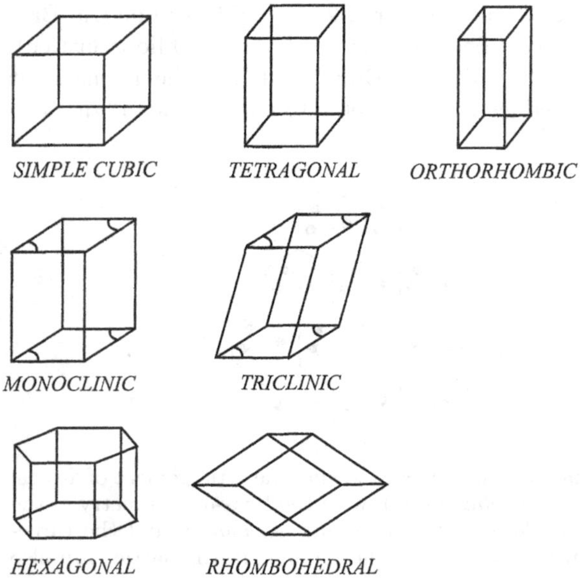

Fig. 2.3. The seven crystal systems in three dimensions.

of inversion only). A good text describing in detail the structures of three-dimensional crystals is that of Wyckoff [8].

The simplest three-dimensional lattice is the simple cubic lattice, consisting of points evenly spaced along all three axes, $\boldsymbol{R} = (\ell d, md, nd)$ for integers ℓ, m, and n, with lattice spacing d. The primitive unit cell is a cube of side d centered on each lattice point, with volume d^3. The simple cubic crystal is rarely found in nature; one element that forms a simple cubic structure is *polonium* (Po).

Two much more common lattice geometries are the face-centered cubic (fcc) and body-centered cubic (bcc) lattices, shown in Fig. 2.4. The face-centered cubic lattice consists of atoms at the eight corner points of a cube, and an additional six atoms at the center of each cube face. A single fcc cube therefore contains a total of four atoms, one-eighth of each corner atom and half of each atom on each face. The fcc lattice is generated by the three vectors $\boldsymbol{a}_1 = (0, d/2, d/2)$, $\boldsymbol{a}_2 = (d/2, 0, d/2)$, and $\boldsymbol{a}_3 = (d/2, d/2, 0)$, where d is the cube side length.

It is simpler, however, to describe the fcc lattice as a simple cubic lattice, with generating vectors along three sides (so $\boldsymbol{a}_1 = (d, 0, 0)$, $\boldsymbol{a}_2 = (0, d, 0)$ and $\boldsymbol{a}_3 = (0, 0, d)$) and a four-atom basis, consisting of the corner atom at $(0, 0, 0)$ and the three atoms on each of the adjacent faces, such as those at $(0.5d, 0.5d, 0)$, $(0, 0.5d, 0.5d)$ and $(0.5d, 0, 0.5d)$.

Elements that have the fcc structure as their natural, lowest-energy structure include copper (Cu), silver (Ag), gold (Au), aluminum (Al), nickel (Ni), palladium (Pd) and platinum (Pt).

The body-centered cubic lattice consists of atoms at each corner point of a cube, and in addition a single atom at the center of the cube. Each cube therefore contains two atoms (one-eighth of each corner atom, and the center atom). The bcc lattice is generated by the three vectors $\boldsymbol{a}_1 = (d/2, d/2, d/2)$, $\boldsymbol{a}_2 = (-d/2, d/2, d/2)$, and $\boldsymbol{a}_3 = (d/2, d/2, -d/2)$. The more commonly used description is however a simple cubic lattice, with generating vectors along the three sides (along x_1, x_2 and x_3), and a two-atom basis consisting of the corner atom at $(0, 0, 0)$ and the center atom at $(0.5d, 0.5d, 0.5d)$.

Elements that have the bcc structure include most of the alkali metals (lithium (Li), sodium (Na), potassium (K), rubidium (Rb) and cesium (Cs)),

(*a*)
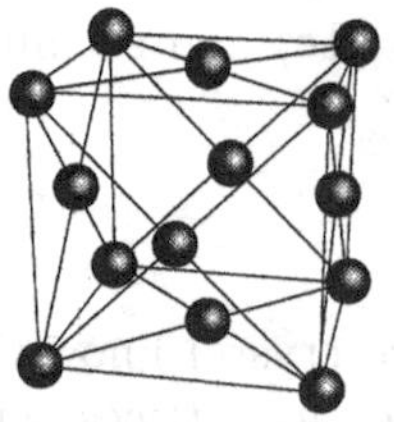
(*b*)
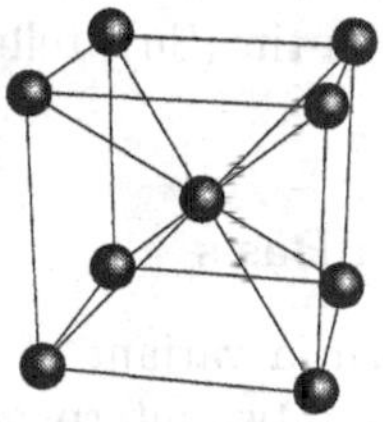

Fig. 2.4. (**a**) Face-centered cubic (fcc) unit cell, and (**b**) body-centered cubic (bcc) unit cell.

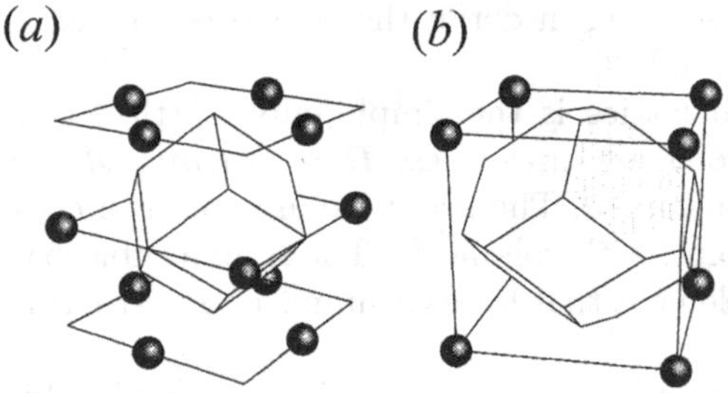

Fig. 2.5. (**a**) Face-centered cubic (fcc) Wigner–Seitz cell, and (**b**) body-centered cubic (bcc) Wigner–Seitz cell.

as well as the transition metals vanadium (V), chromium (Cr), niobium (Nb), molybdenum (Mo), tantalum (Ta) and tungsten (W).

In Fig. 2.5 we display the Wigner–Seitz cells for the fcc and bcc lattices. These are the primitive unit cells for the fcc and bcc lattices with a one-atom basis; the Wigner–Seitz cell for the simple cubic lattice is a cube of side d centered on each lattice point.

The simple hexagonal Bravais lattice forms the basis for another commonly-occurring crystal, the *hexagonal close-packed* (hcp) structure. The simple hexagonal lattice consists of atoms in a triangular pattern in the $x_1 - x_2$ plane. This layer is then repeated by regular displacements along x_3 axis to form the three-dimensional structure. The generating vectors for the simple hexagonal lattice are $\boldsymbol{a}_1 = (a, 0, 0)$, $\boldsymbol{a}_2 = (a/2, \sqrt{3}a/2, 0)$ and $\boldsymbol{a}_3 = (0, 0, c)$, where a is the repeat distance in the plane and c the perpendicular spacing.

The hexagonal close-packed structure is the simple hexagonal lattice with a two-atom basis. It consists of the layered simple hexagonal structure, but with an additional layer interspersed with those already defined. The atoms in the additional layer have the same orientation as in the layers above and below, but are displaced laterally by $\boldsymbol{a}_1/3 + \boldsymbol{a}_2/3$, and vertically by $\boldsymbol{a}_3/2$. The layer planes are known as the ab planes, and the axis perpendicular to the ab plane is the c axis. The hcp crystal is a simple hexagonal lattice with a two-atom basis, one atom at the origin and the second at $(a/2, a/2\sqrt{3}, c/2)$ (see Fig. 2.6). Elements that form the hcp crystal structure include beryllium (Be), magnesium (Mg), a large number of transition metals, such as titanium (Ti), cobalt (Co), zinc (Zn), yttrium (Y), zirconium (Zr), and cadmium (Cd), among others, and materials such as tin (Sn), tellurium (Te) and selenium (Se).

2.3.1 Crystal Structures with a Basis

Carbon, silicon and germanium form a variant of the fcc crystal known as the *diamond structure*; this consists of two interpenetrating fcc lattices, with the origin of the second lattice shifted from that of the first lattice by the vector $(d/4, d/4, d/4)$, where d is the length of the fcc cube.

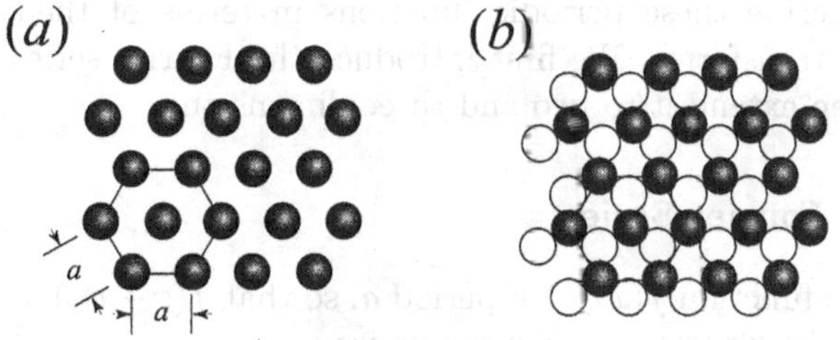

Fig. 2.6. (**a**) First layer of a hexagonal close-packed crystal, with in-plane spacing a, and (**b**) second layer, with atoms at hollows between atoms in first layer.

The compound semiconductor GaAs is in the *zincblende* structure, which also consists of two interpenetrating fcc lattices; one lattice is filled with Ga atoms, and the other with As atoms. This is a common structure for compound materials; AlAs, InAs, GaSb, InSb, GaP, InP and AlP, and a number of others, including SiC, are all in the zincblende structure.

As mentioned above, the hcp structure can be viewed as two interpenetrating simple hexagonal lattices. Some compound elements (ZnS, AlN, GaN, etc.) form the *wurtzite* crystal structure, which consists of two interpenetrating hcp lattices, one lattice filled with one element and the second lattice with the other element. The second hcp lattice is displaced along the c axis by $3/8c$ from the first lattice, with atoms placed directly above. In this arrangement, atoms of one element in the first lattice are equidistant from four atoms of the other element in the second lattice, and vice versa.

2.4 Periodic Functions

When dealing with regular one-dimensional chains, or two and three dimensional crystal lattices, the structure of the solid clearly imposes a strong periodicity on all its physical properties. A periodic function $f(\boldsymbol{r})$ with the periodicity of, say, a three dimensional space lattice, satisfies $f(\boldsymbol{r}+\boldsymbol{R}) = f(\boldsymbol{r})$, for a lattice vector $\boldsymbol{R}$. The ion mass density $\rho(\boldsymbol{r})$ in a crystal , for example, has the periodicity of the crystal space lattice. Note, however, that the *strict* periodicity of a crystal actually only holds when the crystal is in its *ground state*, where no thermal motion perturbs the ion positions; this occurs only at zero temperature.

Many of the microscopic physical properties of a crystal reflect this underlying *translational symmetry* of the crystal: The electron mass density, the total charge density, and so on. Because ions interact with one another, and with the electrons, through forces that depend on these periodic quantities, the interaction force and the interaction potential energy, $\phi(\boldsymbol{r})$, also have the periodicity of the crystal.

It is convenient to describe these periodic functions in terms of their Fourier series and Fourier transforms. We first introduce the Fourier series for one dimension, and then extend it to two and three dimensions.

2.4.1 One-Dimensional Fourier Series

A one-dimensional periodic function $f(x)$ with period a, so that $f(x+na) = f(x)$ for any integer n, can be written as a Fourier series

$$f(x) = \sum_{n=-\infty}^{\infty} F_n \mathrm{e}^{\mathrm{i}2\pi nx/a}. \tag{2.3}$$

For notational convenience, we define the set of points G by $G_n = 2\pi n/a$ for integers n. We can then write (2.3) as a sum over G,

$$f(x) = \sum_G F_G \, \mathrm{e}^{\mathrm{i}Gx}. \tag{2.4}$$

The Fourier coefficients F_G are given by

$$F_G = \frac{1}{a}\int_{-a/2}^{a/2} \mathrm{e}^{-\mathrm{i}Gx} \, f(x) \, \mathrm{d}x. \tag{2.5}$$

This can be verified by multiplying both sides of (2.3) by $\mathrm{e}^{-\mathrm{i}Gx}/a$ and integrating over one period of $f(x)$. The proof is left to the exercises (see Exercise 2.1).

We can write (2.4) in integral form using the Dirac delta function $\delta(k)$, described in Appendix A.3. Using a continuous variable k, we define a new function $F(k)$, the *Fourier transform* of $f(x)$, by

$$F(k) = 2\pi \sum_G F_G \, \delta(k - G). \tag{2.6}$$

$F(k)$ consists of a series of peaks, located at the points $k = G_n = 2\pi n/a$, of varying height F_G. We can relate the original periodic function $f(x)$ to $F(k)$ by

$$f(x) = \frac{1}{2\pi}\int_{-\infty}^{\infty} F(k) \, \mathrm{e}^{\mathrm{i}kx} \, \mathrm{d}k. \tag{2.7}$$

The Fourier transform is defined as a function of k, and the one-dimensional "space" k is known as *reciprocal space.* The set of special points $k = G$ for which the Fourier transform $F(k)$ is non-zero forms a one-dimensional lattice in k-space, with lattice spacing $2\pi/a$. This lattice of points is called the *reciprocal lattice.*

Functions defined in reciprocal space are as useful as those defined in real space. A periodic function $\Phi(k)$ in reciprocal space, with periodicity $\Phi(k + G) = \Phi(k)$, can be written as a series analogous to (2.3),

$$\Phi(k) = \sum_R \mathrm{e}^{\mathrm{i}kR} \phi_R, \tag{2.8}$$

with Fourier coefficients ϕ_R given by

$$\phi_R = \frac{a}{2\pi} \int_{-\pi/a}^{\pi/a} \mathrm{e}^{-\mathrm{i}kR}\, \Phi(k)\, \mathrm{d}k, \tag{2.9}$$

where the points R belong to the real space lattice, $R_n = na$. Defining the real space function $\phi(r)$ by

$$\phi(r) = \sum_R \phi_R\, \delta(r - R), \tag{2.10}$$

we can write the *inverse Fourier transform* as

$$\Phi(k) = \int_{-\infty}^{\infty} \phi(r)\, \mathrm{e}^{-\mathrm{i}kr}\, \mathrm{d}r. \tag{2.11}$$

Note that the Fourier transform and its inverse can be defined for non-periodic as well as for periodic functions.

2.4.2 Two and Three Dimensional Fourier Series: The Reciprocal Lattice

Two and three dimensional Fourier transforms are simple extensions of the one dimensional transform. In one dimension we discussed a periodic function $f(x)$ with period a, and found that the Fourier series involved a set of special points G in reciprocal space, given by $G_n = 2\pi n/a$. In two and three dimensions, for a function $f(\boldsymbol{r})$ that has the periodicity $\boldsymbol{R}$ of a real-space lattice, the Fourier series involves a similar set of special points $\boldsymbol{G}$, which form a regular, periodic reciprocal lattice in k-space. The k-space lattice is intimately related to the real space lattice $\boldsymbol{R}$. In three dimensions, the reciprocal lattice is generated by three vectors $\boldsymbol{b}_1$, $\boldsymbol{b}_2$ and $\boldsymbol{b}_3$, so any reciprocal lattice vector $\boldsymbol{G}$ can be written as a linear combination, $\boldsymbol{G} = \ell \boldsymbol{b}_1 + m\boldsymbol{b}_2 + n\boldsymbol{b}_3$ for integers ℓ, m and n. The generating vectors $\boldsymbol{b}_1$, $\boldsymbol{b}_2$ and $\boldsymbol{b}_3$ are related to the real space generating vectors $\boldsymbol{a}_1$, $\boldsymbol{a}_2$ and $\boldsymbol{a}_3$ through the relations

$$\left.\begin{aligned} \boldsymbol{b}_1 &= \frac{2\pi}{v_c}\, \boldsymbol{a}_2 \times \boldsymbol{a}_3, \\ \boldsymbol{b}_2 &= \frac{2\pi}{v_c}\, \boldsymbol{a}_3 \times \boldsymbol{a}_1, \\ \boldsymbol{b}_3 &= \frac{2\pi}{v_c}\, \boldsymbol{a}_1 \times \boldsymbol{a}_2, \end{aligned}\right\} \tag{2.12}$$

where v_c is the real space unit cell volume, $v_c = \boldsymbol{a}_1 \cdot (\boldsymbol{a}_2 \times \boldsymbol{a}_3)$.

As for the real space lattice, a unit cell for the reciprocal lattice is any volume which, when translated by the reciprocal lattice vectors $\boldsymbol{G}$, fills reciprocal space without overlapping. The reciprocal space Wigner–Seitz cell can be constructed by taking the smallest volume enclosed by the planes bisecting all the vectors $\boldsymbol{G}$; in reciprocal space, the Wigner–Seitz cell is known as the *first Brillouin zone.* The volume of the Brillouin zone is $\boldsymbol{b}_1 \cdot (\boldsymbol{b}_2 \times \boldsymbol{b}_3) = (2\pi)^3/v_c$.

If we now take a periodic real space function $f(\boldsymbol{r})$ with the periodicity of the lattice, the Fourier series can be written as

$$f(\boldsymbol{r}) = \sum_{\boldsymbol{G}} F_{\boldsymbol{G}}\, \mathrm{e}^{\mathrm{i}\boldsymbol{G}\cdot\boldsymbol{r}}, \tag{2.13}$$

with the vectors $\boldsymbol{G}$ running over the reciprocal space lattice. The Fourier components $F_{\boldsymbol{G}}$ are given by

$$F_{\boldsymbol{G}} = \frac{1}{v_c}\int_{v_c} \mathrm{e}^{-\mathrm{i}\boldsymbol{G}\cdot\boldsymbol{r}}\, f(\boldsymbol{r})\,\mathrm{d}\boldsymbol{r}, \tag{2.14}$$

where the integral is over a Wigner–Seitz unit cell volume v_c. We can then define the Fourier transform $F(\boldsymbol{k})$ as a sum of δ-functions,

$$F(\boldsymbol{k}) = (2\pi)^3 \sum_{\boldsymbol{G}} F_{\boldsymbol{G}}\, \delta(\boldsymbol{k} - \boldsymbol{G}). \tag{2.15}$$

The Fourier transform relation is

$$f(\boldsymbol{r}) = \frac{1}{(2\pi)^3}\int F(\boldsymbol{k})\, \mathrm{e}^{\mathrm{i}\boldsymbol{k}\cdot\boldsymbol{r}}\,\mathrm{d}\boldsymbol{k}. \tag{2.16}$$

We see that just as for the one-dimensional case, the three-dimensional Fourier transform of a periodic real space function is zero except for when the wavevector $\boldsymbol{k}$ is equal to a reciprocal space lattice vector $\boldsymbol{G}$.

An analogous set of expressions exists for a function $\Phi(\boldsymbol{k})$ that is periodic in k-space with the periodicity of the reciprocal lattice; we can write it as a Fourier series of *real* space terms as

$$\Phi(\boldsymbol{k}) = \sum_{\boldsymbol{R}} \phi_{\boldsymbol{R}}\, \mathrm{e}^{-\mathrm{i}\boldsymbol{R}\cdot\boldsymbol{k}}, \tag{2.17}$$

with Fourier elements

$$\phi_{\boldsymbol{R}} = \frac{v_c}{(2\pi)^3}\int_{V_c} \mathrm{e}^{\mathrm{i}\boldsymbol{k}\cdot\boldsymbol{R}}\Phi(\boldsymbol{k})\mathrm{d}\boldsymbol{k}, \tag{2.18}$$

where the integral is over the Brillouin zone defined about each reciprocal lattice point $\boldsymbol{G}$, in exactly the manner the Wigner–Seitz cell is defined in real space. The inverse Fourier transform is

$$F(\boldsymbol{k}) = \int f(\boldsymbol{r})\, \mathrm{e}^{-\mathrm{i}\boldsymbol{k}\cdot\boldsymbol{r}}\,\mathrm{d}\boldsymbol{r}. \tag{2.19}$$

A thorough description of the Fourier transform may be found in the text by Champeney [9].

2.4.3 Reciprocal Space Lattices: Some Examples

We now describe the reciprocal space lattices for some of the common Bravais lattices.

The simplest example, the simple cubic lattice with lattice spacing d, has real space generating vectors

$$\left.\begin{aligned} \boldsymbol{a}_1 &= (d,0,0), \\ \boldsymbol{a}_2 &= (0,d,0), \\ \boldsymbol{a}_3 &= (0,0,d). \end{aligned}\right\} \qquad (2.20)$$

The Wigner–Seitz cell has volume $v_c = \boldsymbol{a}_1 \cdot (\boldsymbol{a}_2 \times \boldsymbol{a}_3) = d^3$. The corresponding generating vectors for the reciprocal lattice may be calculated using (2.12), and are parallel to the real space generating vectors,

$$\left.\begin{aligned} \boldsymbol{b}_1 &= (2\pi/d,0,0), \\ \boldsymbol{b}_2 &= (0,2\pi/d,0), \\ \boldsymbol{b}_3 &= (0,0,2\pi/d). \end{aligned}\right\} \qquad (2.21)$$

The first Brillouin zone is thus a cube of side $2\pi/d$ centered on the origin. The simple cubic lattice is therefore its own reciprocal lattice (with a change in scale).

For the fcc lattice, the three real-space generating vectors for the single-atom basis are given in Sect. 2.3:

$$\left.\begin{aligned} \boldsymbol{a}_1 &= (0,d/2,d/2), \\ \boldsymbol{a}_2 &= (d/2,0,d/2), \\ \boldsymbol{a}_3 &= (d/2,d/2,0), \end{aligned}\right\} \qquad (2.22)$$

where d is the cube side length. The Wigner–Seitz cell volume is therefore $v_c = \boldsymbol{a}_1 \cdot (\boldsymbol{a}_2 \times \boldsymbol{a}_3) = d^3/4$; this is as expected, as the simple cube has volume d^3, and as there are four atoms in the fcc cube, the volume is $d^3/4$ per atom. The reciprocal space generating vectors are given by

$$\left.\begin{aligned} \boldsymbol{b}_1 &= (2\pi/d)(-1,1,1), \\ \boldsymbol{b}_2 &= (2\pi/d)(-1,1,-1), \\ \boldsymbol{b}_3 &= (2\pi/d)(1,1,-1). \end{aligned}\right\} \qquad (2.23)$$

These three vectors are precisely those that generate the body-centered cubic lattice, shown in Fig. 2.4. Hence, the reciprocal space lattice of the fcc real space lattice is body-centered cubic.

A similar calculation can be done for the bcc real space lattice, which has a fcc reciprocal space lattice (see Exercise 2.2). Hence the fcc and bcc lattices are complementary.

In Fig. 2.7 we display the first Brillouin zone (reciprocal space Wigner–Seitz cells) for the fcc and bcc real space crystals; we have also marked points of particular symmetry.

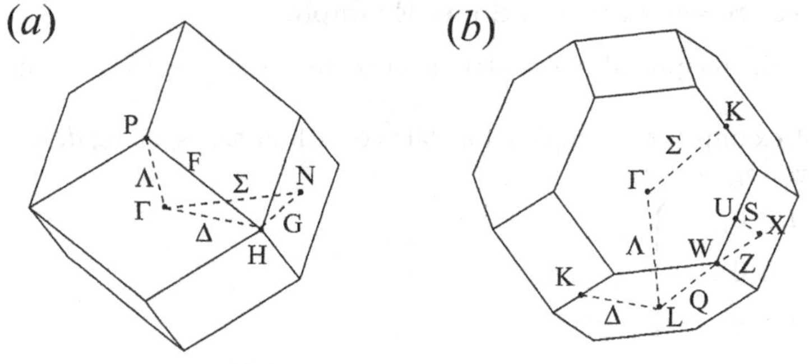

Fig. 2.7. (**a**) Brillouin zone for the bcc real space crystal and (**b**) for the fcc real space crystal, with symmetry points and lines indicated. Note that the Γ point is in the center of the Brillouin zone.

2.5 Bloch's Theorem

A very important theorem, known as Bloch's theorem, states that the solutions of the Hamiltonian equations in a infinite crystal will behave a certain way when the crystal is translated (displaced) by a lattice vector $\boldsymbol{R}$. This theorem can greatly simplify calculational efforts, and sometimes yields qualitative explanations of certain properties of a crystal.

When we solve either a classical or a quantum mechanical problem, the usual approach to is first solve for the normal modes (classical) or the energy eigenstates (quantum). These solutions have a uniform time dependence: For a classical normal mode, such as the vibrational modes of a chain in one dimension, or those of a three dimensional solid, the motion has a uniform frequency ω, and therefore the functional dependence $\phi(\boldsymbol{r})\mathrm{e}^{\mathrm{i}\omega t}$. For an quantum mechanical energy eigenstate, with energy E, the quantum wavefunction has the functional dependence $\phi(\boldsymbol{r})\mathrm{e}^{\mathrm{i}Et/\hbar}$. For economy, we have used the same symbol $\phi(\boldsymbol{r})$ for the spatial dependence in both the classical and quantum solutions.

Bloch's theorem states that spatial part $\phi(\boldsymbol{r})$ of a classical normal mode or of a quantum mechanical energy eigenstate can always be written so that under a real space lattice translation $\boldsymbol{R}$, taken from the set of lattice points $\{\boldsymbol{R}\}$, it transforms as

$$\phi(\boldsymbol{r}+\boldsymbol{R}) = \mathrm{e}^{\mathrm{i}\boldsymbol{q}\cdot\boldsymbol{R}}\phi(\boldsymbol{r}); \tag{2.24}$$

here $\boldsymbol{q}$ is the *wavevector* associated with the solution. As the wavevector $\boldsymbol{q}$ specifies the spatial dependence of the function $\phi(\boldsymbol{r})$, it is usually written as an index, $\phi_{\boldsymbol{q}}(\boldsymbol{r})$. The full proof of this theorem is somewhat involved, and we refer the reader to a text on condensed matter physics, e.g. that of Ziman [10] or Ashcroft and Mermin [1]. We note that we found this type of dependence in Chap. 1, for the N-atom chain.

An important corollary of this theorem is that the eigenfunctions $\phi(\boldsymbol{r})$ can be written in *Bloch form*,

$$\phi_{\boldsymbol{q}}(\boldsymbol{r}) = \mathrm{e}^{\mathrm{i}\boldsymbol{q}\cdot\boldsymbol{r}} f_{\boldsymbol{q}}(\boldsymbol{r}), \tag{2.25}$$

where the function $f_{\boldsymbol{q}}(\boldsymbol{r})$ is periodic with the periodicity of the lattice; its particular form depends on the value of $\boldsymbol{q}$, so we retain this as a subscript.

The proof of this corollary is quite simple, as we can define the function $f(\boldsymbol{r})$ according to (2.25), and write from Bloch's theorem

$$\begin{aligned} \phi_{\boldsymbol{q}}(\boldsymbol{r}+\boldsymbol{R}) &= \mathrm{e}^{\mathrm{i}\boldsymbol{q}\cdot\boldsymbol{R}}\phi_{\boldsymbol{q}}(\boldsymbol{r}) \\ &= \mathrm{e}^{\mathrm{i}\boldsymbol{q}\cdot\boldsymbol{R}}\mathrm{e}^{\mathrm{i}\boldsymbol{q}\cdot\boldsymbol{r}} f(\boldsymbol{r}) \\ &= \mathrm{e}^{\mathrm{i}\boldsymbol{q}\cdot(\boldsymbol{r}+\boldsymbol{R})} f(\boldsymbol{r}). \end{aligned} \tag{2.26}$$

From the definition of $f(\boldsymbol{r})$ from (2.25), applied to $\phi_{\boldsymbol{q}}(\boldsymbol{r}+\boldsymbol{R})$, we have the required periodicity, $f(\boldsymbol{r}) = f(\boldsymbol{r}+\boldsymbol{R})$. An example of a Bloch function is shown in Fig. 2.8.

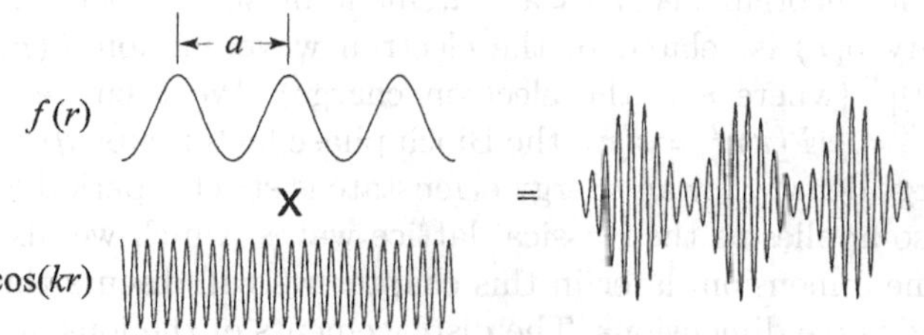

Fig. 2.8. Real part of a Bloch function, made up of the product of a periodic function $f(r)$ with period a and the sinusoidal function $\cos(kr)$.

A second corollary of Bloch's theorem is that the Bloch wavevector $\boldsymbol{q}$ can be restricted to the first Brillouin zone of the reciprocal lattice. This can most easily be seen in one dimension: If we take a one-dimensional crystal with real lattice spacing a, the reciprocal lattice consists of the "vectors" $G_m = m(2\pi/a)$, and the first Brillouin zone (reciprocal space Wigner–Seitz cell) is the set of points g with $-\pi/a < g < \pi/a$. If we take an arbitrary Bloch wavevector q, we can add a lattice vector G to q by writing

$$\phi(x) = f(x)\mathrm{e}^{\mathrm{i}qx} = f(x)\mathrm{e}^{-\mathrm{i}Gx}\mathrm{e}^{\mathrm{i}(q+G)x} = h(x)\mathrm{e}^{\mathrm{i}(q+G)x}, \tag{2.27}$$

with the new function $h(x) = f(x)\mathrm{e}^{-\mathrm{i}Gx}$. Because G is a reciprocal lattice vector, $h(x)$ is a periodic function with the periodicity of the lattice; in other words, the form $h(x)\mathrm{e}^{\mathrm{i}(q+G)x}$ is still in Bloch form, so that $q+G$ is a Bloch wavevector equivalent to q. We can therefore add and subtract whatever lattice vector is necessary to move q into the first Brillouin zone; this is illustrated in Fig. 2.9.

In three dimensions one can use exactly the same argument to show that the wavevector $\boldsymbol{q}$ can be shifted into the first Brillouin zone by adding an arbitrary reciprocal lattice vector $\boldsymbol{G}$ to $\boldsymbol{q}$, and redefining the periodic function,

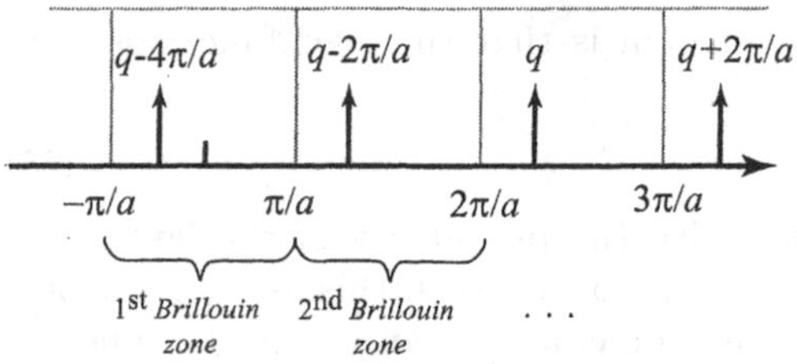

Fig. 2.9. Illustration of how the wavevector q can be shifted by lattice wavevectors $m(2\pi/a)$ until it lies in the first Brillouin zone.

say $f(\boldsymbol{r})$, as $h(\boldsymbol{r}) = f(\boldsymbol{r})\mathrm{e}^{-\mathrm{i}\boldsymbol{G}\cdot\boldsymbol{r}}$. The new Bloch form $h(\boldsymbol{r})\mathrm{e}^{\mathrm{i}(\boldsymbol{q}+\boldsymbol{G})\cdot\boldsymbol{r}}$ can thus be defined so that the Bloch wavevector $\boldsymbol{q}+\boldsymbol{G}$ lies inside the first Brillouin zone.

Bloch's theorem has important implications for the quantum energy eigenstates. When applied to a probability density, such as the electronic charge density $\rho(\boldsymbol{r})$, the theorem becomes a statement of simple periodicity: The charge density $\rho(\boldsymbol{r})$ is related to the electron wavefunction $\Psi(\boldsymbol{r})$ through $\rho(\boldsymbol{r}) = e|\Psi(\boldsymbol{r})|^2$ (where e is the electron charge). We then have $\rho(\boldsymbol{r}+\boldsymbol{R}) = e|\Psi(\boldsymbol{r}+\boldsymbol{R})|^2 = e|\Psi(\boldsymbol{r})|^2 = \rho(\boldsymbol{r})$; the Bloch phase factor $\exp(i\boldsymbol{q}\cdot\boldsymbol{r})$ cancels out, so the charge density for an energy eigenstate is strictly periodic.

Bloch's theorem also applies to the classical lattice waves, which we discussed in Chap. 1 in one dimension; later in this chapter we will discuss the lattice waves in two and three dimensions. The displacements of the ions are strictly real, so the spatial dependence of the displacement field can always be written in the form $\cos(\boldsymbol{q}\cdot\boldsymbol{r})f(\boldsymbol{r})$, with $f(\boldsymbol{r})$ strictly periodic. In the next section we will apply Bloch's theorem to calculate the dispersion relations in two and three dimensions. Note that the theorem only strictly applies to infinite systems, so for a finite solid it is not strictly correct; however, for a "sufficiently large" crystal, approximating the crystal as infinite is still fairly accurate.

2.6 Classical Theory of the Lattice

In Chap. 1, we found the normal mode solutions for the displacement $u(r,t)$ of a crystal in a one-dimensional chain of atoms. Here we will develop the formalism for finding the normal modes for the vector field displacement $\boldsymbol{u}(\boldsymbol{r},t)$ in two and three dimensions, using a somewhat more formal approach than for the one-dimensional description we gave in Chap. 1.

The formalism gives an accurate description of sound waves in crystals; we will find both the non-linear dispersion relations $\omega(\boldsymbol{q})$ that we found in one dimension, and in addition we will discover that the dispersion relation depends on the direction of the wavevector $\boldsymbol{q}$ in the crystal. The theory is *classical* because the atoms are treated as point masses, moving according to the laws of classical mechanics. The forces that determine their interactions are of course best understood using quantum mechanics. In Sect. 2.7.1 we develop the quantum theory of the lattice, in a very similar manner to that used in Chap. 1, using the language of "second quantization".

We will use Hamiltonian mechanics to derive the equations of motion for the atoms in the crystal. We take a crystal with one atom per unit cell, with atomic mass M; our treatment will therefore not apply to crystals with more than one atom in the repeated basis (see Sect. 2.1). Generalizing the arguments we present is fairly straightforward, but complicates the notation significantly. We refer the interested reader to the text by Ziman [11].

We locate the origin of our coordinate system so that when the crystal is at equilibrium, the atom in the jth unit cell is located at the lattice point $\boldsymbol{R}_j$. In order to probe the dynamic behavior of the lattice, we add a *continuous* lattice displacement $\boldsymbol{u}(\boldsymbol{r},t)$, with the displacement $\boldsymbol{u}$ nominally defined at every point $\boldsymbol{r}$ in space (not just the lattice points $\boldsymbol{R}_j$). In the presence of this displacement, the atom in the jth cell will move to $\boldsymbol{r}_j(t) = \boldsymbol{R}_j + \boldsymbol{u}(\boldsymbol{r}_j,t)$, where the displacement $\boldsymbol{u}$ is evaluated at the instantaneous location of the atom. This expression is difficult to evaluate, as it gives the displaced location $\boldsymbol{r}_j$ only through an implicit relation; in the theory of acoustic waves in fluids, the nonlinearity in this relation leads to important effects, apparent when the displacements are larger than the inter-particle spacing in the fluid. It will however prove useful when we discuss the continuum theory of solids, where the atomic nature of the solid is ignored.

For this calculation, we assume the displacements $\boldsymbol{u}$ are very small, and approximate by evaluating the displacement field at the rest position $\boldsymbol{R}_j$ rather than at the instantaneous position $\boldsymbol{r}_j$, so $\boldsymbol{u}(\boldsymbol{r}_j,t) \approx \boldsymbol{u}(\boldsymbol{R}_j,t)$; we will also use the shorthand notation $\boldsymbol{u}_j(t) = \boldsymbol{u}(\boldsymbol{R}_j,t)$. Hence, at time t, the jth atom is located at $\boldsymbol{r}_j(t) = \boldsymbol{R}_j + \boldsymbol{u}_j(t)$.

The total kinetic energy T of the lattice is given by

$$\begin{aligned} T &= \frac{1}{2} M \sum_{j=1}^{N} \left| \frac{\partial \boldsymbol{r}_j}{\partial t} \right|^2 \\ &= \frac{1}{2} M \sum_{j=1}^{N} \left| \frac{\mathrm{d}\boldsymbol{u}_j}{\mathrm{d}t} \right|^2 . \end{aligned} \tag{2.28}$$

The sum over j runs through the N unit cells in the crystal, so for an infinite crystal we take $N \to \infty$.

The total interaction potential energy is in general a function of all the atoms' coordinates, $U = U(\boldsymbol{r}_1, \boldsymbol{r}_2, \ldots \boldsymbol{r}_N)$. For small displacements $\boldsymbol{u}_j$ from the equilibrium position, the interaction can be expanded in a Taylor series about the static positions $\boldsymbol{R}_j$, so that

$$U(\boldsymbol{r}_1, \boldsymbol{r}_2, \ldots \boldsymbol{r}_N) = U(\boldsymbol{R}_1, \boldsymbol{R}_2, \ldots \boldsymbol{R}_N) + \sum_{j=1}^{N} \sum_{\alpha=1}^{3} \frac{\partial U}{\partial r_{j\alpha}} u_{j\alpha}$$
$$+ \sum_{j,k=1}^{N} \sum_{\alpha,\beta=1}^{3} \frac{1}{2} \frac{\partial^2 U}{\partial r_{j\alpha} \partial r_{k\beta}} u_{j\alpha} u_{k\beta} + \ldots \tag{2.29}$$

The derivatives are with respect to the components α of each atom i's coordinates $r_{i\alpha}$, and the derivatives are all evaluated with the atoms at their equilibrium positions $\boldsymbol{R}_j$.

The first term sets the zero of energy and may be dropped. The second term must vanish, as the positions $\boldsymbol{R}_j$ are assumed to be the equilibrium positions; if this sum did not vanish, a net force $\boldsymbol{F} = -\nabla U$ would exist, and the atoms would shift until this force vanished. We thus retain only the second term, dropping the third and higher order derivatives. This is known as the *harmonic approximation*; the higher order terms generate anharmonic, nonlinear responses, and are important for interactions between lattice waves, where they act as perturbations, as will be discussed in Chap. 3. The higher order terms are also necessary to explain the thermal expansion of solids, as was discussed in Chap. 1. Hence we can write the harmonic approximation as

$$U_{\text{harm}}(\boldsymbol{r}_1, \ldots \boldsymbol{r}_N) = \frac{1}{2} \sum_{jk=1}^{N} \sum_{\alpha\beta=1}^{3} \frac{\partial^2 U}{\partial r_{j\alpha} \partial r_{k\beta}} u_{j\alpha} u_{k\beta}. \tag{2.30}$$

We will now define a *tensor*, a 3×3 matrix whose entries are functions of position. A tensor is similar in some ways to a vector. A vector $\boldsymbol{v}$ can be written in component form (v_1, v_2, v_3). If a different coordinate system is chosen, the vector itself does not change, but its components do. A tensor is the same in that it is an entity that does not itself change when the coordinate system changes, but its *entries*, or components, do. A formal description of the properties of tensors appear in the Appendix A.1.2, and we also refer the reader to texts on mathematical physics [3, 4].

We define the tensor function $\Phi(\boldsymbol{R}_i, \boldsymbol{R}_j)$, with tensor elements $\Phi_{\alpha\beta}$, in terms of the curvature of the interaction energy $U(\boldsymbol{r}_1, \ldots \boldsymbol{r}_N)$:

$$\Phi_{\alpha\beta}(\boldsymbol{R}_i, \boldsymbol{R}_j) = \frac{\partial^2 U}{\partial r_{i\alpha} \partial r_{j\beta}}, \tag{2.31}$$

where the derivatives are evaluated with the atoms at their equilibrium positions $\boldsymbol{R}_i$. Note that the tensor is a function of the atom positions $\boldsymbol{R}_i$ and $\boldsymbol{R}_j$, as the derivatives are taken with respect to the components of the displacements of these two atoms.

Our crystal does not change under lattice translations $\boldsymbol{R}$, so the entries in the tensor $\Phi(\boldsymbol{R}_j, \boldsymbol{R}_k)$ should not either. In other words, the tensor should only depend on the difference of the atom positions $\boldsymbol{R}_j - \boldsymbol{R}_k$, so we can write the tensor functions as

$$\Phi_{\alpha\beta}(\boldsymbol{R}_i, \boldsymbol{R}_j) = \Phi_{\alpha\beta}(\boldsymbol{R}_i - \boldsymbol{R}_j). \tag{2.32}$$

Our potential energy can then be written as

$$\begin{aligned} U_{\text{harm}}(\boldsymbol{r}_1, \ldots \boldsymbol{r}_N) &= \frac{1}{2} \sum_{jk=1}^{N} \sum_{\alpha\beta=1}^{3} u_{j\alpha} \Phi_{\alpha\beta}(\boldsymbol{R}_j - \boldsymbol{R}_k) u_{k\beta} \\ &= \frac{1}{2} \sum_{jk} \boldsymbol{u}_j \cdot \Phi(\boldsymbol{R}_j - \boldsymbol{R}_k) \cdot \boldsymbol{u}_k, \end{aligned} \tag{2.33}$$

where we use the tensor inner product to represent the double sum.

2.6.1 Diatomic Interaction Potentials

The expression (2.33) is rather fearsome, so here we work out a somewhat simpler expression. We consider the form for the interaction potential energy when the atoms interact with one another only through a diatomic potential $\phi(\boldsymbol{r})$, i.e. one that involves interactions only between pairs of atoms; we also assume the diatomic potential only depends on the vector separation of the atoms, so $\phi = \phi(\boldsymbol{r}_i - \boldsymbol{r}_j)$. This will be useful when, for example, we consider model solids with atoms connected by springs, as we did in Chap. 1.

The total potential energy is then the sum over all atom pairs, of the form

$$U(\boldsymbol{r}_1, \ldots \boldsymbol{r}_N) = \frac{1}{2} \sum_{ij=1}^{N} \phi(\boldsymbol{r}_i - \boldsymbol{r}_j), \tag{2.34}$$

where the factor of 1/2 is because we are double counting atom pairs. Working out the Taylor expansion for the potential ϕ, we have

$$\begin{aligned} \phi(\boldsymbol{r}_i - \boldsymbol{r}_j) &= \phi(\boldsymbol{R}_i - \boldsymbol{R}_j) + \sum_{\alpha=1}^{3} \left.\frac{\partial \phi}{\partial r_\alpha}\right|_{\boldsymbol{R}_i - \boldsymbol{R}_j} (u_{i\alpha} - u_{j\alpha}) \\ &+ \frac{1}{2} \sum_{\alpha\beta=1}^{3} \left.\frac{\partial^2 \phi}{\partial r_\alpha \partial r_\beta}\right|_{\boldsymbol{R}_i - \boldsymbol{R}_j} (u_{i\alpha} - u_{j\alpha})(u_{i\beta} - u_{j\beta}) \\ &+ \ldots \end{aligned} \tag{2.35}$$

Inserting this in (2.34), and keeping only the second order derivatives, we are left with the equivalent of the harmonic approximation, which may be written

$$\begin{aligned} U_{\text{harm}} &= \frac{1}{4} \sum_{j\,k=1}^{N} \sum_{\alpha\,\beta=1}^{3} (u_{j\alpha} - u_{k\alpha}) \left.\frac{\partial^2 \phi}{\partial r_\alpha \partial r_\beta}\right|_{\boldsymbol{R}_j - \boldsymbol{R}_k} (u_{j\beta} - u_{k\beta}) \\ &= \frac{1}{2} \sum_{j\,k} \sum_{\alpha\,\beta} (u_{j\alpha} u_{j\beta} - u_{j\alpha} u_{k\beta}) \left.\frac{\partial^2 \phi}{\partial r_\alpha \partial r_\beta}\right|_{\boldsymbol{R}_j - \boldsymbol{R}_k} . \end{aligned} \tag{2.36}$$

By inserting the form (2.36) into the general expression (2.33), we can work out the equivalent expressions for the tensor $\Phi(\boldsymbol{R}_j - \boldsymbol{R}_k)$ in terms of the diatomic potential. We find that we can write $\Phi(\boldsymbol{R}_k)$ as (see Exercise 2.7)

$$\left.\begin{aligned}\Phi_{\alpha\beta}(\boldsymbol{R}_k = 0) &= \sum_{\ell=1}^{N} \left.\frac{\partial^2\phi}{\partial r_\alpha \partial r_\beta}\right|_{\boldsymbol{R}_\ell}, \\ \Phi_{\alpha\beta}(\boldsymbol{R}_k \neq 0) &= -\left.\frac{\partial^2\phi}{\partial r_\alpha \partial r_\beta}\right|_{\boldsymbol{R}_k},\end{aligned}\right\} \tag{2.37}$$

Using this tensor, we can write the harmonic approximation to the potential as

$$U_{\text{harm}} = \frac{1}{2}\sum_{jk}\sum_{\alpha\beta} u_{j\alpha}\Phi_{\alpha\beta}(\boldsymbol{R}_j - \boldsymbol{R}_k)u_{k\beta}. \tag{2.38}$$

Symmetries of the Interaction Tensor $\boldsymbol{\Phi}$. The tensor function $\Phi(\boldsymbol{R})$ has certain important symmetries, that are useful in calculations. The function Φ is *inversion symmetric* in $\boldsymbol{R}$, so

$$\Phi_{\alpha\beta}(\boldsymbol{R}) = \Phi_{\alpha\beta}(-\boldsymbol{R}). \tag{2.39}$$

This can be seen by inverting the order of differentiation. Further, for any given $\boldsymbol{R}$, the tensor is symmetric in its indices, so

$$\Phi_{\alpha\beta}(\boldsymbol{R}) = \Phi_{\beta\alpha}(\boldsymbol{R}). \tag{2.40}$$

This symmetry follows from the inversion symmetry of the Bravais lattice [1]. Finally, if we take the $\alpha\beta$ tensor element $\Phi_{\alpha\beta}(\boldsymbol{R})$, and sum over all $\boldsymbol{R}$ in the Bravais lattice, the result is zero:

$$\sum_{j=1}^{N} \Phi_{\alpha\beta}(\boldsymbol{R}_j) = 0. \tag{2.41}$$

This can be seen by examining (2.37), or for a more physical explanation, (2.38): If all the ions in the lattice are given the same displacement $\boldsymbol{u}$, shifting the crystal as a unit, the potential energy U should not change.

Example 2.1: Spherically Symmetric Potential in Two Dimensions. We consider the simple example of a two-dimensional lattice of atoms, interacting through a spherically symmetric, diatomic potential $\phi(r)$, that depends only on the distance r between two atoms. We arrange the atoms in a simple square lattice in the $x_1 - x_2$ plane with equilibrium spacing r_0, and take the interaction potential to be of the Lennard–Jones type,

$$\phi(r) = A\left(\frac{r_0}{r}\right)^{12} - 2A\left(\frac{r_0}{r}\right)^{6}, \tag{2.42}$$

where A is the depth of the potential minimum at the equilibrium spacing r_0 (see Sect. 1.1). This potential is sketched in two dimensions in Fig. 2.10.

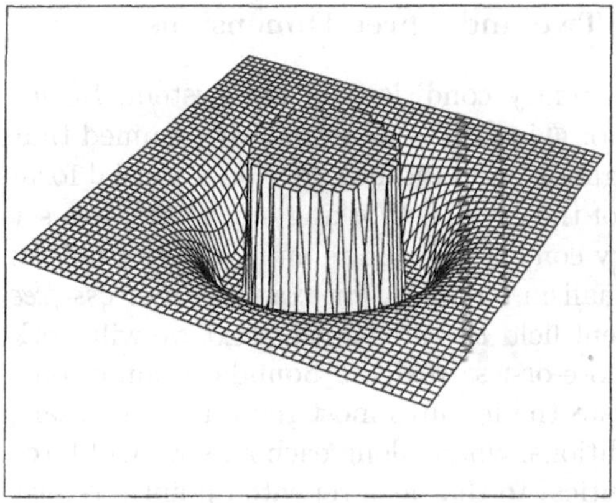

Fig. 2.10. Spherically symmetric Lennard–Jones potential $\phi(r)$.

We only consider nearest-neighbor interactions, so an atom at $\boldsymbol{R}$ only interacts with those at $\boldsymbol{R} \pm r_0\hat{\boldsymbol{x}}_1$ and $\boldsymbol{R} \pm r_0\hat{\boldsymbol{x}}_2$. The 2×2 interaction tensor $\Phi(\boldsymbol{R})$ is then given by

$$\left.\begin{aligned} \Phi(\boldsymbol{0}) &= \frac{A}{r_0^2}\begin{bmatrix} 144 & 0 \\ 0 & 144 \end{bmatrix}, \\ \Phi(\pm r_0\hat{\boldsymbol{x}}_1) &= -\frac{A}{r_0^2}\begin{bmatrix} 72 & 0 \\ 0 & 0 \end{bmatrix}, \\ \Phi(\pm r_0\hat{\boldsymbol{x}}_2) &= -\frac{A}{r_0^2}\begin{bmatrix} 0 & 0 \\ 0 & 72 \end{bmatrix}, \end{aligned}\right\} \tag{2.43}$$

as the reader can verify by evaluating the partial derivatives in (2.37).

The Hamiltonian H for the system can now be formed from the sum of the total kinetic and potential energies, $H = T + U$. Written out, this is

$$H = \frac{1}{2}M\sum_{j=1}^{N}\left|\frac{\partial \boldsymbol{u}_j}{\partial t}\right|^2 + \frac{1}{2}\sum_{jk}\sum_{\alpha\beta} u_{j\alpha}\Phi_{\alpha\beta}(\boldsymbol{R}_j - \boldsymbol{R}_k)u_{k\beta}. \tag{2.44}$$

The Hamiltonian equations of motion then generate a second-order differential equation of motion for each index α of the atomic displacement vector $\boldsymbol{u}_j$,

$$M\frac{\partial^2 u_{j\alpha}}{\partial t^2} = -\sum_{k=1}^{N}\sum_{\beta=1}^{3}\Phi_{\alpha\beta}(\boldsymbol{R}_j - \boldsymbol{R}_k)u_{k\beta}, \tag{2.45}$$

where the component index α runs from 1 to 3 (or 1 to 2 in two dimensions), and the atom index j from 1 to N, the number of atoms in the crystal.

2.6.2 Boundary Conditions in Two and Three Dimensions

We have not yet discussed the boundary conditions in our system. In our development of the interaction tensor Φ in (2.37), we implicitly assumed that every atom has a set of nearest neighbors, and did not develop a special form for Φ for atoms at the boundaries of the crystal. Working out a complete set of equations with realistic boundary conditions is quite an involved problem; we have not yet developed the formalism for what, for example, a *stress-free* boundary means for the displacement field $\boldsymbol{u}$. Later in this text we will work out some simple problems with more-or-less realistic boundary conditions. For now, however, in order to extract the features most relevant to crystals, we will use *periodic* boundary conditions, where along each axis we will force the motion at each point to be identical to that at a repeated point a repeat distance L away, where L is the size of the entire crystal. In two dimensions this is equivalent to wrapping a two-dimensional $L \times L$ sheet into a toroid, with the edges at $x = 0$ and $x = L$ joined together, and the edges at $y = 0$ and $y = L$ joined as well. Each axis then has $\mathcal{N} = L/r_0$ atoms, where r_0 is the atomic spacing, and the sheet has a total of $N = \mathcal{N}^2$ atoms. In three dimensions, the equivalent geometric construction is four dimensional and therefore somewhat hard to picture.

We used periodic boundary conditions for our one-dimensional N-atom chain in Chap. 1, and found that the only change from the infinite chain is to restrict the values of the Bloch wavevector q in the normal mode solutions to $q_m = (2\pi/r_0)m/N$, where the repeat distance was $L = Nr_0$. In two dimensions, as we shall see, the equivalent two-dimensional Bloch wavevector $\boldsymbol{q} = (q_x, q_y)$ will be restricted to the set

$$(q_x, q_y) = \frac{2\pi}{L}(m, n), \tag{2.46}$$

where m and n are integers running from $-\mathcal{N}/2$ to $+\mathcal{N}/2$, and $L = \mathcal{N}r_0$ is the length of the (square) crystal along each axis. The range of q_x and q_y is from $-\pi/r_0$ to $+\pi/r_0$, just as for one dimension. Other than this restriction in the allowed values of the Bloch wavevectors, our expression for the interaction tensor (2.37), and the solutions we derive from it, will remain unchanged.

In three dimensions, applying the same periodic boundary conditions means that the motion is repeated along all three axes. If we again take a repeat distance L, corresponding to a crystalline cube of side L, the values of the Bloch wavevector $\boldsymbol{q} = (q_x, q_y, q_z)$ will be restricted to the set

$$(q_x, q_y, q_z) = \frac{2\pi}{L}(\ell, m, n), \tag{2.47}$$

where (ℓ, m, n) run from $-\mathcal{N}/2$ to $+\mathcal{N}/2$, for $\mathcal{N}$ atoms along each axis of the cube.

Periodic boundary conditions work well when the actual boundaries, and the specific normal modes derived from them, are not important in the calculation. This is typically the case in large systems, where the characteristic wavelength in the displacement field is much smaller than any of the solid's dimensions (which of course depends on the problem); this applies, for example, when calculating the thermal properties at temperatures such that the characteristic thermal wavelength is much smaller than any dimension. In situations where the characteristic wavelength is of the order of, or larger than, at least one of the dimensions, one would like to use boundary conditions that approximate more accurately, or in the best case reproduce exactly, the actual system. As mentioned above, this can present a significant problem. A free boundary, one on which no constraints are placed, is one that is free to move while maintaining zero *stress*, or force, on its surface. This usually results in a set of equations applied to the solid surface that involve both the first and second spatial derivatives of the displacement field $\boldsymbol{u}(\boldsymbol{r}, t)$. A fixed boundary, on the other hand, is one in which no displacement is allowed, and either no *bending* is allowed, in what is known as a clamped boundary, or one where no in-plane stress occurs, known as a *hinged* boundary. In most situations some combination of these occurs, and the exact conditions on the displacement and stress at the boundary must be approximated.

Obtaining accurate solutions in such cases almost always requires the use of numerical approaches. The mixed partial differential equations that result are most often solved using finite-element models, where the solid is approximated by a number of small solid elements joined together, and within each element, the dynamic equations of motion are solved approximately using a numerical partial-differential equation (PDE) solver. The accuracy of such approaches is limited by the number of elements that can be solved simultaneously, and by the approximations used to deal with the (usually poorly known) surface boundary conditions.

In order to proceed, in this chapter we will use periodic boundary conditions, and allow the reader to expand to the more difficult alternatives. We note that if the repeat distance L used along each axis in the periodic system is chosen equal to the actual solid's physical dimension, the total number of normal modes will be correct; however, the specific values, and the specific dependence of frequency on wavevector, $\omega(\boldsymbol{q})$, will only approximate those of the real solid. Higher frequency, shorter wavelength modes will be more accurate than the longer wavelength modes. The broad temperature dependence of quantities such as the specific heat, thermal conductance and so on will be accurately reproduced at higher temperatures, but at very low temperatures, where only a very small number of modes are thermally accessible, discrepancies will appear.

2.6.3 Normal Modes of the Lattice

The set of equations (2.45) is solved by finding the normal modes of the coupled equations, that is, those solutions where every atom has the same harmonic time dependence. These solutions form a complete set, so that any arbitrary motion of the crystal can be written as a linear superposition of the normal modes.

As in Chap. 1, we find the normal modes by taking the same harmonic time dependence for all the atoms,

$$\boldsymbol{u}_j(t) = \boldsymbol{A}_j \mathrm{e}^{\mathrm{i}\omega t}, \tag{2.48}$$

where $\boldsymbol{A}_j$ is the (possibly complex) vector displacement for the jth atom. Inserting this into (2.45) yields a set of $3N$ coupled linear equations:

$$\omega^2 M\, A_{j\alpha} = \sum_{k=1}^{N} \sum_{\beta=1}^{3} \Phi_{\alpha\beta}(\boldsymbol{R}_j - \boldsymbol{R}_k) A_{k\beta}. \tag{2.49}$$

We solve this set of equations by taking the Bloch form for the amplitudes $\boldsymbol{A}_j$ (see Sect. 2.5):

$$\boldsymbol{A}_j = \boldsymbol{A} \mathrm{e}^{\mathrm{i}\boldsymbol{q}\cdot\boldsymbol{R}_j}, \tag{2.50}$$

where $\boldsymbol{q}$ is the wavevector, for the mode with frequency ω; we will write the frequency as $\omega_{\boldsymbol{q}}$ to indicate this relationship. The amplitude and polarization of the mode is given by $\boldsymbol{A}$, with components A_α.

As mentioned in Sect. 2.6.2, the periodic boundary conditions we are using restrict the values of $\boldsymbol{q}$ to a discrete, finite set. For a two dimensional rectangular sheet with length L and $\mathcal{N}$ atoms along each axis, the values of the wavevector are restricted to the set $\boldsymbol{q} = (q_1, q_2) = 2\pi(m/L, n/L)$ with $-\mathcal{N}/2 < m < \mathcal{N}/2$ and $-\mathcal{N}/2 < n < \mathcal{N}/2$. For a three dimensional rectangular solid, similar constraints apply to each of the wavevector components $\boldsymbol{q} = (q_1, q_2, q_3)$. Other than these restrictions, the equations of motion are the same as for an infinite solid; we will henceforth ignore the discretized values for the wavevector except when it is relevant to the discussion, and assume $\boldsymbol{q}$ can be treated as a continuous vector.

Inserting (2.50) in (2.49), the equation for the jth atom takes on the form

$$\omega_{\boldsymbol{q}}^2 M\, A_\alpha = \sum_{k=1}^{N} \sum_{\beta=1}^{3} \Phi_{\alpha\beta}(\boldsymbol{R}_j - \boldsymbol{R}_k) \mathrm{e}^{\mathrm{i}\boldsymbol{q}\cdot(\boldsymbol{R}_k - \boldsymbol{R}_j)}\, A_\beta. \tag{2.51}$$

The expression (2.51) cannot depend on the index j of the atom we started with. Taking $\boldsymbol{R}_\ell = 0$ for some atom ℓ, we can write the equation for $j = \ell$, leaving the equivalent relation

$$\omega_{\boldsymbol{q}}^2 M\, A_\alpha = \sum_{k=1}^{N} \sum_{\beta=1}^{3} \Phi_{\alpha\beta}(\boldsymbol{R}_k) \mathrm{e}^{\mathrm{i}\boldsymbol{q}\cdot\boldsymbol{R}_k}\, A_\beta, \tag{2.52}$$

where we have used the inversion symmetry (2.39).

We now define a new tensor, $\mathsf{D}(\boldsymbol{q})$, as the Fourier transform of the original tensor $\Phi(\boldsymbol{R}_j)$. The components $D_{\alpha\beta}(\boldsymbol{q})$ of the new tensor are given by

$$D_{\alpha\beta}(\boldsymbol{q}) = \sum_{k=1}^{N} \Phi_{\alpha\beta}(\boldsymbol{R}_k) \mathrm{e}^{\mathrm{i}\boldsymbol{q}\cdot\boldsymbol{R}_k}. \tag{2.53}$$

Inserting this form in (2.52) generates the set of equations

$$\omega_{\boldsymbol{q}}^2 M A_\alpha = \sum_{\beta=1}^{3} D_{\alpha\beta}(\boldsymbol{q}) A_\beta. \tag{2.54}$$

The tensor $\mathsf{D}(\boldsymbol{q})$ is known as the *dynamical matrix.*

The set of three equations (2.54) are significantly easier to solve than the original set of $3N$ equations (2.45). They form an *eigenvector-eigenvalue* problem, which can be written

$$M\omega_{\boldsymbol{q}}^2 \begin{bmatrix} A_1 \\ A_2 \\ A_3 \end{bmatrix} = \begin{bmatrix} D_{11} & D_{12} & D_{13} \\ D_{21} & D_{22} & D_{23} \\ D_{31} & D_{32} & D_{33} \end{bmatrix} \begin{bmatrix} A_1 \\ A_2 \\ A_3 \end{bmatrix}, \tag{2.55}$$

where of course $\boldsymbol{A}$ and D are functions of $\boldsymbol{q}$. This type of problem is frequently encountered in normal mode problems; a discussion of how to solve these appears in Appendix A.2.

For each wavevector $\boldsymbol{q}$, there are three distinct eigenfrequencies $\omega_{\boldsymbol{q}}$, and corresponding to each eigenfrequency, an eigenvector $\boldsymbol{A}$ that determines the *polarization* of the mode.

The eigenfrequencies are found by solving the characteristic equation resulting from setting the determinant of the matrix in (2.55) equal to zero. Once these frequencies have been found, the three corresponding polarizations $\boldsymbol{A}(\boldsymbol{q})$ are found by inserting each eigenfrequency $\omega_{\boldsymbol{q}}$ in (1.52). If there are N atoms in the lattice, there will be N distinct values of $\boldsymbol{q}$ (see below), and therefore a total of $3N$ modes for the $3N$ degrees of freedom. In an infinite crystal, there are an infinite number of distinct wavevectors, so $\boldsymbol{q}$ becomes a continuous variable.

We note that the expression (2.53) for the tensor $\mathsf{D}(\boldsymbol{q})$ can be manipulated using the symmetries (2.39–2.41), yielding [1]

$$D_{\alpha\beta}(\boldsymbol{q}) = -2 \sum_{\boldsymbol{R}} \Phi_{\alpha\beta}(\boldsymbol{R}) \sin^2 \frac{\boldsymbol{q}\cdot\boldsymbol{R}}{2}. \tag{2.56}$$

Example 2.2: Two-Dimensional Square Lattice. We now return to our example with atoms on a square lattice, interacting through a spherically symmetric potential (see Sect. 2.1). The atoms are located at the equilibrium points $(x, y) = (\ell r_0, m r_0)$ where r_0 is the equilibrium spacing, and ℓ and m take on integer values. We allow motion only in the $x - y$ plane, with displacement $u_x(\ell, m)$ along the x axis and $u_y(\ell, m)$ along y, for the (ℓ, m) atom. The 2×2 interaction curvature tensor is given in (2.43). We note that the curvature tensor is identical to that found if the atoms are coupled by springs with spring constant $k = 72A/r_0^2$, as shown in Fig. 2.11.

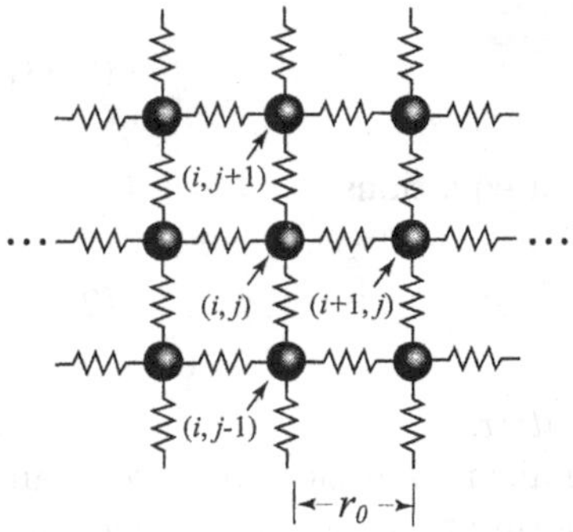

Fig. 2.11. Model for a two-dimensional lattice of atoms of mass M connected by springs with spring constant k; the atoms' rest points are on a square lattice with spacing r_0.

The 2×2 dynamical matrix is given by

$$\mathsf{D}(q_x, q_y) = 288\frac{A}{r_0^2}\begin{bmatrix} \sin^2(q_x d_0/2) & 0 \\ 0 & \sin^2(q_y d_0/2) \end{bmatrix}. \tag{2.57}$$

The normal mode equations for the frequencies $\omega_{\boldsymbol{q}}$ and amplitudes (A_x, A_y) are then given by

$$M\omega_{\boldsymbol{q}}^2 \begin{bmatrix} A_x \\ A_y \end{bmatrix} = 288\frac{A}{r_0^2}\begin{bmatrix} \sin^2(q_x d_0/2) & 0 \\ 0 & \sin^2(q_y d_0/2) \end{bmatrix}\begin{bmatrix} A_x \\ A_y \end{bmatrix}, \tag{2.58}$$

These equations are already separated, as there are no off-diagonal terms in the 2×2 matrix. The two normal mode frequencies, as a function of $\boldsymbol{q}$, are given by

$$\left.\begin{aligned} \omega_{\boldsymbol{q}1} &= \sqrt{\frac{288A}{Mr_0^2}}\sin\frac{q_x d_0}{2}, \\ \omega_{\boldsymbol{q}2} &= \sqrt{\frac{288A}{Mr_0^2}}\sin\frac{q_y d_0}{2}. \end{aligned}\right\} \tag{2.59}$$

The corresponding eigenvectors are $\boldsymbol{A}_1 = (1,0)$ and $\boldsymbol{A}_2 = (0,1)$; regardless of the direction of the $\boldsymbol{q}$-vector, the two polarizations are along $\hat{x}_1$ and $\hat{x}_2$. The eigenvalue solutions $\omega_{\boldsymbol{q}1}$ and $\omega_{\boldsymbol{q}2}$ are plotted in a parametric plot in Fig. 2.12.

We can also plot the solutions as contour plots in the $q_1 - q_2$ plane, where the contour lines correspond to constant values of the frequency $\omega_{\boldsymbol{q}1}$ or $\omega_{\boldsymbol{q}2}$; this is shown in Fig. 2.13. Note that the two plots are for the lower and upper frequency, respectively, so as $\omega_{\boldsymbol{q}2}$ falls below $\omega_{\boldsymbol{q}1}$ as a function of $\boldsymbol{q}$, the contour for the lower frequency changes from one polarization to the other. The vertical contours correspond to polarization along $\hat{q}_x$, while the horizontal contours are for polarization along $\hat{q}_y$. Separating the two frequencies gives a set of horizontal and vertical contours, respectively.

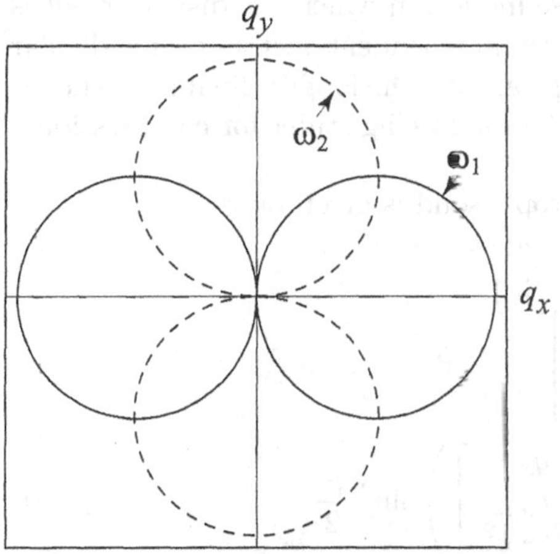

Fig. 2.12. Parametric plot displaying the dispersion relation ω_{q1} (*solid line*) and ω_{q2} (*dashed line*) for a two-dimensional square lattice with a spherically symmetric interaction. The curves are for small wavevector amplitude $|q|$ as a function of the direction of q in the plane; the distance of the curve from the origin is proportional to the frequency.

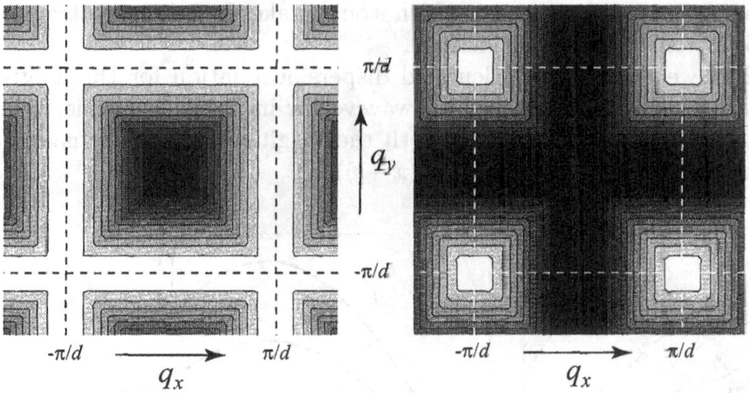

Fig. 2.13. Contour plots for the lower (*left*) and higher (*right*) eigenvalue solutions for the two-dimensional square lattice. The Brillouin zone boundaries corresponding to a square lattice with a real space unit cell of side d are shown as dashed lines.

Example 2.3: Simple Isotropic Model. We now describe an isotropic model that contains many of the aspects of a real solid, but also provides an extremely simple system in which more complex analyses may be pursued. This model is related to the simplest continuum isotropic solid, which is described by only two elastic constants; the isotropic solid will be explored in more detail in Chap. 5. In an isotropic solid, there are three distinct modes: The longitudinal mode, in which the displacement of the atoms is parallel to the wavevector $\boldsymbol{q}$, and the transverse modes, in which the displacement is perpendicular to $\boldsymbol{q}$ (there are two linearly independent vectors perpendicular to $\boldsymbol{q}$, so two transverse modes). In a typical solid the longitudinal frequency is larger than the transverse frequency, the solid being stiffer for compressional than transverse motion.

The dynamic tensor D for the isotropic solid is given by

$$\begin{aligned} \mathsf{D} &= \left(\omega_L^2 \hat{\boldsymbol{q}}\hat{\boldsymbol{q}} + \omega_T^2 (1 - \hat{\boldsymbol{q}}\hat{\boldsymbol{q}})\right) \sin^2 \frac{qa}{2} \\ &= \left(\frac{\omega_L^2}{q^2} \begin{bmatrix} q_x^2 & q_x q_y & q_x q_z \\ q_y q_x & q_y^2 & q_y q_z \\ q_z q_x & q_z q_y & q_z^2 \end{bmatrix} \right. \\ &\quad \left. + \frac{\omega_T^2}{q^2} \begin{bmatrix} q^2 - q_x^2 & -q_x q_y & -q_x q_z \\ -q_y q_x & q^2 - q_y^2 & -q_y q_z \\ -q_z q_x & -q_z q_y & q^2 - q_z^2 \end{bmatrix} \right) \sin^2 \frac{qa}{2}. \end{aligned} \tag{2.60}$$

The frequencies ω_L and ω_T are the maximum frequencies for the longitudinal and transverse polarization normal modes. The two transverse mode frequencies are degenerate (equal) in the isotropic solid. The $\sin^2 qa/2$ gives the shape of the standard dispersion relation. For atomic spacing a, the nominal Brillouin zone boundary would be at $q = |\boldsymbol{q}| = \pi/a$. Note however that we do not specify a crystal structure, which would make the solid anisotropic, so the Brillouin zone is not really defined.

In Fig. 2.14 we display the calculated dispersion relation for the longitudinal polarization as a function of the wavevector in the $x - y$ plane, and also show the dispersion relations for both the longitudinal and transverse polarizations as a function of q for $\boldsymbol{q} = q(\hat{x} + \hat{y})/\sqrt{2}$.

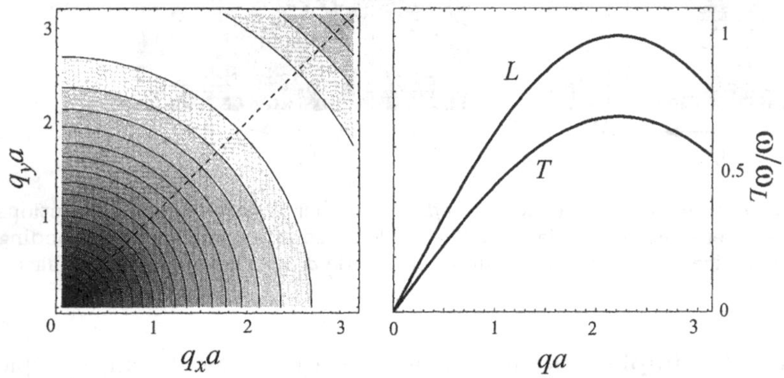

Fig. 2.14. Isotropic solid dispersion relations. (**a**) Contour plot for the longitudinal branch, as a function of wavevector $\boldsymbol{q}$ in the positive quadrant of the $x - y$ plane. (**b**) Dispersion for both the longitudinal and transverse branches, as a function of wavevector amplitude q along $q_x = q_y$, shown by the dotted line on the left. The ratio of transverse to longitudinal polarization frequencies is taken to be $\omega_T/\omega_L = 1/\sqrt{2}$.

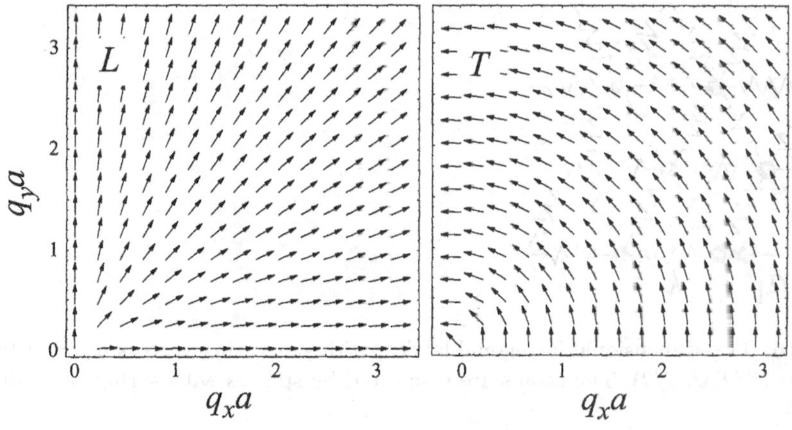

Fig. 2.15. Isotropic solid eigenvectors, for longitudinal (L) polarization (*left*) and transverse (T) polarization (*right*); both are plotted for $\boldsymbol{q}$ in the positive quadrant of the $x - y$ plane, and the vector lengths are normalized.

The dispersion relations plotted in Fig. 2.14 look very similar to that found for the N-dimensional chain. In addition, in this model we can extract the direction of the polarization at each wavevector $\boldsymbol{q}$; this is shown in Fig. 2.15, as a function of wavevector in the $x - y$ plane, for the longitudinal and one of the transverse polarizations; the other transverse polarization is perpendicular to the $x - y$ plane, along z.

Clearly this simple model generates pure longitudinal and pure transverse polarization normal modes, with a dispersion relation closely related to the N-atom chain. We will come back to this model when dealing with topics such as the Debye model for heat capacity.

Example 2.4: Two-Dimensional Hexagonal Lattice. We now modify Example 2.2 by moving the atoms into a hexagonal lattice, shown in Fig. 2.16; the atoms still interact with their nearest neighbors with a spherically symmetric potential $\phi(r)$.

Choosing the origin centered on one atom, the six nearest atoms are at the positions

$$\left.\begin{aligned} \boldsymbol{R}_1 &= (d,0),\ \boldsymbol{R}_2 = (-d,0), \\ \boldsymbol{R}_3 &= (d/2,\sqrt{3}d/2),\ \boldsymbol{R}_4 = (-d/2,-\sqrt{3}d/2), \\ \boldsymbol{R}_5 &= (d/2,-\sqrt{3}d/2),\ \boldsymbol{R}_6 = (-d/2,\sqrt{3}d/2). \end{aligned}\right\} \tag{2.61}$$

To evaluate the various terms in the curvature tensor Ψ, we need the derivatives of $\phi(r) = \phi(\sqrt{x^2+y^2})$ with respect to x and y:

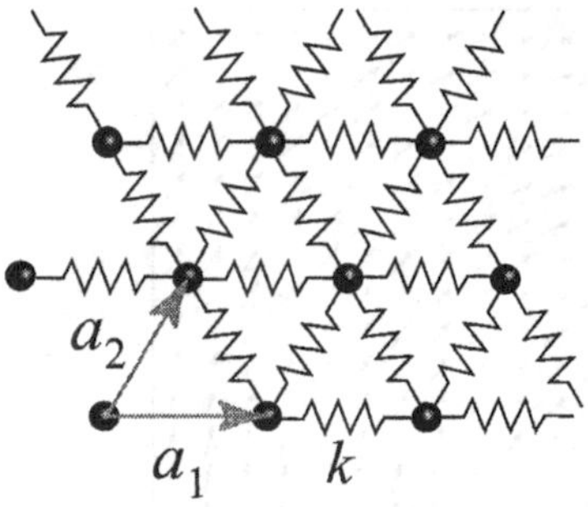

Fig. 2.16. Two-dimensional hexagonal lattice, with generating vectors $\boldsymbol{a}_1 = (d, 0)$ and $\boldsymbol{a}_2 = (d/2, \sqrt{3}d/2)$. The atoms are connected by springs with spring constant $k = \phi''(d)$.

$$\left.\begin{aligned}
\frac{\partial^2\phi}{\partial x^2} &= \frac{\mathrm{d}^2\phi}{\mathrm{d}r^2}\frac{x^2}{r^2} + \frac{\mathrm{d}\phi}{\mathrm{d}r}\left(\frac{1}{r} - \frac{x^2}{r^3}\right), \\
\frac{\partial^2\phi}{\partial y^2} &= \frac{\mathrm{d}^2\phi}{\mathrm{d}r^2}\frac{y^2}{r^2} + \frac{\mathrm{d}\phi}{\mathrm{d}r}\left(\frac{1}{r} - \frac{y^2}{r^3}\right), \\
\frac{\partial^2\phi}{\partial x \partial y} &= \frac{\partial^2\phi}{\partial y \partial x} = \frac{\mathrm{d}^2\phi}{\mathrm{d}r^2}\frac{xy}{r^2} - \frac{\mathrm{d}\phi}{\mathrm{d}r}\frac{xy}{r^3}.
\end{aligned}\right\} \tag{2.62}$$

At the equilibrium spacing $r = d$ we can take $\mathrm{d}\phi/\mathrm{d}r(r = d) = 0$. For the springs shown in Fig. 2.16, we can write $k = \mathrm{d}^2\phi/\mathrm{d}r^2|_d$; if we take the Lennard–Jones potential for ϕ, then we can write $k = 72A/d^2$ in terms of the coupling strength A. In terms of k, the interaction tensor $\Psi(\boldsymbol{R})$ is given by

$$\Psi(\boldsymbol{R} = 0) = 3k \begin{bmatrix} 1 & 0 \\ 0 & 1 \end{bmatrix}, \tag{2.63}$$

and

$$\left.\begin{aligned}
\Psi(\boldsymbol{R}_{1,2}) &= -k \begin{bmatrix} 1 & 0 \\ 0 & 0 \end{bmatrix}, \\
\Psi(\boldsymbol{R}_{3,4}) &= -\frac{k}{4} \begin{bmatrix} 1 & \sqrt{3} \\ \sqrt{3} & 3 \end{bmatrix}, \\
\Psi(\boldsymbol{R}_{5,6}) &= -\frac{k}{4} \begin{bmatrix} 1 & -\sqrt{3} \\ -\sqrt{3} & 3 \end{bmatrix}.
\end{aligned}\right\} \tag{2.64}$$

Using (2.56), the dynamical matrix $\mathsf{D}(q_x, q_y)$ can be written out; it has the form

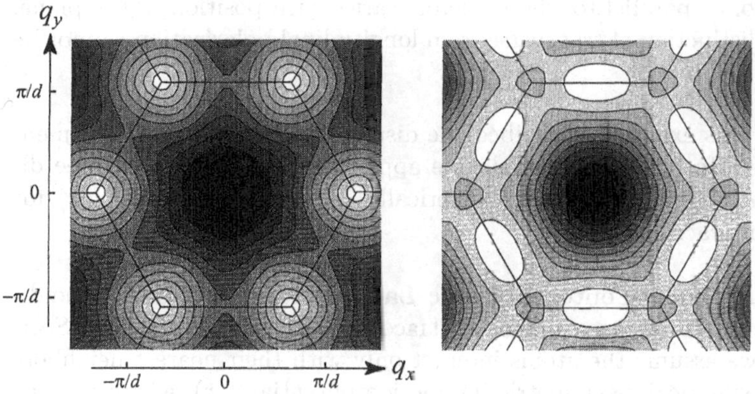

Fig. 2.17. Contour plots of the solutions for the lower (*left*) and upper (*right*) eigenvalues for the two-dimensional hexagonal lattice. The boundaries of the Brillouin zone are shown as solid lines.

$$\mathsf{D}(\boldsymbol{q}) = k \sum_{i=1}^{6} \sin^2 \frac{\boldsymbol{q} \cdot \boldsymbol{R}_i}{2} \hat{\boldsymbol{R}}_i \hat{\boldsymbol{R}}_i, \tag{2.65}$$

using the notation $(\hat{\boldsymbol{R}}_i \hat{\boldsymbol{R}}_i)_{\alpha\beta} = R_{i\alpha} R_{i\beta} / R_i^2$. From the dynamical matrix we can solve for the two eigenvalues $\omega_{\boldsymbol{q}1}$ and $\omega_{\boldsymbol{q}2}$; these are plotted in Fig. 2.17. The left side shows the contours for the lower frequency, the right for the upper, so we do not distinguish between the polarizations.

In Fig. 2.18 we display the polarization vectors, on the left the vectors corresponding to the smaller eigenvalue and on the right those corresponding to the larger. Note that the character of the eigenvalue, whether is is per-

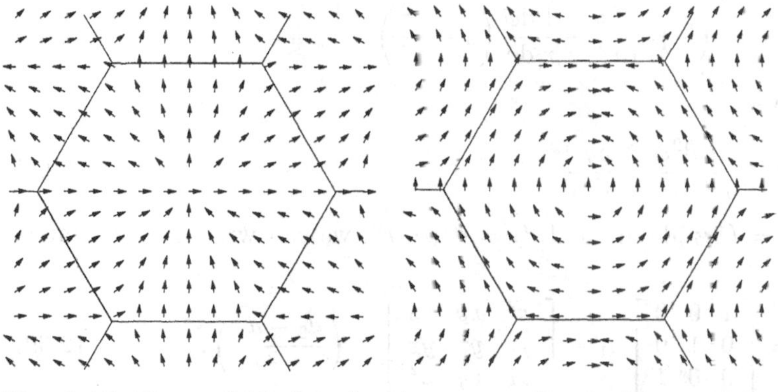

Fig. 2.18. Vector field plots for eigenvectors for the hexagonal lattice, plotted for the same range of $\boldsymbol{q}$ values as for Fig. 2.17. The left side shows the vectors corresponding to the lower eigenvalue and the right side for the upper. The Brillouin zone boundaries are shown in gray.

pendicular to, or parallel to, the $\boldsymbol{q}$-vector, varies with position in the plane, and a clear distinction of transverse from longitudinal polarization cannot be made.

The previous examples described the dispersion relations for two dimensional lattices. In the next example, we apply the formalism to a three dimensional lattice, using the same spherically symmetric potential $\phi(r)$ for the atomic interactions.

Example 2.5: Face–Centered Cubic Lattice. We now calculate the dispersion relations for a three dimensional face-centered cubic crystal (see Sect. 2.3), where we assume the atoms interact only with their nearest neighbors through the spherically-symmetric interaction potential $\phi(r)$, where r is the radial coordinate. The twelve atoms nearest the atom at the cube corner, which we place at the origin $\boldsymbol{R}$=0, are located at

$$\begin{aligned} \boldsymbol{R}_{1-4} &= \frac{d}{2}(\pm\hat{\boldsymbol{x}} \pm \hat{\boldsymbol{y}}), \ \boldsymbol{R}_{5-8} = \frac{d}{2}(\pm\hat{\boldsymbol{x}} \pm \hat{\boldsymbol{z}}), \\ \boldsymbol{R}_{9-12} &= \frac{d}{2}(\pm\hat{\boldsymbol{y}} \pm \hat{\boldsymbol{z}}), \end{aligned} \tag{2.66}$$

where d is the length of the fcc cube side (see Fig. 2.4); the nearest-neighbor distance is $a = d/\sqrt{2}$.

The interaction tensor $\Psi(\boldsymbol{r})$ has the elements $\Psi_{\alpha\beta}(\boldsymbol{r}) = \partial^2\phi/\partial x_\alpha \partial x_\beta$. With $r = (x^2 + y^2 + z^2)^{1/2}$, we have for example

$$\begin{aligned} \Phi_{xx}(\boldsymbol{r}) &= \frac{\partial^2\phi}{\partial x^2} \\ &= \frac{\partial}{\partial x}\left(\frac{x}{r}\frac{\mathrm{d}\phi}{\mathrm{d}r}\right) \\ &= \left(\frac{x}{r}\right)^2 \frac{\mathrm{d}^2\phi}{\mathrm{d}r^2} + \frac{1}{r}\frac{\mathrm{d}\phi}{\mathrm{d}r}\left(1 - \frac{x^2}{r^2}\right). \end{aligned} \tag{2.67}$$

and similarly

$$\Phi_{xy}(\boldsymbol{r}) = \frac{xy}{r^2}\frac{\mathrm{d}^2\phi}{\mathrm{d}r^2} - \frac{xy}{r^3}\frac{\mathrm{d}\phi}{\mathrm{d}r}. \tag{2.68}$$

With $d_1 = (1/r)\mathrm{d}\phi/\mathrm{d}r$ and $d_2 = \mathrm{d}^2\phi/\mathrm{d}r^2$, we can write this in matrix form as

$$\Phi(\boldsymbol{r}) = \begin{bmatrix} 1 & 0 & 0 \\ 0 & 1 & 0 \\ 0 & 0 & 1 \end{bmatrix} d_1 + \begin{bmatrix} x^2 & xy & xz \\ yx & y^2 & yz \\ zx & zy & z^2 \end{bmatrix} \times \left(\frac{d_2 - d_1}{r^2}\right). \tag{2.69}$$

Note that we are not setting $d_1 = 0$, which we did in the previous two examples. The dynamical matrix $\mathsf{D}(\boldsymbol{q})$ is given by the sum (2.56),

$$D_{\alpha\beta}(\boldsymbol{q}) = -2\sum_{\boldsymbol{R}} \Phi_{\alpha\beta}(\boldsymbol{R}) \sin^2 \frac{\boldsymbol{q}\cdot\boldsymbol{R}}{2}, \tag{2.70}$$

where the sum is over the nearest neighbor vectors given in (2.66).

The parameters d_1 and d_2 determine the detailed shape and scale of the interaction matrix elements; the eigenfrequencies $\omega_n(\boldsymbol{q})$ can then be calculated by solving the eigenvalue equation,

$$M\,\omega_n^2(\boldsymbol{q})\,A_\alpha = \sum_{\beta=1}^{3} D_{\alpha\beta}(\boldsymbol{q})\,A_\beta, \tag{2.71}$$

or, written in vector form,

$$M\omega_n^2(\boldsymbol{q})\,\boldsymbol{A} = \mathsf{D}(\boldsymbol{q})\boldsymbol{A}. \tag{2.72}$$

In Fig. 2.19 we display the calculated dispersion curves $\omega_n(\boldsymbol{q})$ for $d_1/M = 3\ \text{THz}^2$ and $d_2/M = 30\ \text{THz}^2$, choices which yield dispersion relations similar to those for lead, an fcc metal (see [12]). The calculations are along different directions in the reciprocal lattice; the first segment from the origin Γ to the Brillouin zone edge $X = (0, 2\pi/a, 0)$, the second from X to a point outside the zone at $(2\pi/a, 2\pi/a, 0)$, and the third from that point back to the zone edge point $K = (3\pi a/2, 3\pi a/2, 0)$, and then returning to the origin Γ; see Sect. 2.1 and Fig. 2.7 for the geometry of the symmetry points. Note the symmetry of the crystal yields degenerate values for the two lower (transverse acoustic) branches along the $\Gamma - X$ line.

Another way to plot the dispersion relations is as a set of contour plots in a plane in reciprocal space; this is shown in Fig. 2.20, for the lowest frequency (transverse) branch for wavevectors $\boldsymbol{q} = (q_x, q_y, 0)$ in the $x - y$ plane.

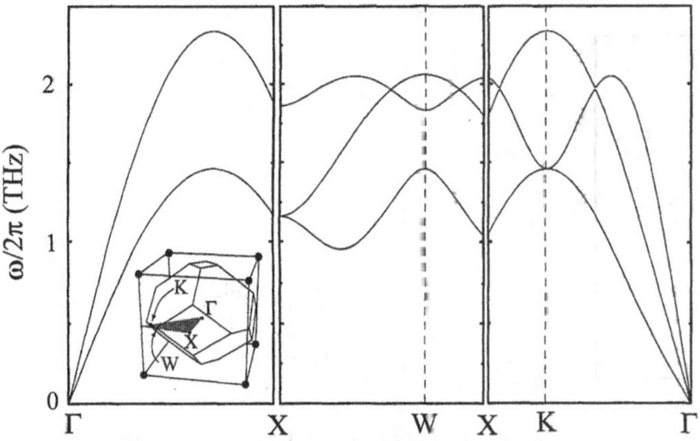

Fig. 2.19. Phonon dispersion relations for a model fcc lattice, using a spherically symmetric potential $\phi(r)$; the potential parameters are chosen to give a dispersion relation similar to that of lead.

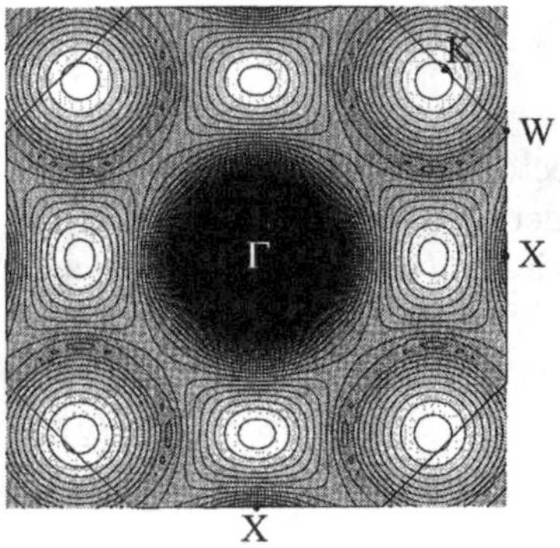

Fig. 2.20. Contour plots for the same model system as in Fig. 2.19. The Brillouin zone symmetry points and the zone edge are marked.

From Fig. 2.20, we can pull out the set of wavevectors that give a fixed value for the mode frequency ω; this is shown, for the same model, in Fig. 2.21 for all three branches, again for wavevectors in the $x-y$ plane. Note that as the longitudinal branch has the highest frequency for a given $\boldsymbol{q}$, the equal frequency contour is achieved for the smallest values of $|\boldsymbol{q}|$. The designations of the longitudinal and transverse branches are somewhat arbitrary, as except for certain specific symmetry points in $\boldsymbol{q}$-space, the polarization is not strictly parallel or perpendicular to the $\boldsymbol{q}$-vector.

Changing the parameters values d_1 and d_2 changes the scale and some of the details of the dispersion relations plotted in these three figures, but the overall form is determined by the symmetry of the crystal; different fcc solids therefore have similar dispersion relations.

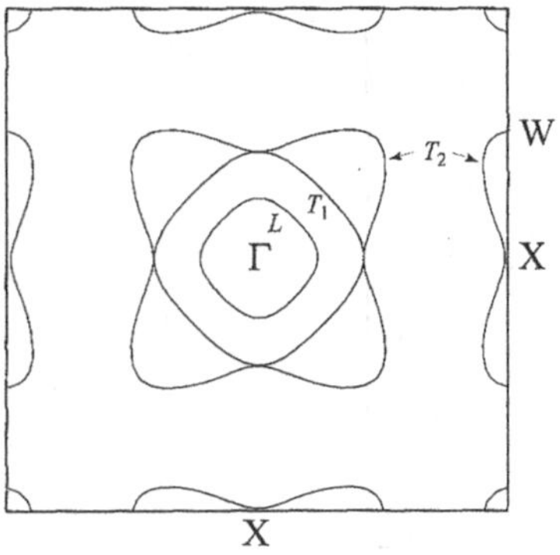

Fig. 2.21. Equal-frequency contours for the three branches in Fig. 2.19.

2.6.4 Optical Phonons

As with the one-dimensional lattice, crystals that include two or more types of atoms in the unit cell will display acoustic modes similar to those for monatomic crystals. The atoms within a unit cell move in the same direction (with different amplitudes of motion). In addition, there are a set of modes where the different atom types move in different directions, also with different amplitudes, a set of modes known as the *optical modes* (see Sect. 1.4). If the atoms have different electrical charges, this motion couples to electromagnetic waves, from which the terminology is derived. The dispersion relations for diatomic and more complex types of crystals include three polarizations for the acoustic modes, and in addition three polarizations for the optical modes. The optical bands are typically separated from the acoustic bands by a range of frequencies in which no modes exist. In Fig. 2.22 we display the dispersion relations for GaAs, a diatomic crystal that forms in the wurtzite crystal structure. The dispersion relations are plotted for lines connecting the various symmetry points in the underlying fcc crystal lattice (see Fig. 2.7).

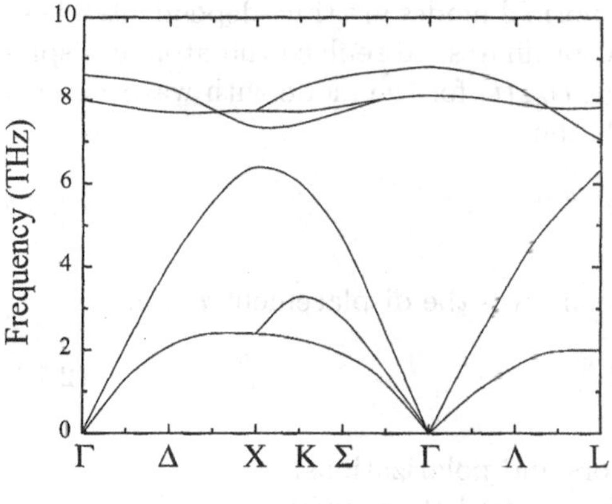

Fig. 2.22. Dispersion relations for GaAs, plotted along the different symmetry directions in the fcc crystal lattice. Adapted from Tamura [13].

2.7 Normal Mode Hamiltonian

In Sect. 2.6 we worked out the Hamiltonian for the crystal lattice, written out in terms of the displacement field $\boldsymbol{u}(\boldsymbol{r},t)$ and its time derivatives, and we found the normal mode solutions of the corresponding equations of motion. If we only consider motion in the normal mode with wavevector $\boldsymbol{q}$ and mode (or polarization) n, then the jth atom in the crystal, with equilibrium position $\boldsymbol{R}_j$, has a vector displacement given by

$$\boldsymbol{u}_j(t) = \boldsymbol{A}_n(\boldsymbol{q})\mathrm{e}^{\mathrm{i}\boldsymbol{q}\cdot\boldsymbol{R}_j}\mathrm{e}^{\mathrm{i}\omega_n(\boldsymbol{q})t}, \tag{2.73}$$

where $\boldsymbol{A}_n(\boldsymbol{q})$ is the displacement vector for the mode and $\omega_n(\boldsymbol{q})$ the frequency. If there are $N = \mathcal{N}_1\mathcal{N}_2\mathcal{N}_3$ atoms in the (three dimensional) solid, there are N distinct values of $\boldsymbol{q}$, each with three values for the polarization index n, and therefore a total of $3N$ distinct normal mode solutions.

The direction of the vector $\boldsymbol{A}_n(\boldsymbol{q})$ is determined by the equations of motion, but the length of $\boldsymbol{A}$ is arbitrary; we can normalize the amplitude, and then build an arbitrary particular solution by superposing normal modes with the appropriate (complex) amplitudes. We therefore separate $\boldsymbol{A}$ into an amplitude and a polarization,

$$\boldsymbol{A}_n(\boldsymbol{q}) = \mathcal{U}_n(\boldsymbol{q})\, \boldsymbol{\varepsilon}_{\boldsymbol{q}n}, \tag{2.74}$$

where $\boldsymbol{\varepsilon}_{\boldsymbol{q}n}$ is the normalized polarization vector and $\mathcal{U}_n(\boldsymbol{q})$ the amplitude. Note that the polarization vectors $\boldsymbol{\varepsilon}_{\boldsymbol{q}n}$ and $\boldsymbol{\varepsilon}_{\boldsymbol{q}n'}$ for different polarization indices n and n' are mutually orthogonal, so

$$\boldsymbol{\varepsilon}_{\boldsymbol{q}n} \cdot \boldsymbol{\varepsilon}_{\boldsymbol{q}n'} = 0 \tag{2.75}$$

(this is a feature of eigenvector-eigenvalue solutions).

As we did with the N-atom chain, we now want to rewrite the solutions (2.73) so that the amplitudes of the normal modes are time-dependent. These amplitudes will form a new set of coordinates, to replace the atomic displacements $\boldsymbol{u}_j$. We define the amplitude $\mathcal{U}_{\boldsymbol{q}n}(t)$ for the mode with wavevector $\boldsymbol{q}$ and polarization index n by the relation

$$\mathcal{U}_{\boldsymbol{q}n}(t) = \frac{1}{\sqrt{N}} \sum_j \boldsymbol{\varepsilon}_{\boldsymbol{q}n} \cdot \boldsymbol{u}_j(t) \mathrm{e}^{-\mathrm{i}\boldsymbol{q}\cdot\boldsymbol{R}_j}. \tag{2.76}$$

The corresponding inverse relation gives the displacement $\boldsymbol{u}_j(t)$,

$$\boldsymbol{u}_j(t) = \frac{1}{\sqrt{N}} \sum_{\boldsymbol{q}n} \mathcal{U}_{\boldsymbol{q}n}(t) \boldsymbol{\varepsilon}_{\boldsymbol{q}n} \mathrm{e}^{\mathrm{i}\boldsymbol{q}\cdot\boldsymbol{R}_j}, \tag{2.77}$$

superposing all the mode wavevectors and polarizations.

To reformulate the Hamiltonian, we need the momentum corresponding to the coordinate $\mathcal{U}_{\boldsymbol{q}n}$. This is given in terms of the single atom momenta $\boldsymbol{p}_j = M\mathrm{d}\boldsymbol{u}_j/\mathrm{d}t$ by

$$\mathcal{P}_{\boldsymbol{q}n}(t) = \frac{1}{\sqrt{N}} \sum_j \boldsymbol{\varepsilon}_{\boldsymbol{q}n} \cdot \boldsymbol{p}_j(t) \mathrm{e}^{\mathrm{i}\boldsymbol{q}\cdot\boldsymbol{R}_j}. \tag{2.78}$$

The Hamiltonian (2.44) can then be re-written in terms of these new coordinates as

$$H = \frac{1}{2M} \sum_{\boldsymbol{q}n} \mathcal{P}_{\boldsymbol{q}n}\mathcal{P}_{-\boldsymbol{q}n} + \frac{1}{2} \sum_{\boldsymbol{q}n} \left(\boldsymbol{\varepsilon}_{\boldsymbol{q}n} \cdot \mathsf{D}(\boldsymbol{q}) \cdot \boldsymbol{\varepsilon}_{-\boldsymbol{q}n}\right) \mathcal{U}_{\boldsymbol{q}n}\mathcal{U}_{-\boldsymbol{q}n}. \tag{2.79}$$

Here the dynamic tensor $\mathsf{D}(\boldsymbol{q})$ is the same as we used for the normal mode solution,

$$\mathsf{D}(\boldsymbol{q}) = \sum_{j=1}^{N} \Phi(\boldsymbol{R}_j)\, \mathrm{e}^{\mathrm{i}\boldsymbol{q}\cdot\boldsymbol{R}_j}. \tag{2.80}$$

We write $\mathcal{U}_{-qn}$ in terms of the complex conjugate of $\mathcal{U}_{qn}$,

$$\mathcal{U}^*_{qn} = \mathcal{U}_{-qn}, \tag{2.81}$$

which follows from (2.76). The Hamiltonian is then

$$H = \frac{1}{2M} \sum_{qn} \mathcal{P}_{qn}\mathcal{P}^*_{qn} + \frac{1}{2}\sum_{qn} (\boldsymbol{\varepsilon}_{qn} \cdot \mathsf{D}(\boldsymbol{q}) \cdot \boldsymbol{\varepsilon}_{qn})\, \mathcal{U}_{qn}\mathcal{U}^*_{qn}. \tag{2.82}$$

We can make a connection with the equivalent one-dimensional expression (1.56), where we found

$$H = \frac{1}{2M} \sum_{n} \mathcal{P}_n\mathcal{P}^*_n + k\sum_{n} (1 - \cos 2\pi n/N)\, \mathcal{U}_n\mathcal{U}^*_n. \tag{2.83}$$

In the one-dimensional case, our interaction potential was that of a system of linear springs with spring constant k and atomic spacing r_0. The indices n in (2.83) are related to the one-dimensional wavevector q through $q = (2\pi/r_0)\, n/N$. The Hamiltonian written in terms of the displacements u_n is given by (1.36) and (1.37), and from these expressions we can identify the terms in the interaction potential by

$$\left.\begin{aligned} \Phi(+d_0) &= -k, \\ \Phi(0) &= 2k, \\ \Phi(-d_0) &= -k. \end{aligned}\right\} \tag{2.84}$$

The corresponding dynamic one-dimensional "tensor" $D(q)$ is given by the analog of (2.80),

$$\begin{aligned} D(q) &= -k\mathrm{e}^{\mathrm{i}qr_0} + 2k - k\mathrm{e}^{-\mathrm{i}qr_0} \\ &= 2k(1 - \cos qr_0) = 2k(1 - \cos 2\pi n/N), \end{aligned} \tag{2.85}$$

yielding the same equation for the Hamiltonian as given by (1.54),

$$H = \frac{1}{2M} \sum_{n=-N/2}^{N/2} \mathcal{P}_n\mathcal{P}^*_n + k \sum_{n=-N/2}^{N/2} (1 - \cos 2\pi n/N)\, \mathcal{U}_n\mathcal{U}^*_n. \tag{2.86}$$

2.7.1 Quantum Operators for Normal Modes

In Chap. 1 we provided a discussion of how the quantum operators for the normal modes of the one dimensional chain are defined (see 1.7); here we do the same for three dimensions. The quantized energy states that appear for each normal mode ω_{qm}, with discrete numbers of energy quanta n_{qm}, are known as phonons, and the collection of these excitations is known as the phonon gas. As we shall see, by including non-linear terms in the Hamiltonian as perturbations, these excitations can interact with one another, and can also interact with electrons and other degrees of freedom in the system.

We have developed an expression for the Hamiltonian in terms of the amplitudes and momenta of the normal modes in (2.82). We now re-write this expression using the quantum mechanical operators $\hat{\mathcal{U}}_{\boldsymbol{q}m}$ and $\hat{\mathcal{P}}_{\boldsymbol{q}m}$,

$$H = \frac{1}{2M} \sum_{\boldsymbol{q}m} \hat{\mathcal{P}}_{\boldsymbol{q}m} \hat{\mathcal{P}}^{\dagger}_{\boldsymbol{q}m} + \frac{1}{2} \sum_{\boldsymbol{q}m} \left(\boldsymbol{\varepsilon}_{\boldsymbol{q}m} \cdot \mathsf{D}(\boldsymbol{q}) \cdot \boldsymbol{\varepsilon}_{\boldsymbol{q}m} \right) \hat{\mathcal{U}}_{\boldsymbol{q}m} \hat{\mathcal{U}}^{\dagger}_{\boldsymbol{q}m}, \tag{2.87}$$

where the vector operator $\hat{\mathcal{U}}^{\dagger}$ is the Hermitian conjugate of $\hat{\mathcal{U}}$.

The commutation relations (1.115) apply to the position and momentum operators, so we have

$$\left.\begin{aligned} \left[\hat{\mathcal{U}}_{\boldsymbol{q}m}, \hat{\mathcal{U}}_{\boldsymbol{q}'m'} \right] &= 0, \\ \left[\hat{\mathcal{P}}_{\boldsymbol{q}m}, \hat{\mathcal{P}}_{\boldsymbol{q}'m'} \right] &= 0, \\ \left[\hat{\mathcal{U}}_{\boldsymbol{q}m}, \hat{\mathcal{P}}_{\boldsymbol{q}'m'} \right] &= i\hbar \delta(\boldsymbol{q} - \boldsymbol{q}') \, \delta_{mm'}. \end{aligned}\right\} \tag{2.88}$$

As we did for the one-dimension system, we can build a set of raising and lowering operators $\hat{a}^{\dagger}_{\boldsymbol{q}m}$ and $\hat{a}_{\boldsymbol{q}m}$,

$$\left.\begin{aligned} \hat{a}^{\dagger}_{\boldsymbol{q}m} &= \sqrt{\frac{M\omega_m(\boldsymbol{q})}{2\hbar}} \hat{\mathcal{U}}^{\dagger}_{\boldsymbol{q}m} - i \frac{1}{\sqrt{2\hbar M \omega_m(\boldsymbol{q})}} \hat{\mathcal{P}}_{\boldsymbol{q}m}, \\ \hat{a}_{\boldsymbol{q}m} &= \sqrt{\frac{M\omega_m(\boldsymbol{q})}{2\hbar}} \hat{\mathcal{U}}_{\boldsymbol{q}m} + i \frac{1}{\sqrt{2\hbar M \omega_m(\boldsymbol{q})}} \hat{\mathcal{P}}^{\dagger}_{\boldsymbol{q}m}. \end{aligned}\right\} \tag{2.89}$$

The Hamiltonian can be written in terms of these operators as

$$\hat{H} = \sum_{\boldsymbol{q}m} \left(\hat{a}_{\boldsymbol{q}m} \hat{a}^{\dagger}_{\boldsymbol{q}m} + \frac{1}{2} \right) \hbar\omega_m(\boldsymbol{q}). \tag{2.90}$$

The eigenstates of this Hamiltonian consist of states with integer numbers $n_{\boldsymbol{q}m}$ of phonons in each mode $\boldsymbol{q}m$, known as *occupation numbers*, written in bracket notation as

$$\left| \ldots n_{\boldsymbol{q}1} \, n_{\boldsymbol{q}2} \, n_{\boldsymbol{q}3} \, n_{\boldsymbol{q}'1} \, n_{\boldsymbol{q}'2} \, n_{\boldsymbol{q}'3} \ldots \right\rangle . \tag{2.91}$$

The Hamiltonian operating on such an eigenstate gives the energy, in the form

$$\begin{aligned} \hat{H} \left| \ldots n_{\boldsymbol{q}1} \, n_{\boldsymbol{q}2} \ldots \right\rangle &= \sum_{\boldsymbol{q}m} \left(n_{\boldsymbol{q}m} + \frac{1}{2} \right) \hbar\omega_{\boldsymbol{q}m} \left| \ldots n_{\boldsymbol{q}1} \, n_{\boldsymbol{q}2} \ldots \right\rangle \\ &= \sum_{\boldsymbol{q}m} E_{\boldsymbol{q}m} \left| \ldots n_{\boldsymbol{q}1} \, n_{\boldsymbol{q}2} \ldots \right\rangle , \end{aligned} \tag{2.92}$$

in terms of the eigenstate energies $E_{\boldsymbol{q}m}$. The occupation number $n_{\boldsymbol{q}m}$ of these eigenstates is the number of *phonons* in the particular mode.

The raising and lowering operators act on these eigenstates in the usual way,

$$\left.\begin{aligned}\hat{a}_{qm}^{\dagger}\,|\ldots n_{qm}\ldots\rangle &= \sqrt{n_{qm}+1}\,|\ldots(n_{qm}+1)\ldots\rangle\,,\\ \hat{a}_{qm}\,|\ldots n_{qm}\ldots\rangle &= \sqrt{n_{qm}}\,|\ldots(n_{qm}-1)\ldots\rangle\,,\end{aligned}\right\}\quad(2.93)$$

where the occupation numbers of all the other states remain unchanged.

By inverting the expressions for the raising and lowering operators, we can write down expressions for the phonon momentum $\hat{\mathcal{P}}_{qm}$ and displacement operators $\hat{\mathcal{U}}_{qm}$, given by

$$\hat{\mathcal{P}}_{qm} = \mathrm{i}\left(\frac{M\hbar\omega_{qm}}{2}\right)^{1/2}\left(\hat{a}_{qm}^{\dagger} - \hat{a}_{-qm}\right), \quad (2.94)$$

and

$$\hat{\mathcal{U}}_{qm} = \left(\frac{2\hbar}{M\omega_{qm}}\right)^{1/2}\left(\hat{a}_{qm}^{\dagger} + \hat{a}_{-qm}\right). \quad (2.95)$$

Note that the expectation value for the momentum and position in a pure energy eigenstate is zero, as the raising and lowering operators generate states orthogonal to the original eigenstate. The energy eigenstates are therefore not eigenstates of either momentum or displacement, as expected.

This operator formalism will prove useful in calculations involving phonons interacting with each other, and for phonons interacting with electrons in the solid.

2.8 Connection to the Classical Continuum Theory of Solids

From the various examples we have worked out (see especially Sect. 2.5), it is apparent that for wavevectors $\boldsymbol{q}$ close to the origin, i.e. for $q = |\boldsymbol{q}|$ small compared to the Brillouin zone edge, the dispersion relation $\omega_m(\boldsymbol{q})$ is linear in q, with $\omega_m(\boldsymbol{q}) = c_m q$. The wave velocity c_m associated with the mode m depends, however, on the direction of the wavevector. Referring to Fig. 2.19, which shows the dispersion relation for a model based on the fcc real space lattice, we see that as we move from the Γ point at the origin to the X symmetry point on the edge of the Brillouin zone, there are two distinct curves $\omega(\boldsymbol{q})$, with a distinct upper (longitudinal) branch, but with the two lower (transverse) branches degenerate, i.e. with equal frequencies. Hence there are two different wave velocities c_ℓ and c_t with $c_\ell > c_t$ (the wave velocities are the slopes of the lines in the figure). Moving from Γ towards K, however, the transverse branches are not degenerate, and there are three wave velocities, with $c_\ell > c_{t1} > c_{t2}$.

The small wavevector behavior of the dispersion relation, with a sound speed that depends on direction, can be reproduced using a continuum description of the solid, as can all the long-wavelength responses. In this section, we will derive the connection between the atomic description of the solid and the continuum theory, which are in exact correspondence for wavelengths much larger than the atomic spacing.

We begin the discussion with the harmonic approximation for the total potential energy of the lattice (2.36) and (2.37),

$$U_{\text{harm}} = \frac{1}{4} \sum_{j,\,k=1}^{N} \sum_{\alpha,\,\beta=1}^{3} (u_{j\alpha} - u_{k\alpha}) \Phi_{\alpha\beta}(\boldsymbol{R}_j - \boldsymbol{R}_k)(u_{j\beta} - u_{k\beta}). \tag{2.96}$$

We treat the displacement field $\boldsymbol{u}$ as a continuous function, $\boldsymbol{u}(\boldsymbol{r})$, with $u_{j\alpha} - u_{k\alpha} = u_\alpha(\boldsymbol{R}_j) - u_\alpha(\boldsymbol{R}_k)$ and so on. For long-wavelength variations in the displacement we can approximate the relative displacement as

$$u_\alpha(\boldsymbol{R}_j) - u_\alpha(\boldsymbol{R}_k) \approx (\boldsymbol{R}_j - \boldsymbol{R}_k) \cdot \nabla u_\alpha(\boldsymbol{r})|_{\boldsymbol{r}=\boldsymbol{R}_k}, \tag{2.97}$$

which written out in component form is

$$u_\alpha(\boldsymbol{R}_j) - u_\alpha(\boldsymbol{R}_k) \approx \sum_{\beta=1}^{3} (R_{j\beta} - R_{k\beta}) \left. \frac{\partial u_\alpha}{\partial x_\beta} \right|_{\boldsymbol{r}=\boldsymbol{R}_k}. \tag{2.98}$$

For sufficiently short-range interactions $\Phi(\boldsymbol{R})$, we can then write the potential energy as

$$\begin{aligned} U_{\text{harm}} &= \frac{1}{4} \sum_{j,\,k=1}^{N} \sum_{\alpha,\,\beta=1}^{3} ((\boldsymbol{R}_j - \boldsymbol{R}_k) \cdot \nabla u_\alpha(\boldsymbol{r})|_{\boldsymbol{r}=\boldsymbol{R}_k}) \\ &\times \Phi_{\alpha\beta}(\boldsymbol{R}_j - \boldsymbol{R}_k) ((\boldsymbol{R}_j - \boldsymbol{R}_k) \cdot \nabla u_\beta(\boldsymbol{r})|_{\boldsymbol{r}=\boldsymbol{R}_k}) \\ &= \frac{v_c}{2} \sum_{k=1}^{N} \sum_{\alpha\,\beta\,\mu\,\nu=1}^{3} \left. \frac{\partial u_\alpha}{\partial x_\mu} \right|_{\boldsymbol{R}_k} E_{\mu\alpha\beta\nu} \left. \frac{\partial u_\beta}{\partial x_\nu} \right|_{\boldsymbol{R}_k}, \end{aligned} \tag{2.99}$$

where we define the fourth-rank tensor $E_{\mu\alpha\beta\nu}$,

$$E_{\mu\alpha\beta\nu} = -\frac{1}{2\,v_c} \sum_{j=1}^{N} R_{j\mu} \Phi_{\alpha\beta}(\boldsymbol{R}_j) R_{j\nu}. \tag{2.100}$$

Here v_c is the volume of the Wigner–Seitz cell for the real space lattice. In the continuum limit, we can replace the sum over k in the potential energy (2.99) as an integral over all space, where the term $V_c = N v_c$ cancels out:

$$U_{\text{harm}} = \frac{1}{2} \sum_{\alpha\,\beta\,\mu\,\nu=1}^{3} \int \frac{\partial u_\alpha}{\partial x_\mu}(\boldsymbol{r}) E_{\mu\alpha\beta\nu} \frac{\partial u_\beta}{\partial x_\nu}(\boldsymbol{r}) \mathrm{d}\boldsymbol{r}. \tag{2.101}$$

The form (2.101) is actually the starting point for Green's method for writing the generalized energy of a solid under strain [14]. Many of the $3^4 = 81$

values $E_{\mu\alpha\beta\nu}$ can be eliminated; for example, the tensor is invariant under the exchange $\alpha \leftrightarrow \beta$ and $\mu \leftrightarrow \nu$, reducing the tensor to 36 independent values.

The derivatives of the displacement $\partial u_\alpha/\partial x_\mu$ can be written as the sum of two very important quantities in continuum mechanics, the *strain tensor* $S_{\alpha\mu}$ and the *rotation tensor* $\omega_{\alpha\mu}$, with

$$S_{\alpha\mu} = \frac{1}{2}\left(\frac{\partial u_\alpha}{\partial x_\mu} + \frac{\partial u_\mu}{\partial x_\alpha}\right), \tag{2.102}$$

and

$$\omega_{\alpha\mu} = \frac{1}{2}\left(\frac{\partial u_\alpha}{\partial x_\mu} - \frac{\partial u_\mu}{\partial x_\alpha}\right). \tag{2.103}$$

The strain tensor is symmetric under interchange of its indices, $S_{\alpha\mu} = S_{\mu\alpha}$, while the rotation tensor is antisymmetric, $\omega_{\alpha\mu} = -\omega_{\mu\alpha}$. Written in matrix form, the strain tensor is

$$S = \begin{bmatrix} \frac{\partial u_x}{\partial x} & \frac{1}{2}\left(\frac{\partial u_x}{\partial y} + \frac{\partial u_y}{\partial x}\right) & \frac{1}{2}\left(\frac{\partial u_x}{\partial z} + \frac{\partial u_z}{\partial x}\right) \\ \frac{1}{2}\left(\frac{\partial u_y}{\partial x} + \frac{\partial u_x}{\partial y}\right) & \frac{\partial u_y}{\partial y} & \frac{1}{2}\left(\frac{\partial u_y}{\partial z} + \frac{\partial u_z}{\partial y}\right) \\ \frac{1}{2}\left(\frac{\partial u_z}{\partial x} + \frac{\partial u_x}{\partial z}\right) & \frac{1}{2}\left(\frac{\partial u_z}{\partial y} + \frac{\partial u_y}{\partial z}\right) & \frac{\partial u_z}{\partial z} \end{bmatrix}. \tag{2.104}$$

The displacement derivative can then be written in terms of the strain and rotation tensors as

$$\frac{\partial u_\alpha}{\partial x_\mu} = S_{\alpha\mu} + \omega_{\alpha\mu}. \tag{2.105}$$

The potential energy (2.101) must be invariant under a rigid rotation of the whole solid. As the strain tensor $S_{\alpha\mu}$ is invariant under such a rotation, while the rotation tensor $\omega_{\alpha\mu}$ is not, the rotational part of the displacement derivative can be eliminated from the potential energy, so we can write

$$U_{\text{harm}} = \frac{1}{2}\sum_{\alpha\,\beta\,\mu\,\nu=1}^{3} \int S_{\alpha\mu}(\boldsymbol{r}) c_{\mu\alpha\beta\nu}\, S_{\beta\nu}(\boldsymbol{r})\mathrm{d}\boldsymbol{r}, \tag{2.106}$$

where the tensor $E_{\mu\alpha\beta\nu}$ has been replaced by the symmetrized elastic tensor $c_{\mu\alpha\beta\nu}$,

$$c_{\mu\alpha\beta\nu} = \frac{1}{4}\left(E_{\mu\alpha\beta\nu} + E_{\alpha\mu\beta\nu} + E_{\mu\alpha\nu\beta} + E_{\alpha\mu\nu\beta}\right). \tag{2.107}$$

As a result of its symmetry under index exchange, $c_{\mu\alpha\beta\nu}$ has at most 21 independent components.

Further simplifications may be made to the elastic tensor c, depending on the symmetries of the underlying crystal lattice. As mentioned in Sect. 2.3, the three-dimensional Bravais lattices can be categorized in seven crystal systems; the *triclinic crystal* class has no symmetries, and therefore has 21 independent components; the *monoclinic crystals* have 13 independent elastic constants, the orthorhombic 9, down to the simplest crystal class, the cubic, with 3 independent constants. A completely isotropic, non-crystalline solid has at most two independent constants, and yields dispersion relations with a longitudinal frequency and a distinct transverse frequency, the latter degenerate for the two transverse polarizations, as discussed in Sect. 2.3.

A very common modification to the tensor notation we have been using reduces the number of indices for the strain tensor $S_{\mu\nu}$ from two to one, where the one index runs from 1 to 6, and the indices for the elastic tensor $c_{\mu\nu\beta\mu}$ from four to two, each of the two indices running from 1 to 6. The double indices $\mu\nu$ in the strain tensor are replaced by the single index α according to the rule

$$xx \to 1,\ yy \to 2,\ zz \to 3,\ zy \to 4,\ zx \to 5,\ xy \to 6. \tag{2.108}$$

At the same time, the off-diagonal components of the strain tensor $S_{\mu\nu}$ are multiplied by a factor of two; this unfortunate, asymmetric definition of the on- and off-diagonal elements is historical, and deeply embedded in the literature, so we shall preserve this convention. In order to distinguish both between the double index and single index form of the strain, and to minimize confusion with the factors of two, we will use the symbol ϵ to indicate the six-element vector form of the strain. The strain components of ϵ are thus defined as

$$\left.\begin{aligned}
\epsilon_1 &= S_{11} = \frac{\partial u_x}{\partial x}, \\
\epsilon_2 &= S_{22} = \frac{\partial u_y}{\partial y}, \\
\epsilon_3 &= S_{33} = \frac{\partial u_z}{\partial z}, \\
\epsilon_4 &= 2S_{23} = \frac{\partial u_y}{\partial z} + \frac{\partial u_z}{\partial y}, \\
\epsilon_5 &= 2S_{13} = \frac{\partial u_x}{\partial z} + \frac{\partial u_z}{\partial x}, \\
\epsilon_6 &= 2S_{12} = \frac{\partial u_y}{\partial x} + \frac{\partial u_x}{\partial y}.
\end{aligned}\right\} \tag{2.109}$$

The corresponding inverse relations are

$$S = \begin{bmatrix} \epsilon_1 & \frac{1}{2}\epsilon_6 & \frac{1}{2}\epsilon_5 \\ \frac{1}{2}\epsilon_6 & \epsilon_2 & \frac{1}{2}\epsilon_4 \\ \frac{1}{2}\epsilon_5 & \frac{1}{2}\epsilon_4 & \epsilon_3 \end{bmatrix}. \tag{2.110}$$

Similarly the elastic tensor is written with two indices,

$$c_{\alpha\beta} = c_{\mu\nu\rho\sigma}, \tag{2.111}$$

with $\mu\nu$ replaced by α, and $\rho\sigma$ replaced by β, using the same rule (2.108). We note that there is no numerical difference between the fourth-rank tensor c and the double index form: The factors of two in the definition of the strain do not change the entries for the elastic tensor c.

To return to our starting point, we can now write the energy of the solid (2.101) in the six-vector form as

$$U_{\text{harm}} = \frac{1}{2}\sum_{\alpha\,\beta=1}^{6} \int \epsilon_\alpha\, c_{\alpha\beta}\, \epsilon_\beta\, \mathrm{d}\boldsymbol{r} \tag{2.112}$$

(the factors of two in the strain definition allow this form for the energy). The energy per unit volume u_E, or energy density, is then

$$\begin{aligned} u_E &= \frac{1}{2}\sum_{\alpha\,\beta=1}^{6} \epsilon_\alpha\, c_{\alpha\beta}\, \epsilon_\beta \\ &= \frac{1}{2}\boldsymbol{\epsilon}\cdot\mathsf{c}\cdot\boldsymbol{\epsilon}. \end{aligned} \tag{2.113}$$

Consider now a small change $\boldsymbol{\delta\epsilon}$ in the strain $\boldsymbol{\epsilon}$. The energy density changes by an amount

$$\delta u_E = \boldsymbol{\delta\epsilon}\cdot(\mathsf{c}\cdot\boldsymbol{\epsilon})\,. \tag{2.114}$$

We can identify the term in the parentheses as the force against which the displacement does work; this term is actually the six-vector form of the *stress tensor* $\boldsymbol{\tau}$, written in terms of the strain and the elastic constants as

$$\tau_\alpha = \sum_{\beta=1}^{6} c_{\alpha\beta}\epsilon_\beta, \qquad (\alpha = 1 \text{ to } 6) \tag{2.115}$$

or in vector form as

$$\boldsymbol{\tau} = \mathsf{c}\cdot\boldsymbol{\epsilon}. \tag{2.116}$$

As for the strain, the six-vector form of the stress is based on a 3×3 stress tensor which we shall write as T. The six-vector elements τ_α are related to the tensor entries $T_{\mu\nu}$ using the same rules (2.108); there are however no factors of two. The tensor form of the strain is related to the tensor form of the stress according to

$$T_{\mu\nu} = \sum_{\alpha\,\beta=1}^{6} c_{\mu\nu\alpha\beta}S_{\alpha\beta}. \tag{2.117}$$

$T_{\mu\nu}$ represents the force per unit area in the direction $\hat{x}_\mu$ on the surface perpendicular to $\hat{x}_\nu$; in other words, T_{xx} represents the force per unit area along $\hat{x}$ on a surface parallel to the $y - z$ plane, while T_{yx} represents the force per unit area along $\hat{y}$ on the same surface. In Chap. 4 we shall explore in much greater detail the definition and use of the stress and strain tensor. We note for reference that the stress-strain relation (2.116) has an equivalent inverse relation,

$$S_{\mu\nu} = \sum_{\alpha\,\beta=1}^{6} s_{\mu\nu\alpha\beta} T_{\alpha\beta}, \tag{2.118}$$

where the *stiffness matrix* s is the matrix inverse of the elastic tensor c, $\mathsf{s} = \mathsf{c}^{-1}$.

There is a six-vector form of the relation (2.118), which is

$$\epsilon_\alpha = \sum_{\beta=1}^{6} s_{\alpha\beta} \tau_\beta, \qquad (\alpha = 1 \text{ to } 6) \tag{2.119}$$

or in vector notation

$$\boldsymbol{\epsilon} = \mathsf{s} \cdot \boldsymbol{\tau}. \tag{2.120}$$

The six-vector form of the stiffness constants $s_{\alpha\beta}$ is related to the tensor form $s_{\mu\nu\rho\sigma}$ using the mapping rules (2.108), with an addition of some factors of 2 and 4:

$$s_{\alpha\beta} = s_{\mu\nu\rho\sigma} \times \begin{cases} 1 \ (\alpha \text{ and } \beta = 1 \text{ to } 3), \\ 2 \ (\alpha \text{ or } \beta = 4 \text{ to } 6), \\ 4 \ (\alpha \text{ and } \beta = 4 \text{ to } 6) \end{cases} \tag{2.121}$$

(see Auld [15]).

In the next chapter, we turn to a discussion of the physics of the phonon gas, calculating its heat capacity, thermal conductance, and interaction properties.

Exercises

2.1 Prove that the Fourier series relations (2.3)–(2.5) are self-consistent.

2.2 Show that the reciprocal lattice to the bcc real space lattice has the fcc structure.

2.3 Calculate the Wigner–Seitz cell volume, reciprocal space lattice vectors and draw the first Brillouin zone for the simple hexagonal real space lattice.

2.4 Show that the fcc crystal lattice can be represented as a triangular lattice, in that using triangular lattice in the plane with generating vectors $\boldsymbol{a}_1$ and $\boldsymbol{a}_2$, find a third generating vector such that the three together generate the fcc crystal.

2.5 Consider atoms in the diamond structure, with a fourfold-symmetric arrangement. Find the angle separating lines drawn from a central atom to two of its neighboring atoms.

2.6 Find the first three terms in the Fourier series for the Fourier transform of the square wave function with $f(x) = 0$ for $2na < x < (2n+1)a$ and $f(x) = A$ for $(2n+1)a < x < (2n+2)a$ for integers n.

2.7 Show that the expression (2.37) can be derived from the more general expression (2.33).

2.8 Find an expression for the dynamical matrix $\Phi(\boldsymbol{q})$ for a simple cubic three-dimensional crystal with one atom per unit cell, with a spherically symmetric nearest-neighbor interaction $\phi(r)$; do not assume that the derivative $\mathrm{d}\phi/\mathrm{d}r$ of the potential is zero at the equilibrium spacing.

2.9 Show that if the wavevector $\boldsymbol{q}$ lies along a symmetry axis of threefold or fourfold symmetry, then the normal mode eigenvectors consist of a vector along $\boldsymbol{q}$, and two other degenerate eigenvectors perpendicular to $\boldsymbol{q}$.

2.10 Work out the values for the 36 elastic constants $c_{\mu\nu}$ for the fcc crystal bound by the spherically symmetric interaction potential $\phi(r)$, described in Sect. 2.5. Note that many of these constants are zero, and the symmetry of the crystal makes many of the non-zero values equal.

3. Properties of the Phonon Gas

In the previous chapter we worked out the formalism for describing the normal modes for vibrations in two- and three-dimensional solids. We showed that the application of quantum mechanics leads to quantization of the energies of the normal modes, and gives a description of a quantum system known as the *phonons*, which in many ways is similar to the quantum excitations of the electromagnetic spectrum, known as *photons*. In this chapter, we begin by deriving the heat capacity associated with the phonon gas, which is the dominant term for insulating solids, but plays a smaller role in metals; we also work out the heat capacity for a nanoscale solid, which should display interesting geometrical effects at very low temperatures. We then turn to a discussion of how to treat phonons as point particles, and discuss phonon scattering and electron-phonon interactions. Finally, we work out the theory for the thermal conductivity of the phonon gas, and conclude with a brief description of the quantum limit to thermal conductance, a theory which has only recently been elucidated and demonstrated experimentally.

3.1 Specific Heat of the Phonon Gas

In Chap. 1 we calculated the specific heat for an N-atom chain. Here we expand the discussion to two- and three-dimensional crystal.

In Sect. 2.7.1, we showed that the energy eigenvalues of the quantum mechanical Hamiltonian for the three dimensional vibrational normal modes can be written in the form

$$E_{\text{tot}} = \sum_{\boldsymbol{q}m} \left(n_{\boldsymbol{q}m} + \frac{1}{2} \right) \hbar\omega_{\boldsymbol{q}m}, \tag{3.1}$$

where each mode $\boldsymbol{q}m$ has occupation $n_{\boldsymbol{q}m}$ and frequency $\omega_{\boldsymbol{q}m}$. In Chap. 2, we showed that each of the modes $\boldsymbol{q}m$ acts as an independent harmonic oscillator, so at non-zero temperature we can calculate the average energy in each mode and then sum to find the total energy of the solid.

The thermally-averaged occupation number $\langle n_{\boldsymbol{q}m} \rangle$ for the mode $\boldsymbol{q}m$ is given by the Bose–Einstein distribution (1.93),

$$\langle n_{\boldsymbol{q}m} \rangle = \frac{1}{\mathrm{e}^{\beta\hbar\omega_{\boldsymbol{q}m}} - 1}, \tag{3.2}$$

where $\beta = 1/k_{\mathrm{B}}T$ is the inverse temperature. The average energy $\langle E_{\boldsymbol{q}m}\rangle$ for this mode is

$$\langle E_{\boldsymbol{q}m}\rangle = \left(\langle n_{\boldsymbol{q}m}\rangle + \frac{1}{2}\right)\hbar\omega_{\boldsymbol{q}m}. \tag{3.3}$$

The total energy E_{tot} of the solid is obtained by summing over all the indices,

$$\langle E_{\mathrm{tot}}\rangle = \sum_{\boldsymbol{q}m}\langle E_{\boldsymbol{q}m}\rangle = \sum_{\boldsymbol{q}m}\left(\frac{1}{\mathrm{e}^{\beta\hbar\omega_{\boldsymbol{q}m}} - 1} + \frac{1}{2}\right)\hbar\omega_{\boldsymbol{q}m}. \tag{3.4}$$

At very high temperatures T, such that $1/\beta = k_{\mathrm{B}}T \gg \hbar\omega_{\mathrm{max}}$, where ω_{max} is the highest frequency in the mode spectrum, every mode has average occupation number $\langle n_{\boldsymbol{q}m}\rangle \approx k_{\mathrm{B}}T/\hbar\omega_{\boldsymbol{q}m}$, and energy $\langle E\rangle \approx k_{\mathrm{B}}T$. With N atoms and 3 polarizations, there are $3N$ modes, and the total thermally-averaged energy is

$$\langle E_{\mathrm{tot}}\rangle = 3Nk_{\mathrm{B}}T. \tag{3.5}$$

This is the classical Dulong–Petit law, which can be arrived at by the purely classical argument that with $3N$ position and $3N$ momentum degrees of freedom, and a thermally-averaged energy of $\frac{1}{2}k_{\mathrm{B}}T$ per degree of freedom, the total energy should be $3Nk_{\mathrm{B}}T$.

If we now lower the temperature, the energy in a mode at frequency ω comprises two parts: A temperature-dependent part that falls exponentially as the temperature is reduced below $\hbar\omega$, and the temperature-independent "zero-point" energy $\frac{1}{2}\hbar\omega$. This is illustrated in Fig. 3.1, showing the thermal energy $\langle E\rangle$ for a single mode at frequency ω, as a function of dimensionless temperature $k_{\mathrm{B}}T/\hbar\omega$. At low temperatures, the contribution of such a mode to the heat capacity becomes negligible, so the higher-frequency modes are said to "freeze out".

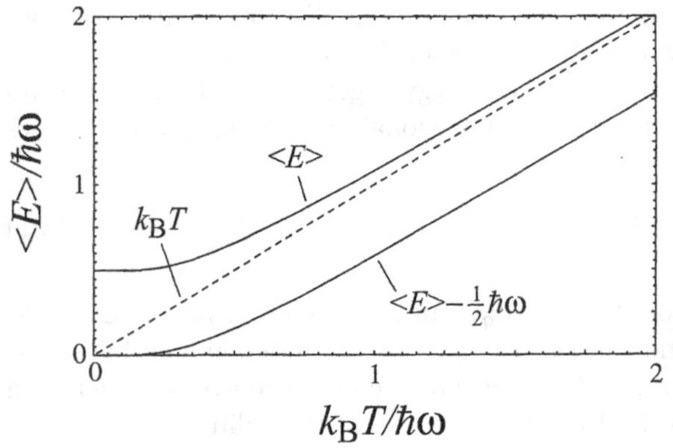

Fig. 3.1. Average energy for a single mode at frequency ω, as a function of temperature, both with and without the zero-point energy, compared to the classical thermal energy $k_{\mathrm{B}}T$.

At moderately low temperatures, such that $k_B T/\hbar$ is less than ω_{max}, but much larger than the frequency difference $\Delta\omega$ between the discrete normal modes, the sum over $\boldsymbol{q}$ in (3.4) may then be approximated as an integral over the first Brillouin zone[1]:

$$\langle E_{tot}\rangle = N\,\frac{v_c}{(2\pi)^3}\sum_m \int_{BZ} \frac{\hbar\omega_{qm}}{e^{\beta\hbar\omega_{qm}}-1}\,d\boldsymbol{q}, \tag{3.6}$$

where $(2\pi)^3/v_c$ is the volume of the Brillouin zone, and $Nv_c = V$ the crystal volume.

This can be transformed into an integral over frequency by using the density of states, $\mathcal{D}(\omega)$, defined as the total number of states per unit crystal volume V, per unit frequency interval $d\omega$. This quantity is defined so that $V\mathcal{D}(\omega)\,d\omega$ is the number of modes in the entire crystal with frequency between ω and $\omega + d\omega$. Using the density of states, we can write down the spectral energy density $u_E(\omega)$, that is, the energy per unit volume per unit frequency interval $d\omega$,

$$u_E(\omega) = \frac{\hbar\omega}{e^{\beta\hbar\omega}-1}\,\mathcal{D}(\omega). \tag{3.7}$$

This is the number occupation for frequency ω, multiplied by the number of distinct states at this frequency (per unit volume). The total energy in the volume V (3.6) is then written

$$\langle E_{tot}\rangle = V\int_0^\infty u_E(\omega)d\omega. \tag{3.8}$$

The specific heat of the system is the derivative of the energy density with temperature, and can be expressed as

$$\begin{aligned} c_V &= \frac{1}{V}\frac{\langle E_{tot}\rangle}{\partial T} = \frac{\partial}{\partial T}\int \frac{\hbar\omega}{e^{\beta\hbar\omega}-1}\,\mathcal{D}(\omega)\,d\omega \\ &= k_B\int (\beta\hbar\omega)^2\,\frac{e^{\beta\hbar\omega}}{(e^{\beta\hbar\omega}-1)^2}\,\mathcal{D}(\omega)\,d\omega. \end{aligned} \tag{3.9}$$

We still need to determine the form for the density of states, $\mathcal{D}(\omega)$. A very useful model was developed by Debye when he first calculated the heat capacity of a solid. Consider the simple isotropic solid, discussed briefly in Sect. 2.3, with one longitudinal branch and two degenerate transverse branches. At small q-vector the dispersion relation corresponds to two sound speeds, c_ℓ and c_t, for the longitudinal and transverse branches, respectively. In the Debye model, we simple take an average sound speed $\bar{c}$, given as the weighted mean of the longitudinal and two transverse sound speeds,

[1] We note that the lower temperature limit for this approximation, $k_B T \sim 2\pi\hbar c/L$, where c is the sound speed and L the dimension of the solid, is of order 0.2 mK for a 1 mm^3 solid, and 20 mK for a 10 µm^3 solid; only for the very smallest objects at the lowest temperatures does the approximation fail. We treat the problem of a nanoscale solid later in this chapter.

$$\frac{1}{\bar{c}^3} = \frac{1}{3}\left(\frac{1}{c_\ell^3} + \frac{1}{c_{t1}^3} + \frac{1}{c_{t2}^3}\right). \tag{3.10}$$

We then assume the dispersion relation $\omega(\boldsymbol{q})$ is strictly linear, with

$$\omega(q) = \bar{c}\,|\boldsymbol{q}|. \tag{3.11}$$

In a real crystal, the range of $\boldsymbol{q}$ is over the first Brillouin zone in the crystal; Debye simplified this by instead taking the range of $\boldsymbol{q}$ to be the volume of a sphere of radius q_D, the Debye wavevector, chosen so that the number of degrees of freedom is preserved.

Each wavevector $\boldsymbol{q}$ within the sphere is associated with a q-space volume $(2\pi)^3/V$, where V is the crystal volume. Hence there are

$$N_D = 3\frac{V}{(2\pi)^3}\frac{4\pi}{3}q_D^3, \tag{3.12}$$

states within the sphere of radius q_D, the factor of 3 to account for the three polarizations. With N atoms in the crystal, there are $3N$ degrees of freedom, so we must have $N_D = 3N$. Hence the radius q_D is given by

$$q_D = \left(6\pi^2\frac{N}{V}\right)^{1/3} = \left(6\pi^2 n\right)^{1/3}, \tag{3.13}$$

with $n = N/V$ the atomic number density. The frequency that corresponds to q_D is the Debye frequency, $\omega_D = \bar{c}q_D$.

We can now calculate the density of states in the Debye model. Take a frequency ω, less than the maximum Debye frequency ω_D; the number of modes N within the corresponding sphere of radius $q = \omega/\bar{c}$ is given by

$$N = 3\frac{V}{(2\pi)^3}\frac{4\pi}{3\bar{c}^3}\omega^3. \tag{3.14}$$

If we increase the frequency slightly by $\mathrm{d}\omega$, the increase in the number of modes $\mathrm{d}N$ is

$$\mathrm{d}N = \frac{\mathrm{d}N}{\mathrm{d}\omega}\mathrm{d}\omega = \frac{3V}{2\pi^2\bar{c}^3}\,\omega^2\,\mathrm{d}\omega. \tag{3.15}$$

Hence the density of states in the Debye model is given by

$$\mathcal{D}(\omega) = \frac{1}{V}\frac{\mathrm{d}N}{\mathrm{d}\omega} = \frac{3}{2\pi^2\bar{c}^2}\omega^2, \tag{3.16}$$

for $\omega \le \omega_D$. For frequencies ω larger than the Debye frequency, we set $\mathcal{D}(\omega) = 0$.

In Fig. 3.2 we show the density of states as a function of frequency, in units of the Debye frequency ω_D. As we shall see, this very simple model gives a fairly good estimate of the low temperature behavior of the specific heat.

We now turn to a discussion of how one calculates the density of states $\mathcal{D}(\omega)$ for a real crystal, with a complex dispersion relation such as that shown

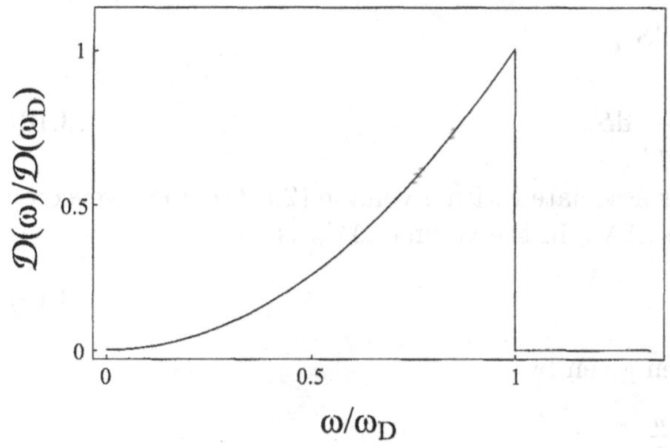

Fig. 3.2. Debye density of states as a function of frequency.

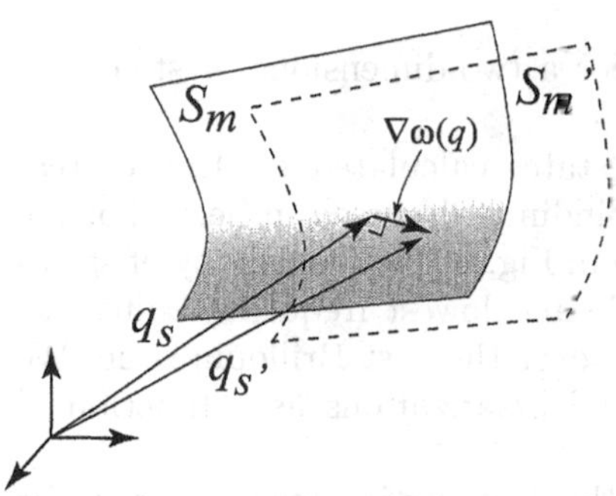

Fig. 3.3. Two surfaces S_m and S_m' defined by the set of points $\{\boldsymbol{q}_s\}$ and $\{\boldsymbol{q}_s^0\}$ that yield frequencies ω_0 and $\omega_0 + \Delta\omega$. The gradient $\nabla_q\omega(\boldsymbol{q})$ is perpendicular to the surface S_m.

in Fig. 2.19. In wavevector space, the set of values $\{\boldsymbol{q}_s\}$ that give a particular frequency ω_0 (so that $\omega_m(\boldsymbol{q}_s) = \omega_0$) form a smooth surface S_m, such as that shown in Fig. 3.3. There is a different surface for each polarization m. At each point on the surface, the local normal to the surface, that is, the vector perpendicular to S_m, is parallel to the vector $\nabla_q \omega_{qm}|_{\boldsymbol{q}_s}$, the gradient taken with respect to $\boldsymbol{q}$. If we then take a slightly different frequency $\omega_0 + \Delta\omega$, defining the surface S_m', the perpendicular distance Δd_m between S_m and S_m' is given to first order by

$$\Delta d_m = \frac{\Delta\omega}{|\nabla_q \omega_{qm}|}. \tag{3.17}$$

The volume ΔV_m (in wavevector space) between the two surfaces is then the integral of this distance over the surface S_m,

$$\Delta V_m = \int_{S_m} \Delta d_m \, \mathrm{d}S_m = \int_{S_m} \frac{\Delta\omega}{|\nabla_q \omega_{\boldsymbol{q}m}|} \mathrm{d}S_m. \tag{3.18}$$

As each normal mode $\boldsymbol{q}m$ is associated with a volume $(2\pi)^3/V$ in wavevector space, the number of states ΔN_m in the volume ΔV_m is

$$\Delta N_m = \frac{V}{(2\pi)^3} \Delta V_m. \tag{3.19}$$

The density of states is then given by

$$\mathcal{D}(\omega) = \sum_m \frac{1}{V} \frac{\Delta N_m}{\Delta\omega} = \frac{1}{(2\pi)^3} \sum_m \int_{S_m} \frac{\mathrm{d}S_m}{|\nabla_q \omega_{\boldsymbol{q}m}|}. \tag{3.20}$$

An equivalent expression can be derived for a two-dimensional system (see Exercise 3.2).

In Fig. 3.4 we display the density of states calculated for the fcc real space crystal with spherically symmetric binding, discussed in Sect. 2.5; the dispersion relation for this crystal is shown in Fig. 2.19. The density of states is shown separately for the highest, middle and lowest frequency solutions, as a function of frequency, each integrated over the first Brillouin zone. We also show the total density of states for all polarizations as a function of frequency.

For a solid with optical phonon modes, the density of states is non-zero for frequencies in both the acoustic and the optical bands. In Fig. 3.5 we display the calculated density of states for GaAs, using the dispersion relations shown in Fig. 2.22.

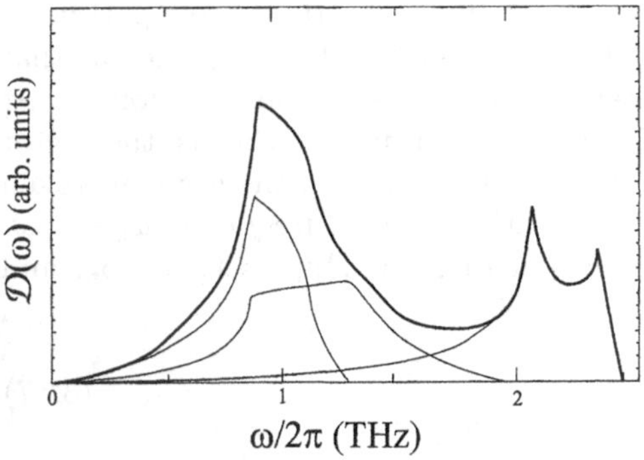

Fig. 3.4. Density of states for the model crystal discussed in Sect. 2.5. The corresponding dispersion relations are shown in Fig. 2.19. We show the density of states for each of the three frequencies (high, middle and low) separately, and for the total density of states $\mathcal{D}(\omega)$.

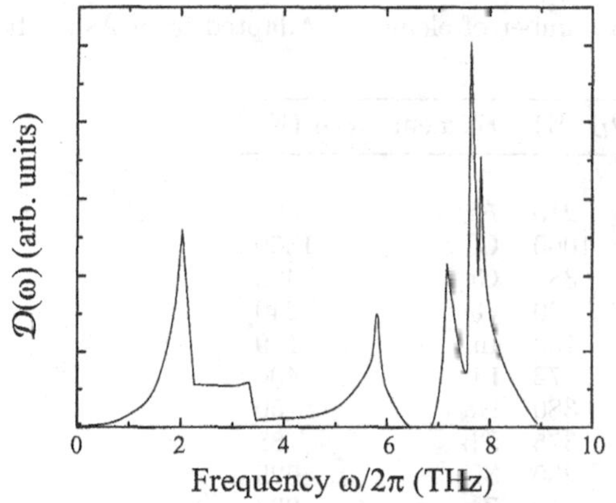

Fig. 3.5. Density of states for a GaAs crystal. The vertical axis is in arbitrary units. The acoustic bands contribute to energies up about 6.5 THz, and the optical bands contribute from just under 7 THz to about 9 THz. The gap between the two bands is about 0.5 THz wide. Adapted from Tamura [13].

The density of states for the model fcc crystal, as well as for that calculated for GaAs, include sharp peaks. These are due to the large number of states that appear at an energy when one of the bands in the dispersion relation has a local maximum. The group velocity at that energy is then zero, i.e. the slope of the dispersion relation is zero, and the contribution to the integral in (3.20) is then very large. These singularities are known as *van Hove singularities*, and appear in the density of states for phonons as well as in that for electrons.

Given the form for the density of states, we can now return to (3.9) in order to calculate the specific heat for a solid. We begin by calculating the specific heat for the Debye model.

Example 3.1: Debye Model for Three-Dimensional Crystal. We first calculate the specific heat using the simplified model of Debye for the density of states, (3.16). At temperature T, the specific heat is given by (3.9),

$$\begin{aligned} c_V &= k_B \int_0^{\omega_D} (\beta\hbar\omega)^2 \frac{e^{\beta\hbar\omega}}{(e^{\beta\hbar\omega}-1)^2} \frac{3}{2\pi^2}\frac{\omega^2}{\bar{c}^3}\,d\omega \\ &= \frac{3}{2\pi^2}\frac{k_B}{\bar{c}^3}\left(\frac{k_BT}{\hbar}\right)^3 \int_0^{x_D} x^4 \frac{e^x}{(e^x-1)^2}\,dx, \end{aligned} \qquad (3.21)$$

where we define the integration variable $x = \beta\hbar\omega$, and x_D is evaluated at the Debye frequency ω_D. It is convenient to define the *Debye temperature* Θ_D,

$$\Theta_D = \frac{\hbar\omega_D}{k_B} = \frac{\hbar}{k_B\bar{c}}\left(6\pi^2 n\right)^{1/3}. \qquad (3.22)$$

In Table 3.1 we display the Debye temperature for a number of elements.

Table 3.1. Debye temperature for a number of elements. Adapted from Ashcroft and Mermin [1].

Element	Θ_D (K)	Element	Θ_D (K)	Element	Θ_D (K)
Al	394	Ag	215	Ar	92
Au	170	Be	1000	C	1860
Cd	120	Co	385	Cr	460
Cu	315	Fe	420	Ga	240
Ge	360	Hg	100	In	129
K	91	Kr	72	Li	400
Mg	318	Mo	380	Na	150
Ne	75	Ni	375	Pb	88
Pd	275	Pt	230	Si	625
Sn (white)	170	W	310	Zn	234

At very low temperatures $T \ll \Theta_D$, $x_D \to \infty$ and the integral may be evaluated exactly:

$$\begin{aligned} c_V &= \frac{2\pi^2}{5}\left(\frac{k_B T}{\hbar \bar{c}}\right)^3 k_B \\ &= \frac{12\pi^4}{5} n \left(\frac{T}{\Theta_D}\right)^3 k_B, \qquad (T \ll \Theta_D) \end{aligned} \tag{3.23}$$

in terms of the atom density n and the Debye temperature Θ_D.

We see that at low temperatures, the Debye model yields a specific heat that scales as T^3; this is the most accurate limit of this model, as the thermal phonons then have the longest wavelengths (smallest wavevectors), where the approximation that the dispersion relation is linear, a critical assumptions of the Debye model, actually holds. This result agrees well with measurements of the specific heat of insulators at the lowest temperatures, and the scaling with number density n and Debye temperature Θ_D is fairly accurate as well. Note that for metals, the contribution to the specific heat from electrons dominates that of the phonons, so the Debye theory is much more difficult to test.

At very high temperatures, such that $T \gg \Theta_D$, we have $x_D \ll 1$. We can then approximate the integrand in (3.21) for small x as

$$x^4 \frac{e^x}{(e^x - 1)^2} \approx x^2, \tag{3.24}$$

so that the integration yields $x_D^3/3 = (\hbar\omega_D/k_B T)^3/3$, and the specific heat is approximately

$$c_V = 3nk_B. \qquad (T \gg \Theta_D), \tag{3.25}$$

the Dulong–Petit law.

At intermediate temperatures $T \sim \Theta_D$, the integral (3.21) may be calculated numerically. The result is shown in Fig. 3.6.

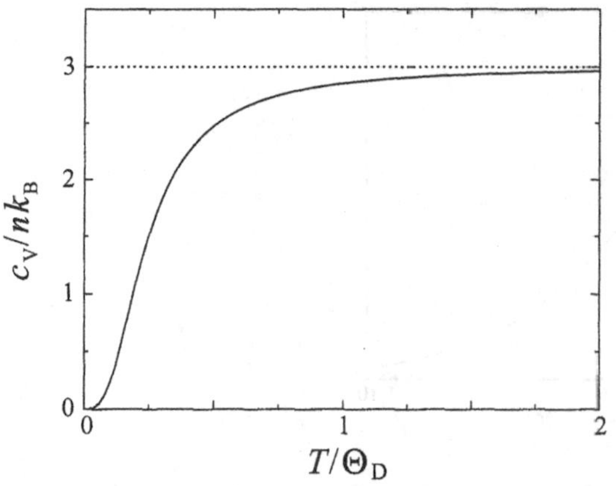

Fig. 3.6. The Debye model for the heat capacity, where n is the number of ions per unit volume, and the Debye temperature Θ_D is defined in the text. The dashed line at $c_V = 3nk_B$ is the classical Dulong–Petit result.

3.1.1 Primary Thermal Phonon Wavelength

The T^3 temperature dependence of the Debye model at low temperatures can be understood by considering what is happening with the thermal distribution of phonons as a function of frequency. The average spectral energy density $u_E(\omega)$ in the Debye model (3.7) is given by

$$u_E(\omega) = \frac{3\hbar V}{2\pi^2 c^3} \frac{\omega^3}{e^{\beta\hbar\omega} - 1}, \tag{3.26}$$

ignoring the zero-point energy. This distribution is plotted in Fig. 3.7, showing the peak in the distribution that occurs at the phonon frequency ω_T, or equivalently the primary phonon wavelength λ_T,

$$\lambda_T = \frac{2\pi c}{\omega_T} = \frac{hc}{2.821 k_B T}. \tag{3.27}$$

Equivalently we can define the primary thermal phonon wavevector, $q_T = 2\pi/\lambda_T = 2.821 k_B T/\hbar c$.

The energy of the solid is given by the average energy per mode, which scales as T, and by the number of modes contributing to the energy, which scales as the volume of wavevector space within the sphere bounded by the thermal wavevector q_T. This volume scales as $q_T^3 \sim T^3$, so we find the total energy scaling as T^4. The specific heat then scales as T^3, as given by Debye. This scaling argument works as long as the thermal wavevector is much smaller than the Debye wavevector, $q_T \ll q_D$; for higher temperatures, the modes are ultimately limited to a sphere of radius q_D, and the energy then scales as T, yielding the constant Dulong–Petit specific heat.

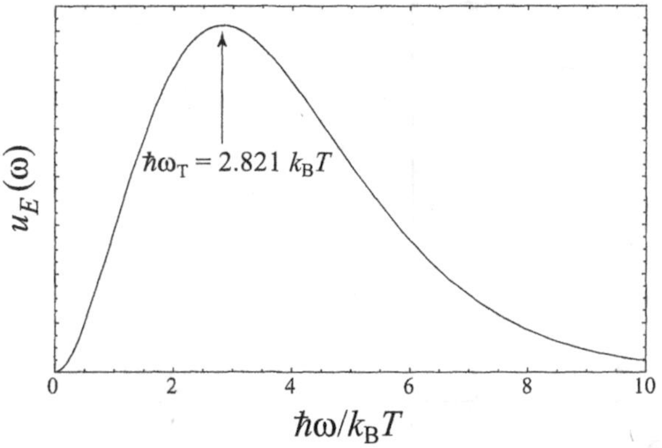

Fig. 3.7. Distribution of thermal phonon energy as a function of phonon frequency, at a fixed temperature T. The vertical axis is the spectral energy density $u_E(\omega)$ at frequency ω, as in (3.26).

Example 3.2: Specific Heat for the Model fcc Crystal. We now calculate the specific heat for the model fcc crystal discussed in Sect. 2.5. The dispersion relation is shown in Fig. 2.19, and the density of states is shown in Fig. 3.4. The specific heat is given by (3.9),

$$\begin{aligned} c_V &= k_B \int_0^{\omega_{max}} (\beta\hbar\omega)^2 \frac{e^{\beta\hbar\omega}}{(e^{\beta\hbar\omega}-1)^2} \mathcal{D}(\omega)\, d\omega \\ &= k_B \left(\frac{k_B T}{\hbar}\right) \int_0^{x_{max}} \mathcal{D}\left(\frac{k_B T x}{\hbar}\right) x^2 \frac{e^x}{(e^x-1)^2}\, dx, \end{aligned} \tag{3.28}$$

where again $x = \beta\hbar\omega$, and x_{max} is evaluated at the maximum frequency ω_{max}. In Fig. 3.8 we display the numerically-integrated result for (3.28), along with the result for the Debye model for comparison. Note that while there is a rather striking difference between the actual dispersion relation and that of the Debye model, in the thermal averaging involved in calculating the heat capacity, the sharp features disappear.

Example 3.3: Heat Capacity of a Thin Bar. We now look at how the geometry of a solid can affect its heat capacity. We will consider a long thin nanometer-scale bar, that has square cross-section with width and thickness a, and has length $L \gg a$. The bar has N atoms, and we assume the width and thickness are sufficiently large that they include several atoms along each dimension. The bar is isotropic, with a single sound speed c for both longitudinal and transverse polarizations; this assumption is highly improbable, but simplifies the calculation. We will ignore the dispersion related to the atomic nature of the solid, as in the Debye model.

We assume the displacement field $\boldsymbol{u}(\boldsymbol{r}, t)$ satisfies the linear, three-dimensional wave equation

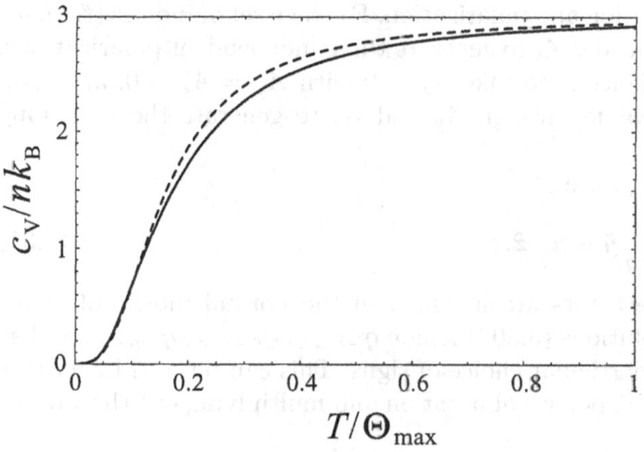

Fig. 3.8. The calculated specific heat for the model fcc crystal as a function of temperature, in units of the maximum frequency $\hbar\omega_{max}/k_B$, shown as a solid line. The specific heat is plotted in units nk_B, where the Dulong- Petit result is $c_V/nK_B = 3$. We also display the Debye model as a dotted line, with a temperature dependence very close to the more complete model.

$$\frac{\partial^2 \boldsymbol{u}}{\partial t^2} = c^2 \nabla^2 \boldsymbol{u}; \tag{3.29}$$

this is the simplest possible wave equation for a solid (see Chap. 7). We place one corner of the bar at the origin, and the diametrically opposite corner at $(x, y, z) = (L, a, a)$, as sketched in Fig. 3.9. We choose boundary conditions such that the derivative of the displacement $\boldsymbol{u}$ on the surface of the bar is equal to zero, in order to simulate stress-free boundaries (see Chap. 7).

The normal modes that satisfy the wave equation and the boundary conditions have displacement along x given by

$$u_x(x, y, z, t) = A_x \cos(\pi \ell x/L) \cos(\pi m y/a) \cos(\pi n z/a) e^{i\omega_{\ell m n} t}, \tag{3.30}$$

with displacements u_y and u_z similar to this, with the amplitude A_x replaced by the amplitudes A_y and A_z, respectively. The indices ℓ, m, and n are non-negative integers, with at least one index larger than zero, and A_x, A_y and A_z

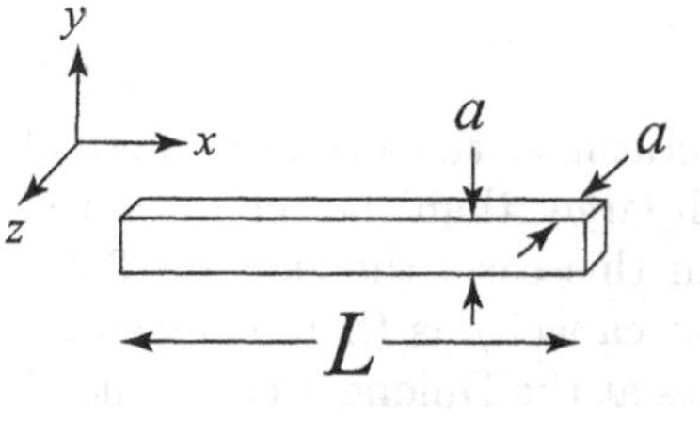

Fig. 3.9. Orientation and geometry of thin bar with length L and width and thickness a.

determine the normalization and polarization. For each set of indices (ℓ, m, n), we can choose the amplitudes A_i to generate three independent polarizations; a particularly simple choice is to take $A_x \neq 0$ with $A_y = A_z = 0$, and then permute the non-zero value through A_y and A_z to generate the remaining polarizations.

The wavevectors $\boldsymbol{q}_{\ell mn}$ are

$$\boldsymbol{q}_{\ell mn} = \ell\frac{\pi}{L}\hat{\boldsymbol{x}} + m\frac{\pi}{a}\hat{\boldsymbol{y}} + n\frac{\pi}{a}\hat{\boldsymbol{z}}. \tag{3.31}$$

Note that several wavevectors are included in the normal mode solutions; along with $\boldsymbol{q}_{\ell mn}$, the solutions (3.30) include $\boldsymbol{q}_{-\ell\, m\, n}$, $\boldsymbol{q}_{\ell\, -m\, n}$, $\boldsymbol{q}_{\ell\, m\, -n}$, and so on, superposed with a particular choice of signs. This can be seen by writing out the normal mode in exponential notation and multiplying out the various products.

The normal mode frequency is given by the wave equation (3.29) as

$$\omega_{\ell mn} = c\,|\boldsymbol{q}_{\ell mn}| = \pi c\sqrt{\ell^2/L^2 + m^2/a^2 + n^2/a^2}. \tag{3.32}$$

The normal mode solutions have no limit on the magnitude of the wavevector $\boldsymbol{q}$. As in the Debye model, we will artificially restrict the range of wavevectors q that satisfy (3.29) to limit the number of normal modes to $3N$, the number of degrees of freedom. The range of $\boldsymbol{q}$ is over the positive octant $q_x, q_y, q_z \geq 0$; once we take one value in this volume, the normal mode includes wavevectors from the other octants, but in a particular combination fixed by the first wavevector in order to yield the displacement (3.30). If we include all solutions with wavevector less than q_D, we have a total volume $\pi q_D^3/6$ in wavevector space. Each wavevector is associated with a volume $\Delta V = \pi^3/a^2 L$ in wavevector space (see (3.31)), so for $q_D \gg \pi/a$, we have $a^2 L q_D^3/2\pi^2$ solutions (counting three polarizations for each wavevector). Given that our bar has $3N$ degrees of freedom, q_D is given by

$$q_D = (6\pi^2 N/a^2 L)^{1/3} = (6\pi^2 n)^{1/3}, \tag{3.33}$$

in terms of the atom density n, just as with the Debye model (3.13). The heat capacity C_V is given by

$$C_V = 3k_{\mathrm{B}} \sum_{\substack{\boldsymbol{q} \\ q < q_D}} (\beta\hbar\omega_{\boldsymbol{q}})^2 \frac{\mathrm{e}^{\beta\hbar\omega_{\boldsymbol{q}}}}{(\mathrm{e}^{\beta\hbar\omega_{\boldsymbol{q}}} - 1)^2} \tag{3.34}$$

At very high and moderately high temperatures, the primary thermal phonon frequency ω_T (see Sect. 3.1.1) is much larger than the separation in frequency $\Delta\omega$ between the discrete modes; in this case, the sum in (3.34) can be approximated by an integral. The heat capacity is then as given by the Debye model, starting at high temperatures at the Dulong–Petit value of $3Nk_{\mathrm{B}}$, and as the temperature is reduced to below the Debye temperature Θ_D, scaling as T^3.

As the temperature is further reduced, the thermal phonon frequency ω_T, or equivalently the thermal phonon wavevector q_T, falls to a value comparable to that associated with a unit transverse wavevector, i.e.

$$q_T = \frac{2.821 k_B T}{\hbar c} \sim \frac{\pi}{a}. \tag{3.35}$$

Below this temperature, these transverse modes are not thermally accessible, and the bar effectively becomes one-dimensional, with only the longitudinal modes $\boldsymbol{q} = (\pi \ell / L, 0, 0)$ contributing to the heat capacity. The sum in (3.34) is then a one-dimensional sum, and in converting to an integral the equivalent density of states is constant (see Exercises). For these low temperatures, then, the heat capacity is found to scale linearly with temperature T.

At the very lowest temperatures, the primary phonon wavevector falls below the smallest non-trivial wavevector,

$$q_T \sim \frac{\pi}{L}. \tag{3.36}$$

There is then only a single wavevector (and three polarizations) contributing to the sum in the heat capacity, with wavevector $\boldsymbol{q}_0 = (\pi/L, 0, 0)$ and frequency $\omega_0 = c\pi/L$. The heat capacity in this limit is then given by

$$C_V = 3k_B \left(\beta \hbar \omega_0\right)^2 \frac{e^{\beta \hbar \omega_0}}{\left(e^{\beta \hbar \omega_0} - 1\right)^2} \tag{3.37}$$

$$\approx 3k_B \left(\beta \hbar \omega_0\right)^2 e^{-\beta \hbar \omega_0}, \tag{3.38}$$

exponentially dependent on the inverse temperature $\beta = 1/k_B T$.

The entire temperature dependence is shown in Fig. 3.10, calculated for a system with $10 \times 10 \times 100$ atoms. The horizontal axis is the dimensionless temperature $a k_B T / \hbar c$, and the vertical axis is the dimensionless heat capacity

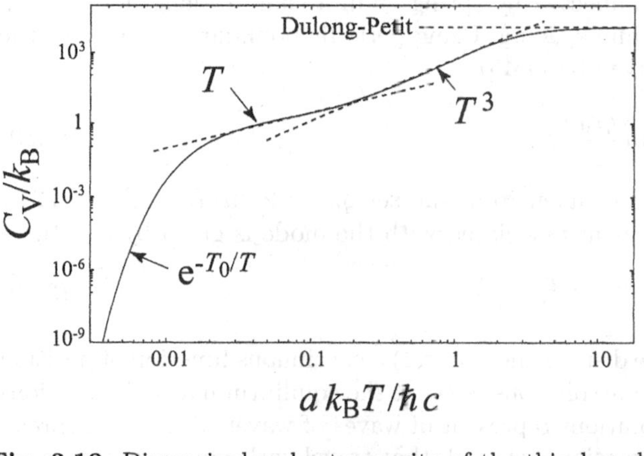

Fig. 3.10. Dimensionless heat capacity of the thin bar sketched in Fig. 3.9, as a function of dimensionless temperature; parameters are as defined in the text.

C_V/k_B. Note that at high temperatures the vertical axis approaches the Dulong–Petit limit of $N = 10^4$; the temperature scale T_0 for the exponential dependence is $T_0 = \hbar\omega_0/k_\mathrm{B}$.

Our bar, with $10 \times 10 \times 100$ atoms, will have dimensions of about $1 \times 1 \times 100\,\mathrm{nm}^3$. The cross-over from the T^3 dependence to the linear T dependence occurs roughly at the temperature $T \approx \hbar c/ak_\mathrm{B} \approx 40$ K. The cross-over from linear to exponential dependence then occurs at $T \approx \hbar c/Lk_\mathrm{B} \approx 400$ mK. We note, however, that this calculation assumes the bar is completely thermally isolated from its environment; any physical link will allow thermal conductance and additional heat capacity, so the dependence calculated here will not hold.

3.2 Phonons as Particles: Position, Momentum and Scattering

In the previous chapter, we developed the quantum description of the normal modes, and found that the quantum eigenstates were ones with definite phonon numbers n in the normal modes $\boldsymbol{q}m$. For static problems, such as the previous section where we calculated the heat capacity of a solid, this picture is quite useable. However, if we are interested in transport properties or in interactions of phonons, the picture of phonons as excitations that are spread throughout the entire crystal is not a very useful one. Here we will develop an alternative picture, where we create superpositions of the normal modes that look and act more like point particles; in so doing, the new states will no longer be energy eigenstates of the Hamiltonian, but will approximate them closely, with an energy spread δE much smaller than the energy E.

We develop this picture with a one-dimensional model; extrapolating to three dimensions is straightforward. Consider a one dimensional system with N atoms of mass M, connected by springs k in a chain of length $L = Nr_0$, where r_0 is the equilibrium spacing. Using periodic boundary conditions, the dispersion relation is given by (1.45),

$$\omega(q) = 2\sqrt{\frac{k}{M}}\left|\sin\frac{qr_0}{2}\right|, \tag{3.39}$$

where the values of q are taken from the set $q_m = 2\pi m/L$, with $-N/2 < m \leq N/2$. The displacement associated with the mode is given by (1.42),

$$u_q(x,t) = \frac{1}{\sqrt{N}}\,\mathrm{e}^{\mathrm{i}qx - \mathrm{i}\omega(q)t}, \tag{3.40}$$

where we have made the displacement $u(x,t)$ a continuous function of position x, giving the normal mode solutions $u_n(t)$ at the equilibrium atomic positions $x = r_n = nr_0$. These solutions represent of waves of wavelength $2\pi/q$, spread throughout the one-dimensional crystal, that travel with phase velocity $c_\phi = \omega(q)/q$.

Consider now building a superposition of the modes, known as a *wave packet*, according to the recipe

$$\begin{aligned} u(x,t) &= \frac{L}{2\pi}\sum_q \mathcal{A}(q)u_q(x,t) \\ &= \frac{L}{2\pi}\frac{1}{\sqrt{N}}\sum_q \mathcal{A}(q)\mathrm{e}^{\mathrm{i}qx-\mathrm{i}\omega(q)t}. \end{aligned} \tag{3.41}$$

The amplitude $\mathcal{A}(q)$ is an envelope function that picks out the modes in a range Δq about a central value q_0. Assume that our chain is very long, and that the range Δq is much larger than the discrete spacing in q of $2\pi/L$, but smaller than the Brillouin zone width $2\pi/r_0$; we can then replace the sum in (3.41) by an integral and write it as

$$u(x,t) = \frac{1}{\sqrt{N}}\int_{-\infty}^{\infty} \mathcal{A}(q)\,\mathrm{e}^{\mathrm{i}qx-\mathrm{i}\omega(q)t}\,\mathrm{d}q. \tag{3.42}$$

At any fixed time, say at $t = 0$, (3.42) has the form of an inverse Fourier transform, with $u(x,0)$ the spatial transform of $\mathcal{A}(q)$. A general property of such transforms is that if $\mathcal{A}$ has a spread Δq in wavevector space, then the wave packet $u(x,0)$ will have a spread Δx in real space, such that

$$\Delta x\,\Delta q \geq \frac{1}{2}. \tag{3.43}$$

If we take a spread in q of 1% of the Brillouin zone, $\Delta q \approx 0.01(2\pi/r_0)$, then the spread of the wave packet in physical space will be $\Delta x \approx 8r_0$. Hence, by making the packet energy, or frequency, spread out by 1%, we have reduced the real-space size of the mode from the length L of the chain to that of a few atomic spacings. An example is sketched in Fig. 3.11.

If the dispersion relation $\omega(q)$ is linear, with $\omega = c_\phi q$, the wave packet will travel in x at the phase velocity c_ϕ. For the dispersion relation (3.39), this holds for small q, but at large q the relation is strongly non-linear. In that case, any individual Fourier component with wavevector q will travel at the local phase velocity $c_\phi = \omega/q$, but the packet as a whole travels at the *group velocity* $c_g = \mathrm{d}\omega/\mathrm{d}q$.

This can be seen as follows: Assume $\mathcal{A}(q)$ is sharply peaked about q_0; we then expand the dispersion relation $\omega(q)$ in a Taylor series about this value, so that

$$\omega(q) \approx \omega(q_0) + \frac{\mathrm{d}\omega}{\mathrm{d}q}(q-q_0) = \omega_0 + c_g(q-q_0). \tag{3.44}$$

In the wave packet expression (3.42), we then make the approximation

$$u(x,t) \approx \frac{1}{\sqrt{N}}\,\mathrm{e}^{\mathrm{i}t(c_g q_0-\omega_0)}\int \mathcal{A}(q)\,\mathrm{e}^{\mathrm{i}q(x-c_g t)}\,\mathrm{d}q. \tag{3.45}$$

The integral part of this expression is $u(x-c_g t, 0)$, the wave packet at $t = 0$, displaced along x by $c_g t$; the remaining exponential is a simple phase factor.

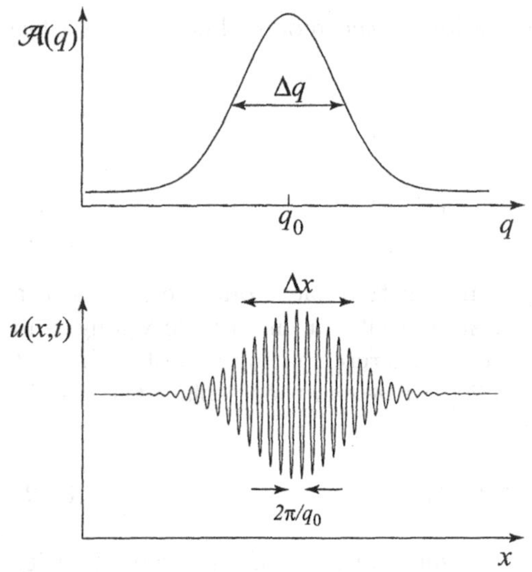

Fig. 3.11. Example of a wave packet envelope function $\mathcal{A}(q)$ and the corresponding spatial wave packet $u(x, t)$.

Hence we see that, to this order of approximation, the wave packet travels undistorted at speed c_g, as promised. Note however that the approximation (3.45) is not exact, and that at long times the wave packet will spread out in real space; the spreading is more severe for packets that initially contain a small range Δq, and much slower for those with larger spreads. A specific example of building such a wave packet is left to the exercises.

In three dimensions, the construction of the wave packet is the same. We build the packet using a three-dimensional envelope $\mathcal{A}(\boldsymbol{q})$, superposing modes in a volume v_q in q-space about a central wavevector $\boldsymbol{q}_0$. Typically the superposed modes would all have the same polarization $\boldsymbol{\varepsilon}_{\boldsymbol{q}_0 n}$, so the construction is given by

$$\boldsymbol{u}(\boldsymbol{r}, t) = \boldsymbol{\varepsilon}_{\boldsymbol{q}_0 n} \int \mathcal{A}(\boldsymbol{q})\, \mathrm{e}^{\mathrm{i}\boldsymbol{q}\cdot\boldsymbol{r} - \mathrm{i}\omega(\boldsymbol{q})t}\, \mathrm{d}\boldsymbol{q}. \tag{3.46}$$

The spread in real space of the wave packet $\boldsymbol{u}$ follows the same rules as for the one dimensional packets; the spread Δx_α along each axis x_α is related to the spread Δq_α by $\Delta x_\alpha \Delta q_\alpha \geq 1/2$.

3.2.1 Phonon Momentum and Energy

Now that we have built a picture of a phonon as a particle, we can work out the particle's momentum and energy. These are evaluated from the central value $\boldsymbol{q}_0$ of the wave packet in (3.46). The energy E of a single phonon with central wavevector $\boldsymbol{q}_0$ and polarization index n is

$$E = \hbar\omega_{q_0 n}, \tag{3.47}$$

as expected from the quantum mechanical description of the lattice. There is an uncertainty ΔE in the energy due to the spread in the wavevectors that make up the wave packet; this is approximately given by

$$\left(\frac{\Delta E}{\hbar}\right)^2 \approx \sum_{\alpha=1}^{3}\left(\left.\frac{\partial\omega_{qn}}{\partial q_\alpha}\right|_{q_0}\Delta q_\alpha\right)^2, \tag{3.48}$$

in terms of the spread Δq_α along each wavevector direction.

The momentum $\boldsymbol{P}$ associated with a wave packet can be evaluated from the quantum mechanical raising and lowering operators given in (2.89), evaluated at the central wavevector $\boldsymbol{q}_0$,

$$\boldsymbol{P} = \hbar\boldsymbol{q}_0. \tag{3.49}$$

Note that this is a momentum in the sense that it is usually conserved in interactions between phonons and between phonons and electrons (see below). Such interactions preserve momentum only in a "weak" sense, as in addition to the various particles involved, the crystal as a whole can change its momentum. For example, an electron with momentum $\boldsymbol{p}$ that emits a phonon with wavevector $\boldsymbol{P}$ can end up with final momentum $\boldsymbol{p}' = \boldsymbol{p} - \boldsymbol{P}$, conserving momentum in the "strong" sense. This type of scattering is known as a *normal* scattering process. The electron can also end up with momentum

$$\boldsymbol{p}' = \boldsymbol{p} - \boldsymbol{P} - \hbar\boldsymbol{G}, \tag{3.50}$$

where $\boldsymbol{G}$ is a reciprocal lattice vector, and $\hbar\boldsymbol{G}$ the corresponding momentum absorbed by the crystal (with a + sign, this would be the momentum added to the interaction by the crystal). These types of processes are known as *Umklapp* scattering processes.

The spread in the wave packet momentum is related to the spread in wavevector,

$$\Delta P_\alpha = \hbar\Delta q_\alpha. \qquad (\alpha = 1 \text{ to } 3) \tag{3.51}$$

We thus find that the position and momentum uncertainties are related by Heisenberg's uncertainty principle:

$$\Delta x_\alpha \Delta P_\alpha \geq \hbar/2. \qquad (\alpha = 1 \text{ to } 3) \tag{3.52}$$

The dispersion relation gives us the relation between momentum and energy. For long-wavelength, small wavevector packets, $\omega_{qm} \approx c_\phi|\boldsymbol{q}|$, and we find

$$E = c_\phi\,|\boldsymbol{P}|, \tag{3.53}$$

the same relation as is found for photons. At small q the phase and group velocities of the packet are the same; larger values of $\boldsymbol{q}$ will have different values for the two velocities, so here we specify the use of the phase velocity.

3.3 Phonon Scattering from Point Defects

The simplest crystal, from the point of view of phonon transport, is a perfect, infinite crystal with no defects, i.e. no deviations from compositional or crystalline order. In such a crystal, a phonon wave packet could travel without disturbance, although suffering spread due to the nonlinearity of the dispersion relation. At zero temperature, a normal mode vibration in such a crystal would exist forever without damping.

It is extremely difficult to achieve such perfection in reality. While extremely high chemical purity can be achieved, ensuring that only very small numbers of unwanted elements remain in a crystal, it is much more difficult to eliminate isotopic variations, that is, variations in the atomic mass of the primary element, due to its naturally occurring isotopes. The isotopic variations create small variations in the local atomic mass, breaking the perfect crystalline symmetry. The easiest way to calculate the effect these imperfections have on phonons is to treat them as a perturbation, which in quantum mechanics translates to calculating the rate at which phonons scatter off the mass variations; this is shown schematically in Fig. 3.12.

We can calculate the isotopic scattering rate using a fairly simple approach; this approach can be approximately applied to chemical variations in crystal composition as well, although the approximations hold less well in the latter case. We can write the Hamiltonian for the crystal as (see Sect. 2.6)

$$\hat{H} = \sum_{j=1}^{N} \frac{\hat{p}_j^2}{2M_j} + \frac{1}{2} \sum_{jk} \sum_{\alpha\beta} \hat{u}_{j\alpha} \Phi_{\alpha\beta}(\boldsymbol{R}_j - \boldsymbol{R}_k) \hat{u}_{k\beta}^{\dagger}, \tag{3.54}$$

with local atomic mass M_j, atomic momentum operator $\hat{\boldsymbol{p}}_j$ and displacement operator $\hat{\boldsymbol{u}}_j$. The interactions between atoms are in the interaction tensor $\Phi(\boldsymbol{R}_j - \boldsymbol{R}_k)$.

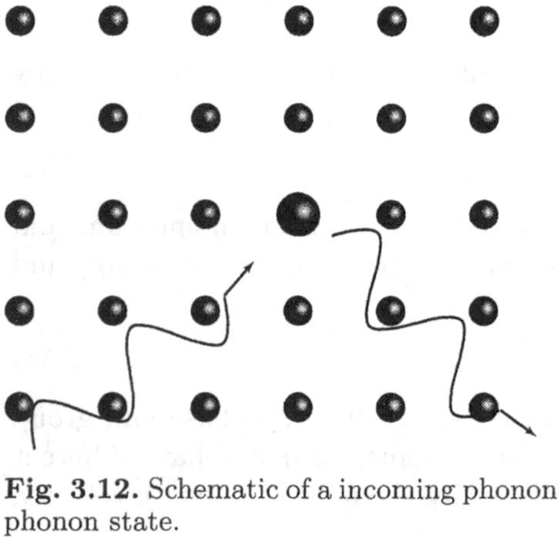

Fig. 3.12. Schematic of a incoming phonon scattering off a mass defect into another phonon state.

The majority of the atoms in the crystal have mass $M_j = M$. Isotopic mass variations cause fluctuations in the local mass M_j, but do not affect the interaction potential. If these variations are small, then we may write the isotopic mass $M_j = M - \Delta M_j$ in terms of the average mass M, and then make the approximation $1/(M - \Delta M_j) \approx (1/M)(1 + \Delta M_j/M)$. We then separate the small terms, involving $\Delta M/M$, as a perturbation term from the perfect crystal Hamiltonian. For a single isotope in the crystal at the position l, the Hamiltonian can be written as

$$\hat{H} = \hat{H}_0 + \Delta\hat{H}_l, \tag{3.55}$$

with

$$\hat{H}_0 = \sum_{j=1}^{N} \frac{\hat{p}_j^2}{2M} + \frac{1}{2}\sum_{jl}\sum_{\alpha\beta} \hat{u}_{j\alpha}\Phi_{\alpha\beta}(\boldsymbol{R}_j - \boldsymbol{R}_k)\hat{u}_{k\beta}^{\dagger}, \tag{3.56}$$

representing the (perfect) crystal Hamiltonian and

$$\Delta\hat{H}_l \approx \frac{\Delta M_l}{M}\frac{\hat{p}_l^2}{2M} \tag{3.57}$$

giving the perturbation. In a real crystal there will be variations in a large fraction f (of order $f \approx 10\%$) of the atoms, with mass variations of order $\Delta M/M \approx 10^{-3}$.

The perturbation causes transitions between the energy eigenstates of the unperturbed Hamiltonian, i.e. between the normal modes $\boldsymbol{q}m$, so that phonons will scatter from $|\boldsymbol{q}m\rangle$ to $|\boldsymbol{q}'m'\rangle$. We calculate the rate Γ for transitions between two such eigenstates, each with definite numbers of phonons (see Sect. 2.7.1). Starting with an initial phonon state $|i\rangle$, the scattering rate $\Gamma_{i\to f}$ to a final state $|f\rangle$ may be calculated using *Fermi's golden rule*,

$$\Gamma_{i\to f} = \frac{2\pi}{\hbar}\left|\langle f|\,\Delta\hat{H}\,|i\rangle\right|^2 \delta(E_f - E_i) \tag{3.58}$$

(see a textbook on introductory quantum mechanics, such as Schiff [16] or Cohen–Tannoudji [7] for a discussion of this fundamental result from scattering theory).

The term $\langle f|\,\Delta\hat{H}\,|i\rangle$, which has units of energy, is known as the *scattering matrix element*, and the δ-function ensures that energy is conserved in this *elastic* (energy-conserving) scattering process.

We would like to write the perturbation Hamiltonian in terms of the normal mode momenta $\hat{\mathcal{P}}_{qn}(t)$. The momentum for the lth atom is given by the inverse of (2.78),

$$\hat{\boldsymbol{p}}_l = \frac{1}{\sqrt{N}}\sum_{\boldsymbol{q}n} \hat{\mathcal{P}}_{\boldsymbol{q}n}\boldsymbol{\varepsilon}_{\boldsymbol{q}n}\mathrm{e}^{-\mathrm{i}\boldsymbol{q}\cdot\boldsymbol{R}_l}. \tag{3.59}$$

The square amplitude can then be written

$$\hat{p}_l^2 = \frac{1}{N} \sum_{qn} \sum_{q'n'} \hat{\mathcal{P}}_{qn}^{\dagger} \hat{\mathcal{P}}_{q'n'} \left(\varepsilon_{qn} \cdot \varepsilon_{q'n'} \right) \mathrm{e}^{\mathrm{i}(q-q') \cdot R_l} . \tag{3.60}$$

For simplicity we position the scattering site at $\boldsymbol{R}_l = 0$.

Next we write (3.60) in terms of the raising and lowering operators, which act directly on the energy eigenstates of the unperturbed Hamiltonian. We can re-write the product $\hat{\mathcal{P}}_{qn}^{\dagger} \hat{\mathcal{P}}_{q'n'}$ using (2.94),

$$\begin{aligned} \hat{\mathcal{P}}_{qn}^{\dagger} \hat{\mathcal{P}}_{q'n'} &= \left(\frac{M^2 \hbar^2 \omega_{qn} \omega_{q'n'}}{4} \right)^{1/2} \left(\hat{a}_{qn} - \hat{a}_{-qn}^{\dagger} \right) \left(\hat{a}_{q'n'}^{\dagger} - \hat{a}_{-q'n'} \right) \\ &= \frac{M\hbar}{2} \left(\omega_{qn} \omega_{q'n'} \right)^{1/2} \\ &\times \left(\hat{a}_{qn} \hat{a}_{q'n'}^{\dagger} - \hat{a}_{-qn}^{\dagger} \hat{a}_{q'n'}^{\dagger} - \hat{a}_{qn} \hat{a}_{-q'n'} + \hat{a}_{-qn}^{\dagger} \hat{a}_{-q'n'} \right) \end{aligned} \tag{3.61}$$

We see that the perturbation $\Delta\hat{H}$ involves the simultaneous manipulation of two phonon states; this is as expected, as the scattering process should remove a phonon from one wavevector state and put it in another, which is accomplished by the destruction of a phonon in the initial state and the creation of a different phonon in the final state. Given that the scattering is elastic, with conservation of energy as given by (3.58), the terms that involve two lowering operators or two raising operators, which destroy and create two phonons respectively, must vanish, so the relevant terms in the perturbation are

$$\begin{aligned} \Delta\hat{H}_l &= \frac{\hbar}{N} \frac{\Delta M_l}{4M} \sum_{qn} \sum_{q'n'} \left(\omega_{qn} \omega_{q'n'} \right)^{1/2} \left(\varepsilon_{qn} \cdot \varepsilon_{q'n'} \right) \\ & \left(\hat{a}_{qn} \hat{a}_{q'n'}^{\dagger} + \hat{a}_{-qn}^{\dagger} \hat{a}_{-q'n'} \right) . \end{aligned} \tag{3.62}$$

Hence the matrix element in the scattering rate couples states with one phonon fewer in the state $\boldsymbol{q}n$ or the state $-\boldsymbol{q}'n'$, and one more in the state $\boldsymbol{q}'n'$ or $-\boldsymbol{q}n$, respectively. If initially there are $m_{\boldsymbol{q}n}$ phonons in $\boldsymbol{q}n$, and the final state has $m_{\boldsymbol{q}'n'}$ phonons, the matrix elements yield factors of the form

$$\langle f | \, \hat{a}_{qn} \hat{a}_{q'n'}^{\dagger} \, | i \rangle = \left[m_{qn} (m_{q'n'} + 1) \right]^{1/2} . \tag{3.63}$$

In the sum for the momentum $\hat{p}_l^2$ there are two such terms, as the q-vectors run through both signs. The scattering rate is then given by

$$\begin{aligned} \Gamma_{i \to f} &= \frac{\pi\hbar}{2N^2} \frac{\Delta M_l^2}{M^2} \left(\omega_{qn} \omega_{q'n'} \right) \left(\varepsilon_{qn} \cdot \varepsilon_{q'n'} \right)^2 \\ &\times \; m_{qn} (m_{q'n'} + 1) \delta (E_f - E_i) . \end{aligned} \tag{3.64}$$

This is the rate for scattering of any one of the $m_{\boldsymbol{q}n}$ phonons in the state $\boldsymbol{q}n$ to the final state $|f\rangle$. We will calculate the rate for a single such phonon, so henceforth we set $m_{\boldsymbol{q}n} = 1$.

Summing over all possible final states $|f\rangle$, the total rate for scattering a single phonon from $|i\rangle$ is given by

$$\Gamma = \frac{\pi\hbar}{2N^2}\frac{\Delta M_l^2}{M^2}\omega_{qn}^2 \sum_{q'n'} (\varepsilon_{qn}\cdot\varepsilon_{q'n'})^2 (m_{q'n'}+1)\delta(E_f - E_i). \tag{3.65}$$

We now assume the final states are in thermal equilibrium with the crystal at temperature T, so the occupation of the final states is given by the Bose–Einstein distribution (1.99). The polarizations are then equally occupied; we replace[2] the sum over polarization n', including the inner product $(\varepsilon_{qn}\cdot\varepsilon_{q'n'})^2$, with the value 1/3. We can then write the remaining sum over q' as an integral over frequency,

$$\begin{aligned}\Gamma &= \frac{\pi\hbar}{2N^2}\frac{1}{3}\frac{\Delta M_l^2}{M^2}\omega_{qn}^2 \\ &\times V\int \mathrm{d}\omega_{q'n'}\,\mathcal{D}(\omega_{q'n'})\left(\frac{1}{\mathrm{e}^{\beta\hbar\omega_{q'n'}}-1}+1\right)\delta(E_f-E_i) \\ &= \frac{\pi V}{6N^2}\frac{\Delta M_l^2}{M^2}\omega_{qn}^2\,\frac{\mathcal{D}(\omega_{qn})}{1-\mathrm{e}^{-\beta\hbar\omega_{qn}}},\end{aligned} \tag{3.66}$$

The δ-function forces the evaluation of the integrand at the frequency $\omega_{q'n'} = \omega_{qn}$. The crystal volume V appears because $\mathcal{D}(\omega)$ is the density of states per unit volume.

Taking the Debye form for the density of states, which is quite accurate for low frequencies ω_{qn} (see Sect. 3.1),

$$D(\omega) = \frac{3}{2\pi^2\bar{c}^3}\omega^2, \tag{3.67}$$

and summing over all scattering sites, with fractional density f and therefore a total number of scattering sites NVf, with an average mass variation ΔM, we find the scattering rate for the single phonon

$$\Gamma = \frac{1}{4\pi n\bar{c}^3} f\left(\frac{\Delta M}{M}\right)^2 \frac{\omega^4}{1-\mathrm{e}^{-\beta\hbar\omega}}. \tag{3.68}$$

At very low temperatures $k_\mathrm{B}T \ll \hbar\omega$, the thermal factor goes to 1, and the rate approaches the limit

$$\Gamma = \frac{1}{4\pi n\bar{c}^3} f\left(\frac{\Delta M}{M}\right)^2 \omega^4. \qquad (k_\mathrm{B}T/\hbar\omega \to 0) \tag{3.69}$$

In this limit the scattering rate goes as ω^4, similar to that found for Rayleigh scattering of light, a limit approached when the wavelength is much larger than the physical dimensions of the scattering center. Writing the radial frequency ω as $\nu = \omega/2\pi$, this is

$$\Gamma = \frac{(2\pi)^3}{2n\bar{c}^3} f\left(\frac{\Delta M}{M}\right)^2 \nu^4. \qquad (k_\mathrm{B}T/h\nu \to 0) \tag{3.70}$$

The first ratio in (3.70) is of order $10^9\ \mathrm{s}^{-1}/\mathrm{THz}^4 = 10^{-39}\ \mathrm{s}^3$.

---

[2] This sum is exactly 1/3 for a cubic crystal, but can differ in other crystal classes.

At high temperatures, or low frequencies, with $k_B T \gg \hbar\omega$, the rate instead goes as

$$\Gamma = \frac{1}{4\pi n \bar{c}^3} f \left(\frac{\Delta M}{M}\right)^2 (k_B T/\hbar)\, \omega^3. \qquad (k_B T/\hbar\omega \gg 1) \tag{3.71}$$

Note however that this temperature effect is much smaller than that due to anharmonic phonon-phonon interactions, to be discussed in Sect. 3.4.

The discussion above was couched in the context of a monatomic crystal, but applies equally to diatomic and more complex crystals. A treatment of the technologically important materials GaAs and InSb appears in the paper by Tamura [13]. In the low-frequency limit for GaAs at $T = 0$, Tamura finds that the scattering rate is given by

$$\begin{aligned} \Gamma &= 4.64 \times 10^{-45} \omega^4 \mathrm{s}^{-1} \\ &= 7.38 \times 10^{-42} \nu^4 \mathrm{s}^{-1}, \end{aligned} \tag{3.72}$$

where the frequency ν is measured in Hertz. In Fig. 3.13 we show the calculated isotopic scattering rates for GaAs and Si as a function of frequency ν, for both the Rayleigh scattering limit and for theoretical predictions that take the actual phonon dispersion into account.

We note that the mean free path Λ, that is, the average distance a phonon can travel before undergoing an isotopic scattering event, is given by

$$\Lambda = \frac{c_g}{\Gamma}. \tag{3.73}$$

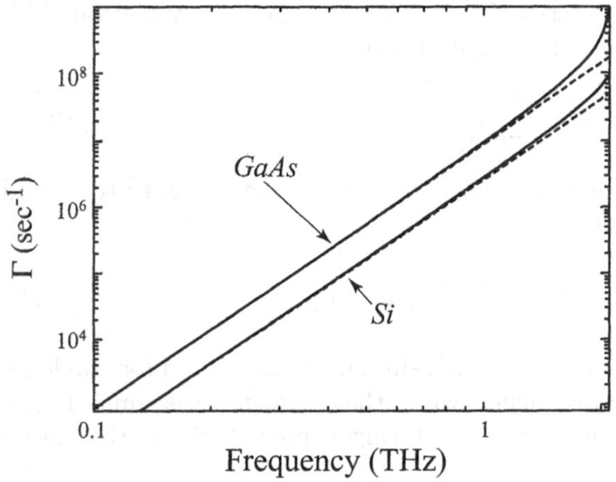

Fig. 3.13. Elastic scattering rate in GaAs and Si due to isotopic mass variations in a crystal, for a mass fluctuation $f\,\Delta M/M$ of 1.97×10^{-4} for GaAs and 2.02×10^{-4} for Si. The dashed lines indicate the low-frequency Rayleigh limit of $7.38 \times 10^{-42}\nu^4$ Hz^{-3} for GaAs and $2.43 \times 10^{-42}\nu^4$ Hz^{-3} for Si. Adapted from Tamura [13], [17].

In low temperature Si, for example, for frequencies in the range of 100 GHz, the lifetime is of order 5 ms, and the mean free path is of order 10 m! Note that the temperatures to achieve this limit are below ~ 1 K. By comparison, the low-temperature mean free path for a 1 THz phonon is of order 2 mm. It should be clear that phonon scattering from the crystal surfaces, which we have not discussed, will dominate for small structures at low temperatures.

At more elevated temperatures, or for high-frequency phonons, phonon-phonon scattering is a much more important process than isotopic scattering; we discuss this process next.

3.4 Phonon–Phonon Interactions

A very important source of scattering of phonons comes from an approximation we made very early in our discussion: The harmonic approximation to the interaction potential. A quantum theory for phonon-phonon interactions, due to this approximation, has been developed by a number of authors (see [18, 19, 20, 21, 22, 23, 24]).

Referring back to (2.29), the full Taylor expansion of the potential energy of the lattice contains terms proportional to the cubic and higher powers of the relative displacement $\boldsymbol{u}_j$. The third-order term in the expansion of the Hamiltonian in (2.29) is given by

$$H_3 = \frac{1}{3!}\sum_{jk\ell}\sum_{\alpha\beta\gamma} u_{j\alpha}u_{k\beta}u_{\ell\gamma}\frac{\partial^3 U}{\partial r_{j\alpha}\partial r_{k\beta}\partial r_{\ell\gamma}}, \tag{3.74}$$

where the derivatives are with respect to each atom's position $\boldsymbol{r}_i$, evaluated at the equilibrium positions $\boldsymbol{R}_i$. Following the same procedure as we did for the harmonic approximation, we write the displacements $\boldsymbol{u}_i$ in terms of the normal mode solutions $\mathcal{U}_{qn}(t)$, where the third order term acquires the form (see Ziman [11])

$$\begin{aligned} H_3 &= \frac{1}{3!}\frac{1}{N}\sum_{q_1n_1}\sum_{q_2n_2}\sum_{q_3n_3} \varepsilon_{q_1n_1}\varepsilon_{q_2n_2}\varepsilon_{q_3n_3} \vdots \mathsf{F}(\boldsymbol{q}_1\boldsymbol{q}_2\boldsymbol{q}_3) \\ &\times \delta(\boldsymbol{G}-\boldsymbol{q}_1-\boldsymbol{q}_2-\boldsymbol{q}_3)\,\mathcal{U}_{q_1n_1}\mathcal{U}_{q_2n_2}\mathcal{U}_{q_3n_3}. \end{aligned} \tag{3.75}$$

The term ε_{qn} is the normalized polarization vector for the mode $\boldsymbol{q}n$. The tensor F is a third-rank tensor, and is related to the Fourier transform of the third-order derivative of U by

$$F_{\alpha\beta\gamma}(\boldsymbol{q}_1\boldsymbol{q}_2\boldsymbol{q}_3) = \sum_{jk=1}^{N} \mathrm{e}^{\mathrm{i}\boldsymbol{q}_2\cdot\boldsymbol{R}_j+\mathrm{i}\boldsymbol{q}_3\cdot\boldsymbol{R}_k}\,\frac{\partial^3 U}{\partial x_{i\alpha}\partial x_{j\beta}\partial x_{k\gamma}}. \tag{3.76}$$

This tensor is the third-order equivalent of the dynamic tensor $\mathsf{D}(\boldsymbol{q})$ we obtained in the harmonic approximation, (2.53). In (3.75), the symbol $\vdots$ signifies a triple inner product of the components of three polarization vectors ε_{qn} with F, so

$$\begin{aligned}\varepsilon_{q_1n_1}\varepsilon_{q_2n_2}\varepsilon_{q_3n_3} &\vdots \mathsf{F}(q_1q_2q_3)\\ &= \sum_{\alpha\beta\gamma=1}^{3} \varepsilon_{q_1n_1,\alpha}\varepsilon_{q_2n_2,\beta}\varepsilon_{q_3n_3,\gamma}F_{\alpha\beta\gamma}(q_1q_2q_3). \quad (3.77)\end{aligned}$$

Finally we note the presence of the δ-function in (3.75); this restricts the possible combinations of $\boldsymbol{q}_1$, $\boldsymbol{q}_2$ and $\boldsymbol{q}_3$ to those that add to zero, or that add to a reciprocal lattice vector $\boldsymbol{G}$. We shall discuss the implications of this restriction below.

To proceed, we replace the displacements $\mathcal{U}_{qn}$ with the appropriate combination of raising and lowering operators for the photon number eigenstates, as given by (2.95),

$$\hat{\mathcal{U}}_{qn} = \left(\frac{2\hbar}{M\omega_{qn}}\right)^{1/2} \left(\hat{a}_{qn}^{\dagger} + \hat{a}_{-qn}\right). \quad (3.78)$$

Placed in (3.75), we find the perturbation to the harmonic Hamiltonian has the form

$$\begin{aligned}\hat{H}_3 &= \frac{1}{3!}\frac{1}{NV}\sum_{q_1n_1}\sum_{q_2n_2}\sum_{q_3n_3}\mathcal{F}(q_1n_1, q_2n_2, q_3n_3)\delta(G - q_1 - q_2 - q_3)\\ &\times (\hat{a}_{q_1n_1}^{\dagger} + \hat{a}_{-q_1n_1})(\hat{a}_{q_2n_2}^{\dagger} + \hat{a}_{-q_2n_2})(\hat{a}_{q_3n_3}^{\dagger} + \hat{a}_{-q_3n_3}). \quad (3.79)\end{aligned}$$

The term $\mathcal{F}(q_1n_1, q_2n_2, q_3n_3)$ involves the prefactors in (3.78) and the triple inner product of the eigenvectors and the tensor F, given by (3.77). The perturbation also involves a triple product of raising and lowering operators, which when multiplied out, will couple initial and final states that differ in their phonon occupation numbers for up to three different modes (or possibly by three phonons in a single mode).

To calculate the scattering rate between number eigenstates of the harmonic Hamiltonian, we proceed as we did for the calculation of the isotopic scattering rate. The rate $\Gamma_{i\to f}$ for scattering from an initial state $|i\rangle$ to a particular final state $|f\rangle$ is given by (3.58),

$$\Gamma_{i\to f} = \frac{2\pi}{\hbar}\left|\langle f|\,\hat{H}_3\,|i\rangle\right|^2 \delta(E_f - E_i). \quad (3.80)$$

The total scattering rate from an initial state $|i\rangle$ is then (3.80) summed over all final states,

$$\Gamma_i = \frac{2\pi}{\hbar}\sum_f\left|\langle f|\,\hat{H}_3\,|i\rangle\right|^2 \delta(E_f - E_i). \quad (3.81)$$

In (3.81) and the Hamiltonian (3.79), the δ-functions in energy and wavevector ensure conservation of energy and momentum, respectively. The δ-function $\delta(\boldsymbol{G}-\boldsymbol{q}_1-\boldsymbol{q}_2-\boldsymbol{q}_3)$ conserves momentum in a "weak" manner: the

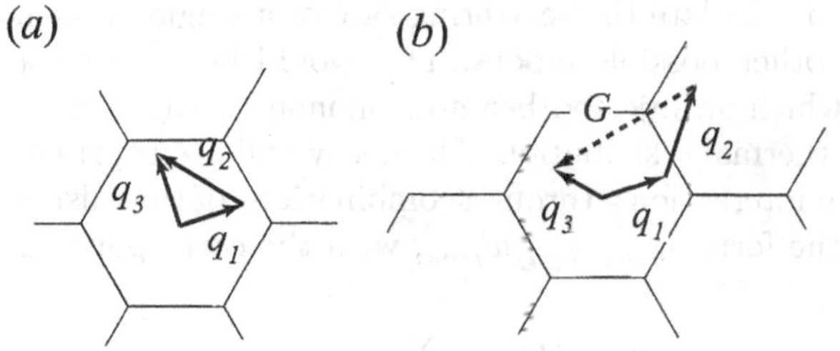

Fig. 3.14. (**a**) Three-phonon normal scattering process, with $1 + 2 \to 3$, where momentum is strictly conserved; also shown is the Brillouin zone, for a hexagonal crystal, for reference. (**b**) *Umklapp* scattering process, involving a reciprocal lattice vector $\boldsymbol{G}$.

q-vectors involved in the scattering process must either sum to zero ("strong" momentum conservation), or can sum to a reciprocal lattice vector $\boldsymbol{G}$ ("weak" conservation). Processes that involve a non-zero reciprocal lattice vector correspond to the crystal as a whole absorbing the momentum $\hbar\boldsymbol{G}$, so that after such a process the crystal's center of mass acquires a (infinitesimal) velocity $\hbar\boldsymbol{G}/NM$ for N total atoms of mass M. These processes are known as *Umklapp* processes, while those for which $\boldsymbol{G} = 0$ are called *normal* processes; we will also use the notation U-process and N-process. In Fig. 3.14 we display these two types of scattering. We note that for *Umklapp* scattering to occur, the initial phonon wavevector (and energy) must be quite large, in the range of one-half to nearly equal to the Brillouin zone wavevector; correspondingly such processes are important only for very high frequency phonons, or at temperatures significantly larger than the Debye temperature, so that such states have non-zero thermal populations.

The energy and momentum conservation rules strongly limit what processes are allowed; one involving the destruction of three phonons, through a product of lowering operators of the form

$$\hat{a}_{q_1 n_1} \hat{a}_{q_2 n_2} \hat{a}_{q_3 n_3}, \tag{3.82}$$

or creation of three phonons from the product of three raising operators, will not be allowed, because in the former case the final state is lower in energy than that of the initial state by $\hbar(\omega_{q_1 n_1} + \omega_{q_2 n_2} + \omega_{q_3 n_3})$, while in the latter case the final state is higher in energy by the same amount.

A process that would be allowed would, for example, be one where two phonons are destroyed and one created, which involves a term $\hat{a}_{q_1 n_1} \hat{a}_{q_2 n_2} \hat{a}^{\dagger}_{q_3 n_3}$, with momentum and energy conservation implying

$$\boldsymbol{q}_1 + \boldsymbol{q}_2 - \boldsymbol{q}_3 = 0 \text{ or } \boldsymbol{G}, \tag{3.83}$$

and

$$\omega_{q_1 n_1} + \omega_{q_2 n_2} - \omega_{q_3 n_3} = 0. \tag{3.84}$$

A common problem is to calculate the scattering rate of a single phonon in the mode $\boldsymbol{q}_1 n_1$ into all other possible modes. This would be of interest in calculating the rate at which an injected beam of phonons decays (eventually) into an equilibrium thermal distribution. The decay will occur primarily through phonon-phonon interactions, through combinations of the raising and lowering operators of the form $\hat{a}_{\boldsymbol{q}_1 n_1}\hat{a}^\dagger_{\boldsymbol{q}_2 n_2}\hat{a}^\dagger_{\boldsymbol{q}_3 n_3}$, with the corresponding conservation equations

$$\left.\begin{array}{rcl} \boldsymbol{q}_1 - \boldsymbol{q}_2 - \boldsymbol{q}_3 &=& 0 \text{ or } \boldsymbol{G}, \\ \hbar\omega_{\boldsymbol{q}_1 n_1} - \hbar\omega_{\boldsymbol{q}_2 n_2} - \hbar\omega_{\boldsymbol{q}_3 n_3} &=& 0. \end{array}\right\} \tag{3.85}$$

An interesting point is that N-processes, of the form shown in Fig. 3.14(a), do not remove momentum or energy from the phonon gas: A beam of phonons in the mode $\boldsymbol{q}n$ decaying through N-processes will generate phonons in other modes $\boldsymbol{q}'n'$, but these other phonons will continue to transport the energy and momentum of the original beam. A thermal current of phonons, in particular, will continue to transport the same amount of energy at the same rate, even in the presence of N scattering processes; only U-processes lead to the decay of thermal currents.

In a solid with elastic isotropy, as discussed in Sect. 2.3, the dispersion relations $\omega(\boldsymbol{q})$ are such that the longitudinal mode has a higher frequency than the two degenerate transverse modes (see Fig. 3.15). In an isotropic solid, therefore, energy and momentum conservation place strong limits on the possible decay processes. A single transverse (T) phonon cannot decay through processes of the form of (3.85), generating two final state phonons, while conserving both energy and momentum [20]. The only available decay process is that in which two initial transverse phonons decay into a third, that is, $T + T \to T$ or $T + L \to T$. Such processes are extremely rare unless there are significant numbers of thermal phonons, so that $k_B T \geq \hbar\omega$, or the

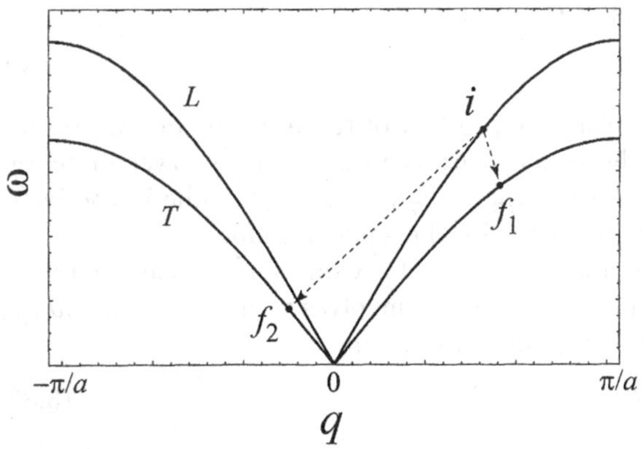

Fig. 3.15. Example of a initial longitudinal phonon i decaying to two final transverse phonon states f_1 and f_2.

injected phonon density is extremely high. The same considerations show that a longitudinal branch (L) phonon cannot decay into two L phonons.

The primary spontaneous decay processes in isotropic materials are therefore of the form

$$\begin{aligned} L &\rightarrow L+T, \text{ and} \\ L &\rightarrow T+T. \end{aligned} \tag{3.86}$$

Tamura has calculated the rates for these decay processes [21], with a single phonon in the initial state, and no phonons initially in the final state (this is therefore a zero temperature calculation). A longitudinal phonon with wavevector $\boldsymbol{q}$ and polarization j decays into two phonons, one with wavevector $\boldsymbol{q}'$ and polarization j', the other with wavevector $\boldsymbol{q}'' = \boldsymbol{q} - \boldsymbol{q}'$, and polarization j'', at a rate

$$\Gamma_{qj} = \frac{\pi\hbar}{8N\omega_{qj}} \sum_{q'} \sum_{j'j''} \frac{|M|^2}{\omega_{q'j'}\omega_{q-q'j''}} \delta(\omega_{qj} - \omega_{q'j'} - \omega_{q-q'j''}). \tag{3.87}$$

The matrix element $|M|^2$ includes all the dependence of the perturbation term $\hat{H}_3$ on q-vector and energy; the δ-function forces conservation of energy. The sum is over all possible final states, with the final wavevector $\boldsymbol{q}'$ set by momentum conservation; note that polarization need not be conserved, as long as the anharmonic matrix element $|M|^2$ connecting the various polarizations is non-zero.

Obtaining numerical results from (3.87) involves a fairly complicated manipulation of the matrix element $|M|^2$, with a number of simplifying approximations. In the low-frequency limit, as was first described by Klemens [18], the decay rate is found to scale as the initial phonon frequency to the fifth power, $\Gamma \sim \omega^5$. We note that the frequency dependence is more rapid than that for Rayleigh scattering from isotopic mass variations, which scales as ω^4 (see Sect. 3.3). However, for typical materials, the overall magnitude for isotopic scattering is significantly larger than that for zero-temperature phonon-phonon scattering; a phonon is much more likely to undergo several isotopic scattering events before it experiences anharmonic decay. In Fig. 3.16 we display the anharmonic decay rates calculated for phonons in Si and GaAs crystal, from [21].

In anisotropic crystals, the rule that one transverse mode phonon cannot decay into two phonons through the anharmonic perturbation does not hold. Calculations for the decay rate of transverse phonons in anisotropic crystals have been published by Tamura and Maris [22, 24]. The predictions are that both fast transverse and slow transverse phonons can decay, but at average rates significantly lower than for longitudinal phonons. There are large regions of wavevector space where transverse phonon decay remains forbidden; however, there are also certain directions for the wavevectors in which the transverse phonons decay faster than the longitudinal phonons. Details of the calculations may be found in the papers cited above.

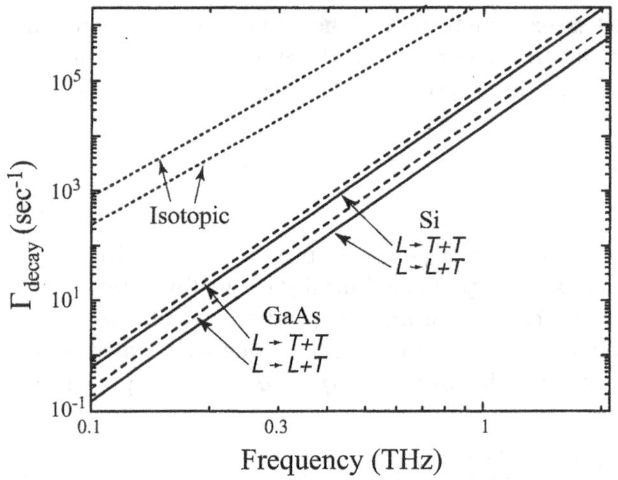

Fig. 3.16. Zero temperature anharmonic decay rates for longitudinal phonons in Si (*solid lines*) and GaAs (*dashed lines*). The rates have been separately calculated for the processes $L \to L+T$ and $L \to T+T$. Also shown are the Rayleigh scattering rates from isotopic mass variations for Si and GaAs (*dotted lines*). Adapted from Tamura [22].

3.4.1 Phonon–Phonon Interactions: Temperature Dependence

Tamura's expression (3.87) is calculated for the decay rate of a single initial phonon, in the absence of phonons in the final states; it therefore applies in the limit $k_{\rm B}T \ll \hbar\omega$. Non-zero occupation numbers affect the rate calculation through the action of the raising and lowering operators; a decay process with $\boldsymbol{q}_1 j_1$ decaying to $\boldsymbol{q}_2 j_2$ and $\boldsymbol{q}_3 j_3$ has a rate $\Gamma_{1\to 2,3}$ proportional to the factors

$$\Gamma_{1\to 2,3} \propto n_{\boldsymbol{q}_1 j_1} \left(n_{\boldsymbol{q}_2 j_2} + 1\right) \left(n_{\boldsymbol{q}_3 j_3} + 1\right). \tag{3.88}$$

With a single phonon in $\boldsymbol{q}_1 n_1$ and no phonons in the final state, this factor is unity. With $\mathcal{N}_1$ phonons in the initial state, and no others, the rate is proportional to $\mathcal{N}_1$, and the population decays in a exponential manner, with a time dependence $\mathrm{e}^{-t/\tau}$, with a well-defined lifetime $\tau = 1/\Gamma_{1\to 2,3}$.

If however we are at a non-zero temperature T, the thermal populations of states 2 and 3 are given by the Bose–Einstein distribution,

$$n_{2,3} = \frac{1}{\mathrm{e}^{\beta\hbar\omega_{2,3}} - 1}, \tag{3.89}$$

with $\beta = 1/k_{\rm B}T$. As the temperature increases, the occupation number increases as well, and we find a strong temperature dependence for the decay rate $\Gamma_{1\to 2,3}$. The overall frequency and temperature dependence for Γ is found to scale as ωT^4; this is known as the *Landau–Rumer* effect, after the authors who worked out the detailed dependence (see [25]).

3.4.2 Higher Order Phonon-Phonon Interactions

The third-order term in the Taylor expansion of the potential energy led us to a perturbation Hamiltonian in which three phonons interact. The fourth-order term in the expansion will similarly lead to four-phonon interactions, and so on; for crystals which are well approximated by the harmonic approximation, these higher-order terms in general get smaller as the order gets larger.

It is also possible to generate fourth and higher-order phonon processes simply from the third-order term. These can be generated through a *second-order perturbation*, where the initial state $|i\rangle$ transforms through a three-phonon process to a short-lived intermediate state $|\alpha\rangle$, and a second three-phonon process leads to the final state $|f\rangle$. Such a sequence is illustrated in Fig. 3.17, along with the three and four-phonon direct processes.

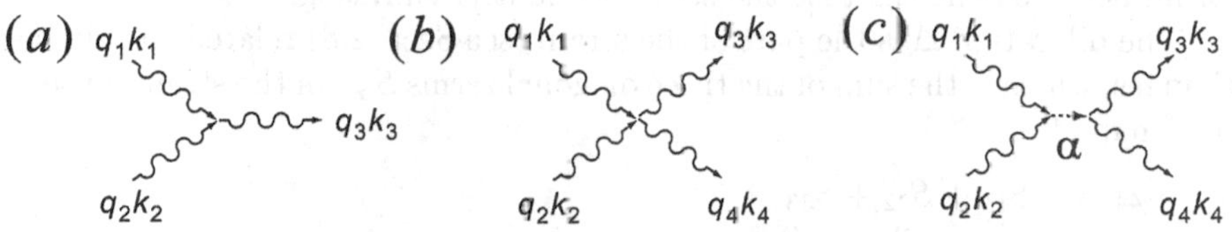

Fig. 3.17. Feynman diagrams for (**a**) a three-phonon interaction from first-order perturbation theory using the cubic term in the potential expansion, (**b**) a four-phonon interaction from first-order perturbation and the quartic term in the expansion, and (**c**) a four-phonon process from second-order perturbation theory involving the cubic term in the expansion.

The rate for the second-order process is given by the expression

$$\Gamma_{i\to f} = \frac{2\pi}{\hbar}\left|\sum_{\alpha\neq i,f}\frac{\langle i|\,\hat{H}_3\,|\alpha\rangle\,\langle\alpha|\,\hat{H}_3\,|f\rangle}{E_i - E_\alpha}\right|^2 \delta(E_i - E_f), \tag{3.90}$$

with the sum over all intermediate states α, and the corresponding total rate for transitions out of the initial state $|i\rangle$ is given by the sum of (3.90) over all final states $|f\rangle$. The constraints on second-order processes are weaker than on first-order processes, in that the intermediate state $|\alpha\rangle$ does not need to have the same energy as the initial or final states; the momentum must however satisfy

$$\boldsymbol{q}_1 + \boldsymbol{q}_2 + \boldsymbol{q}_\alpha = \boldsymbol{G}. \tag{3.91}$$

Interactions occurring through second-order processes can be combined to form four-, five- and higher-number phonon processes, relying only on the cubic term in the potential expansion. Note however that the second-order process (3.90) involves the product of two matrix elements of $\hat{H}_3$, and is therefore significantly weaker than the first-order perturbation involving only a single such matrix element. Similarly, a five-phonon process involves a product of three matrix elements, and so on.

3.4.3 Akhiezer Effect

At high temperatures, above the range controlled by the Landau–Rumer effect, a process distinct from the quantum phonon-phonon interaction process can cause the decay of an initial non-equilibrium phonon population. We will discuss this effect in the context of an injected phonon population, all with the same wavevector and frequency. At high temperatures, the source of interaction between the (assumed low-frequency) injected phonons and the (usually) higher-frequency thermal phonons is due to the change in the elastic constants of the solid that occurs when the crystal undergoes a *dilatation* Δ. This is analogous to, although distinct from, the change in tone of a violin string heard as one tightens the string while it is vibrating.

The dilatation Δ is the part of the strain (see Sect. 2.8) related to volume changes, which is the sum of the three diagonal terms $S_{\alpha\alpha}$ in the strain tensor; see Chap. 4:

$$\begin{aligned}\Delta &= S_{11} + S_{22} + S_{33} \\ &= \frac{\partial u_1}{\partial x_1} + \frac{\partial u_2}{\partial x_2} + \frac{\partial u_3}{\partial x_3}.\end{aligned} \tag{3.92}$$

For a uniform compression, this yields the simple result $\Delta = \delta V/V$, i.e. the dilatation is equal to the fractional change in volume. This dilatation changes the elastic constants in the crystal

The change in the elastic constants changes the normal mode frequencies of the crystal, so that a normal mode at frequency ω_0 and wavevector $\boldsymbol{q}$ will shift to $\omega(\boldsymbol{q})$ according to the relation

$$\omega(\boldsymbol{q}) = \omega_0(\boldsymbol{q})\,(1 - \gamma(\boldsymbol{q})\Delta)\,. \tag{3.93}$$

Here the wavevector-dependent constant $\gamma(\boldsymbol{q})$ that relates the dilatation to the frequency change is known as Grüneisen's constant . This constant also relates the temperature dependence of the thermally-induced dilatation $\Delta(T)$ to the specific heat $c_V(\boldsymbol{q})$ associated with the normal mode with wavevector $\boldsymbol{q}$:

$$\frac{\mathrm{d}\Delta(T)}{\mathrm{d}T} = \frac{1}{B}\sum_{\boldsymbol{q}} c_V(\boldsymbol{q})\gamma(\boldsymbol{q}), \tag{3.94}$$

Here B is the *bulk modulus* of the solid, defined as

$$B = -\left.\frac{\partial P}{\partial \ln V}\right|_T. \tag{3.95}$$

The bulk modulus is equal to the uniform, hydrostatic pressure P needed to effect a unit fractional change in the volume V.

To simplify the treatment, we will ignore the $\boldsymbol{q}$−dependence of γ and the specific heat c_V, so that we can write

$$\frac{\mathrm{d}\Delta}{\mathrm{d}T} \approx \frac{\gamma}{B}\, c_V(T). \tag{3.96}$$

The temperature dependence of the dilatation is related to the coefficient of thermal expansion $\alpha = \partial \ln L/\partial T$ by

$$\frac{\mathrm{d}\Delta}{\mathrm{d}T} = 3\,\alpha\,, \tag{3.97}$$

(see (3.92)). In the Debye model, all the normal mode frequencies are proportional to the Debye temperature Θ_D, so the Grüneisen constant in the Debye model is given by

$$\gamma = -\frac{\partial(\ln \Theta_D)}{\partial(\ln V)}. \tag{3.98}$$

From (3.22), this then yields the simple formula

$$\gamma = \frac{1}{3}. \qquad \text{(Debye model)} \tag{3.99}$$

In Table 3.2 we display values for Grüneisen's constant, the coefficient of linear expansion and the bulk modulus for a number of monatomic crystals.

We now wish to describe the Akhiezer effect. A low-frequency strain wave, with non-zero dilatation Δ, will cause periodic, local variations in the frequency of the normal modes of the crystal, where a positive dilatation reduces the normal mode frequency. These modes are occupied by thermal phonons, nominally in equilibrium at temperature T. A drop in the frequency means that the instantaneous mode occupation will be lower than that dictated by the equilibrium temperature, so that the *effective* local temperature for that mode is reduced. These temperature variations, reductions in T for positive dilatation and increases for negative dilatation, drive a diffusive motion of thermal phonons from volumes of negative dilatation to those with positive dilatation, and thus remove energy from the strain wave, causing it to decay. This is known as the *Akhiezer effect.*

The time constant τ_{th} for the relaxation of the thermal phonons is given by

$$\tau_{\mathrm{th}} = \frac{3\kappa}{c_V \bar{c}^2}, \tag{3.100}$$

where κ is the thermal conductivity and $\bar{c}$ the average velocity used in the Debye model (3.10). This time-delayed relaxation causes inelastic relaxation, treated using Zener's model for anelastic solids (see Chap. 8), with a decay rate Γ_i for the injected phonons given by

$$\Gamma_i = \frac{\Delta E}{\bar{E}} \frac{\omega^2 \tau_{\mathrm{th}}}{1+\omega^2 \tau_{\mathrm{th}}^2}. \tag{3.101}$$

Table 3.2. Values for Grüneisen's constant γ, the coefficient of linear expansion α, and the bulk modulus B for a number of monatomic crystals. The coefficient of linear expansion is in units of 10^{-6}/K, the bulk modulus is in units of 10^{11} N/m^2. All values are at room temperature. Adapted from Ashcroft and Mermin [1] and Kittel [26].

Element	γ	α 10^{-6}/K	B 10^{11} N/m^2
Li	-	45	0.12
Na	-	71	0.068
K	-	83	0.032
Cr	-	5.0	1.90
Mo	-	5.0	2.73
W	1.62	4.5	3.23
Fe	1.60	11.7	1.68
Co	1.87	13.7	1.91
Ni	1.88	12.8	1.86
Cu	1.96	17.0	1.37
Ag	2.40	18.9	1.01
Au	-	13.9	1.73
Al	-	23.6	0.72
Ga	-	19.7	0.57
In	-	-7.5 ($\parallel$), 50 ($\perp$)	0.41
C (diamond)	-	1.0	5.45
Si	-	2.5	0.99
Ge	-	5.7	0.77
Sn (white)	-	21.9	1.11

Here ΔE is the difference between the relaxed and the unrelaxed elastic modulus caused by the dilatation, and $\bar{E}$ is the average elastic modulus. A thermodynamic argument (see [27]) shows that the change in elastic modulus is given approximately by

$$\Delta E = \gamma^2 c_V T. \tag{3.102}$$

One should in principle distinguish between the Grüneisen's constants appearing for different wavevectors $\boldsymbol{q}$, but this result captures the principal effects. The decay rate may therefore be written using the phase velocity relation $v_\phi^2 = \bar{E}/\rho$,

$$\Gamma_i = \frac{\gamma^2 c_V T}{\rho v_\phi^2} \frac{\omega^2 \tau_{\text{th}}}{1 + \omega^2 \tau_{\text{th}}^2}. \tag{3.103}$$

This expression holds in the limit $\omega\tau_{\text{th}} < 1$, where the relaxation of the thermal phonons is faster than the oscillation frequency of the strain wave. In the opposite limit where the relaxation is slower, direct phonon-phonon interactions between the thermal phonons and the strain must be considered, which involves the second- and third-order terms in the Hamiltonian directly;

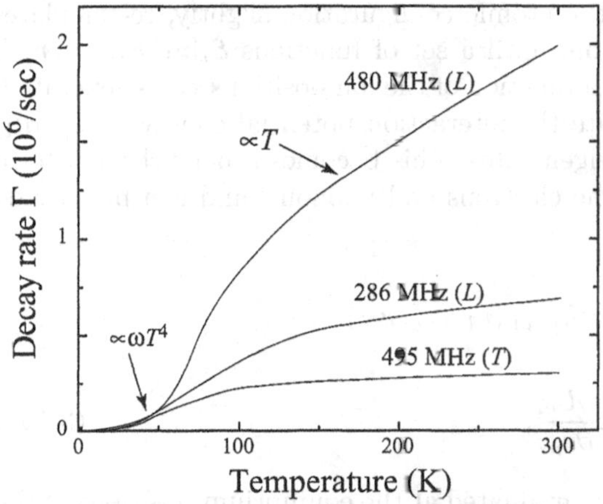

Fig. 3.18. Measured decay rates for different phonon frequencies as a function of temperature; for longitudinal (L) and transverse (T) phonons. The T^4 dependence from the Landau–Rumer effect is observed at the lowest temperatures; at higher temperatures, the decay rate is approximately proportional to T, from the Akhiezer effect. Adapted from Mason and Bateman [29].

in other words, a quantum treatment is required. These calculations are significantly more complicated, and relaxation rates are not so easily derived. A discussion of the detailed theory may be found in Klemens' paper on this subject [28], treating quantum effects at high temperatures.

Measurements of the decay rates for longitudinal and transverse phonons in Si are shown in Fig. 3.18. The T^4 dependence at low temperatures, in the Landau–Rumer regime, as well as the scaling with T at higher temperatures, in the Akhiezer regime, are both clearly visible.

3.5 Electron–Phonon Scattering

A very important scattering process for phonons appears in solids that have free electrons, i.e. in metals or semiconductors. The electrons in a metal or a semiconductor are coupled to the normal mode vibrations through what is in principle a simple mechanism: A lattice vibration causes a change in the atomic charge density, which couples to the electrons through their mutual electrostatic interaction. Calculating the magnitude of the electron-lattice interaction is not entirely straightforward; the approach relies on the fact that the electrons have small mass, and can therefore respond very rapidly to changes in the configuration of the comparatively slowly moving ions. This is the *Born–Oppenheimer* approximation. In this context, one starts by choosing a spatial configuration $\boldsymbol{r}_i = \boldsymbol{R}_i + \boldsymbol{u}_i$ for the ions, and then solves for the resulting electron eigenstates $\Phi(\boldsymbol{r})$ and corresponding electron energy eigenvalues $\mathcal{E}_j$. One then changes the atomic configuration slightly, re-calculates the electron energies, and so on, until a set of functions $\mathcal{E}_j[\boldsymbol{r}_1, \boldsymbol{r}_2, \ldots, \boldsymbol{r}_N]$, giving the electron energies as a function of the ion positions $\boldsymbol{r}_i$, is generated.

In this way we can calculate the interaction potential energy $U_{\text{el-ph}}[\boldsymbol{u_i}]$ that determines the electron eigenstates. This becomes a perturbation term in the Hamiltonian for both the electrons and phonons, and can be written in operator form as

$$\begin{aligned} \Delta\hat{H}_{\text{el-ph}} &= \sum_{i=1}^{N} \hat{\boldsymbol{u}}_i \cdot \nabla_{\boldsymbol{r}_i} U_{\text{el-ph}}(\boldsymbol{r}_1, \ldots, \boldsymbol{r}_N) \\ &= \sum_{i=1}^{N} \sum_{\alpha=1}^{3} \hat{u}_{i\alpha} \frac{\partial U_{\text{el}}}{\partial r_{i\alpha}}, \end{aligned} \tag{3.104}$$

where the partial derivatives are evaluated at the equilibrium positions of the atoms, and the $\hat{\boldsymbol{u}}_i$ are the ion displacement quantum operators. The scattering rate from the unperturbed eigenstates can be calculated for both the phonons and the electrons once the dependence $U_{\text{el}}(\boldsymbol{r_i})$ has been determined.

The simplest model for the electron-phonon interaction is provided by the scalar deformation potential theory, where only the dilatational part of the lattice vibration interacts with the electrons. We have already defined the strain from a lattice vibration in (2.102); the dilatation Δ is related to the strain as given by (3.92).

A plane wave with wavevector $\boldsymbol{q}$ and polarization index m has ion displacements $\boldsymbol{u}(\boldsymbol{r}, t) = (1/\sqrt{N})\mathcal{U}_{\boldsymbol{q}m}(t)\boldsymbol{\varepsilon}_{\boldsymbol{q}m}\mathrm{e}^{\mathrm{i}\boldsymbol{q}\cdot\boldsymbol{r}}$; the corresponding dilatation is given by

$$\Delta = \mathrm{i}\frac{1}{\sqrt{N}}(\boldsymbol{q} \cdot \boldsymbol{\varepsilon}_{\boldsymbol{q}m})\mathcal{U}_{\boldsymbol{q}m}\mathrm{e}^{\mathrm{i}\boldsymbol{q}\cdot\boldsymbol{r}}. \tag{3.105}$$

A wave with transverse polarization has $\boldsymbol{\varepsilon}_{\boldsymbol{q}m} \perp \boldsymbol{q}$, and therefore has zero dilatation. Only a longitudinally polarized wave can couple to the electrons in this model. Furthermore, the coupling is proportional to the phonon wavevector q. Writing the amplitude in terms of the raising and lowering operators, the dilatation operator is

$$\hat{\Delta} = \mathrm{i}\left(\frac{2\hbar}{NM\omega_{\boldsymbol{q}m}}\right)^{1/2} (\boldsymbol{q} \cdot \boldsymbol{\varepsilon}_{\boldsymbol{q}m})\mathrm{e}^{\mathrm{i}\boldsymbol{q}\cdot\boldsymbol{r}} \left(\hat{a}^{\dagger}_{\boldsymbol{q}m} + \hat{a}_{-\boldsymbol{q}m}\right). \tag{3.106}$$

When the dilatation scatters an electron, we need to add to the terms in (3.106) a factor that removes an electron from one state, and a factor that creates one in another, the pair acting as a scattering element. The eigenstates for the electrons are similar to those of the phonon normal modes, with solutions of the Bloch form (2.25). Electrons can therefore be identified by their energy $\mathcal{E}$ and by their Bloch wavevector $\boldsymbol{k}$. The creation and destruction of electrons is performed by raising and lowering operators analogous to those

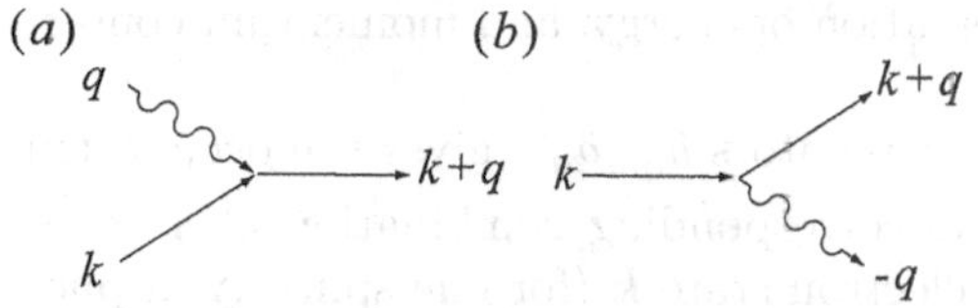

Fig. 3.19. Feynman diagrams for (a) an electron with state $\boldsymbol{k}$ absorbing a phonon with wavevector $\boldsymbol{q}$ and scattering into a state $\boldsymbol{k}+\boldsymbol{q}$, and (b) emitting a phonon with wavevector $-\boldsymbol{q}$ and scattering into the state $\boldsymbol{k}+\boldsymbol{q}$.

for the phonons, which for electrons are written $c_{\boldsymbol{k}}^{\dagger}$ and $c_{\boldsymbol{k}}$ respectively. We note however that the commutation relations for electrons differ from those for phonons, as electrons are fermions, obeying the Pauli exclusion principle, while phonons are bosons.

A scattering process involves an electron either emitting or absorbing a phonon. After this process, the system will have the same total momentum as before. The scattering process can however include *Umklapp* scattering, where the crystal lattice as a whole absorbs a momentum $\hbar\boldsymbol{G}$, with $\boldsymbol{G}$ a reciprocal lattice vector (see Sect. 3.4). Hence, if the electron has initial wavevector $\boldsymbol{k}$, and the phonon has wavevector $\boldsymbol{q}$, the electron will scatter into a state with wavevector $\boldsymbol{k} \pm \boldsymbol{q} - \boldsymbol{G}$ (+ for absorption, − for emission). For the sake of simplicity, here we will ignore *Umklapp* processes, and thus take $\boldsymbol{G} = 0$. The two processes are sketched in Fig. 3.20.

We will write the strength of the deformation potential as D, which has units of energy; in a metal D is given by $2E_F/3$ in terms of the Fermi energy E_F, while in a semiconductor D is typically in the range of 5–15 eV [11]; screening makes the metal coupling weaker than in a semiconductor. The interaction Hamiltonian is given by the the dilatation operator $\hat{\Delta}$ multiplied by the coupling strength D, with the addition of the appropriate electron raising and lowering operators, so

$$\Delta\hat{H} = -\mathrm{i}D\sum_{\boldsymbol{k},\boldsymbol{q}}\left(\frac{\hbar}{2NM\omega(\boldsymbol{q})}\right)^{1/2} q\left(\hat{a}_{\boldsymbol{q}}\,\hat{c}_{\boldsymbol{k}+\boldsymbol{q}}^{\dagger}\hat{c}_{\boldsymbol{k}} - \hat{a}_{\boldsymbol{q}}^{\dagger}\hat{c}_{\boldsymbol{k}-\boldsymbol{q}}^{\dagger}\hat{c}_{\boldsymbol{k}}\right). \quad (3.107)$$

The first term scatters an electron from $\boldsymbol{k}$ to $\boldsymbol{k}+\boldsymbol{q}$ by absorbing a phonon of wavevector $\boldsymbol{q}$, and the second term scatters an electron from $\boldsymbol{k}$ to $\boldsymbol{k}-\boldsymbol{q}$ by emitting a phonon $\boldsymbol{q}$.

The perturbation $\Delta\hat{H}$ couples the state with an electron with wavevector $\boldsymbol{k}$ and $n_{\boldsymbol{q}m}$ phonons in the state $\boldsymbol{q}m$, to the state with an electron with wavevector $\boldsymbol{k}' = \boldsymbol{k}+\boldsymbol{q}$ and $n_{\boldsymbol{q}m}+1$ phonons in the state $\boldsymbol{q}m$, a phonon in the state $\boldsymbol{q}m$ having been emitted. The rate for this phonon emission process is given by Fermi's golden rule:

$$\Gamma_{\boldsymbol{q}-} = \frac{2\pi}{\hbar}\left|\left\langle \boldsymbol{k}-\boldsymbol{q}, n_{\boldsymbol{q}m}+1\left|\Delta\hat{H}\right|\boldsymbol{k}, n_{\boldsymbol{q}m}\right\rangle\right|^2 \delta(E_{\boldsymbol{k}} - \hbar\omega_{\boldsymbol{q}m} - E_{\boldsymbol{k}-\boldsymbol{q}}), (3.108)$$

where the δ-function ensures conservation of energy, and momentum conservation set by setting $\boldsymbol{k} = \boldsymbol{k} - \boldsymbol{q}$.

For phonons, the combination of operators $\hat{a}^\dagger_{\boldsymbol{q}m}\hat{a}_{\boldsymbol{q}m}$ gives the occupation number $n_{\boldsymbol{q}m}$ for the state $\boldsymbol{q}m$. The corresponding combination $\hat{c}^\dagger_{\boldsymbol{k}}\hat{c}_{\boldsymbol{k}}$ gives the occupation probability for the electron state $\boldsymbol{k}$ (for one spin). At a phonon temperature T_{ph}, the phonon occupation is given by the Bose–Einstein distribution,

$$n(\varepsilon_{\boldsymbol{q}}) = \frac{1}{\mathrm{e}^{\varepsilon_q/k_{\mathrm{B}}T_{\mathrm{ph}}} - 1} = \frac{1}{\mathrm{e}^{\beta_{\mathrm{ph}}\varepsilon_q} - 1}. \tag{3.109}$$

For the electrons, at the electron temperature T_{el}, the corresponding distribution is given by the Fermi distribution $f(E)$, with

$$f(E) = \frac{1}{\mathrm{e}^{(E-\mu)/k_{\mathrm{B}}T_{\mathrm{el}}} + 1} = \frac{1}{\mathrm{e}^{\beta_{\mathrm{el}}(E-\mu)} + 1}, \tag{3.110}$$

where μ is the chemical potential for the electron distribution. At zero temperature, all the states below the energy μ are occupied with unit probability, while all states above μ are empty. This energy is known as the Fermi energy E_F, so at zero temperature we have $\mu = E_F$. At non-zero temperature, the chemical potential μ falls slightly below E_F, and marks the energy at which the occupation probability is 1/2; for energies well below μ the occupation approaches 1, while for energies well above it the occupation tends to zero. For energies in the range $\mu \pm k_{\mathrm{B}}T$, the occupation falls smoothly from 1 to 0.

Writing the phonon energy as ε_q, we can write the rate (3.108) as

$$\Gamma_{\boldsymbol{q}-} = \frac{2\pi}{\hbar}\mathcal{M}^2\,(1 - f(E_{\boldsymbol{k}-\boldsymbol{q}}))\,(n_{\boldsymbol{q}\ell} + 1)\,\delta(E_{\boldsymbol{k}} - \varepsilon_q - E_{\boldsymbol{k}-\boldsymbol{q}}), \tag{3.111}$$

where the square matrix element $\mathcal{M}^2$ is given by

$$\mathcal{M}^2 = \frac{\hbar q}{2NMc_\ell}D^2 \equiv \mathcal{M}_0^2\,q. \tag{3.112}$$

The phonon frequencies and occupation numbers all refer to the longitudinal polarization; transverse polarization phonons, as discussed above, have zero dilatation and do not couple to the electrons in the scalar deformation model.

Following the same reasoning, we can find the rate $\Gamma_{\boldsymbol{q}+}$ for electrons scattering from $\boldsymbol{k}$ to $\boldsymbol{k}+\boldsymbol{q}$ by absorbing a phonon of wavevector $\boldsymbol{q}$. This is given by (3.108), with $-\boldsymbol{q}$ replaced by $\boldsymbol{q}$, and $n_{\boldsymbol{q}m}+1$ replaced by $n_{\boldsymbol{q}m}$,

$$\Gamma_{\boldsymbol{q}+} = \frac{2\pi}{\hbar}\mathcal{M}^2\,(1 - f(E_{\boldsymbol{k}+\boldsymbol{q}}))\,n_{\boldsymbol{q}\ell}\,\delta(E_{\boldsymbol{k}} + \varepsilon_q - E_{\boldsymbol{k}+\boldsymbol{q}}). \tag{3.113}$$

These expressions for the scattering rates can be used to calculate various electron-phonon processes. In the following example, we calculate the expression for power flow between an electron gas and a phonon gas, each at different temperatures T_{el} and T_{ph}.

Example 3.4: Electron–Phonon Coupling in a Low-Temperature Metal. Here we examine a particular problem of great importance in small volumes of metal at very low temperatures: The thermal decoupling of the electrons in the metal from the phonons. This decoupling is quite pronounced at temperatures below 1 K, and allows the electrons and phonons to acquire different temperatures $T_{\rm el}$ and $T_{\rm ph}$, respectively. When electrical power is dissipated in the electron gas, they can heat significantly above the temperature of the crystal lattice. This so-called "hot-electron" effect was first measured by Roukes [30], and later confirmed in a number of other systems (see e.g. Wellstood [31]).

We are interested in using the scattering rates (3.111) and (3.113) to calculate the power flow between the electrons and phonons in a volume V. An electron loses an energy $\varepsilon_q = \hbar\omega_{q\ell}$ when it emits a longitudinal phonon of wavevector $\boldsymbol{q}$. The power flow is this energy times the rate (3.111), $P_e = \varepsilon_q \Gamma_{\boldsymbol{q}-}$. This process can involve phonons of all wavevectors $\boldsymbol{q}$, so we should sum the power over all phonon wavevectors; the electron gas as a whole loses power at a rate given by summing this result over all possible electron initial states, multiplied by the initial occupation probability $f(E)$. Hence the total power emission is given by

$$P_e = 2\sum_{\boldsymbol{k}}\sum_{\boldsymbol{q}} \varepsilon_q\, f(E_{\boldsymbol{k}})\, \Gamma_{\boldsymbol{q}-}. \tag{3.114}$$

The factor of two is to account for the two electron spin states. Writing the sum over $\boldsymbol{k}$ as an integral over energy, multiplied by the electron density of electron states per unit volume, $\mathcal{D}_{\rm el}(E)$, and by the volume V, and writing the sum over $\boldsymbol{q}$ as an integral over $\boldsymbol{q}$ divided by the density of states in $\boldsymbol{q}$-space $(2\pi)^3/V$, this is

$$\begin{aligned} P_e &= \frac{2\pi}{\hbar}\int V\,\mathcal{D}_{\rm el}(E)\,{\rm d}E\frac{V}{(2\pi)^3}\int {\rm d}^3q\,\mathcal{M}_0^2\,q\,\varepsilon_q \\ &\times\; f(E)\,(1-f(E_{\boldsymbol{k}-\boldsymbol{q}}))\,(n_{\boldsymbol{q}\ell}+1)\,\delta(E-\varepsilon_q-E_{\boldsymbol{k}-\boldsymbol{q}}). \end{aligned} \tag{3.115}$$

Similarly we can calculate the power *absorption* P_a by the electron gas, due to absorption processes at the rate (3.113), which yields

$$\begin{aligned} P_a &= \frac{2\pi}{\hbar}\int \mathcal{D}_{\rm el}(E)\,{\rm d}E\frac{V^2}{(2\pi)^3}\int {\rm d}^3q\,\mathcal{M}_0^2\,q\,\varepsilon_q \\ &\times\; f(E)\,(1-f(E_{\boldsymbol{k}+\boldsymbol{q}}))\,n_{\boldsymbol{q}\ell}\,\delta(E+\varepsilon_q-E_{\boldsymbol{k}+\boldsymbol{q}}). \end{aligned} \tag{3.116}$$

The net power flow out of the electrons is $P = P_e - P_a$. Following Wellstood [32], this power can then be separated into the sum $P = P_0 + P_T$, where P_0 is the power emitted when the phonon gas is at zero temperature and P_T the power for non-zero phonon temperature. The powers P_0 and P_T are given by

$$\begin{aligned} P_0 &= \frac{2\pi}{\hbar}\frac{V^2}{(2\pi)^3}\int \mathcal{D}_{\rm el}(E)\,\mathrm{d}E \int \mathrm{d}^3q\,\mathcal{M}_0^2\,q\,\varepsilon_q \\ &\times\; f(E)\,(1-f(E_{\boldsymbol{k}+\boldsymbol{q}}))\,\delta(E-\varepsilon_q-E_{\boldsymbol{k}+\boldsymbol{q}}), \end{aligned} \tag{3.117}$$

and

$$\begin{aligned} P_T &= \frac{2\pi}{\hbar}\frac{V^2}{(2\pi)^3}\int \mathcal{D}_{\rm el}(E)\,\mathrm{d}E \int \mathrm{d}^3q\,\mathcal{M}_0^2\,q\,\varepsilon_q\,n_{q\ell} \\ &\times\; f(E)\,(1-f(E_{\boldsymbol{k}+\boldsymbol{q}})) \\ &\times\; (\delta(E-\varepsilon_q-E_{\boldsymbol{k}+\boldsymbol{q}})-\delta(E+\varepsilon_q-E_{\boldsymbol{k}+\boldsymbol{q}}))\,. \end{aligned} \tag{3.118}$$

The integrals for the emission and absorption of power involve a product of Fermi functions $f(E)$ and $1-f(E\pm\varepsilon_q)$, which at low temperatures forms a sharply peaked function centered at $E=E_F$; we can then evaluate the density of states $\mathcal{D}_{\rm el}(E)$ at the Fermi energy and pull it out of the integral. The integral over wavevectors $\boldsymbol{q}$ involves the delta function that conserves energy; this can be evaluated using the identity

$$\int \delta(g(\boldsymbol{q}))\mathrm{d}^3q = \int \frac{\mathrm{d}S_q}{|\nabla_q g(\boldsymbol{q})|}, \tag{3.119}$$

where the surface integral on the right side of the equation is over the surface defined by the zeroes of the function g, i.e. the points $\boldsymbol{q}$ where $g(\boldsymbol{q})=0$.

The function g in (3.111) is the electron energy $E(\boldsymbol{k}-\boldsymbol{q})$, given in metals by the usual parabolic form $E(k)=\hbar^2k^2/2m^*$, with m^* the reduced electron mass, so we have

$$\begin{aligned} \nabla_q E(\boldsymbol{k}-\boldsymbol{q}) &= \frac{\hbar^2(\boldsymbol{k}-\boldsymbol{q})}{m^*} - \hbar c_\ell\,\frac{\boldsymbol{q}}{q^2} \\ &\approx \frac{\hbar^2 k_F}{m^*}\,\hat{k}. \end{aligned} \tag{3.120}$$

The zeroes of $g(\boldsymbol{q})$ are on a sphere of radius k_F centered on $\boldsymbol{k}$. At low temperatures only the parts of this sphere near the origin $\boldsymbol{q}=0$ contribute significantly, so we approximate the sphere by a plane tangent to the sphere (see Fig. 3.20). The surface integral becomes $\mathrm{d}S_q=2\pi q\mathrm{d}q$, integrated over

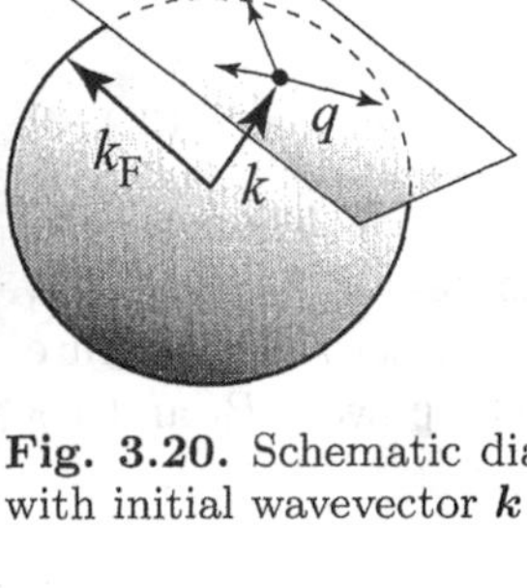

Fig. 3.20. Schematic diagram showing scattering sheet available for an electron, with initial wavevector $\boldsymbol{k}$ emitting a phonon with wavevector $\boldsymbol{q}$.

the surface perpendicular to $\boldsymbol{k}$. Finally, we approximate the phonon energy for small phonon wavevectors q by $\varepsilon_q = \hbar c_\ell q$.

With these approximations, the integral for P_0 becomes

$$P_0 = \frac{V^2 \mathcal{D}_{\rm el}(E_F) \mathcal{M}_0^2 c_\ell}{2\pi\hbar v_F} \int \mathrm{d}E \int \mathrm{d}q\, q^3 f(E)\,(1 - f(E - \varepsilon_q)), \tag{3.121}$$

and for P_T

$$\begin{aligned} P_T &= \frac{V^2 \mathcal{D}_{\rm el}(E_F) \mathcal{M}_0^2 c_\ell}{2\pi\hbar v_F} \int \mathrm{d}E \int \mathrm{d}q\, q^3 \times \\ &\times\ n_{q\ell}\,(f(E - \varepsilon_q) - f(E + \varepsilon_q)), \end{aligned} \tag{3.122}$$

where we have used $v_F = k_F/m^*$.

At low temperatures $k_B T_{\rm el} \ll E_F$, the Fermi function goes to zero for energies a few times $k_B T_{\rm el}$ above the chemical potential $\mu \approx E_F$, so the integral over energy can be extended to $\pm\infty$. Similarly the integral over q can be extended to infinity. The integrals can then be evaluated in closed form to yield

$$P_0 = \Sigma V T_{\rm el}^5, \tag{3.123}$$

and

$$P_T = \Sigma V T_{\rm ph}^5, \tag{3.124}$$

where the coupling constant Σ is given by

$$\Sigma = \frac{\mathcal{D}_{\rm el}(E_F)\, V\, \mathcal{M}_0^2 k_B^5}{2\pi\hbar^5 c_\ell^3 v_F} \Gamma(5)\zeta(5), \tag{3.125}$$

where $\Gamma(5) = 4!$ is the gamma function and $\zeta(5) \approx 1.037$ is the Riemann zeta function. The net power is then

$$P = \Sigma V (T_{\rm el}^5 - T_{\rm ph}^5). \tag{3.126}$$

For typical metals, the interaction potential $D = 2E_F/3$ [11], so $\mathcal{M}_0^2 = 2\hbar E_F^2/9NMc_\ell$ (M is the atomic mass). The density of states per unit volume at the Fermi level is related to the electronic specific heat $c_{\rm el}$ through

$$c_{\rm el} = \frac{\pi^2}{3} \mathcal{D}_{\rm el}(E_F)\, k_B^2\, T_{\rm el} \equiv \gamma_{\rm el} T_{\rm el}, \tag{3.127}$$

(see e.g. Ashcroft and Mermin [1]). The matrix element $\mathcal{M}_0^2$ is given by

$$\begin{aligned} \mathcal{M}_0^2 &= \frac{\hbar}{2NMc_\ell} D^2 \\ &= \frac{\hbar}{2\rho V c_\ell} \frac{4E_F^2}{9}, \end{aligned} \tag{3.128}$$

and we can write

$$\Sigma = \frac{1}{3\pi^3} \frac{\gamma_{\rm el}\, E_F^2 k_B^3}{\hbar^4 \rho c_\ell^4 v_F} \Gamma(5)\,\zeta(5). \tag{3.129}$$

For typical metals, Σ is of order $0.1-1\times10^9$ W/m^3-K^5. For a thin film of metal with dimensions of $1\times10\times0.01$ μm^3, a dissipated power of 1 pW (10^{-12} W) will increase the electron temperature from $T_{\rm el}=0$ to 400 mK, with the phonons still at absolute zero, a rather spectacular temperature increase for such a small power. This decoupling of the electrons from phonons means that achieving truly low electron temperatures, below 100 mK, requires fairly large metal volumes, due to the weak 1/5-power dependence of the electron temperature on both power and volume.

In the previous example we saw how the theory for the scalar deformation potential can be used to calculate the effective thermal coupling between electrons and phonons in a metal. In semiconductors, the deformation potential theory also applies, although the much weaker screening, caused by the lower electron density, changes the theory somewhat. The interaction energy is still taken to be proportional to the dilatation Δ, but in addition there is a non-scalar component; the perturbation Hamiltonian ΔH can then be written [11]

$$\Delta H = D_1\Delta + D_2\left(\hat{k}\cdot\mathsf{S}\cdot\hat{k} - \frac{1}{3}\Delta\right), \tag{3.130}$$

where S is the strain tensor (2.102), with components

$$S_{\alpha\beta} = \frac{1}{2}\left(\frac{\partial u_\alpha}{\partial x_\beta} + \frac{\partial u_\beta}{\partial x_\alpha}\right), \tag{3.131}$$

in terms of the local atomic displacement $\boldsymbol{u}$ (assumed a continuous function of position). The vectors $\hat{k}$ are the unit vectors in the direction of the electron wavevector $\boldsymbol{k}$.

In a uniform dilatational displacement Δ, the strain S has no off-diagonal components, and the diagonal components are all equal. The perturbation is then the same form as for a metal (with a different interaction strength D_1). If (say) there is only one diagonal strain component, or the strain includes off-diagonal components, the coupling will be a function of the direction of the electron wavevector. In semiconductors such as Si and Ge, the conduction band valleys, where electrons (the negative charge carriers) have their minimum energies, lie at non-zero $\boldsymbol{k}$ along the axes of cubic symmetry (x, y and z). In a strained crystal of Si or Ge, then, the electrons in different valleys shift in energy by different amounts; the electrons then redistribute in the various valleys to keep the Fermi level the same. As the different valleys then participate to different degrees in carrying an electrical current, and they have different electrical resistance, the resistance of the crystal as a whole becomes strain-dependent. This effect is the source of the strong *piezoresistance* in these indirect semiconductors.

3.5.1 Piezoelectric Interaction

Piezoelectric Materials. Materials that are not inversion-symmetric, that is, ones where inverting all the atomic positions from $\boldsymbol{R}_i$ to $-\boldsymbol{R}_i$ does not return the original crystal, and at the same time include ions with different equilibrium charges, will often display an effect known as *piezoelectricity*. This effect can be understood using Fig. 3.21.

At equilibrium, the four-atom chain in Fig. 3.21 has charges $q_i = \pm 2q$ at $x_i = \pm d/2$ and $q_i = \mp q$ at $x_i = \pm d$. The net polarization, defined as

$$P = \sum_{i=1}^{4} q_i x_i, \tag{3.132}$$

is equal to zero. When an compressional force F is applied to both ends of the chain, the springs connecting the charges compress; for simplicity, we assume the central spring is rigid, while the outer springs have spring constant k. The outer charges move in by equal amounts $\delta x = F/k$, and yield a net polarization $P = 2q\,\delta x$. Note that this polarization is associated with an internal electric field as well (assumed weak). Hence an external force generates both a displacement, and a polarization and electric field. Similarly, when an external electric field E is applied, the outside charges move in by equal amounts $\delta x' = qE/k$, yielding a polarization $P = 2(q^2/k)E$; in addition to the polarization, an internal force, or stress, is generated.

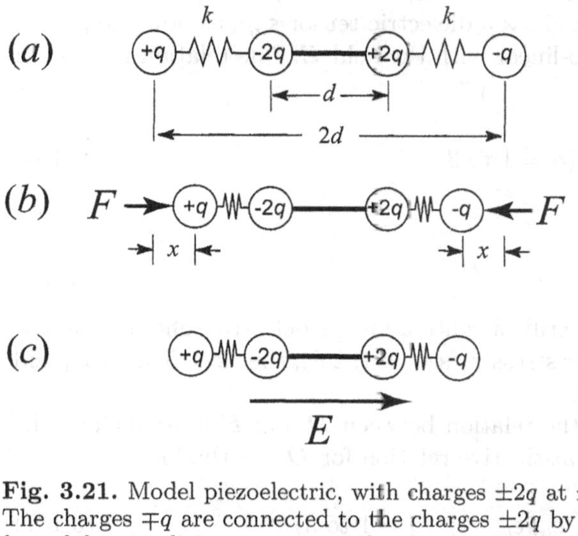

Fig. 3.21. Model piezoelectric, with charges $\pm 2q$ at $\pm d/2$, and charges $\mp q$ at $\pm d$. The charges $\mp q$ are connected to the charges $\pm 2q$ by springs with spring constant k, and for simplicity we assume the charges $\pm 2q$ are connected by rigid rods (the argument also works if one takes very stiff springs). The equilibrium polarization is zero, but external forces and electric fields generate non-zero polarizations. Figure adapted from Auld [15].

We therefore find that this system has its displacement, polarization, force and electric field forming a coupled set, each generating the other in turn. If we take the force and the electric field as the independent variables, we can write these relations in the form

$$\left.\begin{aligned} \delta x &= (1/k)F + (q/k)E, \\ P &= (2q/k)F + (2q^2/k)E. \end{aligned}\right\} \qquad (3.133)$$

Note that the choice of dependent and independent variables is arbitrary; we could equally well have chosen the displacement δx and the polarization P.

In a real piezoelectric solid, the same type of behavior is found: The lack of inversion symmetry in the charge distribution, along with differing displacements of one charge type compared to the other, generates displacements (strains) and polarizations from externally applied forces (stresses) and electric fields. The constitutive relations corresponding to (3.133) for a three-dimensional solid are usually written in terms of the *electric displacement* $\boldsymbol{D}$, instead of the polarization $\boldsymbol{P}$. The electric displacement is defined through

$$\boldsymbol{D} = \varepsilon_0 \boldsymbol{E} + \boldsymbol{P}, \qquad (3.134)$$

where $\varepsilon_0 = 8.85 \times 10^{-12}$ F/m is the permittivity of vacuum, and $\boldsymbol{P}$ the polarization density. In a linear, isotropic dielectric solid, with relative permittivity ε_r and dielectric constant $\varepsilon = \varepsilon_r \varepsilon_0$, the polarization density is given by $\boldsymbol{P} = (\varepsilon_r - 1)\varepsilon_0 \boldsymbol{E}$, and the electric displacement is simply $\boldsymbol{D} = \varepsilon \boldsymbol{E}$.

In a crystalline non-piezoelectric solid, the corresponding linear relation between $\boldsymbol{D}$ and $\boldsymbol{E}$ involves the 3×3 dielectric tensor $\varepsilon_{\alpha\beta}$, as now the polarization $\boldsymbol{P}$ is not ncessarily co-linear with the field $\boldsymbol{E}$. The relation is then, in component form,

$$D_\alpha = \sum_{\beta=1}^{3} \varepsilon_{\alpha\beta} E_\beta, \qquad (\alpha = 1 \text{ to } 3) \qquad (3.135)$$

or in tensor form,

$$\boldsymbol{D} = \varepsilon \cdot \boldsymbol{E}. \qquad (3.136)$$

The variables used to describe a continuum piezoelectric solid are the six-vector strain $\boldsymbol{\epsilon}$, the six-vector stress $\boldsymbol{\tau}$ (see Sect. 2.8), the electric field $\boldsymbol{E}$, and the electric displacement $\boldsymbol{D}$.

In a piezoelectric solid, the relation between $\boldsymbol{D}$ and $\boldsymbol{E}$ must include the effect of the strain $\boldsymbol{\epsilon}$. The constitutive relation for $\boldsymbol{D}$ has the form

$$D_\alpha = \sum_{\beta=1}^{3} \varepsilon^{\epsilon}_{\alpha\beta} E_\beta + \sum_{\mu=1}^{6} e_{\alpha\mu}\epsilon_\mu, \qquad (\alpha = 1 \text{ to } 3) \qquad (3.137)$$

where ε^ϵ is the form for the dielectric tensor when the strain is chosen as the independent variable. We have introduced a new matrix e, which has 3 rows and 6 columns, and is known as the piezoelectric stress matrix. This matrix gives the relation between the electric displacement $\boldsymbol{D}$ and the strain ϵ.

The corresponding relation for the stress τ is

$$\tau_\mu = -\sum_{\beta=1}^{3} e_{\beta\mu} E_\beta + \sum_{\nu=1}^{6} c_{\mu\nu}^E \epsilon_\nu, \qquad (\mu = 1 \text{ to } 6) \tag{3.138}$$

with the elastic constant c^E relating the stress and strain when the electric field is treated as an independent variable. Note the interchange of indices for the piezoelectric stress constant matrix e that appears in (3.138). We will indicate the transpose by using an underline, so that $e_{\beta\mu} = \underline{e}_{\mu\beta}$.

These relations can be written in matrix form as

$$\left.\begin{aligned} \boldsymbol{D} &= \varepsilon^\epsilon \cdot \boldsymbol{E} + \mathsf{e} \cdot \epsilon, \\ \tau &= -\underline{\mathsf{e}} \cdot \boldsymbol{E} + \mathsf{c}^E \cdot \epsilon, \end{aligned}\right\} \tag{3.139}$$

where $\underline{\mathsf{e}}$ the transpose of the 3×6 matrix e, yielding a 6×3 matrix.

There are a number of other forms for the piezoelectric relations (3.139); for example, if we choose the electric field $\boldsymbol{E}$ and the stress τ as the independent parameters, the relations are

$$\left.\begin{aligned} \boldsymbol{D} &= \varepsilon^\tau \cdot \boldsymbol{E} + \mathsf{d} \cdot \tau, \\ \epsilon &= \underline{\mathsf{d}} \cdot \boldsymbol{E} + \mathsf{s}^E \cdot \tau, \end{aligned}\right\} \tag{3.140}$$

where ε^τ is the dielectric tensor with the stress τ as the independent parameter. These relations involve a different piezoelectric strain matrix d. The terms $d_{\alpha\nu}$ form a 3×6 tensor, related to the piezoelectric constants $e_{\alpha\nu}$ through the relation

$$e_{\alpha\nu} = \sum_{\mu=1}^{6} d_{\alpha\mu}\, c_{\mu\nu}^E. \qquad (\alpha = 1 \text{ to } 3,\ \nu = 1 \text{ to } 6) \tag{3.141}$$

The dielectric tensors are related by

$$\varepsilon_{\alpha\beta}^\epsilon = \varepsilon_{\alpha\beta}^\tau - \sum_{\mu\,\nu=1}^{6} d_{\alpha\mu}\, c_{\mu\nu}^E\, \underline{d}_{\nu\beta}. \qquad (\alpha, \beta = 1 \text{ to } 3) \tag{3.142}$$

Yet another form is

$$\left.\begin{aligned} \boldsymbol{E} &= \beta^\epsilon \cdot \boldsymbol{D} - \mathsf{h} \cdot \epsilon, \\ \tau &= -\underline{\mathsf{h}} \cdot \boldsymbol{D} + \mathsf{c}^{\mathrm{D}} \cdot \epsilon, \end{aligned}\right\} \tag{3.143}$$

where β^ϵ is the inverse of the permittivity tensor ε, h is another piezoelectric matrix, defined by

$$\mathsf{h} = \beta^\epsilon \cdot \mathsf{e}, \tag{3.144}$$

and the elastic tensor c^{D} is given by

$$\mathsf{c}^{\mathrm{D}} = \underline{\mathsf{h}} \cdot \mathsf{e} + \mathsf{c}^E. \tag{3.145}$$

Piezoelectric Electron–Phonon Interaction. We now work out the equivalent of the deformation potential, discussed in Sect. 3.5, but with the electron-phonon coupling mediated by the piezoelectric interaction. We will work this out for an insulating solid, with a very small number of charge carriers. For solids that have large numbers of carriers, the carriers will act to screen out the piezoelectric fields, significantly complicating the calculation.

For the insulating solid, Gauss' law for the electric displacement with no free charge carriers is

$$\nabla \cdot \boldsymbol{D} = \rho_{\text{free}} \approx 0 \tag{3.146}$$

where ρ_{free} is the free charge density and is taken to be zero. As this is essentially an electrostatic problem, we can then take $\boldsymbol{D} = 0$. The simplest choice for the set of piezoelectric stress-strain equations is then that given by (3.143), which with (3.146) takes the form

$$\left.\begin{aligned} \boldsymbol{E} &= -\mathsf{h} \cdot \boldsymbol{\epsilon}, \\ \boldsymbol{\tau} &= \mathsf{c}^{\mathrm{D}} \cdot \boldsymbol{\epsilon}. \end{aligned}\right\} \tag{3.147}$$

Electrons will interact with the electric field, with the electrostatic potential ϕ given by $\boldsymbol{E} = -\nabla\phi$.

Consider a transverse polarization phonon travelling along $\hat{z}$, with wavevector $q\hat{z}$, polarized along $\hat{y}$. The displacement field $\boldsymbol{u}$ has the form

$$\boldsymbol{u} = \mathcal{A}\,\hat{y}\,\mathrm{e}^{\mathrm{i}qz-\mathrm{i}\omega t}, \tag{3.148}$$

with amplitude $\mathcal{A}$. The strain $\boldsymbol{\epsilon}$ associated with the displacement is (see Chap. 2)

$$\left.\begin{aligned} \epsilon_1 &= \epsilon_2 = \epsilon_3 = 0, \\ \epsilon_4 &= \mathrm{i}q\,\mathcal{A}\,\hat{y}\,\mathrm{e}^{\mathrm{i}qz-\mathrm{i}\omega t}, \\ \epsilon_5 &= \epsilon_6 = 0. \end{aligned}\right\} \tag{3.149}$$

The electric field corresponding to the strain (3.149) is given by (3.143). We will assume that the piezoelectric matrix h has zeroes in the fourth column except for h_{14}, which corresponds to the matrix for zincblende crystals such as GaAs. The electric field is then

$$\boldsymbol{E} = \mathrm{i}\,h_{14}\,q\,\mathcal{A}\,\hat{z}\,\mathrm{e}^{\mathrm{i}qz-\mathrm{i}\omega t}. \tag{3.150}$$

The electrostatic potential is then given by

$$\phi = h_{14}u_y = h_{14}\,\mathcal{A}\,\mathrm{e}^{\mathrm{i}qz-\mathrm{i}\omega t}, \tag{3.151}$$

and is proportional to the displacement u_y, i.e. to the amplitude $\mathcal{A}$.

We can now proceed by analogy with the deformation potential calculation. We can write from (3.151) the form for the operator $\hat{\phi}$, in terms of the raising and lowering operators that represent the displacement $\boldsymbol{u}$. The potential ϕ interacts with the electrons with an energy $e\phi$, so the perturbation Hamiltonian $\Delta\hat{H}$ is given by

$$\begin{aligned}\Delta\hat{H} &= -e\hat{\phi} \\ &= -ie\,h_{14}\sum_{k,q}\left(\frac{\hbar}{2NM\omega(q)}\right)^{1/2}\left(\hat{a}_q\,\hat{c}^\dagger_{k+q}\hat{c}_k - \hat{a}^\dagger_q\hat{c}^\dagger_{k-q}\hat{c}_k\right). \end{aligned} \quad (3.152)$$

The rate for electrons to scatter from $\boldsymbol{k}$ to $\boldsymbol{k}' = \boldsymbol{k} - \boldsymbol{q}$ by emitting a phonon with wavevector $\boldsymbol{q}$ is given by (3.111), with matrix element $\mathcal{M}_P{}^2$ given by

$$\mathcal{M}_P{}^2 = \frac{\hbar}{2NMc_\ell q}\,(e\,h_{14})^2. \quad (3.153)$$

From the equivalent form (3.112), we see that in all the expressions for scattering rates, we can make the substitution

$$D^2 \to \frac{e^2 h_{14}^2}{q^2} \quad (3.154)$$

to obtain the equivalent expressions for the piezoelectric scattering. Piezoelectric scattering becomes very strong in the limit $q \to 0$

3.6 Phonon Thermal Conductivity

In Sect. 3.1, we derived the formula for the equilibrium specific heat of the phonon gas in the solid. We then went through a detailed discussion of the primary types of scattering that phonons undergo in a bulk solid, from isotopic scattering and phonon-phonon interactions, to scattering that involves electrons in conducting solids (metals and semiconductors). We are now ready to discuss phonon transport in a bulk solid; we shall do this in the context of thermal transport, the conduction of heat by phonons.

The thermal conductivity κ of a solid is defined as the thermal power, or heat, transmitted through a volume of solid for a unit cross-sectional area, length, and temperature difference. In an insulating solid, thermal power is transmitted due to the imbalance in the number density and energy distribution of the phonons at two ends of the solid, generated by imposing a temperature difference, typically at the ends of a long rod. In Fig. 3.22 we display the spectral energy density $u_E(\omega)$ for two different temperatures T and $T + \Delta T$ imposed at opposite ends of a rod of length L.

At every frequency, the energy density at the higher temperature is larger than at the lower temperature. Due to the difference in the spectral density, the net flow of thermal energy will be from the end at $T > T'$ to the end at T'. In either case, for small temperature differences $\Delta T = T - T'$, the net thermal power ΔP will be proportional to ΔT; the *thermal conductance* G_{th} is then defined as the ratio of power to temperature difference, $G_{\text{th}} = \Delta P/\Delta T$.

In the process of transmitting the thermal energy, phonons from each end will undergo two types of scattering processes: Elastic processes, such as normal isotope and boundary scattering, where the phonon energy remains

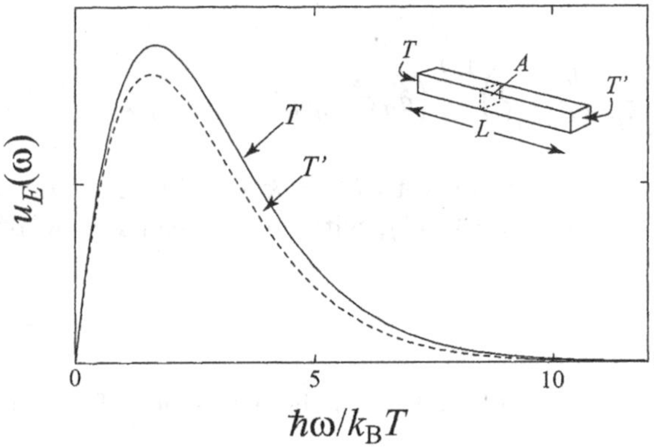

Fig. 3.22. Spectral energy density $u_E(\omega)$ for two ends of a rod of length L, cross-sectional area A, with one end at temperature T, the other at $T' < T$.

the same, and inelastic processes, such as *Umklapp* isotope scattering, scattering with other phonons, scattering with electrons, etc., processes where the phonon energy is changed. We therefore distinguish the mean free path for elastic scattering, Λ_{el}, from the inelastic mean free path Λ_{inel}; these can be written in terms of scattering rates, $\Lambda_{\mathrm{el}} = \bar{c}/\Gamma_{\mathrm{el}}$, and $\Lambda_{\mathrm{inel}} = \bar{c}/\Gamma_{\mathrm{inel}}$, with $\bar{c}$ the average sound speed used in the Debye model. The overall mean free path is determined by the total scattering rate $\Gamma_{\mathrm{el}} + \Gamma_{\mathrm{inel}}$, or equivalently the mean free path

$$\frac{1}{\Lambda} = \frac{1}{\Lambda_{\mathrm{el}}} + \frac{1}{\Lambda_{\mathrm{inel}}}. \tag{3.155}$$

The elastic mean free path is usually temperature-independent, as it is usually dominated by geometric factors such as sample size or isotopic distribution. The inelastic mean free path is strongly temperature dependent, as it is determined by processes such as *Umklapp* scattering and phonon-phonon interactions, which are strong functions of frequency, and therefore of temperature. At low temperatures, typically $\Lambda_{\mathrm{el}} < \Lambda_{\mathrm{inel}}$, while at high temperatures $\Lambda_{\mathrm{el}} > \Lambda_{\mathrm{inel}}$. We are interested in the thermal transport through a rod of characteristic cross-sectional diameter D; note that typically the rod length L is larger than D. The relation of the rod length L to the mean free path determines three different transport regimes:

$$\left.\begin{array}{rl} \text{Ballistic:} & L < \Lambda_{\mathrm{el}}, \\ \text{Non-equilibrium:} & \Lambda_{\mathrm{el}} < L < \Lambda_{\mathrm{inel}}, \\ \text{Diffusive:} & \Lambda_{\mathrm{inel}}, \Lambda_{\mathrm{el}} < L. \end{array}\right\} \tag{3.156}$$

In the ballistic regime, phonons do not scatter while passing through the length L, so the phonon distribution at any point in the rod is the superposition of the distributions from both ends, at temperatures T and T'. We assume the phonons are transmitted in normal modes of the rod that satisfy the transverse boundary conditions, so that no scattering occurs on the rod boundaries. In the "non-equilibrium" regime, phonons undergo elastic scattering while passing through the rod, randomizing their momentum, but their energy distribution is unchanged. Finally, in the diffusive regime, the phonons undergo both elastic and inelastic scattering, and in the limit of many such scattering events will achieve local thermal equilibrium at each point in the rod.

In the diffusive limit, where Λ_{el} and Λ_{inel} are both small compared to L, the local phonon number distribution $n_{qm}(\boldsymbol{r})$ at each point $\boldsymbol{r}$ in the solid can be described by a well-defined local temperature $T(\boldsymbol{r})$, and a corresponding average phonon energy $E(T)$. An average phonon travelling with velocity $\boldsymbol{v}$ carries this energy, and as it diffuses through the solid, its average energy must change to maintain local thermal equilibrium. Then the local energy will be constant, $\partial E/\partial t = 0$, and the *total* derivative will satisfy

$$\frac{\mathrm{d}E}{\mathrm{d}t} = \boldsymbol{v}\cdot\nabla E = \boldsymbol{v}\cdot\nabla T\,\frac{\partial E}{\partial T}, \tag{3.157}$$

where ∇T is the gradient of the local temperature $T(\boldsymbol{r})$. The derivative of E with respect to temperature is simply the heat capacity per phonon, or $c_V/\langle n\rangle$ in terms of the specific heat c_V and the average phonon number $\langle n\rangle$. The average energy ΔE transported per scattering length Λ is the rate (3.157) times the average time between scattering events, $\Lambda/\bar{c}$. The corresponding average thermal current per phonon, $\langle \boldsymbol{j}_U\rangle$, is given by

$$\begin{aligned}\langle \boldsymbol{j}_U\rangle &= \langle \boldsymbol{v}\Delta E\rangle \\ &= \left\langle \boldsymbol{v}\boldsymbol{v}\cdot\nabla T\,\frac{c_V}{\langle n\rangle}\frac{\Lambda}{\bar{c}}\right\rangle \\ &= \frac{1}{3}\frac{c_V}{\langle n\rangle}\,\bar{c}\,\Lambda\nabla T,\end{aligned} \tag{3.158}$$

where the average of the dyadic $\boldsymbol{v}\boldsymbol{v}$ is $\bar{c}^2/3$ times the identity tensor 1. The total thermal current density $\boldsymbol{J}_U$ is the thermal energy transported per unit cross-sectional area, equal to the current per phonon, times the average phonon number density $\langle n\rangle$. In the diffusive limit, therefore, we find

$$\begin{aligned}J_U &= \frac{1}{3}c_V\,\bar{c}\,\Lambda\nabla T \\ &= \kappa\nabla T,\end{aligned} \tag{3.159}$$

where we define the *thermal conductivity* $\kappa = (1/3)c_V\,\bar{c}\,\Lambda$. For a rod of length L and cross-sectional area A, the thermal conductance is related to the thermal conductivity by $G_{\text{th}} = \kappa A/L$.

3.6.1 The Boltzmann Equation

A more detailed understanding of non-equilibrium transport of phonons is best approached through the Boltzmann transport equation. This is a three-dimensional integro-differential equation, that describes the spatial and temporal dependence of the local phonon density function $n(\boldsymbol{r}, \boldsymbol{q}, t)$, in terms of diffusion and scattering effects (we will drop the mode index m to simplify the notation; we also explicitly express the dependence of the number density on the wavevector $\boldsymbol{q}$). The distribution is usually non-thermal, i.e. the distribution in the wavevector $\boldsymbol{q}$ is not described by a simple local temperature $T(\boldsymbol{r})$. A full description of the theory for this equation may be found in e.g. Ashcroft and Mermin [1].

The density $n(\boldsymbol{r}, \boldsymbol{q}, t)$ evolves due to the motion of phonons, and due to scattering between modes. A basic assumption is that the phonons can be treated both as point particles and as having specific wavevectors $\boldsymbol{q}$: The particles are built from the wave packet constructions discussed in Sect. 3.2. In the absence of scattering, the number density at $\boldsymbol{r}$ with wavevector $\boldsymbol{q}$ at time t is related to that at time $t - \Delta t$ by

$$n(\boldsymbol{r}, \boldsymbol{q}, t) = n(\boldsymbol{r} - \dot{\boldsymbol{r}}\,\Delta t, \boldsymbol{q} - \dot{\boldsymbol{q}}\,\Delta t, t - \Delta t), \tag{3.160}$$

where the position is evaluated at the point the phonons came from, as is the wavevector. We can write the velocity $\dot{\boldsymbol{r}} = \boldsymbol{v}_{\boldsymbol{q}} = \nabla_q \omega(\boldsymbol{q})$, in terms of the group velocity for mode $\boldsymbol{q}$. In the limit $\Delta t \to 0$, (3.160) can be written

$$\frac{Dn}{Dt}(\boldsymbol{r}, \boldsymbol{q}, t) = \frac{\partial n}{\partial t}(\boldsymbol{r}, \boldsymbol{q}, t) + \boldsymbol{v}_{\boldsymbol{q}} \cdot \nabla_r n(\boldsymbol{r}, \boldsymbol{q}, t) + \dot{\boldsymbol{q}} \cdot \nabla_q n(\boldsymbol{r}, \boldsymbol{q}, t), \tag{3.161}$$

the usual expression for the total time derivative of $n(\boldsymbol{r}, \boldsymbol{q}, t)$. Note that this expression includes the term $\dot{\boldsymbol{q}} \cdot \nabla_q n$, appearing if there is a driving force that changes the wavevector $\boldsymbol{q}$. The Akhiezer effect (discussed in Sect. 3.4.3) could be included through a term of this sort, as could any external effect that changes $\boldsymbol{q}$ (such as an externally applied strain). In most situations, however, $\dot{\boldsymbol{q}}$ can safely be set to zero.

The Boltzmann equation balances the total time derivative Dn/Dt with the rate of transfers in and out of the state $\boldsymbol{q}$ due to scattering. This balance is thus expressed as

$$\frac{\partial n}{\partial t} = -\boldsymbol{v} \cdot \nabla_r n(\boldsymbol{r}, \boldsymbol{q}, t) - \dot{\boldsymbol{q}} \cdot \nabla_q n(\boldsymbol{r}, \boldsymbol{q}, t) + \left[\frac{\partial n_{\boldsymbol{q}}}{\partial t}\right]_{\text{scattering}}. \tag{3.162}$$

Clearly the meat of this equation is in the scattering term. As an example, we work out what this term would look like for phonons interacting with one another through the cubic (three phonon) perturbation term in the Hamiltonian discussed in Sect. 3.4. The rate for a phonon $\boldsymbol{q}$ colliding with a phonon $\boldsymbol{q}'$ and producing a phonon $\boldsymbol{q}''$ is proportional to the product of the occupation terms $n_{\boldsymbol{q}} n_{\boldsymbol{q}'}(1 + n_{\boldsymbol{q}''})$; we will write the proportionality constant, also known as the *kernel*, as $K_{\boldsymbol{q}\boldsymbol{q}'}^{\boldsymbol{q}''}$. The other possible interactions involving these three phonon states are $\boldsymbol{q} \to \boldsymbol{q}' + \boldsymbol{q}''$, $\boldsymbol{q}'' \to \boldsymbol{q} + \boldsymbol{q}'$, and so on. Summing over the distinct possibilities yields the scattering rate (see Ziman [11])

$$\left[\frac{\partial n_{\boldsymbol{q}}}{\partial t}\right]_{\text{scattering}} = \int\int \mathrm{d}\boldsymbol{q}'\mathrm{d}\boldsymbol{q}''$$
$$\Big[\left(n_{\boldsymbol{q}}n_{\boldsymbol{q}'}(1+n_{\boldsymbol{q}''}) - (1+n_{\boldsymbol{q}})(1+n_{\boldsymbol{q}'})n_{\boldsymbol{q}''}\right) K_{\boldsymbol{q}\boldsymbol{q}'}^{\boldsymbol{q}''} \tag{3.163}$$
$$+\tfrac{1}{2}\left(n_{\boldsymbol{q}}(1+n_{\boldsymbol{q}'})(1+n_{\boldsymbol{q}''}) - (1+n_{\boldsymbol{q}})n_{\boldsymbol{q}'}n_{\boldsymbol{q}''}\right) K_{\boldsymbol{q}}^{\boldsymbol{q}'\boldsymbol{q}''}\Big] .$$

This is clearly a rather horrendous equation: The distribution $n(\boldsymbol{r},\boldsymbol{q},t)$ must be chosen so that it solves the differential equation (3.162), with the scattering term (3.163), all self-consistently.

Some standard simplifications are usually applied, rather than trying to solve this problem directly. Typically the Boltzmann equation can be solved assuming a static distribution, so $\partial n/\partial t = 0$. Furthermore, one assumes only small deviations from equilibrium, so that one can write $n(\boldsymbol{r},\boldsymbol{q}) = n^0(\boldsymbol{r},\boldsymbol{q}) + \delta n(\boldsymbol{r},\boldsymbol{q})$, where $n^0(\boldsymbol{r},\boldsymbol{q})$ is the local distribution at thermal equilibrium, using a local temperature $T(\boldsymbol{r})$, given by

$$n^0(\boldsymbol{r},\boldsymbol{q}) = \left[\exp\left(\frac{\mathcal{E}}{k_{\mathrm{B}}T(\boldsymbol{r})}\right) - 1\right]^{-1}, \tag{3.164}$$

where $\mathcal{E} = \hbar\omega(\boldsymbol{q})$. The term $\delta n(\boldsymbol{r},\boldsymbol{q})$ is the deviation from n^0, assumed small. Scattering terms of the form (3.163) can then be linearized by dropping terms of order δn^2 and higher.

A great simplification can be made by using the *relaxation time approximation*, where the scattering term is replaced by a term of the form

$$\begin{aligned}\left[\frac{\partial n}{\partial t}\right]_{\text{scattering}} &= -\frac{1}{\tau}\left(n(\boldsymbol{r},\boldsymbol{q}) - n^0(\boldsymbol{r},\boldsymbol{q})\right)\\ &= -\frac{\delta n(\boldsymbol{r},\boldsymbol{q})}{\tau},\end{aligned} \tag{3.165}$$

representing a uniform relaxation for all out-of-equilibrium terms with a single relaxation time τ. With this approximation we can write the Boltzmann equation as

$$\frac{Dn}{Dt}(\boldsymbol{r},\boldsymbol{q},t) = -\frac{n(\boldsymbol{r},\boldsymbol{q}) - n^0(\boldsymbol{r},\boldsymbol{q})}{\tau}. \tag{3.166}$$

Following Marder [6], we integrate both sides with respect to time to find

$$n(\boldsymbol{r},\boldsymbol{q},t) = \int_{-\infty}^{t} \frac{\mathrm{e}^{-(t-t')/\tau}}{\tau} n^0(\boldsymbol{r}(t'),\boldsymbol{q}(t'),t')\,\mathrm{d}t'. \tag{3.167}$$

This can be integrated by parts to yield

$$\begin{aligned} n &= n^0 - \int_{-\infty}^{t} \frac{\mathrm{e}^{-(t-t')/\tau}}{\tau}\frac{\mathrm{d}n^0}{\mathrm{d}t'}\,\mathrm{d}t' \\ &= n^0 - \int_{-\infty}^{t} \mathrm{d}t'\,\frac{\mathrm{e}^{-(t-t')/\tau}}{\tau}\left[\dot{\boldsymbol{r}}\cdot\nabla_r + \dot{\boldsymbol{q}}\cdot\nabla_q\right] n^0. \end{aligned} \tag{3.168}$$

We simplify by taking $\dot{\boldsymbol{q}} = 0$, and assume τ is much shorter than any other time scale. Then

$$n = n^0 - \tau\, \boldsymbol{v_q} \cdot \nabla_r n^0. \tag{3.169}$$

Using the form for the equilibrium density (3.164), we can write

$$\nabla_r n^0 = \mathcal{E}\, \frac{\nabla T}{T} \frac{\partial n^0}{\partial \mathcal{E}}. \tag{3.170}$$

so that

$$n = n^0 - \tau\, \mathcal{E}\, \boldsymbol{v_q} \cdot \frac{\nabla T}{T} \frac{\partial n^0}{\partial \mathcal{E}}. \tag{3.171}$$

Hence we see that the solution n of the Boltzmann equation, in this approximation, differs from the local equilibrium n^0 by a term that is proportional to the derivative of n^0 with respect to energy, $\partial n^0/\partial \mathcal{E}$.

Using the solution of the form (3.169), we can calculate, for example, the thermal current $\boldsymbol{j}_U$, using the formula

$$\begin{aligned} \boldsymbol{j}_U &= \int \mathrm{d}\boldsymbol{q} \mathcal{E} \boldsymbol{v_q} n(\boldsymbol{r}, \boldsymbol{q}, t) \\ &= -\tau \int \mathrm{d}\boldsymbol{q} \frac{\mathcal{E}^2}{T} \frac{\partial n^0}{\partial \mathcal{E}} \boldsymbol{v_q} \boldsymbol{v_q} \cdot \nabla T \\ &= \kappa \nabla T, \end{aligned} \tag{3.172}$$

where the integral is over the first Brillouin zone.

We can now pull out an expression for the thermal conductivity; the integral depends on the direction of $\boldsymbol{q}$ only through the phonon velocity, which in a cubic crystal averages (for low temperatures) to $\bar{c}^2/3$, leaving the integral over the magnitude of $\boldsymbol{q}$,

$$\begin{aligned} \kappa &= -\tau \frac{\bar{c}^2}{3} \int \mathrm{d}q \frac{\mathcal{E}^2}{T} \frac{\partial n^0}{\partial \mathcal{E}} \\ &= \frac{1}{3} \Lambda \bar{c} c_V, \end{aligned} \tag{3.173}$$

the same expression we derived earlier. While this just shows that a very sophisticated approach can return the same result as a more simple-minded one, the Boltzmann formalism allows the calculation of parameters such as the thermal conductivity in situations with large non-equilibrium distributions, and without some of the approximations that are otherwise needed.

In the theory we have described for the thermal conductivity, the phonons are assumed to scatter within the volume of the solid often enough that a well-defined temperature exists at each point, and the local phonon distribution is assumed to be in equilibrium at that temperature. This requires that the phonons undergo frequent momentum scattering, where the $\boldsymbol{q}$ vector of the phonons changes in a scattering event, and that they as well undergo frequent energy scattering, where the phonons change frequency through inelastic collisions. In both types of scattering, the overall energy and momentum are conserved; the change in a given phonon's energy and momentum is taken up either by the solid as a whole, or by another phonon (or an electron).

Another source of inelastic scattering can occur when phonons scatter from a solid's surfaces, which as we shall see in our discussion of continuum mechanics, usually cause changes in both polarization and q-vector. For objects with diameters in the 1-100 nm range, the low temperature mean free paths for both isotopic and phonon-phonon scattering are significantly larger than the solid diameter, so surface scattering completely dominates in determining the mean free path. Surface scattering has been shown to be the dominant effect in the thermal transport of macroscopic rods at low temperatures; in a beautiful set of experiments by Klitsner [33, 34, 35], the phonon transport in highly polished rods of Si, with cross-sectional area $A = 0.5 \times 0.5\ \text{cm}^2$, in the temperature range 0.05–1 K, was found to correspond to a thermal conductance determined by ballistic phonon transport. More than 95% of the phonons would scatter specularly from the highly polished surfaces of the rod, without any change in energy. In this limit the thermal conductance was limited by the heat generating and detecting contacts at the ends of the rod, and the energy-equilibrating mean free path was found to be as high as 10-15 cm, corresponding to the contact spacing. The phonons in the temperature range of this experiment have peak thermal frequencies in the range from 5 to 100 GHz. Roughening the surfaces slightly was found to significantly increase the rate of diffuse, energy-equilibrating surface scattering, and the thermal conductance became limited by this rate; the inverse energy-equilibrating mean free path in rough samples was found to $\Lambda^{-1} = 2.08\ \text{cm}^{-1}$, corresponding to the *Casimir limit* for diffuse scattering, given by $\overline{\Lambda} = 1.12\sqrt{A}$ for cross-sectional area A.

In nanoscale rods, similar types of measurements on very smooth rods with diameters of order 0.1-0.2 μm have also shown mean free paths that significantly exceed the diameter of the rods; mean free paths for such rods have been measured in excess of 3-10 μm, more typical of the length of the rod than its diameter [36, 37, 38]. In these and in the larger scale systems, then, the temperature within the rod is not well defined, and we must turn to the quantum theory of thermal conductance, which we now describe.

3.6.2 Quantum of Thermal Conductance

In nanoscale insulating wires with specular scattering, or a small amount of diffusive scattering, it has recently been discovered that there is a quantum limit to the thermal *conductance* through the wire. This limit is associated with the fact that each vibrational normal mode in a solid can, at low temperatures, only carry a limited amount of heat.

If we consider a nanoscale rod, connected at both ends to much larger volume solids, the normal modes will consist of a continuum of values for the wavevector along the axis of the rod, but discretized for the wavevector components along the two axes perpendicular to the rod. A simple model

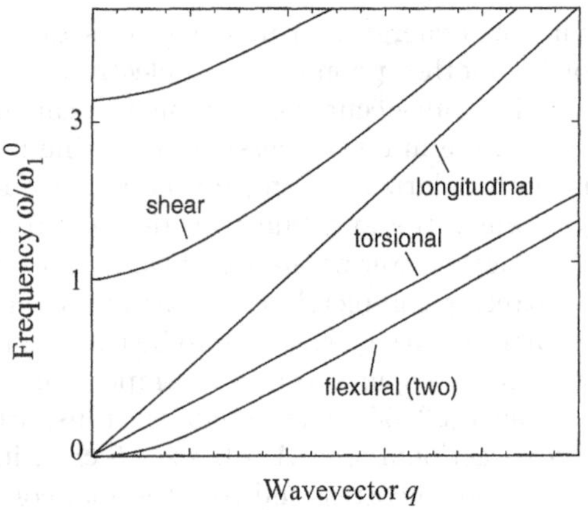

Fig. 3.23. Model normal modes for a rod connected to infinite reservoirs at either end. The two degenerate flexural modes, the longitudinal, torsional and shear modes are shown, as a function of wavevector q. Adapted from Rego and Kirczenow [39].

was described in Sect. 3.3; here we take the limit where the length L goes to infinity, but retain the discretization along the other two axes. We can therefore write the allowed modes in the form

$$\boldsymbol{u}_{qm} = \boldsymbol{A}_{qm}\, \mathrm{e}^{\mathrm{i}qx} \phi_{qm}(y, z), \tag{3.174}$$

where we take the x-axis along the length of the rod, y and z perpendicular to the length, ϕ_{qm} is the mode shape perpendicular to x, given wavevector q along x and mode index m for the perpendicular mode. $\boldsymbol{A}_{qm}$ is the polarization wavevector. Each mode has frequency $\omega_m(q)$; in Fig. 3.23 we display the q-dependence of different normal modes. There are four modes whose frequencies go to zero as the wavevector q goes to zero; one longitudinal, one torsional, and two flexural modes. The first two have a linear dispersion relation, with $\Omega \propto q$, while the two flexural modes have frequencies that scale with q^2 (see Chap. 8).

We can associate a transmittance $\mathcal{T}_m(\omega)$ with each mode m, which ranges from 0 for no transmission to unity for perfect transmission. In a mode with unit transmissivity, every phonon is transmitted, while for a transmissivity less than one, there is some probability that a phonon will scatter into another mode, such that only the fraction $\mathcal{T}_m(\omega)$ will make it through. For a beam with square ends connected to the bulk solid, the abrupt coupling leads to a complex frequency dependence for the transmissivity of any given transverse mode m [39, 40]. Unity coupling for the longitudinal mode can however be achieved for a gradual transition from the bulk to the beam, with the optimal shape found for a catenoidal beam [39].

Each mode has energy $\mathcal{E}_m(q) = \hbar\omega_m(q)$; at thermal equilibrium, the occupation of this mode is $n^0(\mathcal{E})$, given by the Bose–Einstein distribution (3.164).

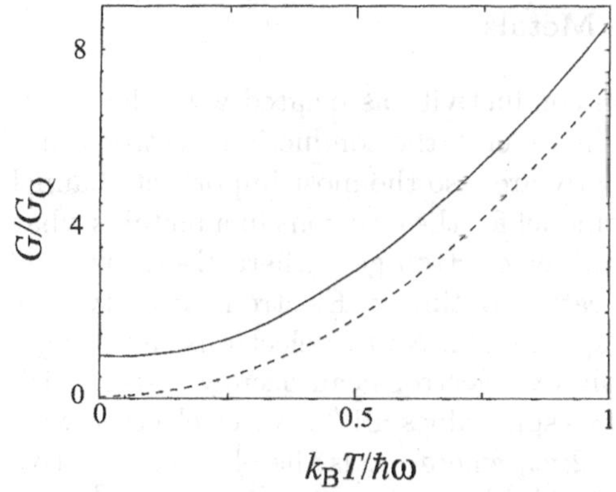

Fig. 3.24. Temperature dependence for the quantum-limited thermal conductance (*solid line*), for one transverse mode, compared to the bulk dependence, proportional to T^2. The thermal conductance is in units of the quantum of conductance given by (3.177).

After some calculation [39, 41], it can be shown that the thermal conductance G_m associated with the transverse mode index m is given by

$$G_m = \frac{1}{2\pi}\int_{\omega_m^0}^{\infty} \mathcal{T}_m(\omega)\hbar\omega\frac{\partial n^0}{\partial T}\mathrm{d}\omega \tag{3.175}$$

where $h = 6.6262 \times 10^{-34}$ J-s is Planck's constant. The lower limit of the integral is the lowest frequency ω_m^0 in the band m. Summing over all modes, and using the form for the distribution n^0, the total thermal conductance is

$$G = \sum_m \frac{k_B^2 T}{h}\int_{x_m}^{\infty} \mathcal{T}_m(k_B T x)\frac{x^2 e^x}{(e^x - 1)^2}\mathrm{d}x, \tag{3.176}$$

with lower limit $x_m = \hbar\omega_m^0/k_B T$. At very low temperatures, the only modes that have non-zero occupation are the four modes that do not "cut off", that is, those whose frequency goes to zero at long wavelengths. Each of these modes contribute a minimum conductance G_Q, proportional to T, given by

$$G_Q = \frac{\pi^2 k_B^2 T}{3h}. \tag{3.177}$$

Note that this conductance does not depend on any geometric factors. The total low-temperature conductance for a single beam is then four times G_Q. In Fig. 3.24 we show the quantum theory compared to the bulk theory, the latter having a low-temperature limit of $G_{mathrmbulk} \propto T^3$, so that $G_{\mathrm{bulk}}/G_Q \propto T^2$.

The theory for the quantum limit of thermal conductance was verified in a beautiful experiment by Schwab and Roukes [37]; the experiment is described in Sect. 9.5.

3.6.3 Thermal Conductivity in Metals

We have only described the thermal conductivity associated with the phonons, applicable to insulating solids. In a metal, the conduction electrons that give a metal its low electrical resistivity are also the most important channel for thermal transport. The simplest model for the electrons in a metal is what is known as the Debye model for the free electron gas, where electrons scatter with a single, uniform average scattering time τ. Electrons are a type of particle known as fermions, with a spin of 1/2. No two electrons can occupy the same quantum state, so that only two electrons can occupy a state with wavevector $\boldsymbol{k}$, each with one of the two spin values $\pm 1/2$. A free electron with wavevector $\boldsymbol{k}$ has energy $E = \hbar^2 k^2/2m_e$, where m_e is the electron effective mass (usually closely approximated by the free mass, $m_e = 9.11 \times 10^{-31}$ kg). At $T = 0$, the lowest energy electron states, with the smallest wavevectors $\boldsymbol{k}$, are uniformly occupied up to the *Fermi wavevector* k_F, with Fermi energy $E_F = \hbar^2 k_F^2/2m_e$. This sphere of occupied states in k-space is known as the Fermi sphere. At non-zero temperatures, the occupation of states with wavevector length near k_F and energy near E_F is smeared out, with occupation probability $f(E)$ given by the Fermi–Dirac distribution function,

$$f(E) = \frac{1}{e^{\beta(E-\mu)} + 1}, \tag{3.178}$$

where $\beta = 1/k_B T$ is the inverse temperature. Here μ is the chemical potential, given by the Fermi energy E_F at $T = 0$ and decreasing slightly with temperature.

The Fermi energy and Fermi velocity are determined by the electron number density n_e. For a volume $V = L^3$ with periodic boundary conditions, the allowed wavevectors $\boldsymbol{k}$ are the same as those for phonons, and must come from the set $\boldsymbol{k} = (2\pi n/L, 2\pi m/L, 2\pi \ell/L)$, for integers n, m and ℓ. For a sphere of radius k_F in wavevector space, with k-space volume $v_F = 4\pi k_F^3/3$, there are $N = 2(L/2\pi)^3\, v_F$ quantum states (the factor of two for the two possible spin values). In order to accommodate a number density $n_e = N/V$ electrons, the Fermi wavevector and energy are thus related to n_e by

$$\left.\begin{aligned} k_F &= (3\pi^2 n_e)^{1/3}, \\ E_F &= \frac{\hbar^2}{2m_e}(3\pi^2 n_e)^{2/3}. \end{aligned}\right\} \tag{3.179}$$

We can also define the Fermi velocity $v_F = \hbar k_F/m_e$.

Thermal transport in pure metals is dominated by the thermal conductance of the electrons. The thermal conductivity of an electron gas is closely related to its electrical conductivity, as both are affected, in certain limits, by the same scattering mechanisms. The electrical conductivity of a free electron gas with a mean free path $\Lambda_e = v_F \tau$ is given by

$$\sigma = \frac{n_e e^2 \tau}{m_e} = \frac{n_e e^2 \Lambda_e}{m_e v_F}. \tag{3.180}$$

The thermal conductivity of the electrons is then given by a formula similar to that for the phonons,

$$\kappa_e = \frac{1}{3} c_{\rm el}\, v_F\, \Lambda_e, \tag{3.181}$$

where the electron specific heat $c_{\rm el}$ for a free electron gas is given by

$$c_{\rm el} = \frac{\pi^2}{2} n_e k_{\rm B} \left(\frac{k_{\rm B} T}{E_F} \right) \tag{3.182}$$

(see e.g. Ashcroft and Mermin [1]).

We see that the electrical and thermal conductivity are proportional to the mean free path Λ_e, which allows us to write a famous relation known as the *Wiedemann–Franz law*, relating the electrical and thermal conductivities:

$$\frac{\kappa_e}{\sigma} = \frac{\pi^2}{3} \left(\frac{k_{\rm B}}{e} \right)^2 T = LT. \tag{3.183}$$

The factor L is called the *Lorenz number* and has the value $L = 2.44 \times 10^{-8}$ W $- \Omega/\mathrm{K}^2$ for a free electron gas. This relation holds very well for many metals, both dirty and clean, at high temperatures as well as at very low temperatures. At intermediate temperatures, the scattering process affects the electrical and thermal transport somewhat differently, so the relation does not hold as accurately. If the Lorenz number is treated as a parameter, temperature-independent but adjusted for each material, the relation is somewhat more accurate.

The characteristic temperature dependence of the electrical and thermal conductivity in a good metal is shown in Fig. 3.25; the deviation from the Lorenz formula is in the temperature range above the peaks in conductivity.

At intermediate temperatures, of order $0.01-1\,\Theta_D$ (when Θ_D is the Debye temperature discussed in Sect. 3.1), the Wiedemann–Franz law does not apply as well; the Lorenz number effectively becomes temperature-dependent, achieving the value given by the free electron gas model (3.182) at high ($T > \Theta_D$) and at low temperatures ($T < 0.01\Theta_D$). In many metals, in between these temperatures, it is found that the value of L falls by about 30-40%, with a minimum achieved at around $0.1\,\Theta_D$. The reason for this is that *normal* phonon scattering dominates in this temperature range, which removes the energy of the electrons more effectively than their momentum. Hence the effective mean free path for thermal conduction becomes shorter than that for electrical conduction; a thorough discussion may be found in Ziman [11].

By comparing the size of the thermal conductivity of the electron gas with that of the phonons in the lattice, we see that the electron conduction

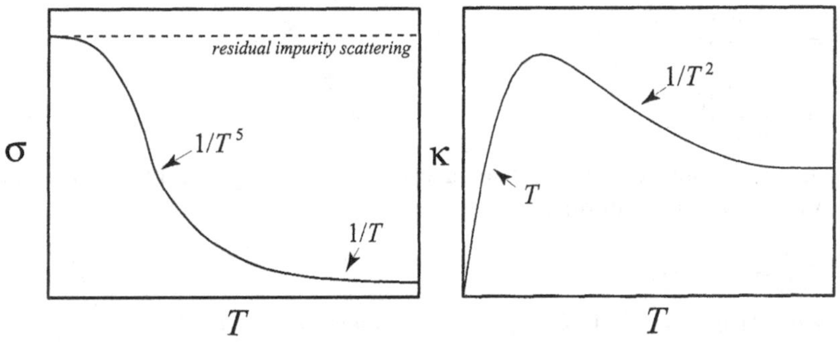

Fig. 3.25. Characteristic electrical and thermal conductivity of a good metal over the three important temperature ranges, with the functional dependence on temperature as indicated. The $1/T^5$ and $1/T^2$ regions for the electrical and thermal conductivity above the peak values are where *normal* phonon scattering controls the electron mean free path, while above that *Umklapp* scattering takes over.

dominates in a good metal; the phonons provide comparable thermal conduction only in a very dirty metal or in a semimetal. In Fig. 3.26 we display the relative thermal conductance for a pure metal, a single crystal insulator, and a dirty metal, whose thermal conductances are arranged in that order. Note that the two metals are polycrystalline, and the thermal conductance of both the electrons and the phonons is dominated by impurity scattering in the metal alloy.

The data shown in Fig. 3.26 are for bulk materials, except that for the single-crystal quartz, which at the lowest temperatures is in the limit where

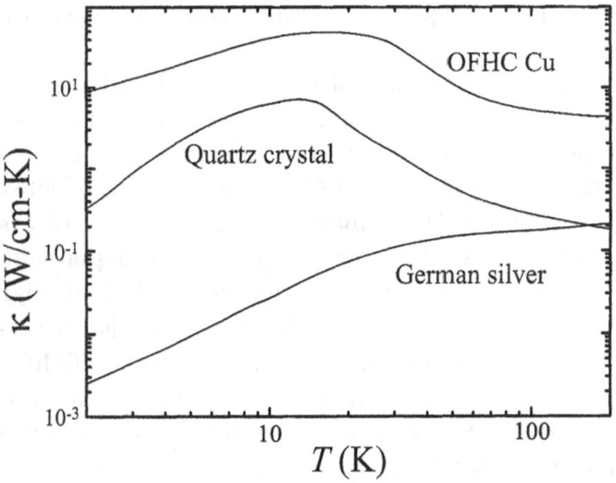

Fig. 3.26. Thermal conductivity for oxygen-free, high conductivity, (OFHC) annealed copper, single-crystal quartz, and German silver, the latter an alloy of copper, zinc and nickel. Data adapted from [42].

the phonon mean free path Λ is of the order of the cross-sectional dimensions D of the quartz, and is therefore in the Casimir limit for phonons described above. In thin metal films, the smallest conductor dimension D will also determine the electron mean free path Λ_e at low enough temperatures, once the mean free path as determined by elastic and inelastic scattering becomes larger than D. The electrical and thermal conductivities in the Casimir limit for electrons are then given by

$$\left.\begin{aligned} \sigma &\approx \frac{n_e\, e^2 D}{m_e\, v_F}, \\ \kappa_e &\approx \frac{1}{3}\, c_{el}\, v_F\, D. \end{aligned}\right\} \tag{3.184}$$

Exercises

3.1 Referring to Sect. 3.2, take the wave packet envelope function to be $\mathcal{A}(q) = \exp(-a^2(q-q_0)^2)$; find the corresponding real space wave packet $u(x,0)$. With $q_0 = \pi/2r_0$, find the phase and group velocity for the wave packet, using the dispersion relation (3.39).

3.2 Derive an expression for the density of states per unit area L^2 in two dimensions; show that it is given by

$$\mathcal{D}(\omega) = \frac{1}{4\pi^2} \sum_\alpha \int_{\mathcal{L}} \frac{d\ell}{|\nabla\omega_\alpha(q)|}, \tag{3.185}$$

where $\mathcal{L}$ is a line in two-dimensional q-space of points with equal frequency, and $d\ell$ a line segment.

3.3 Using the expression (3.185), find an expression for the two dimensional equivalent density of states in the Debye model, and find an expression for the specific heat per unit area L^2.

3.4 Calculate the heat capacity for three atoms of mass M connected in a line by springs k; assume the center of mass is at rest. Sketch the temperature dependence on a log–log plot.

3.5 Using the wave equation given for a isotropic solid in Example 3.3, plot the dispersion relation $\omega(q)$ in the limit where the length of the rod L is much larger than the width or thickness, treating q, the wavevector along L, as a continuous variable. Plot the relations for the lowest four modes, i.e. transverse wavevectors.

3.6 For the long thin bar in Example 3.3, calculate the very low temperature dependence of the heat capacity in the limit $T \ll hc/4L$; what is the primary functional dependence?

3.7 Compare the temperature dependence of the specific heat of a $100\times100\times 100\ \text{nm}^3$ block of Cu to that of Si, at temperatures below 100 K. Assume no free carriers in the Si.

3.8 Consider the thermal conductance G of a rod of length 1 μm and width and thickness 100 nm. Compare the thermal conductance for a rod of Cu to a rod of Si, assuming bulk values for the thermal conductivity for each. Now compare the conductance of the Cu rod to that of Si, using the quantum of thermal conductance for the conduction in Si. Sketch all three on a log-log plot of conductance as a function of temperature.

3.9 Using the wave equation given for a isotropic solid in Example 3.3, taking the length L to go to infinity, find an expression for the thermal conductance in the quantum limit, as a function of temperature, for each set of transverse indices along y and z. Assume a single polarization.

4. Stress and Strain

In the first three chapters, we have given a fairly simple description of a solid from the atomic point of view, where we explicitly brought out the connection between the individual atoms, their interactions, and the resulting properties of the solid as a whole. In this chapter, we begin a discussion of the formal theory for the continuum solid; we have already introduced the important concepts of stress and strain, and the elastic moduli that connect them. Here we describe these concepts, and the mathematical theory used to handle them, in more detail.

4.1 Relative Displacement and Strain

Classical dynamics, as covered in most modern courses on introductory mechanics, is a description of the laws that govern the motion of objects without any internal degrees of freedom: The objects involved are assumed to be infinitely rigid. The motion of a rigid object, subjected to various external forces, is then completely described by the position of the object's center of mass, and the orientation of the object with respect to a fixed set of coordinate axes.

As we have discussed, the bonds between the atoms that make up a solid are not infinitely rigid, and many of the thermodynamic properties of insulators are based on the flexibility of these bonds. Rigid body mechanics is therefore only an approximation to the actual motion of the solid, as it ignores all the internal motion when an object is acted upon by external forces: These external forces will cause deformation, or *strain*, of the object.

When an object is strained, a point P, fixed in the object, will shift to a different point P'. This shift can occur due to the rigid translation and rotation of the object, but here we are interested in shifts due to strain in the object. Hence we will assume that the object as a whole does not move or rotate. We could equivalently use a coordinate system that is fixed to the center of mass of the object, and assume the resulting non-inertial effects (due to the object's acceleration and rotation) are small enough or slow enough that they can be ignored.

Referring to Fig. 4.1, the vector $\boldsymbol{r}$ locates the initial position P of our point, and $\boldsymbol{r}'$ the final position P', both with respect to the center of mass

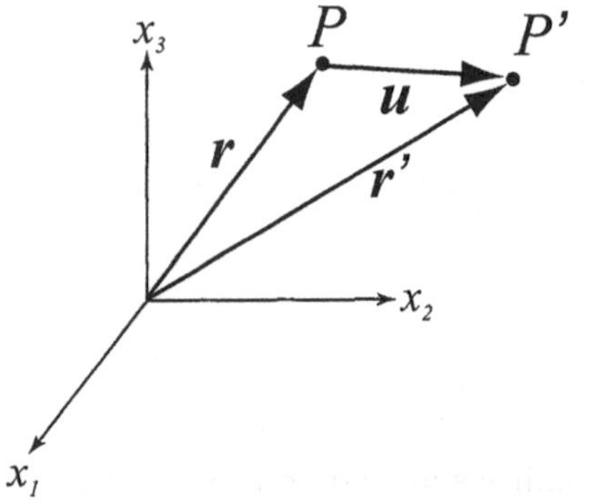

Fig. 4.1. The point P defined by position vector $\boldsymbol{r}$ displaces to the point P' as given by the vector $\boldsymbol{r}^0$, under the relative displacement $\boldsymbol{u}$.

of the object. The strain-induced displacement of P' with respect to P is the vector $\boldsymbol{u} = \boldsymbol{r}' - \boldsymbol{r}$, the same vector we used to described the displacement of atoms in a solid. The displacement vector $\boldsymbol{u}$ varies with position $\boldsymbol{r}$ within the body, and is time-dependent, so we write it as $\boldsymbol{u}(\boldsymbol{r}, t)$. The center of mass, which is the origin of our coordinate system, is assumed to have zero strain-induced displacement, so that $\boldsymbol{u}(\boldsymbol{0}, t) = \boldsymbol{0}$.

Consider now two points P and Q, Q spaced a small distance from P, so that P is at $\boldsymbol{r}$ and Q at $\boldsymbol{r} + \Delta\boldsymbol{r}$. Under strain P displaces to P', located at $\boldsymbol{r} + \boldsymbol{u}(\boldsymbol{r})$, and Q displaces to Q', located at $\boldsymbol{r} + \Delta\boldsymbol{r} + \boldsymbol{u}(\boldsymbol{r} + \Delta\boldsymbol{r})$, as shown in Fig. 4.2. The position of Q with respect to P is $\Delta\boldsymbol{r}$, and the position of

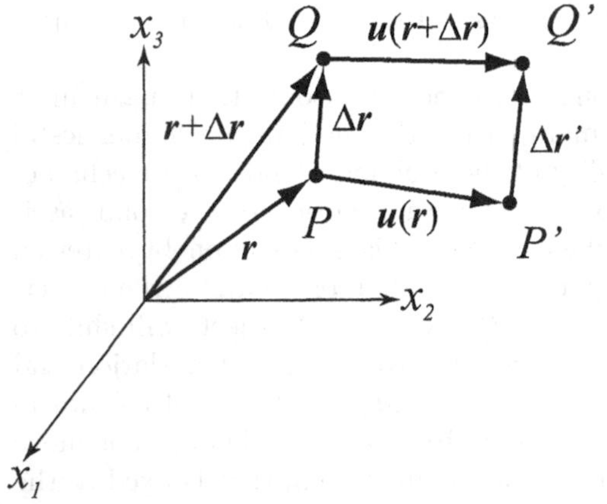

Fig. 4.2. Points P and Q, initially separated by $\Delta\boldsymbol{r}$, shift under strain to P' and Q' separated by $\Delta\boldsymbol{r}'$.

Q' relative to P' is $\Delta\boldsymbol{r}' = \Delta\boldsymbol{r} + \boldsymbol{u}(\boldsymbol{r}+\Delta\boldsymbol{r}) - \boldsymbol{u}(\boldsymbol{r})$. We now introduce the strain-induced *relative displacement* of Q with respect to P as the vector $\Delta\boldsymbol{u} = \boldsymbol{u}(\boldsymbol{r}+\Delta\boldsymbol{r}) - \boldsymbol{u}(\boldsymbol{r})$. The position of Q' with respect to P' is then $\Delta\boldsymbol{r}' = \Delta\boldsymbol{r} + \Delta\boldsymbol{u}$.

In the limit where Q and P are infinitesimally close, so that $\Delta\boldsymbol{r}$ is a differential vector, we can expand each component u_i of the displacement $\boldsymbol{u}(\boldsymbol{r}+\Delta\boldsymbol{r})$ in a vector Taylor series about the point $\boldsymbol{r}$:

$$u_i(\boldsymbol{r}+\Delta\boldsymbol{r}) = u_i(\boldsymbol{r}) + \sum_{j=1}^{3} \frac{\partial u_i}{\partial x_j} \Delta r_j + \text{ higher order terms.} \tag{4.1}$$

Dropping the higher order terms, we can write the relative displacement $\Delta\boldsymbol{u}$ of Q with respect to P in component form as

$$\Delta u_i = \sum_{j=1}^{3} \frac{\partial u_i}{\partial x_j} \Delta r_j. \tag{4.2}$$

Given the set of partial derivatives $\partial u_i/\partial x_j$ as a function of position $\boldsymbol{r}$ within the body, we can calculate the strain-induced relative displacement of any two points by integrating these derivatives in the appropriate way; later we will work out an example of such a calculation. These derivatives can be assembled into a tensor D, whose components are given by $D_{ij} = \partial u_i/\partial x_j$. This tensor is a special class of tensors known as *dyadic* tensors, with the general form $\mathsf{C} = \boldsymbol{AB}$ whose indices are $C_{ij} = A_i B_j$, for two vectors $\boldsymbol{A}$ and $\boldsymbol{B}$; we have already used dyadic tensors in Chaps. 2 and 3.

In this case the tensor D has the vector $\boldsymbol{A}$ as the *gradient operator* ∇, so that $\mathsf{D} = \nabla\boldsymbol{u}$. Written out in tabular form, D is

$$\mathsf{D} = \begin{bmatrix} \frac{\partial u_1}{\partial x_1} & \frac{\partial u_1}{\partial x_2} & \frac{\partial u_1}{\partial x_3} \\ \frac{\partial u_2}{\partial x_1} & \frac{\partial u_2}{\partial x_2} & \frac{\partial u_2}{\partial x_3} \\ \frac{\partial u_3}{\partial x_1} & \frac{\partial u_3}{\partial x_2} & \frac{\partial u_3}{\partial x_3} \end{bmatrix}. \tag{4.3}$$

The tensor D is the basis for the definition of the strain tensor, which we defined but did not justify in Chap. 3. We first rewrite D by breaking it up into two pieces, a symmetric and an antisymmetric piece:

$$\mathsf{D} = \mathsf{S} + \Omega, \tag{4.4}$$

$$\mathsf{S} = \frac{1}{2}(\nabla\boldsymbol{u} + (\nabla\boldsymbol{u})^T), \tag{4.5}$$

$$\Omega = \frac{1}{2}(\nabla\boldsymbol{u} - (\nabla\boldsymbol{u})^T), \tag{4.6}$$

where T indicates the transpose. The separation of D into separate symmetric and antisymmetric components is unique, as is shown by Shames [43]. The tensor S is the *strain tensor*, already defined in Sect. 2.8, and Ω is the *rotation tensor*, not to be confused with the coordinate transformation rotation tensor R. Written out in component form, the strain tensor is

$$S_{ij} = \frac{1}{2}\left(\frac{\partial u_i}{\partial x_j} + \frac{\partial u_j}{\partial x_i}\right), \tag{4.7}$$

and in tabular form

$$\mathsf{S} = \begin{bmatrix} \frac{\partial u_1}{\partial x_1} & \frac{1}{2}\left(\frac{\partial u_1}{\partial x_2} + \frac{\partial u_2}{\partial x_1}\right) & \frac{1}{2}\left(\frac{\partial u_1}{\partial x_3} + \frac{\partial u_3}{\partial x_1}\right) \\ \frac{1}{2}\left(\frac{\partial u_1}{\partial x_2} + \frac{\partial u_2}{\partial x_1}\right) & \frac{\partial u_2}{\partial x_2} & \frac{1}{2}\left(\frac{\partial u_2}{\partial x_3} + \frac{\partial u_3}{\partial x_2}\right) \\ \frac{1}{2}\left(\frac{\partial u_1}{\partial x_3} + \frac{\partial u_3}{\partial x_1}\right) & \frac{1}{2}\left(\frac{\partial u_2}{\partial x_3} + \frac{\partial u_3}{\partial x_2}\right) & \frac{\partial u_3}{\partial x_3} \end{bmatrix}. \tag{4.8}$$

The rotation tensor is

$$\Omega_{ij} = \frac{1}{2}\left(\frac{\partial u_i}{\partial x_j} - \frac{\partial u_j}{\partial x_i}\right), \tag{4.9}$$

and in tabular form

$$\Omega = \begin{bmatrix} 0 & \frac{1}{2}\left(\frac{\partial u_1}{\partial x_2} - \frac{\partial u_2}{\partial x_1}\right) & \frac{1}{2}\left(\frac{\partial u_1}{\partial x_3} - \frac{\partial u_3}{\partial x_1}\right) \\ \frac{1}{2}\left(\frac{\partial u_2}{\partial x_1} - \frac{\partial u_1}{\partial x_2}\right) & 0 & \frac{1}{2}\left(\frac{\partial u_2}{\partial x_3} - \frac{\partial u_3}{\partial x_2}\right) \\ \frac{1}{2}\left(\frac{\partial u_3}{\partial x_1} - \frac{\partial u_1}{\partial x_3}\right) & \frac{1}{2}\left(\frac{\partial u_3}{\partial x_2} - \frac{\partial u_2}{\partial x_3}\right) & 0 \end{bmatrix}. \tag{4.10}$$

The tensor $\Omega(\boldsymbol{r})$ gives the local rotation of the volume element at the point $\boldsymbol{r}$. Note that Ω does not represent the rigid body rotation of the solid as a whole, as such overall rotations do not give rise to relative displacements $\boldsymbol{u}$; later we will look at an example that illustrates the role of the tensor Ω.

Both S and Ω behave as second-order tensors under coordinate transformations. Under a rotation of the coordinate system given by transformation R, the strain tensor S′ in the new coordinate system has components that are given in terms of those in the old coordinate system by

$$S'_{ij} = \sum_{m=1}^{3}\sum_{n=1}^{3} R_{im} S_{mn} R_{nj} \qquad (i, j = 1 \text{ to } 3) \tag{4.11}$$

(see Sect. A.1). The rotation tensor Ω transforms in the same way.

We now look at two simple examples, that will help to illustrate the geometric meaning of the strain and rotation tensors.

Example 4.1: Simple Strain. As a first example, we apply a tensile force to each end of a bar, causing it to lengthen, as shown in Fig. 4.3. We ignore

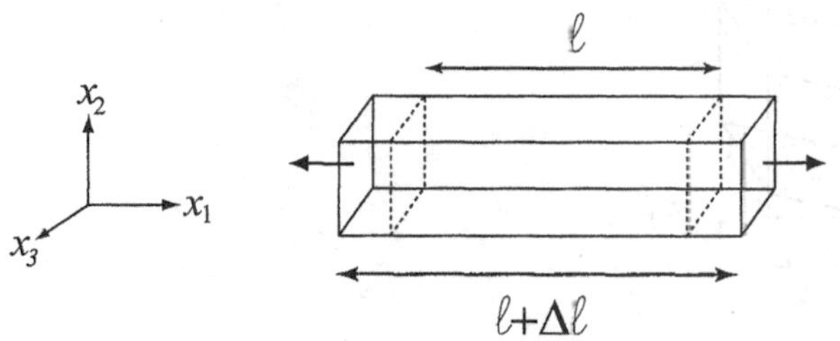

Fig. 4.3. Bar subjected to a pure strain, increasing its length from ℓ to $\ell + \Delta\ell$.

the strains in the direction perpendicular to the tension, so that this is a one-dimensional problem; we will discuss the complete problem later. The bar, of original length ℓ along the x_1 axis, will extend due to the force to a length $\ell' = \ell + \Delta\ell$.

The fractional change in length s is given by $s = (\ell' - \ell)/\ell = \Delta\ell/\ell$. The final length can thus be written as $\ell' = (1 + s)\ell$. Any differential element of initial length $\mathrm{d}x$ will, as it is subject to the same tension as the whole bar, lengthen by the same multiplier, to $(1 + s)\mathrm{d}x$. Placing the origin at the center of the bar, and integrating all the differential elements $\mathrm{d}x$ at initial (unstrained) positions from 0 to x, we find that an element at x shifts its position to $(1 + s)x$. The displacement u of an element originally at x is thus $u(x) = (1 + s)x - x = sx$. Note that the displacement depends on the choice of origin, while the fractional change s, as it is defined as a difference, does not.

We can now calculate the strain. In one dimension the strain is simply $S = \mathrm{d}u/\mathrm{d}x$, so in this example the strain is $S = s = \Delta\ell/\ell$, equal to the fractional change in length (not an uncommon situation, as we shall see!). Using our formal three-dimensional definition, with the vector displacement $\boldsymbol{u}(\boldsymbol{r}) = sx\hat{\boldsymbol{x}}_1$ (where x is the initial position along x_1, and $\hat{\boldsymbol{x}}_1$ is the unit vector along x_1), the strain tensor is

$$\mathsf{S} = \begin{bmatrix} \Delta\ell/\ell & 0 & 0 \\ 0 & 0 & 0 \\ 0 & 0 & 0 \end{bmatrix}. \tag{4.12}$$

The rotation tensor Ω is uniquely zero (no rotational strain).

Example 4.2: Pure Shear Strain. We now consider the case of pure *shear* strain, where the solid is strained as shown in Fig. 4.4: The two plane surfaces perpendicular to x_1 are shifted by $\Delta\ell$ along x_2, keeping the planes parallel to one another without changing their separation. The fractional displacement is $s = \Delta\ell/\ell$, the ratio of the shear displacement to the distance between the shear planes. The displacement vector that gives the strain shown in the figure is $\boldsymbol{u} = (0, sx_1, 0)$. The strain tensor is then given by

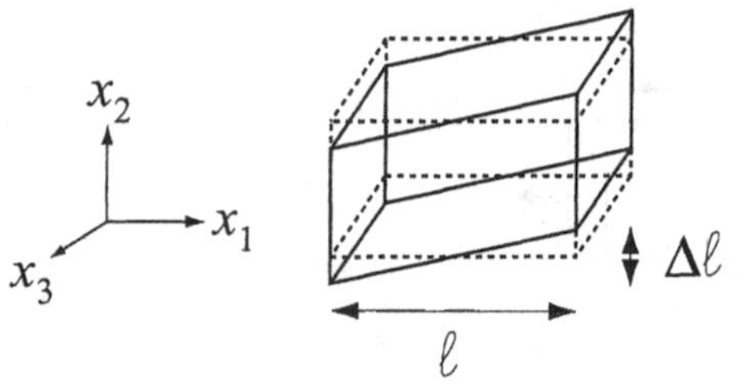

Fig. 4.4. Bar under shear strain.

$$\mathsf{S} = \begin{bmatrix} 0 & s/2 & 0 \\ s/2 & 0 & 0 \\ 0 & 0 & 0 \end{bmatrix} . \tag{4.13}$$

The rotation tensor Ω is given by

$$\Omega = \begin{bmatrix} 0 & s/2 & 0 \\ -s/2 & 0 & 0 \\ 0 & 0 & 0 \end{bmatrix} . \tag{4.14}$$

The motion depicted in Fig. 4.4 consists of both shear strain and a rotation. A slightly different displacement will yield shear strain only, with no rotation: $\boldsymbol{u} = (-\frac{1}{2}sx_2, \frac{1}{2}sx_1, 0)$. As the reader can verify, the strain tensor (4.13) remains unchanged for this displacement, while the rotation tensor is now zero. The reader is encouraged to try to sketch what shape a cube would acquire under this displacement.

Example 4.3: Pure Bending. Our last example deals with the problem of a long beam, with its length aligned along the x-axis, bent (or strained) in the $x - y$ plane by a torque about the z axis. The center of the beam is at the origin, and it has length $2L$ and width and thickness w. The center of each end of the beam moves upwards by Δy, assumed small in comparison to L, as shown in Fig. 4.5.

The displacement vector for the bent beam has the form

$$\boldsymbol{u}(x, y, z) = \left(-\frac{2xy}{L^2} \Delta y, \frac{x^2}{L^2} \Delta y, 0 \right) . \tag{4.15}$$

Consider the line through the geometric cross-sectional center of the beam, known as the center line. For the unstrained beam, this is the line $y = z = 0$. When the beam is subjected to the torque along z, the y displacement of the center line at position x is given by $\delta y = (\Delta y / L^2)x^2$; for small δy, this is the equation for the arc of a circle of radius $R = L^2/2\Delta y$. Now consider the planes perpendicular to the center line in the unstrained beam: Under the

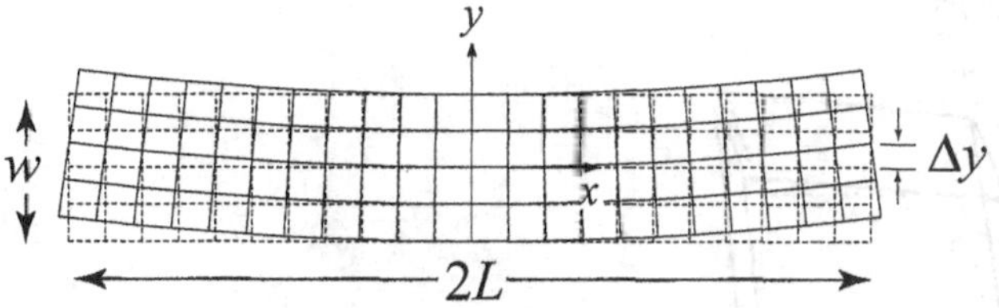

Fig. 4.5. Beam of length L, width and thickness t bent in the $x-y$ plane so that either end moves up by Δy. Solid and dotted lines are for the bent and relaxed beam, respectively.

displacement (4.15), these remain perpendicular in the strained beam. This is characteristic of pure bending.

The strain tensor corresponding to the displacement (4.15) is

$$\mathsf{S} = \begin{bmatrix} -2\frac{y}{L^2}\Delta y & 0 & 0 \\ 0 & 0 & 0 \\ 0 & 0 & 0 \end{bmatrix}. \tag{4.16}$$

Surprisingly, the beam shown in Fig. 4.5 is in a state of pure strain along x! The reason the beam bends, rather than extending as in the earlier example with an extensional force, is because the strain depends on y, so that a line parallel to, but above, the center line, with $y > 0$, contracts along x, but a line with $y < 0$, below the center line, is extended (stretched) along x. The result is that the beam bends to accommodate this height-dependent strain, as shown in the figure.

The rotation tensor for this example is

$$\Omega = \begin{bmatrix} 0 & -2\frac{x}{L^2}\Delta y & 0 \\ 2\frac{x}{L^2}\Delta y & 0 & 0 \\ 0 & 0 & 0 \end{bmatrix}. \tag{4.17}$$

In this case, we see that the rotation tensor is non-zero, as can be seen from the rotation of the small cells drawn in Fig. 4.5.

We now sketch out the geometrical interpretation of the various elements in the strain tensor.

The diagonal elements S_{11}, S_{22}, and S_{33} are all related to linear changes in the respective dimensions of the solid, as can be seen from Examples 4.1 and 4.3.

The off-diagonal elements $S_{12} = S_{21}$, $S_{13} = S_{31}$ and $S_{23} = S_{32}$ are related to angular distortions in the solid. This can be seem from Fig. 4.6, showing an element in pure shear strain, with no rotation, so $\Omega = 0$ (see Example 4.2). For $s = \Delta\ell/\ell$, the strain tensor is

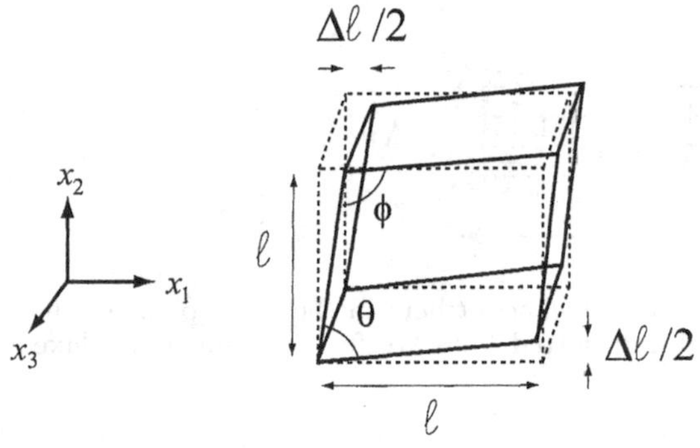

Fig. 4.6. Definition of the shear angles θ and ϕ

$$\mathsf{S} = \begin{bmatrix} 0 & s/2 & 0 \\ s/2 & 0 & 0 \\ 0 & 0 & 0 \end{bmatrix} . \tag{4.18}$$

The angle θ changes under shear from $\theta = \pi/2$ to $\theta = \tan^{-1}(\ell/\Delta\ell) = \tan^{-1}(1/s)$. For small strains, we have $s = \Delta\ell/\ell \ll 1$, so we can write $\theta \cong \pi/2 - s$. The change in angle is then $\Delta\theta \cong -s = -2S_{12}$. By similar considerations, we can show that the angle ϕ increases by $\Delta\phi = 2S_{12}$.

By similar arguments, for shear in the $x_1 - x_3$ plane, we can show that $2S_{13}$ is the angular change between the $x_1 - x_2$ and $x_2 - x_3$ planes, and for shear in the $x_2 - x_3$ plane, $2S_{23}$ gives the angular change between the $x_1 - x_2$ and $x_1 - x_3$ planes.

4.2 Principal Axes of Strain

The strain tensor is a symmetric, real tensor, and has three real eigenvalues and eigenvectors at each point in the solid. These may be found by solving the equation

$$\mathsf{S} \cdot \boldsymbol{A} = \lambda \boldsymbol{A}, \tag{4.19}$$

for the three eigenvalues λ and eigenvectors $\boldsymbol{A}$ (see Sect. A.2). The eigenvalues that satisfy this equation are the roots of the characteristic equation

$$\lambda^3 + \Sigma_1\lambda^2 + \Sigma_2\lambda + \Sigma_3 = 0, \tag{4.20}$$

in terms of the three strain invariants Σ_i, given by

$$\left.\begin{aligned} \Sigma_1 &= S_{11} + S_{22} + S_{33}, \\ \Sigma_2 &= S_{11}S_{22} + S_{11}S_{33} + S_{22}S_{33} - S_{12}^2 - S_{13}^2 - S_{23}^2, \\ \Sigma_3 &= S_{11}S_{22}S_{33} - S_{11}S_{23}^2 - S_{22}S_{13}^2 - S_{33}S_{12}^2 + 2S_{12}S_{13}S_{23}. \end{aligned}\right\} \tag{4.21}$$

The values of these invariants do not change if the coordinate system is rotated. Once we solve (4.11) for the three eigenvalue solutions λ_i, these can then be inserted in (4.19) to find the corresponding three eigenvectors $\boldsymbol{s}_i$.

These three eigenvectors $\boldsymbol{s}_i$ are mutually orthogonal, and the coordinate axes can then be chosen to be aligned with them. These axes are known as the *principal axes* of the strain tensor; if the strain tensor is written in this coordinate system, it will be purely diagonal, with only the terms $S_{ii} \neq 0$, at that point, and the diagonal terms will be equal to the eigenvalues λ_i.

Example 4.4: Principal Axes for a Strained Object. Let's take a displacement $\boldsymbol{u}$ for a solid that has the form

$$\boldsymbol{u} = \begin{pmatrix} 7x/4 + 3y/4 + z/2\sqrt{2} \\ 3x/4 + 7y/4 + z/2\sqrt{2} \\ x/2\sqrt{2} + y/2\sqrt{2} + 5z/2 \end{pmatrix}, \tag{4.22}$$

written in the "laboratory" coordinate system. The corresponding strain tensor is given by

$$\mathsf{S} = \begin{bmatrix} \frac{7}{4} & \frac{3}{4} & \frac{1}{2\sqrt{2}} \\ \frac{3}{4} & \frac{7}{4} & \frac{1}{2\sqrt{2}} \\ \frac{1}{2\sqrt{2}} & \frac{1}{2\sqrt{2}} & \frac{5}{2} \end{bmatrix}. \tag{4.23}$$

We wish to solve for the eigenvalues and eigenvectors for this tensor. The eigenvalue equation (4.19) may be written in the form

$$(\mathsf{S} - \lambda 1) \cdot \boldsymbol{A} = 0, \tag{4.24}$$

where 1 is the identity matrix. This equation has a nontrivial solution only if the expression $\mathsf{S} - \lambda 1$ has zero determinant. Inserting the form for S, and evaluating the determinant, we find

$$\lambda^3 - 6\lambda^2 + 11\lambda - 6 = 0. \tag{4.25}$$

By comparing with (4.20), we can read off the stress invariants, $\Sigma_1 = -6$, $\Sigma_2 = 11$ and $\Sigma_3 = -6$.

The solutions to the cubic equation turn out (miraculously) to be the three roots $\lambda_1 = 1$, $\lambda_2 = 2$ and $\lambda_3 = 3$. Inserting these solutions one by one into the set of equations given by (4.24), we find the following relations for the respective eigenvectors $\boldsymbol{A}_1$, $\boldsymbol{A}_2$, and $\boldsymbol{A}_3$:

$$\boldsymbol{A}_1 = \alpha \begin{pmatrix} 1/\sqrt{2} \\ -1/\sqrt{2} \\ 0 \end{pmatrix}, \tag{4.26}$$

$$A_2 = \beta \begin{pmatrix} 1/2 \\ 1/2 \\ -1/\sqrt{2} \end{pmatrix}, \tag{4.27}$$

$$A_3 = \gamma \begin{pmatrix} 1/2 \\ 1/2 \\ 1/\sqrt{2} \end{pmatrix}, \tag{4.28}$$

where α, β and γ are adjustable constants; to normalize the eigenvectors, i.e. to make them have unit length, these can all be set equal to 1. These eigenvectors are all mutually orthogonal, and the coordinate axes can be aligned with them by first rotating by 45° about the x_3-axis and then by 45° about the new x_1'-axis.

We note that in this example, the displacement vector $\boldsymbol{u}$ only contains terms to first order in the position, so that the strain tensor S is constant throughout the solid. Therefore, for this example, the principal axes are the same at all points within the solid. In general this will not be the case, and the principal axes will vary as one moves through the solid.

All of our examples have involved known displacements, from which the strain could be calculated. Normally, one starts with known stresses (applied by the various forces and constraints on the solid), from which the strain tensor S is calculated. In order to calculate the corresponding displacement $\boldsymbol{u}$, however, we must then integrate the relations given in (4.5). This is often a nontrivial problem, as is shown in the following example.

Example 4.5: Displacement From Strain. Consider a solid bar, with dimensions $L \times w \times t$ in the unstrained state. The bar is then strained, with a strain tensor given by

$$\mathsf{S} = \begin{bmatrix} -2dx_2 + 6d^2x_1^2 & -4dx_1 & 0 \\ -4dx_1 & 0 & 0 \\ 0 & 0 & 0 \end{bmatrix}. \tag{4.29}$$

The constant d is related to the amplitude of the strain, and has units of $(\text{length})^{-1}$ in order to make the strain dimensionless.

We want to construct the displacement vector $\boldsymbol{u}(\boldsymbol{r})$ from the strain tensor. The third column of S is all zeros, so the derivatives of $\boldsymbol{u}$ with x_3 vanish. The third row of S is also zero, so all derivatives of u_3 vanish. Hence we know $u_3 = c$, where c is a constant, and we know that u_1 and u_2 are not functions of x_3.

S_{11} can be integrated to give $u_1 = -2dx_1x_2 + 2d^2x_1^3 + a$, where a is a constant of integration. S_{22} is zero, so u_2 does not depend on x_2. The term S_{12} can then be integrated to yield $u_2 = -dx_1^2 + b$ where b is a constant.

Hence we have the displacement vector

$$u(x_1, x_2, x_3) = \begin{pmatrix} -2dx_1x_2 + 2d^2x_1^3 + a \\ -dx_1^2 + b \\ c \end{pmatrix}. \tag{4.30}$$

We note that at this point we must choose the origin of our coordinate system, which of course cannot be determined from the strain tensor. In the undistorted solid the usual choice is at the center of mass. Here we must make an assumption, which will be that the distortion does not change the density of the solid (actually one only need assume that the final density is constant throughout the distorted volume). We know that the original center of mass is at the origin, so for the final center of mass to remain at the origin we require

$$\int_{-L/2}^{L/2} \int_{-w/2}^{w/2} \int_{-t/2}^{t/2} u(x_1, x_2, x_3)\, dx_1\, dx_2\, dx_3 = 0. \tag{4.31}$$

Solving this equation yields $a = 0$, $b = -dw^3/12$, and $c = 0$. A sketch of the undistorted and distorted solid is shown in Fig. 4.7. Note the close similarity to Example 4.3, although the state of strain is not exactly the same.

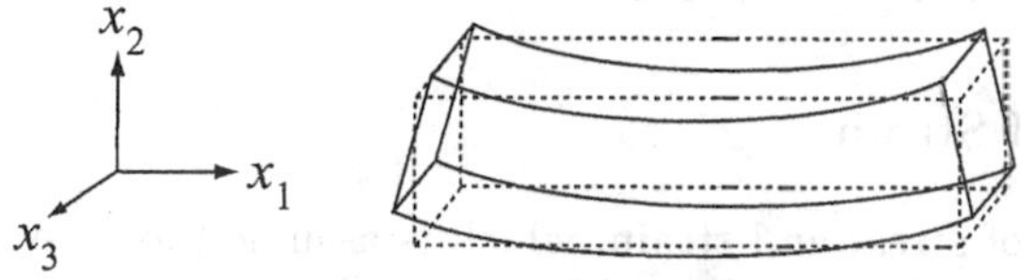

Fig. 4.7. Calculated distortion of the beam in Example 4.5.

4.3 Superposition

We now briefly discuss the delicate issue of the superposition of strain fields. In any linear system, such as for electromagnetic and gravitational fields in vacuum, a given source term (such as a charge distribution, or a mass distribution for gravity) leads to a given response, in the form of an electrostatic or gravitational field. If a problem consists of a combination of two charge distributions (or two masses), the resulting field is the algebraic sum of the fields from each charge distribution by itself: This is the principle of superposition, and results from the linearity of the corresponding equations. One would think that strain in a solid would act the same way, so that the total strain resulting from two sequential strains should be the algebraic (tensor) sum of the two strains. The difficulty lies in the fact that strain is defined as a function of position within an object, and the strain acts to change the position. This leads to an intrinsic nonlinearity in the response, so that linear superposition only holds in the limit of small strains.

Consider a displacement field $\boldsymbol{u}(\boldsymbol{r})$, and another displacement field $\boldsymbol{u}'(\boldsymbol{r})$, each considered as independent. Their superposition, for example the result from applying first $\boldsymbol{u}$ and then $\boldsymbol{u}'$, should just be $\boldsymbol{u}_{tot}(\boldsymbol{r}) = \boldsymbol{u}(\boldsymbol{r}) + \boldsymbol{u}'(\boldsymbol{r})$. However, the second displacement, occurring after the first rather than independent of it, should be evaluated as $\boldsymbol{u}'(\boldsymbol{r}+\boldsymbol{u}(\boldsymbol{r}))$, accounting for the result of the first displacement. Note that we also must be careful about the sequence: Applying $\boldsymbol{u}'(\boldsymbol{r})$ first, and then applying $\boldsymbol{u}(\boldsymbol{r} + \boldsymbol{u}'(\boldsymbol{r}))$, can lead to a different result. The sequence of strains does not necessarily commute.

Expanding the argument of the second displacement field in a Taylor series,

$$u_i'(\boldsymbol{r} + \boldsymbol{u}(\boldsymbol{r})) = u_i'(\boldsymbol{r}) + \sum_{j=1}^{3} \frac{\partial u_i'}{\partial x_j} u_j(\boldsymbol{r}) + \text{ higher order terms}, \tag{4.32}$$

we see that the superposed displacements add linearly only if the second and higher order terms may be neglected. If this is an accurate approximation, then the superposed displacements will add linearly; furthermore, in this limit, the superposed strain fields will commute, so that the sequence in which (small) strains are imposed is not important.

4.3.1 Eulerian Definition of Strain

The linearity of the equations of stress and strain, which is assumed for the majority of this text, are predicated on the linearity of the relation between strain and relative displacement, given by (4.7). The derivation of the exact expression between strain and displacement however has nonlinear terms, which vanish only in the limit of small displacements.

If one has two points P and Q, initially separated by $\mathbf{d}\boldsymbol{r}$, which under strain are separated by $\mathbf{d}\boldsymbol{r}'$ (see Fig. 4.8), then the change in the *squared* distance can be shown to be given by [44]

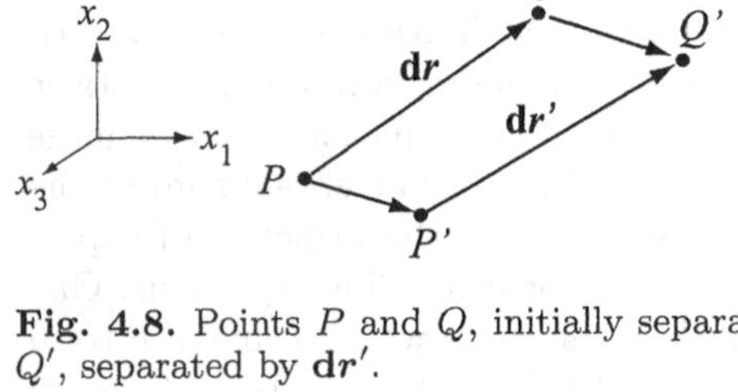

Fig. 4.8. Points P and Q, initially separated by $\mathbf{d}\boldsymbol{r}$, shift under strain to P' and Q', separated by $\mathbf{d}\boldsymbol{r}'$.

$$dr'^2 - dr^2 = 2 \sum_{i,j=1}^{3} \eta_{ij} dr_i dr_j, \tag{4.33}$$

with the Eulerian strain tensor η_{ij} defined by

$$\eta_{ij} = \frac{1}{2} \left(\frac{\partial u_i}{\partial x_j} + \frac{\partial u_j}{\partial x_i} + \sum_{k=1}^{3} \frac{\partial u_k}{\partial x_i} \frac{\partial u_k}{\partial x_j} \right). \tag{4.34}$$

The Eulerian strain tensor η_{ij} reduces to the conventional strain tensor S_{ij} in the limit where the product of derivatives of the relative displacements, the third term in (4.34), is much smaller than the derivatives themselves, which holds for either very gradual or for very small displacements $\boldsymbol{u}(\boldsymbol{r})$. For large displacements, or rapid changes in the displacement, the strain will no longer be linearly related to displacement, and the theory then becomes much more complex. We will not discuss this limit of the theory further.

4.4 The Stress Tensor

We have seen how to describe the state of strain of an object in terms of the position-dependent vector displacement $\boldsymbol{u}(\boldsymbol{r})$. From this displacement we defined the displacement tensor D, the strain tensor S, and the rotation tensor Ω. We now describe how the connection is made between strain and the forces acting on a deformable object.

The description of the forces acting on a solid object, and their distribution in the volume of the solid, are somewhat more complex than the description of the strain. In classical mechanics, it is usually sufficient to consider the force, as a vector, acting at the center of mass of the body, and add to this a torque, or couple, acting about the center of mass. Elastic objects, which can deform under the action of forces, are more difficult to handle, as one must consider the location and distribution of forces on the surfaces and within the volume of the object.

In this section we will introduce the concepts for dealing with the surface stresses and the body forces acting on an object. We will for the time being only treat static forces that are in equilibrium, so that no acceleration occurs. A discussion of how to treat non-equilibrium, and therefore time-dependent situations, will appear in Chap. 7.

Example 4.6: Bar Under Tension. Consider a bar with cross-sectional area A in static equilibrium, with equal and opposite tensile (extensional) forces F applied to each end (see Fig. 4.9). *Tensile stress* is defined as the tensile force exerted on a surface, per unit area of the surface. We will represent stress with the symbol T. The tensile stress in the bar is $T = F/A$. Now imagine dividing the bar in two with a plane perpendicular to the line of force, i.e. perpendicular to the x_1 axis. In order to maintain equilibrium on

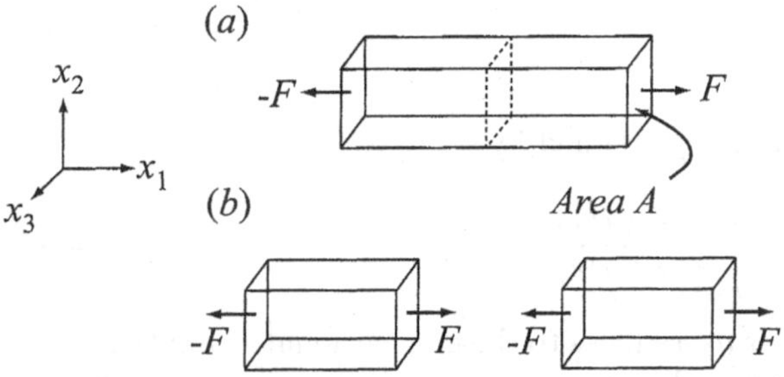

Fig. 4.9. Example 4.6. (**a**) A bar under tensile stress F/A applied on the end surfaces, when divided in two (**b**) reveals internal tensile stress of the same magnitude.

either side of the division, there must exist internal forces on the two surfaces of the cut, with equal magnitudes F but opposite directions (see Fig. 4.9). The bar must therefore have a uniform distribution of such internal forces through its volume, and a uniform internal stress $T = F/A$.

Example 4.7: Bar Under Shear. Now consider a situation where a cube of cross-section A is subjected to a shear force, applied by forces F and $-F$ on two opposing faces of the cube, shown in Fig. 4.10a. The forces are assumed to be uniform over the faces of the cube. *Shear stress* is defined as the in-plane force divided by the area of the surface to which the force is applied. In this case the shear stress is again $T = F/A$. There is no net force applied to the bar, but there is a net torque, applied about the x_3 axis. In order to achieve static equilibrium, we must add two additional forces F on two other faces, in opposite directions, to reduce this torque to zero. This is shown in Fig. 4.10(b).

A rigid body under this revised force distribution is static. However, just as for the tensile force situation, there are internal stresses in the cube, which we wish to evaluate. We cut the cube with an imaginary plane to expose two internal surfaces, cutting first parallel to the top surface of the cube, perpendicular to the x_2 axis, as shown in Fig. 4.11. The exposed surfaces

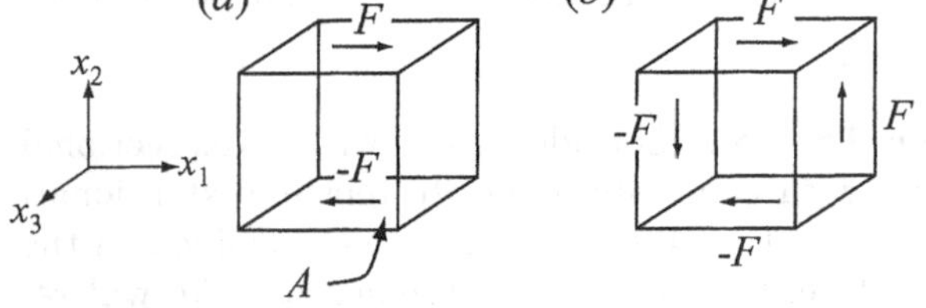

Fig. 4.10. (**a**) Forces F and $-F$ applied to two opposing faces, generating no net force but non-zero torque, and (**b**) applied to four faces to establish static equilibrium.

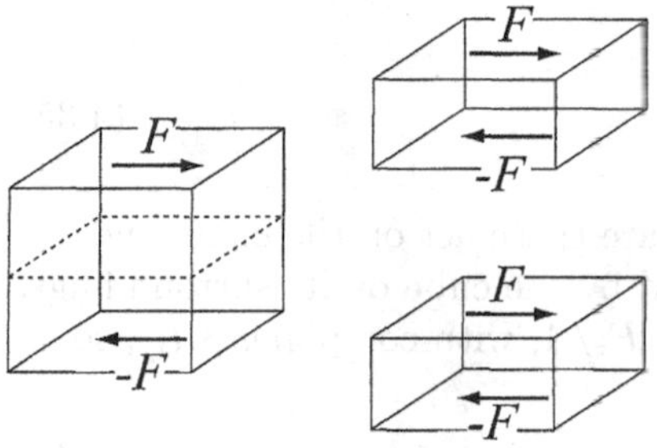

Fig. 4.11. Shear stress is uniform through the solid.

must have equal and opposite forces F and $-F$ as shown, directed along x_1, and therefore must have shear stresses $T = F/A$. Hence the shear stress along the x_1 axis, in the plane perpendicular to the x_2 axis, is uniform through the cube. Similarly, we can cut the cube with a plane perpendicular to x_1, and we find that the shear stress in this plane is also uniform through the cube, with magnitude $T = F/A$.

We now need to formalize our discussion a bit. Consider a differential solid cube, taken from the volume of an arbitrary solid, with its edges aligned with the coordinate axes, as shown in Fig. 4.12; the cube has sides of length $d\ell$, and faces with area $A = d\ell^2$. The cube is acted upon by forces from the surrounding solid; we assume the cube is small enough that the forces do not vary through its volume, and we will ignore any body forces (i.e. those not applied through the surfaces); we will discuss the effect of body forces later.

Each face i of the cube is associated with a vector force $\boldsymbol{F}_i$, with i running from 1 to 6; $\boldsymbol{F}_1$ is the force acting on surface 1, the surface perpendicular to x_1 at $x_1 = d\ell$, $\boldsymbol{F}_4$ is acting on surface 4, perpendicular to x_1 at $x_1 = 0$, and so on. Each of these forces may be written out in component form, with

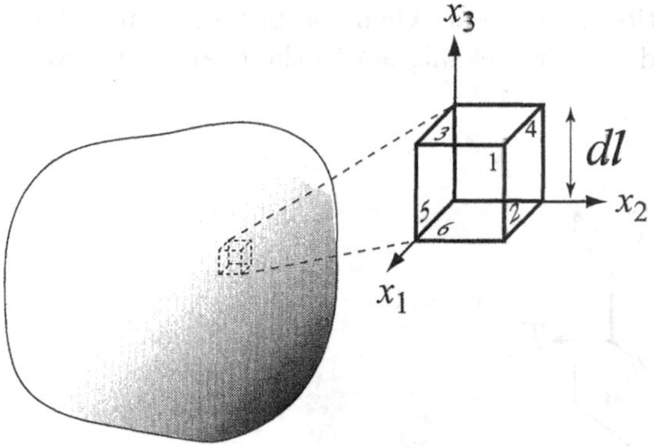

Fig. 4.12. Differential cube of side $d\ell$, taken from interior of a solid.

$$\begin{aligned} \boldsymbol{F}_i &= F_{i1}\,\hat{\boldsymbol{x}}_1 + F_{i2}\,\hat{\boldsymbol{x}}_2 + F_{i3}\,\hat{\boldsymbol{x}}_3 \\ &= \sum_{j=1}^{3} F_{ij}\,\hat{\boldsymbol{x}}_j \qquad (i = 1 \text{ to } 6). \end{aligned} \tag{4.35}$$

Note that we emphasize that these forces are those act on the cube, and are not the (equal and opposite) forces exerted by the cube on its surroundings.

We can now define a vector stress $\boldsymbol{t}_i = \boldsymbol{F}_i/A$, with components t_{ij}, so

$$\boldsymbol{t}_i = \sum_{j=1}^{3} t_{ij}\,\hat{\boldsymbol{x}}_j \qquad (i = 1 \text{ to } 6). \tag{4.36}$$

There are $3 \times 6 = 18$ values t_{ij}. However, because the cube is very small, the forces do not vary with position on the cube surface, so we have the following identities between forces on opposite faces of the cube:

$$\left.\begin{aligned} \boldsymbol{F}_4 &= -\boldsymbol{F}_1 \\ \boldsymbol{F}_5 &= -\boldsymbol{F}_2 \\ \boldsymbol{F}_6 &= -\boldsymbol{F}_3, \end{aligned}\right\} \tag{4.37}$$

with similar identities for the stress vectors $\boldsymbol{t}_i$. Hence we have only nine (not eighteen) independent stress vector components, the t_{ij} with i and j running from 1 to 3. We collect these components in a matrix, and define the *stress tensor* T as the tabulation

$$\mathsf{T} = \begin{bmatrix} t_{11} & t_{12} & t_{13} \\ t_{21} & t_{22} & t_{23} \\ t_{31} & t_{32} & t_{33} \end{bmatrix}, \tag{4.38}$$

identifying the element T_{ij} of the stress tensor with the jth component of the ith stress vector, $T_{ij} = t_{ij}$. In Fig. 4.13 we display the various elements of the stress tensor and their corresponding stress vector components.

We also define a sign convention for the stresses. This choice of sign varies among the different treatments of solid mechanics, so that one must always verify what sign convention was chosen when comparing results. The sign convention is easily defined for surfaces aligned to the coordinate axes;

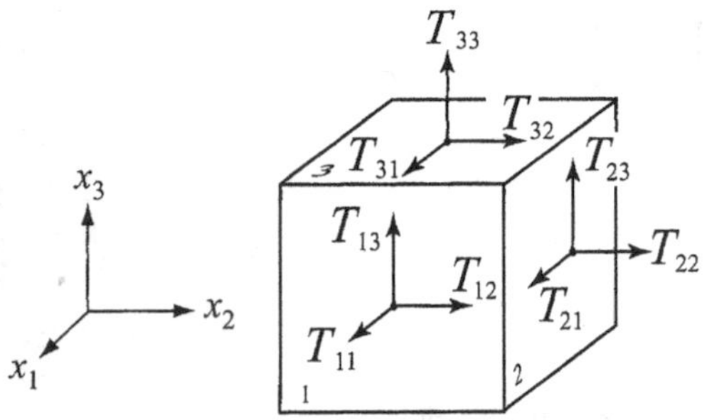

Fig. 4.13. Elements T_{ij} of the stress tensor T.

the sign for an arbitrary surface can be calculated using the transformation procedure that we will define below.

We first define the *outward surface normal* of a surface as the vector pointing away from the interior of the solid element; in Fig. 4.12, the surface normals of faces 1, 2 and 3 point along their respective coordinate axes, while the surface normals for faces 4, 5 and 6 point in the opposite direction of the axis unit vectors.

Our sign convention is then defined as follows: *A stress component acting on an element surface is positive if it points in the same direction as the corresponding coordinate axis, and if the surface normal points in the same direction as its corresponding coordinate axis. Conversely, if the surface normal points in the opposite direction, then the stress component is positive if it also points in the opposite direction.* We can write this using vector notation. With outward surface normal $\hat{\boldsymbol{n}}$ of the surface α, aligned perpendicular to the x_α axis, so that $\hat{\boldsymbol{n}} = n\hat{\boldsymbol{x}}_\alpha$ with $n = \pm 1$, and the stress components $T_{\alpha i}$, the component of the stress along x_i is positive if $nT_{\alpha i} > 0$, and negative if $nT_{\alpha i} < 0$.

Note that the definition is for stress acting *on* a volume element's surface, not for the volume's responding stress. Hence the stresses T_{ij} shown in Fig. 4.13, which act on the surfaces of the cube, are all positive as drawn. On the opposing faces, the positive stress components would all be reversed from the directions shown in Fig. 4.13, as the surface normal vectors point in the direction opposite to the corresponding coordinate axes. For example, in Fig. 4.14, which shows a section of the cube perpendicular to the x_3 axis, the stresses as drawn are all positive.

Given the stress tensor T at a point, the stress vectors for the faces of a differential cube centered on that point can be reconstructed. The stress vector on face 1 is just $\boldsymbol{t}_1 = T_{11}\hat{\boldsymbol{x}}_1 + T_{12}\hat{\boldsymbol{x}}_2 + T_{13}\hat{\boldsymbol{x}}_3$; that for face 4, opposite face 1, is $\boldsymbol{t}_4 = -T_{11}\hat{\boldsymbol{x}}_1 - T_{12}\hat{\boldsymbol{x}}_2 - T_{13}\hat{\boldsymbol{x}}_3$, and so on. Clearly care must be taken to preserve the sign convention.

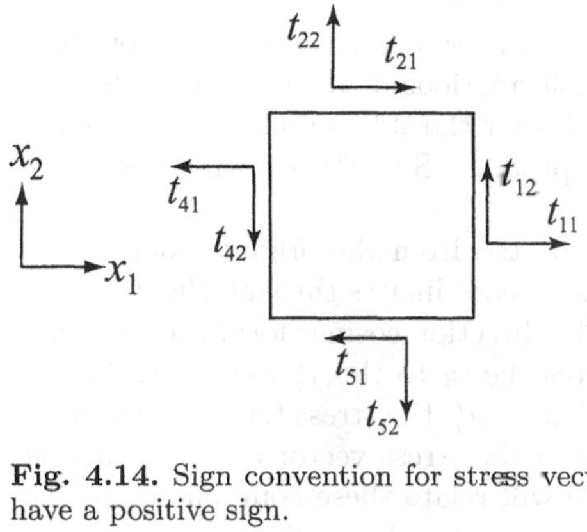

Fig. 4.14. Sign convention for stress vectors; all the indicated stress components have a positive sign.

4.4.1 Symmetry of the Stress Tensor

The static force balance of our cube is automatically satisfied by the assumption that the surface forces are equal and opposite on opposite faces. However, we do not yet have such a guarantee for the torque. Consider the situation shown in Fig. 4.14. The torque about the x_3 axis involves only the stress components shown in the figure, and is given by $N_3 = (\mathrm{d}\ell/2)\,(\mathrm{d}\ell)^2\,(T_{21} - T_{51} - T_{12} + T_{42}) = (\mathrm{d}\ell)^3(T_{21} - T_{12})$, taking into account the moment arm $\mathrm{d}\ell/2$ and the area $(\mathrm{d}\ell)^2$ over which the stress is applied. This torque, if non-zero, will give an angular acceleration $\ddot{\theta}$ of the solid element, where θ is the rotational angle about the x_3 axis. The acceleration is related to the torque by the rotational moment of inertia about the x_3 axis, $I_3 = \rho(\mathrm{d}\ell)^5/6$ for a cube of side $\mathrm{d}\ell$ with mass density ρ, rotating about its center (see a text on classical mechanics, for instance Marion and Thornton [45]). The angular equation of motion is then

$$I_3 \frac{\mathrm{d}^2\theta}{\mathrm{d}t^2} = N_3, \tag{4.39}$$

which yields

$$\frac{\mathrm{d}^2\theta}{\mathrm{d}t^2} = \frac{6}{\rho\,(\mathrm{d}\ell)^2}(T_{21} - T_{12}). \tag{4.40}$$

In the limit where the differential element's length $\mathrm{d}\ell \to 0$, the acceleration becomes infinite unless $T_{12} = T_{21}$. Hence, to achieve static equilibrium, we require the components T_{12} and T_{21} to be equal. By similar considerations of the torque about the other two coordinate axes, you can show that $T_{ij} = T_{ji}$, i.e. the stress tensor T must be symmetric.

4.5 Properties of the Stress Tensor

4.5.1 Transformations of the Stress Tensor

Our next step is to show that the stress tensor T is actually a tensor, that is, it transforms under coordinate transformations as a second rank tensor, according to (A.12). We did not show this for the strain tensor, as the strain tensor is defined as a standard dyadic product, $\mathsf{S} = \nabla \boldsymbol{u} + (\nabla \boldsymbol{u})^T$, which is automatically a tensor.

We define a set of coordinates x_i', rotated from the original coordinates x_i. The x_i' coordinates are related to the x_i coordinates through the direction cosines, defined as $\cos(\theta_{ij}) = \hat{\boldsymbol{x}}_i' \cdot \hat{\boldsymbol{x}}_j$. The direction cosines form the elements R_{ij} of the rotation tensor R that rotates the x_i to the x_i' (see Sect. A.1.1). Expressed in terms of the primed coordinates x_i', the stress tensor T will have components T_{ij}', giving the component of the stress vector along x_j', acting on the surface perpendicular to $\hat{\boldsymbol{x}}_i'$. We will relate these components to the components T_{ij} expressed in terms of the unprimed coordinates x_i.

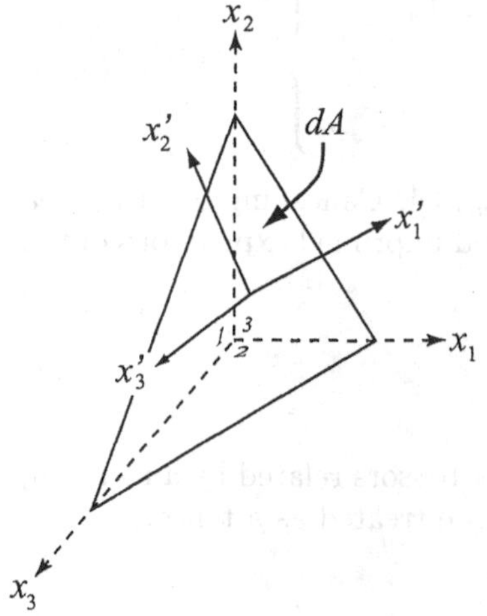

Fig. 4.15. Construction of differential tetrahedron to demonstrate tensor nature of T.

We construct a differential tetrahedron, with three sides formed by the planes perpendicular to the original x_1, x_2, and x_3 axes, and the fourth side formed by the surface normal to the x'_m axis, where m is any one of the three axes 1, 2 or 3. This construction, for $m = 1$, is shown in Fig. 4.15. The tilted surface has area dA, so the area of the 1 plane of the tetrahedron (perpendicular to $\hat{\boldsymbol{x}}_1$) is $\mathrm{d}A_1 = \mathrm{d}A \cos(\theta_{m1}) = R_{m1}\mathrm{d}A$, that of the 2 plane (perpendicular to $\hat{\boldsymbol{x}}_2$) is $\mathrm{d}A_2 = \mathrm{d}A \cos(\theta_{m2}) = R_{m2}\mathrm{d}A$, and that of the 3 plane (perpendicular to $\hat{\boldsymbol{x}}_3$) is $\mathrm{d}A_3 = \mathrm{d}A \cos(\theta_{m3}) = R_{m3}\mathrm{d}A$, where the R_{ij} are the components of the rotation tensor R.

The force acting on the plane dA perpendicular to $\hat{\boldsymbol{x}}'_m$ is given by $\boldsymbol{F}' = \mathrm{d}A(T'_{m1}\hat{\boldsymbol{x}}'_1 + T'_{m2}\hat{\boldsymbol{x}}'_2 + T'_{m3}\hat{\boldsymbol{x}}'_3)$. The force $\boldsymbol{F}_j$ on the j plane ($j = 1$, 2 or 3), perpendicular to the x_j axis, is given by $F_j = \mathrm{d}A_j(T_{j1}\hat{\boldsymbol{x}}_1 + T_{j2}\hat{\boldsymbol{x}}_2 + T_{j3}\hat{\boldsymbol{x}}_3)$. We can express the forces $\boldsymbol{F}_j$ in terms of the primed coordinates as well,

$$\boldsymbol{F}_j = \sum_{k=1}^{3} F'_{jk}\hat{\boldsymbol{x}}'_k, \tag{4.41}$$

with the F'_{jk} components related to the F_{jk} through a vector (first-order tensor) transformation,

$$F'_{ij} = \sum_{j=1}^{3} R_{jk}F_{ik} \tag{4.42}$$

(see Sect. A.1.1).

The net force on the tetrahedron must vanish if we are to maintain a static balance. Equating each component i of the force $\boldsymbol{F}'$ along $\hat{\boldsymbol{x}}'_i$ to the sum of the components i of the forces $\boldsymbol{F}'_j$, we have

$$\left.\begin{aligned} F'_i &= T'_{mi}\,\mathrm{d}A = F'_{1i} + F'_{2i} + F'_{3i} = \textstyle\sum_{j=1}^{3} F'_{ji} \\ &= \textstyle\sum_{j=1}^{3}\sum_{k=1}^{3} R_{ik}\,F_{jk} \\ &= \textstyle\sum_{j=1}^{3}\sum_{k=1}^{3} R_{ik}\,T_{jk}R_{mj}\,\mathrm{d}A, \end{aligned}\right\} \tag{4.43}$$

where the last step follows because $\mathrm{d}A_j = R_{mj}\mathrm{d}A$. Cancelling out the area $\mathrm{d}A$, we find the relation between the primed and unprimed expressions of the stress tensor,

$$T'_{mi} = \sum_{j=1}^{3}\sum_{k=1}^{3} R_{mj}T_{jk}R_{ik} = \mathsf{R}\cdot\mathsf{T}\cdot\mathsf{R}^T, \tag{4.44}$$

This is exactly the relation for two second-rank tensors related by a rotation, and concludes our proof that the stress T can be treated as a tensor.

4.5.2 Stress on an Arbitrary Surface

Given the stress tensor T, we would like to evaluate the vector stress $\boldsymbol{t}'$ acting on any surface, either one aligned to the coordinate axes, or one at an arbitrary orientation. The construction is identical to the one we used to prove that the stress tensor transforms correctly under coordinate transformation. Consider again the differential tetrahedron, redrawn in Fig. 4.16. The unit vector normal to the surface $\mathrm{d}A$ is $\hat{\boldsymbol{n}} = n_1\hat{\boldsymbol{x}}_1 + n_2\hat{\boldsymbol{x}}_2 + n_3\hat{\boldsymbol{x}}_3$, and it forms direction cosines a_1, a_2 and a_3 with the three coordinate axes x_1, x_2 and x_3 respectively, where $a_i = n_i$. A force $\boldsymbol{F}'$ acts on A, and is balanced by the forces $\boldsymbol{F}_i$ on the other three surfaces.

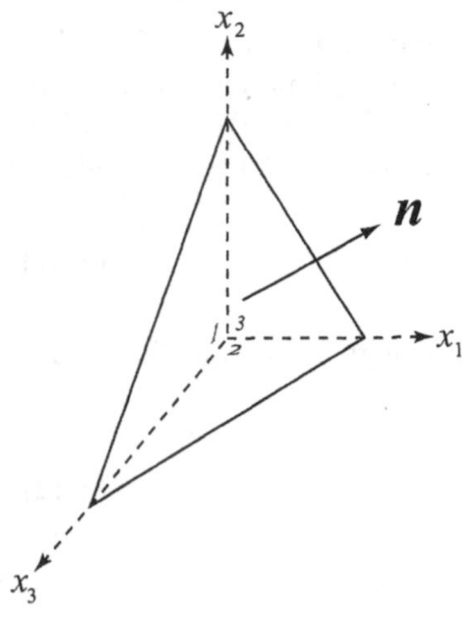

Fig. 4.16. Differential tetrahedron used to find stress on arbitrary surface with normal $\boldsymbol{n}$.

The areas of the three planes $i = 1$, 2 and 3 are given by $\mathrm{d}A_i = n_i \mathrm{d}A$. The force $\boldsymbol{F}'$ on the surface $\mathrm{d}A$ must balance the forces $\boldsymbol{F}_i$ on each surface i, so we must have the component balance

$$\left.\begin{aligned} F_1' &= T_{11}\mathrm{d}A_1 + T_{21}\mathrm{d}A_2 + T_{31}\mathrm{d}A_3, \\ F_2' &= T_{12}\mathrm{d}A_1 + T_{22}\mathrm{d}A_2 + T_{32}\mathrm{d}A_3, \\ F_3' &= T_{13}\mathrm{d}A_1 + T_{23}\mathrm{d}A_2 + T_{33}\mathrm{d}A_3. \end{aligned}\right\} \tag{4.45}$$

Dividing out the common area terms, we arrive at an expression for the three components of the stress vector $\boldsymbol{t}' = \boldsymbol{F}'/\mathrm{d}A$ acting on the surface $\mathrm{d}A$:

$$t_1' = T_{11}n_1 + T_{21}n_2 + T_{31}n_3, \tag{4.46}$$
$$t_2' = T_{12}n_1 + T_{22}n_2 + T_{32}n_3, \tag{4.47}$$
$$t_3' = T_{13}n_1 + T_{23}n_2 + T_{33}n_3. \tag{4.48}$$

Using the symmetry of the tensor T, we can invert the indices and write (4.46-4.48) in a more compact form:

$$t_i' = \sum_{j=1}^{3} T_{ij}n_j, \tag{4.49}$$

or

$$\boldsymbol{t}' = \mathsf{T} \cdot \hat{\boldsymbol{n}}, \tag{4.50}$$

relating the stress vector $\boldsymbol{t}'$ on an arbitrary surface with unit normal $\hat{\boldsymbol{n}}$ to the stress tensor T.

4.5.3 Principal Axes of Stress

We have already shown that the stress tensor T is symmetric. The stress tensor is therefore guaranteed to have three real eigenvalues t_1, t_2 and t_3, and three corresponding eigenvectors $\boldsymbol{e}_1$, $\boldsymbol{e}_2$ and $\boldsymbol{e}_3$. These can be obtained by solving the characteristic equation as in Sect. 4.2 (see Sect. A.2). The eigenvalue-eigenvector equation for the stress tensor is

$$\mathsf{T} \cdot \boldsymbol{e}_i = t\boldsymbol{e}_i, \tag{4.51}$$

where the three values of t that can solve this equation correspond to the eigenvalues t_1, t_2 and t_3. This leads in the usual way to the secular equation

$$t^3 + \Sigma_1 t^2 + \Sigma_2 t + \Sigma_3 = 0, \tag{4.52}$$

using the three stress invariants Σ_i, given by

$$\left.\begin{aligned} \Sigma_1 &= T_{11} + T_{22} + T_{33}, \\ \Sigma_2 &= T_{11}T_{22} + T_{11}T_{33} + T_{22}T_{33} - T_{12}^2 - T_{13}^2 - T_{23}^2, \\ \Sigma_3 &= T_{11}T_{22}T_{33} - T_{11}T_{23}^2 - T_{22}T_{13}^2 - T_{33}T_{12}^2 + 2T_{12}T_{13}T_{23}. \end{aligned}\right\} \tag{4.53}$$

The stress invariants do not change their values under coordinate rotations.

Just as for the strain tensor, the three roots t_1, t_2, t_3 of the secular equation are known as the principal values of the stress tensor. Given the three eigenvalues, we can solve for the corresponding eigenvectors $\boldsymbol{e}_1$, $\boldsymbol{e}_2$ and $\boldsymbol{e}_3$, which are then the principal axes of the stress tensor. If we choose our coordinate system to coincide with the principal axes, then in this coordinate system the stress tensor is diagonal,

$$\mathsf{T} = \begin{bmatrix} t_1 & 0 & 0 \\ 0 & t_2 & 0 \\ 0 & 0 & t_3 \end{bmatrix} . \tag{4.54}$$

A plane perpendicular to one of these axes, say the axis e_i, will only experience normal stress, that is, there will be no shear stress on such a plane. The vector stress on such a plane is therefore given by $\boldsymbol{t} = t_i \boldsymbol{e}_i$. Note that as the stress tensor of course varies throughout a solid, the principal values and principal axes vary from point to point, so that the orientation of the planes of pure normal stress will vary as one moves through the solid.

Example 4.8: Principal Axes for a Bar Under Tensile Stress. Consider again the bar under pure tensile stress, described in Sect. 4.6. We have reproduced the figure below, Fig. 4.17. The vector stress on the left side of the bar is given by $\boldsymbol{t} = (-F/A, 0, 0)$, while that on the right side is $\boldsymbol{t} = (F/A, 0, 0)$. If we section the bar at some point with a plane perpendicular to the x_1 axis, then the side of the section to the right will experience the stress $\boldsymbol{t} = (-F/A, 0, 0)$, while that to the left side of the section will have the vector stress $\boldsymbol{t} = (F/A, 0, 0)$. Hence the planes perpendicular to x_1 are principal planes of the stress tensor, and we can write the stress tensor at any point within the volume of the bar as

$$\mathsf{T} = \begin{bmatrix} F/A & 0 & 0 \\ 0 & 0 & 0 \\ 0 & 0 & 0 \end{bmatrix} . \tag{4.55}$$

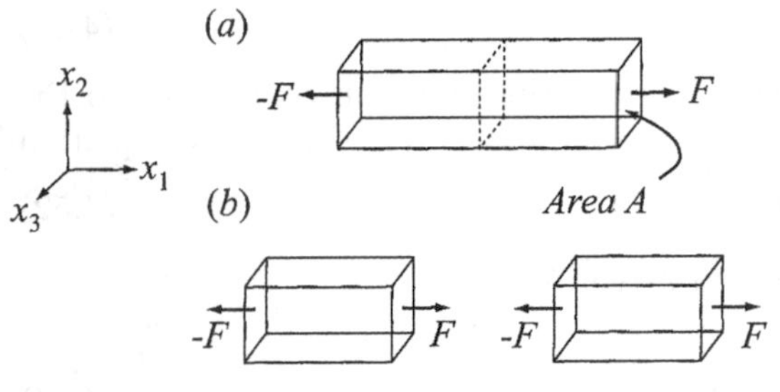

Fig. 4.17. Example 4.8.

Example 4.9: Principal Axes for a Bar Under Shear Stress. We now revisit the example of the bar under pure shear, given in Sect. 4.7. In Fig. 4.18 we display the geometry with a choice of coordinate axes. Every differential volume element experiences a shear stress in the $x_1 - x_2$ plane, a situation known as one of pure *plane stress.* We can write the stress tensor by inspection, as only the components $T_{12} = T_{21}$ are non-zero:

$$\mathsf{T} = \begin{bmatrix} 0 & F/A & 0 \\ F/A & 0 & 0 \\ 0 & 0 & 0 \end{bmatrix}. \tag{4.56}$$

We wish to find the principal axes and the stress along them. The three stress invariants are $\Sigma_1 = 0$, $\Sigma_2 = -(F/A)^2$, and $\Sigma_3 = 0$. The secular equation has the form

$$t^3 - (\frac{F}{A})^2\, t = 0, \tag{4.57}$$

which has the solutions $t_1 = F/A$, $t_2 = -F/A$, and $t_3 = 0$. The corresponding unit length eigenvectors are $\boldsymbol{e}_1 = (1/\sqrt{2}, -1/\sqrt{2}, 0)$, $\boldsymbol{e}_2 = (1/\sqrt{2}, 1/\sqrt{2}, 0)$, and $\boldsymbol{e}_3 = (0, 0, 1)$. If we rotate our original coordinate system by 45° counter-clockwise about the x_3 axis, we will be aligned with the principal axes, and the stress tensor in this new system will have the form

$$\mathsf{T}' = \begin{bmatrix} F/A & 0 & 0 \\ 0 & -F/A & 0 \\ 0 & 0 & 0 \end{bmatrix}. \tag{4.58}$$

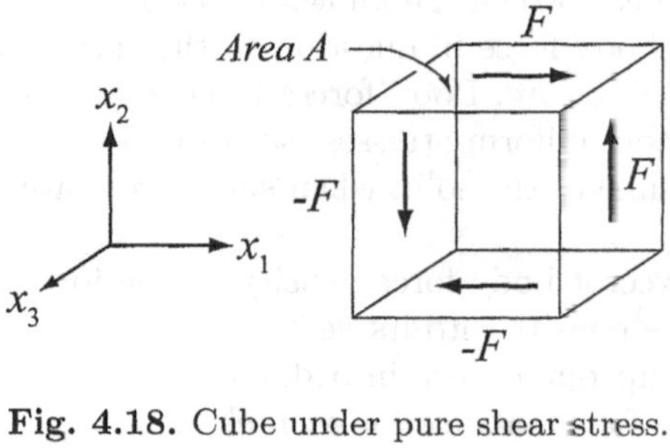

Fig. 4.18. Cube under pure shear stress.

4.6 Inhomogeneous Stresses and Arbitrary Shapes

The discussion so far has been restricted to simple differential cubes, with uniform homogeneous stresses, in the absence of body forces and torques. We now remove these constraints in order to handle more general problems.

We first consider how to handle non-uniform and inhomogeneous stress. If a (non-differential) cube of side L has a non-uniform, inhomogeneous vector stress $\boldsymbol{t}(\boldsymbol{r})$ applied to its surfaces, we can divide the cube into many smaller differential cubes, of side $d\ell$, sufficiently small that the variations in the vector stress can be neglected, to first order. When the small cubes share a surface with the larger cube, the shared surface takes on the same, local, surface stress. When the smaller cubes share an (internal) common surface, the stress exerted on one cube's face must be equal and opposite to that on its neighbor's face, exactly in the manner that we discussed for determining the internal stress and strain through the use of imaginary sections (see Sect. 4.6 and 4.7).

Hence, all the results given in Sect. 4.4 can be applied to each differential cube. The stress tensor T, defined for each differential cube at its center $\boldsymbol{r}$, now becomes a function of position within the larger, non-uniformly stressed cube, $\mathsf{T}(\boldsymbol{r})$. The stress eigenvalues, principal axes, and stress invariants are all now position-dependent, defined for each differential cubic element, varying smoothly with position within the solid.

The extension to problems with shapes other than simple cubes is handled in exactly the same fashion; the volume is divided up into differential cubic elements, and the surface stress applied to the surfaces of the object are merely distributed as tensile and shear stresses on the surfaces of the differential elements.

4.7 Body Forces

We have so far only discussed surface forces. We would also like to include forces that act on the volume of our object, which are known as *body forces*. Body forces are treated by adding the body force in question to the surface stresses in order to determine static equilibrium. Body forces in general can only be balanced by inhomogeneous, non-uniform stresses, so we expect to find stresses that vary through the volume of the solid when such forces are present.

Consider an object subjected to a vector body force density $\boldsymbol{f}$, the force per unit volume. We break up the force density into its vector components, $\boldsymbol{f} = f_1\hat{\boldsymbol{x}}_1 + f_2\hat{\boldsymbol{x}}_2 + f_3\hat{\boldsymbol{x}}_3$. We divide up our object into differential cubic elements, each of side $d\ell$, and write the force balance for each element.

Consider a particular element, shown in cross-section in Fig. 4.19. The stress will now vary with position, so we expand the stress tensor in a Taylor series as a function of position within the solid. The force balance along each coordinate axis for a cube of side $d\ell$ is given by

$$\begin{aligned} x_1 : \quad & \left(T_{11} + \frac{\partial T_{11}}{\partial x_1}\,d\ell - T_{11}\right)d\ell^2 + \left(T_{21} + \frac{\partial T_{21}}{\partial x_2}\,d\ell - T_{21}\right)d\ell^2 \\ & \left(T_{31} + \frac{\partial T_{31}}{\partial x_3}\,d\ell - T_{31}\right)d\ell^2 + f_1 d\ell^3 = 0, \end{aligned} \tag{4.59}$$

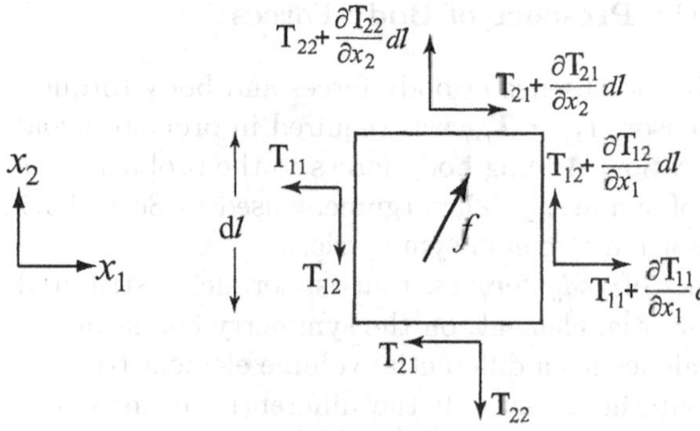

Fig. 4.19. Variation of stress due to presence of body force density $\boldsymbol{f}$.

$$x_2: \quad \left(T_{12} + \frac{\partial T_{12}}{\partial x_1}\,\mathrm{d}\ell - T_{12}\right)\mathrm{d}\ell^2 + \left(T_{22} + \frac{\partial T_{22}}{\partial x_2}\,\mathrm{d}\ell - T_{22}\right)\mathrm{d}\ell^2$$

$$\left(T_{32} + \frac{\partial T_{32}}{\partial x_3}\,\mathrm{d}\ell - T_{32}\right)\mathrm{d}\ell^2 + f_2\mathrm{d}\ell^3 = 0, \tag{4.60}$$

$$x_3: \quad \left(T_{13} + \frac{\partial T_{13}}{\partial x_1}\mathrm{d}\ell - T_{13}\right)\mathrm{d}\ell^2 + \left(T_{23} + \frac{\partial T_{23}}{\partial x_2}\mathrm{d}\ell - T_{23}\right)\mathrm{d}\ell^2$$

$$\left(T_{33} + \frac{\partial T_{33}}{\partial x_3}\mathrm{d}\ell - T_{33}\right)\mathrm{d}\ell^2 + f_3\mathrm{d}\ell^3 = 0, \tag{4.61}$$

where the stresses are multiplied by the surface area $(\mathrm{d}\ell)^2$ and the body force density by the volume $(\mathrm{d}\ell)^3$ of the differential cube. Eliminating common terms and simplifying, we find the result

$$\left.\begin{aligned}
\frac{\partial T_{11}}{\partial x_1} + \frac{\partial T_{21}}{\partial x_2} + \frac{\partial T_{31}}{\partial x_3} + f_1 = 0,\\
\frac{\partial T_{12}}{\partial x_1} + \frac{\partial T_{22}}{\partial x_2} + \frac{\partial T_{32}}{\partial x_3} + f_2 = 0,\\
\frac{\partial T_{13}}{\partial x_1} + \frac{\partial T_{23}}{\partial x_2} + \frac{\partial T_{33}}{\partial x_3} + f_3 = 0.
\end{aligned}\right\} \tag{4.62}$$

The stress tensor is assumed to retain its symmetry $T_{ij} = T_{ji}$ (see following section), so we can write this equation in a more compact form:

$$\sum_{j=1}^{3} \frac{\partial T_{ji}}{\partial x_j} + f_i = 0 \quad (i = 1 \text{ to } 3). \tag{4.63}$$

This can be written in vector form as

$$\nabla \cdot \mathsf{T}(\boldsymbol{r}) + \boldsymbol{f}(\boldsymbol{r}) = 0. \tag{4.64}$$

This differential equation can then be integrated to find the stress T that will balance the body force density $\boldsymbol{f}$.

4.7.1 Torque Balance in the Presence of Body Forces

Previously we showed that, in the absence of body forces and body torques, the symmetry of the stress tensor, $T_{ij} = T_{ji}$, was required in order to avoid having infinite angular acceleration. Adding body forces to the problem does not change the requirement of symmetry. The argument used in Sect. 4.4.1 will show that the stress tensor must remain symmetric.

Here we consider the effect of *body torques*, that is, torques distributed through the volume of a differential element, on the symmetry condition.

We consider the torque balance for a differential volume element,treating the balance for rotation about the x_3 axis. If the differential element has a body torque density g_3 about the x_3 axis, then the total torque on the element is $G_3 = g_3(\mathrm{d}\ell)^3$, for a volume $(\mathrm{d}\ell)^3$. The body torque must be added to the torque N_3 from the surface forces. The angular acceleration $\ddot{\theta}$ then satisfies the dynamic equation adapted from (4.40),

$$I_3 \frac{\mathrm{d}^2\theta}{\mathrm{d}t^2} = (\mathrm{d}\ell)^3(T_{21} - T_{12} + g_3), \tag{4.65}$$

with moment of inertia for the cube given by $I_3 = \rho(\mathrm{d}\ell)^5/6$. The angular acceleration is infinite in the limit $\mathrm{d}\ell \to 0$, unless there is a torque balance, now given by the relation $T_{21} - T_{12} + g_3 = 0$. Hence the presence of a body torque *does* cause an asymmetry in the stress tensor, making $T_{21} \neq T_{12}$. There are a small number of systems in which pure body torques can be generated, such as ferroelectric and ferromagnetic materials that include a permanent electric polarization and magnetization, respectively [15]. We will however assume that the body torque densities are always negligible in comparison to the typically much larger stresses associated with mechanical deformation, and therefore we will always take the stress tensor to be symmetric.

Example 4.10: Supported Cube. We consider an example consisting of a cube of side L with body force density $\boldsymbol{f} = \rho g \hat{\boldsymbol{x}}_3$, due to gravity g. Our coordinate system has its origin at the center of the top surface of the cube, with gravity along x_3. The cube is supported on its bottom surface (at $x_3 = L$) by a uniform vector stress $\boldsymbol{t} = -t\hat{\boldsymbol{x}}_3$, as sketched in Fig. 4.20. For static equilibrium, we require that the support force tL^2 be equal the weight, so $tL^2 = \rho g L^3$, or $t = \rho g L$. Hence the stress on the bottom surface of the cube is not in balance with the stress on the top surface (assumed zero), as expected from examining either (4.63) or (4.64).

We can calculate the stress distribution in the cube. We first do this by using intuitive considerations. Consider a section of the cube, taken at distance z from the top surface, with thickness $\mathrm{d}z$ (see Fig. 4.21). The section has mass $dm = \rho g L^2 dz$, so the stress on the bottom surface must exceed that on the top by $dt = \rho g dz$. We can then integrate the stress from the top surface, where $t = 0$:

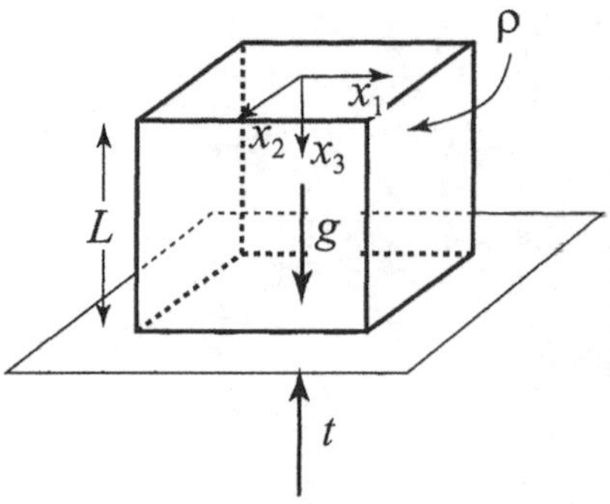

Fig. 4.20. Cube of side L and density ρ, acted upon by gravity while supported from the bottom by a uniform stress t.

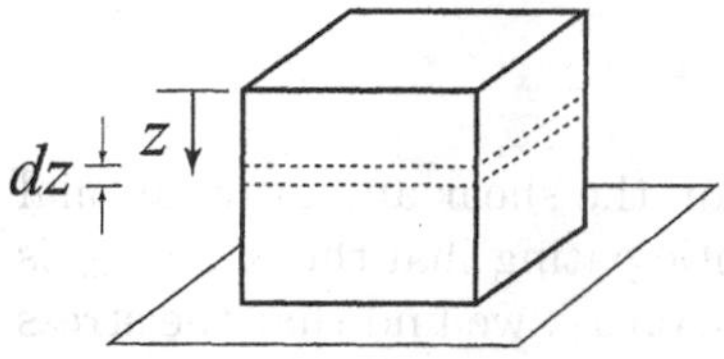

Fig. 4.21. Definition of height z and thickness dz for a supported cube.

$$
\begin{aligned}
t(z) &= \int_0^z \frac{\mathrm{d}t}{\mathrm{d}z}\,\mathrm{d}z \\
&= \int_0^z \rho g\,\mathrm{d}z \\
&= \rho g z,
\end{aligned}
\tag{4.66}
$$

yielding the correct result at the cube bottom where $z = L$.

We can do the same calculation using the formal relations (4.63). The equation along x_3 is

$$
\frac{\partial T_{31}}{\partial x_1} + \frac{\partial T_{32}}{\partial x_2} + \frac{\partial T_{33}}{\partial x_3} + \rho g = 0 \tag{4.67}
$$

The symmetry implies that the derivatives with respect to x_1 and x_2 must vanish, leaving us with a straightforward integration for T_{33},

$$
T_{33}(z) = -\rho g z, \tag{4.68}
$$

identical to (4.66) when the stress tensor is projected onto the bottom surface.

Example 4.11: Cube Supported by Shear Stress. We now consider the situation where the same cube, of density ρ and acted upon by gravity, is supported by equal shear stresses applied to the two faces perpendicular to the x_1 axis, in the $\hat{x}_3$ direction, as shown in Fig. 4.22. The overall force

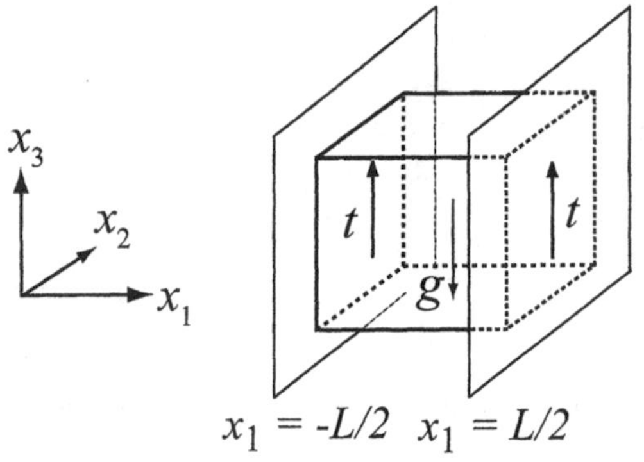

Fig. 4.22. Example 4.11.

balance is the same, where to avoid acceleration the shear and gravitational forces must balance. Referring to (4.67), and anticipating that the stress T_{33} is zero, and assuming that there is no dependence on x_2, we find that the stress $T_{13}(x_1) = \rho g x_1 + c$, where c is a constant of integration. The symmetry of the stress tensor gives $T_{31}(x_1) = T_{13}(x_1) = \rho g x_1 + c$ as well. Symmetry with respect to x_1 implies that $T_{13}(-L/2) = T_{13}(L/2)$, so that $c = 0$. Projecting the stress T_{13} on the end surfaces, we find an in-plane stress $t = -\rho g L/2$ at $x_1 = -L/2$ and $t = \rho g L/2$ at $x_1 = L/2$, for a total supporting force $F = 2tL^2 = \rho g L^3$ as expected. Note that the negative stress on the left side is in accordance with the sign convention discussed in Sect. 4.4.

In this chapter, we have introduced the formal description for both the stress and strain tensors. We have shown how the strain tensor is related to the displacements within a solid, and given examples of how to go from the displacement to an expression for the strain, and also how to do the reverse, taking the strain tensor and calculating the displacement. We then showed how, in a system in static equilibrium, the stress tensor can be calculated from the forces acting on the object, both from forces applied to the surfaces of the object, and from body forces acting within its volume. In the next chapter, we describe the formal theory that connects the stress to the strain, known as *elasticity theory.*

Exercises

4.1 Given the displacement vector $\boldsymbol{u} = (x_1^2 x_2, x_3, x_3^2)^T$, find expressions for the strain tensor S and rotation tensor Ω.

4.2 Show that the rotations given in Example 4.4 indeed rotate the laboratory axes into the eigenvectors in that example. Find the corresponding rotation tensor. Find the form for the tensor T in the eigenvector coordinate frame by applying the rotation tensor to the tensor as given in the example.

4.3 A displacement vector field with the dependence shown in Example 4.4, which has the form

$$u_i = \sum_{j=1}^{3} T_{ij} x_j, \tag{4.69}$$

where T is a constant matrix, is known as an *affine deformation*. The strain tensor for such a deformation is clearly constant through the solid. Show that a plane in a solid remains a plane after such a deformation, and that a straight line remains a straight line.

4.4 A solid bar with dimensions $L \times w \times w$ is stretched along its length to a final length $1.1L$. During the deformation its total volume remains fixed. Find the displacement vector field $\boldsymbol{u}$ and the strain tensor S.

4.5 A cylindrical rod of length L and radius r is twisted along its cylindrical axis to give a total twist angle of ϕ. Find the displacement vector field $\boldsymbol{u}$ and strain tensor S.

4.6 A solid is subjected to stresses as shown in Fig. 4.23. Indicate the indices for each of the stress components, and whether the stresses should be positive or negative.

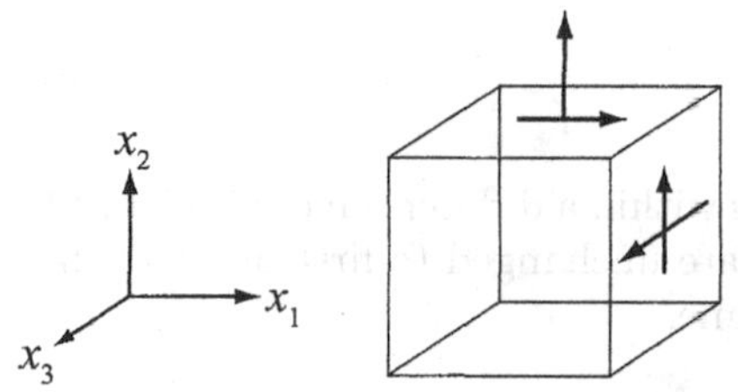

Fig. 4.23. Exercise 2.1

4.7 A uniform compressive force F is applied to the two surfaces of a cube perpendicular to the x_1 axis, and a uniform tensile force $2F$ applied to the surfaces perpendicular to the x_2 axis. Write down the stress tensor, for a cube of side L.

4.8 The cube of Exercise 2.2 is in addition subjected to a shear force F on the plane perpendicular to the x_3 axis, in the $\hat{x}_2$ direction. Write down the stress tensor.

4.9 A solid is subjected to the stress given by (4.70). Find the three stress invariants and the three principal values of the stress. Solve for the directions of the three principal axes.

$$\mathsf{T} = \begin{bmatrix} 1 & 1 & 0 \\ 1 & 1 & 0 \\ 0 & 0 & 3 \end{bmatrix} N/m^2. \tag{4.70}$$

4.10 A solid is subjected to the stress given by (4.71). Find the expression for the stress tensor if the coordinate axes are rotated by 60° counterclockwise about the x_3 axis.

$$\mathsf{T} = \begin{bmatrix} 2 & 1 & 0 \\ 1 & 0 & 0 \\ 0 & 0 & 4 \end{bmatrix} N/m^2. \tag{4.71}$$

4.11 A solid undergoes a stress such that the stress tensor is a constant p multiplied by the identity tensor, i.e.

$$\mathsf{T} = p \begin{bmatrix} 1 & 0 & 0 \\ 0 & 1 & 0 \\ 0 & 0 & 1 \end{bmatrix} . \tag{4.72}$$

Show that this is the form of the stress tensor under all rotations. This type of stress in known as *hydrostatic stress.*

4.12 A solid is stressed according to the tensor given by (4.73). Show that for this form, the stress tensor is invariant under rotations about the x_3 axis. Is it also invariant for rotations about other axes?

$$\mathsf{T} = \begin{bmatrix} a & 0 & 0 \\ 0 & a & 0 \\ 0 & 0 & b \end{bmatrix} . \tag{4.73}$$

4.13 Show that the properties of the stress within a differential cubic element (symmetry, homogeneity and uniformity) are unchanged to first order in the presence of a body torque $\boldsymbol{\nu}$ per unit volume.

5. Elasticity Relations

In 1676 Robert Hooke published his observations on the extensions of metal springs under tension, as an anagram at the end of a lecture describing a new type of telescope. The anagram, "*ceiiinosssttuu*", when decoded, yields the Latin phrase *ut tensio sic vis*. In 1678 he published a full lecture on the topic, summarized as "The Power of any Spring is in the same proportion with the Tension thereof", or that extension is proportional to strain [46]. This one scientific study, which is now enshrined as Hooke's law, is the only for which Hooke is so honored. His extensive work in mechanics, astronomy, geology, optics, and other matters (such as the incidental invention of the universal joint), go mostly unmentioned.

Hooke's formulation of the relation between stress and strain holds for many materials for small strains, typically up to of order a few tenths of a percent. Deviations begin to appear for larger strains, as one begins to generate dislocations and other irreversible changes in the material. Plastic deformation, brittle fracture, and hysteresis, all represent deviations from Hooke's law.

We begin this chapter by roughly classifying the various types of elastic response of materials to stress, and then turn to a formulation of the linear theory, known as the "generalized Hooke's law". We end the chapter by listing the form and specific values of the elastic moduli for a small number of pure materials

5.1 Classes of Elastic Response

Materials are generally classified as responding to stress with a strain that is either uniquely determined by the stress, i.e. is in a $1:1$ relation with the stress, or is multiply valued and hysteretic. Materials that fall in the former class are known as *elastic* materials, while the latter are known as *anelastic* materials. These two classes of response are sketched in Fig. 5.1. The looped, hysteretic curve characteristic of an anelastic response depends, in most anelastic materials, on the time scale over which the stress is applied: If a stress is applied suddenly to an anelastic material, such as a glass or rubber, there is a rapid response in the strain, followed by a slow relaxation towards the ultimate equilibrium strain. The size of the hysteresis loop shown in

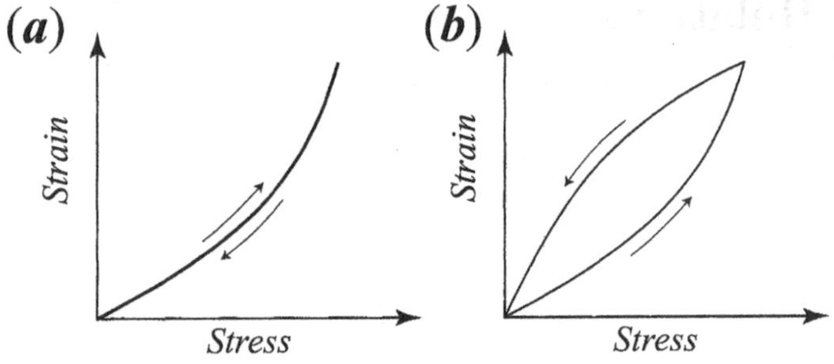

Fig. 5.1. (**a**) Perfectly elastic response, (**b**) anelastic response.

Fig. 5.1 therefore depends on the time over which the cyclic strain is applied. For a very rapidly varying strain the loop is large, as the stress is always far from the ultimate equilibrium, while for a very slowly varying strain the loop is small, as the strain is always close to equilibrium.

Elastic materials are further classified as *elastic linear* and *elastic nonlinear*, where the former applies if stress and strain are strictly proportional to one another, and the latter if not. All materials are somewhat nonlinear for large strains, although the deviation from linearity is often quite small. Elastic materials always recover their original shape and size when the strain is released.

Ductile materials, such as metals, are elastic for small strains, but can be deformed permanently with sufficient imposed strain. These typically have a linear elastic response for low strains, then exhibit *plastic deformation*, typified by a nonlinear and hysteretic response, where in addition permanent changes in the shape and size are exhibited. The state of strain in the object thus depends not only on the existing applied strain, but depends on the history of the material as well. An example is shown in Fig. 5.2. Cycles of plastic deformation such as the one shown in this figure typically lead to *strain hardening*, where the elastic response becomes stiffer with the load cycling, until the material fails.

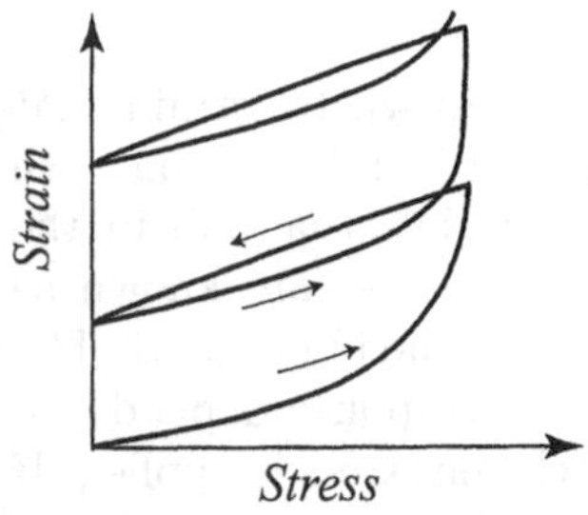

Fig. 5.2. Stress-strain curves for a solid undergoing elastic and plastic deformation.

A large class of materials, are *brittle*, where the response is essentially linear elastic up to the breaking or *fracture point*, beyond which the material fails catastrophically. Typically a crack begins at an irregularity that supports larger-than-average strain. Measured brittle fracture points vary widely for nominally identical samples; the detailed mechanics of brittle fracture depend on crack propagation, which can be arrested or aided by the presence of small numbers of defects. Many of the covalently bonded materials, such as diamond, silicon and germanium, fall into the class of brittle materials.

Some organic materials, especially some types of rubber, as well as proteins such as *titin*, exhibit repeatable, anelastic behavior, where the load cycle returns to the start point but exhibits hysteresis and strong nonlinearity. An example is shown in Fig. 5.3, where the elongation of a single titin molecule is shown as a function of the applied stress. These measurements are taken using an atomic force microscope, where a cantilevered force sensor is attached to one end of the molecule of titin. The other end of the molecule is attached to a plate whose distance from the cantilever is increased at a constant rate, and the force imposed on the cantilever recorded as a function of the strain on the molecule (see [47, 48]). The remarkable sawtoothed hysteresis results from the unravelling of pieces of tightly coiled protein, repeated segments of which make up the titin molecule.

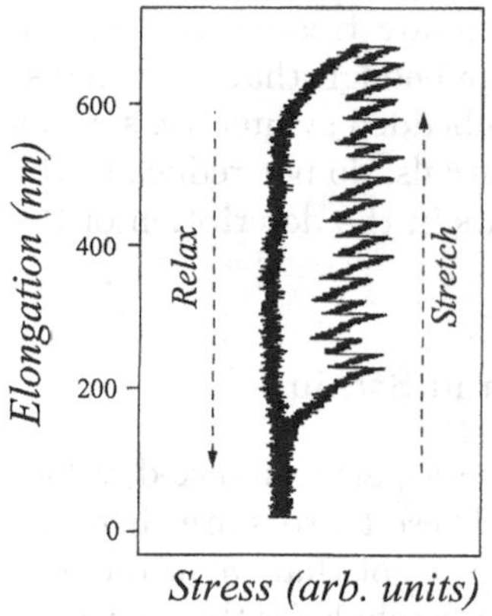

Fig. 5.3. Load curve for a single molecule of titin. Figure courtesy M. Viani and P. Hansma.

5.2 Linear Elastic Response

We now turn to a description of the formal theory for linear elastic materials. In Sect. 2.8 we described the elements of this theory, in which the energy associated with a strain is intimately related to the form for the interatomic potential. From consideration of the form for this energy, we defined the six-vector form for the stress, in (2.115).

The most general linear relation relating stress T to strain S is given by

$$T_{ij} = \sum_{k=1}^{3}\sum_{l=1}^{3} \alpha_{ijkl} S_{kl}, \tag{5.1}$$

where the elastic moduli α_{ijkl} are constants. As written, the α_{ijkl} form a fourth-rank tensor, with a coordinate transformation that can be extrapolated from the formulations given in Sect. A.1.2. The values of the α_{ijkl} can vary from point to point within a solid; we will however always assume the material is *homogeneous*, so that the elastic moduli are independent of position within the solid. Note that this does not restrict the shape of the object, merely its material makeup.

There are $3^4 = 81$ distinct values in the moduli α_{ijkl}. However, the stress tensor T and the strain tensor S are both symmetric, as discussed in Chap. 4. This imposes symmetries on the elastic moduli, so that α_{ijkl} must be symmetric under interchanges of i and j as well as k and l. This reduces the number of independent values to 36. Further reductions (to a minimum of two) can be made if the material has certain symmetries; the minimum of two is achieved for an *isotropic solid*, with full rotational and inversion symmetry, as we discuss in Sect. 5.3. Many macroscopic materials such as polycrystalline metals, disordered organic solids, and some glasses are treated as isotropic materials when the relevant length scales are large enough that the microscopic variations average out. Materials with an embedded symmetry, such as wood, crystalline solids, or organized organic materials, do not reduce to the simplest form, and one must retain more constants in the description of the stress–strain relation.

5.2.1 Six-Vector Representation of Stress and Strain

The tensors T and S are symmetric, so the nine separate indexed values T_{ij} and S_{ij} include only six independent values. These tensors are therefore commonly written as vectors with six elements, a concept that we introduced in Chap. 2. To avoid confusion, we will use different symbols for the six-vector as opposed to the tensor form for both the stress and the strain.

The six-vector form for the stress will be written as τ_i, and the strain as ϵ_i. The correspondence between the tensor indices and the six-vector indices are defined for the stress tensor by

$$\left.\begin{aligned} &\tau_1 = T_{11}, \quad \tau_2 = T_{22}, \quad \tau_3 = T_{33}, \\ &\tau_4 = T_{12}, \quad \tau_5 = T_{13}, \quad \tau_6 = T_{23}. \end{aligned}\right\} \tag{5.2}$$

Written out in array format, this is

$$\mathsf{T} = \begin{bmatrix} \tau_1 & \tau_6 & \tau_5 \\ \tau_6 & \tau_2 & \tau_4 \\ \tau_5 & \tau_4 & \tau_3 \end{bmatrix} . \tag{5.3}$$

An equivalent form is defined for the strain tensor, with an important (and annoying) additional factor of two in some of the terms:

$$\left.\begin{aligned} &\epsilon_1 = S_{11} = \frac{\partial u_1}{\partial x_1}, \quad \epsilon_2 = S_{22} = \frac{\partial u_2}{\partial x_2}, \quad \epsilon_3 = S_{33} = \frac{\partial u_3}{\partial x_3}, \\ &\epsilon_4 = 2S_{23} = \frac{\partial u_2}{\partial x_3} + \frac{\partial u_3}{\partial x_2}, \quad \epsilon_5 = 2S_{13} = \frac{\partial u_1}{\partial x_3} + \frac{\partial u_3}{\partial x_1}, \\ &\epsilon_6 = 2S_{12} = \frac{\partial u_1}{\partial x_2} + \frac{\partial u_2}{\partial x_1}. \end{aligned}\right\} \tag{5.4}$$

Written out in array format, this is

$$\mathsf{S} = \begin{bmatrix} \epsilon_1 & \epsilon_6/2 & \epsilon_5/2 \\ \epsilon_6/2 & \epsilon_2 & \epsilon_4/2 \\ \epsilon_5/2 & \epsilon_4/2 & \epsilon_3 \end{bmatrix} . \tag{5.5}$$

In (5.4), we have also given the definition in terms of the displacement vector $\boldsymbol{u}$. The factor of two in the definitions for the stress six-vectors is historical, and is too engrained to risk an attempt to change it.

5.2.2 Elastic Stiffness

Using the definitions above, we can re-write the elasticity relation (5.1) as one between the stress and strain six-vectors, using a two-dimensional, 6 × 6 elasticity matrix with elements c_{ij}:

$$\tau_i = \sum_{j=1}^{6} c_{ij}\, \epsilon_j \qquad (i = 1 \text{ to } 6). \tag{5.6}$$

The constants c_{ij} are called the *elastic stiffness* coefficients. In *Systéme International* (SI) units, the constants c_{ij} have dimensions of $\mathrm{N/m^2}$. Note that in this notational form the c_{ij} do not form a tensor, as they do not transform correctly under coordinate transformations. One can however relate the constants c_{ij} to the linear coefficients α_{ijkl}, and thereby find the transformation rules; we leave this as an exercise. We note that the vector form of the linear response can be written as

$$\tau = \mathsf{c} \cdot \epsilon. \tag{5.7}$$

5.3 Orthotropic and Isotropic Materials

We now explore how the symmetries of the material can be used to simplify the number of independent elastic moduli. The form given by (5.6) allows 36 independent values for the c_{ij}, as we had found for the moduli α_{ijkl}. Many materials can be described by significantly fewer constants.

We first assume that our material is *inversion symmetric* about the $x_1 - x_2$ plane, that is, it does not change for a mirror reflection through the $x_1 - x_2$ plane. The relations between the stress and strain must then also remain unchanged under this transformation. A mirror inversion about this plane, to a new set of axes x_i', is given by the coordinate transformation tensor

$$\mathsf{R} = \begin{bmatrix} 1 & 0 & 0 \\ 0 & 1 & 0 \\ 0 & 0 & -1 \end{bmatrix}. \tag{5.8}$$

We apply this transformation to the stress and strain tensors, so that $\mathsf{T}' = \mathsf{R}\mathsf{T}\mathsf{R}^T$, and similarly for the stress tensor, $\mathsf{S}' = \mathsf{R}\mathsf{S}\mathsf{R}^T$. The inversion symmetry gives $\mathsf{T}' = \mathsf{T}$, and $\mathsf{S}' = \mathsf{S}$. In the six-vector form, this forces the equalities

$$\left.\begin{aligned} \tau_1' &= \tau_1, \quad \tau_2' = \tau_2, \quad \tau_3' = \tau_3, \\ \tau_4' &= -\tau_4, \quad \tau_5' = -\tau_5, \quad \tau_6' = \tau_6. \end{aligned}\right\} \tag{5.9}$$

Identical relations hold for the strain six-vector. Writing out the linear relation between the stress and strain tensors, in the two coordinate systems, we find the relations for τ_1 and τ_1' in terms of the strain six-vector ϵ in the original axes:

$$\tau_1 = c_{11}\epsilon_1 + c_{12}\epsilon_2 + c_{13}\epsilon_3 + c_{14}\epsilon_4 + c_{15}\epsilon_5 + c_{16}\epsilon_6, \tag{5.10}$$

$$\tau_1' = c_{11}'\epsilon_1 + c_{12}'\epsilon_2 + c_{13}'\epsilon_3 - c_{14}'\epsilon_4 - c_{15}'\epsilon_5 + c_{16}'\epsilon_6. \tag{5.11}$$

As the material itself is assumed invariant under this rotation, the constants c_{ij}' and c_{ij} must be the same. Comparing the expressions (5.10) and (5.11), and using (5.9), we see that we must have $c_{14} = c_{15} = 0$. Similar comparisons may be made for the other five stress components in the two coordinate systems, which yields a matrix of elastic moduli with the form

$$\mathsf{c} = \begin{bmatrix} c_{11} & c_{12} & c_{13} & 0 & 0 & c_{16} \\ c_{21} & c_{22} & c_{23} & 0 & 0 & c_{26} \\ c_{31} & c_{32} & c_{33} & 0 & 0 & c_{36} \\ 0 & 0 & 0 & c_{44} & c_{45} & 0 \\ 0 & 0 & 0 & c_{54} & c_{55} & 0 \\ c_{61} & c_{62} & c_{63} & 0 & 0 & c_{66} \end{bmatrix}. \tag{5.12}$$

This is the form for the moduli in a material with a single mirror symmetry.

We now assume that our material is inversion symmetric for the $x_2 - x_3$ and $x_1 - x_3$ planes as well. Applying the same type of reasoning for these inversion symmetries, we find that the matrix of elastic moduli simplifies to

$$\mathsf{c} = \begin{bmatrix} c_{11} & c_{12} & c_{13} & 0 & 0 & 0 \\ c_{21} & c_{22} & c_{23} & 0 & 0 & 0 \\ c_{31} & c_{32} & c_{33} & 0 & 0 & 0 \\ 0 & 0 & 0 & c_{44} & 0 & 0 \\ 0 & 0 & 0 & 0 & c_{55} & 0 \\ 0 & 0 & 0 & 0 & 0 & c_{66} \end{bmatrix}. \tag{5.13}$$

A material that is mirror symmetric about all three planes, $x_1 - x_2$, $x_2 - x_3$ and $x_1 - x_3$, is known as an *orthotropic* material, and will have elastic constants having the form given by (5.13); all monatomic cubic crystals, including those with fcc and bcc crystal lattices, as well as the diamond structure (see Chap. 2), are orthotropic. This simplified form therefore applies to a large number of materials. We note that certain common semiconductor materials are not mirror symmetric, such as GaAs, GaN and other III – V and II – VI binary compounds.

We now proceed from the elastic constants for an orthotropic material to the limit of a homogeneous, isotropic material. We add to the three mirror symmetries a symmetry for independent 90° rotations about the x_1 and x_3 axes, and finally an independent 45° rotation about the x_3 axis. Materials that are symmetric for the three mirror symmetries, and these three particular rotations, are mechanically equivalent to an isotropic material, that is, one that is the same when viewed in any orientation. The elastic moduli for these types of material have the form

$$\mathsf{c} = \begin{bmatrix} c_{11} & c_{12} & c_{12} & 0 & 0 & 0 \\ c_{12} & c_{11} & c_{12} & 0 & 0 & 0 \\ c_{12} & c_{12} & c_{11} & 0 & 0 & 0 \\ 0 & 0 & 0 & c_{44} & 0 & 0 \\ 0 & 0 & 0 & 0 & c_{44} & 0 \\ 0 & 0 & 0 & 0 & 0 & c_{44} \end{bmatrix}, \tag{5.14}$$

where $c_{44} = (c_{11} - c_{12})/2$; there are therefore only two independent elastic constants.

The two elastic constants required to describe the entire matrix date back to the mid-19th century. The moduli are traditionally referred to as the Lamé constants λ and μ, introduced by G. Lamé (1798-1870), and are defined as

$$\lambda = c_{12}, \tag{5.15}$$
$$\mu = c_{44} = (c_{11} - c_{12})/2. \tag{5.16}$$

Using these constants the elastic moduli may be written in matrix form as

$$\mathsf{c} = \begin{bmatrix} 2\mu + \lambda & \lambda & \lambda & 0 & 0 & 0 \\ \lambda & 2\mu + \lambda & \lambda & 0 & 0 & 0 \\ \lambda & \lambda & 2\mu + \lambda & 0 & 0 & 0 \\ 0 & 0 & 0 & \mu & 0 & 0 \\ 0 & 0 & 0 & 0 & \mu & 0 \\ 0 & 0 & 0 & 0 & 0 & \mu \end{bmatrix}. \tag{5.17}$$

The number of materials that meet the conditions of true isotropy are actually quite small. Polycrystalline materials, viewed on a size scale much larger than the crystalline grain size, are in general isotropic. Hence most metals and disordered plastics are isotropic on size scales of 100 μm or larger. However, the manner in which the materials are fabricated quite often introduces some anisotropy. An example is sheet or plate metal, which acquires an asymmetry due to the rolled extrusion process; many polycrystalline thin films, due to in-plane stress, also exhibit anisotropy. However, the simplifications awarded by the assumption of isotropy are so great that one quite often uses this model, even for materials that are quite anisotropic.

Example 5.1: Stress for an Extended Bar. Consider a bar made of an isotropic material with Lamé constants λ and μ, whose length along x_1 has been increased by an external stress from ℓ to $\ell + \Delta\ell$. What is the stress required to achieve this extension?

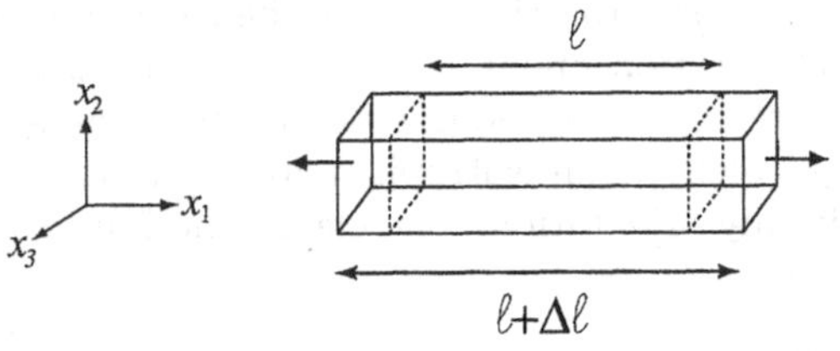

Fig. 5.4. Example 5.1.

The strain six-vector consists of a single non-zero entry, with $\epsilon_1 = \Delta\ell/\ell$; this was discussed in Example 4.1. The elastic moduli for an isotropic material have the form given by (5.17). Therefore the stress six-vector has the form

$$\tau = ((2\mu + \lambda), \lambda, \lambda, 0, 0, 0)\ \frac{\Delta\ell}{\ell}. \tag{5.18}$$

The stress τ_1 must be applied normal to the ends of the bar, along the x_1 axis, i.e. one primarily pulls along the length with a stress proportional to the strain (as discovered by Hooke). We however find that we must in addition apply additional extensional stresses τ_2 and τ_3 along the x_2 and x_3 axes, in order to prevent a reduction in the bar's transverse dimensions when τ_1 is applied.

Example 5.2: Stress for an Sheared Bar. Consider now a bar made of an isotropic material under pure shear, as shown in Fig. 5.5. The bar is sheared along the x_1 axis in the x_2 direction, for a total shear of $\Delta\ell$ over a length ℓ; this was discussed in Sect. 4.2.

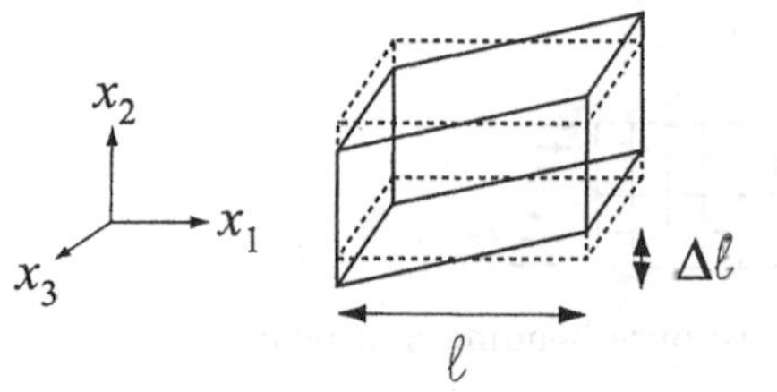

Fig. 5.5. Example 5.2.

The strain six-vector has the form $\epsilon = (0,0,0,0,0,\Delta\ell/\ell)$ (note one must take care with the factor of two). As for Example 5.1, the stress is given by (5.17), so the stress six-vector is given by $\tau = (0,0,0,0,0,\mu\Delta\ell/\ell)$. For shear displacement, one needs to apply a shear force in the planes normal to the x_1 axes, in the x_2 direction, with magnitude proportional to the total shear strain. From the symmetry of the stress tensor, an equal shear force normal to x_2, in the x_1 direction, is also required

Example 5.3: Bar under Pure Bending. Our last example is the same as in Example 4.3. A beam of length L is bent in the $x-y$ plane, shown in Fig. 5.6, under a strain six-vector given by

$$\epsilon = \left(-2\frac{y}{L^2}\,\Delta y, 0, 0, 0, 0, 0\right). \tag{5.19}$$

The corresponding stress for an isotropic material is given by

$$\tau = -2\frac{y}{L^2}\,\Delta y\;(2\mu+\lambda, \lambda, \lambda, 0, 0, 0)\,. \tag{5.20}$$

The stress, applied to either end of the beam, is compressive on the upper side and tensile on the lower side, as sketched in Fig. 5.7. Additional compressive and tensile stress must be applied along the y and z surfaces, τ_2 and τ_3, to obtain pure strain along x, as in Example 5.1.

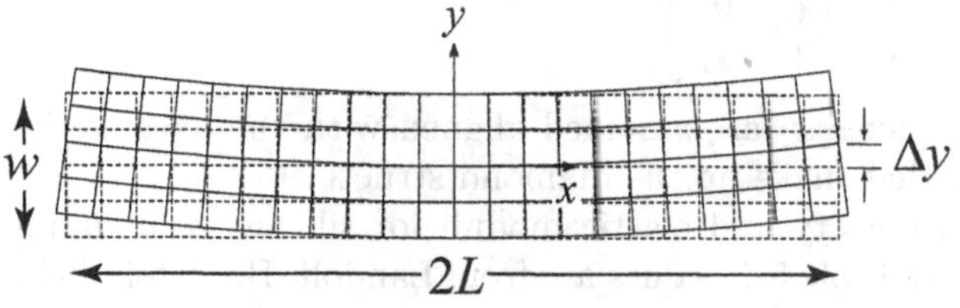

Fig. 5.6. Beam of length L, width and thickness t bent in the $x-y$ plane so that either end moves up by Δy. Solid and dotted lines are for the bent and relaxed beam, respectively.

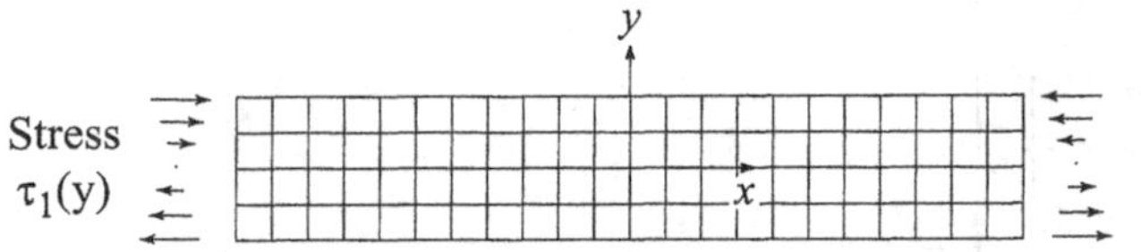

Fig. 5.7. Stress distribution needed to achieve pure bending in a beam.

5.4 Crystalline Materials

Many materials are not isotropic or orthotropic, but display more restrictive symmetries. Crystalline materials display a wealth of different symmetry classes, and correspondingly a range of forms for the elastic moduli. A summary of the types of symmetries, and the corresponding form for the elastic moduli, appears in Mason's treatise on piezoelectric materials [49]; a more comprehensive and material-specific compendium can be found in Landolt–Börnstein [50]. Here we cite the symmetries and corresponding form for the elastic stiffness matrix for a few common, semiconducting crystals.

Diamond Structure. Crystals formed from elements in column IV of the periodic table typically have what is known as the diamond structure, with carbon in its lowest energy state crystallized in the eponymous gem. The diamond structure consists of two interpenetrating face-centered cubic (fcc) lattices with origins offset by 1/4 of the cube diagonal; see Kittel [26] or Ashcroft and Mermin [1]. Silicon, germanium and carbon all have the diamond crystal structure, with T_h symmetry in the Schönflies notation, or $m3$ in the Hermann–Mauguin system. The form of the elastic stiffness coefficients is that given for the cubic system, (5.13),

$$\mathsf{c} = \begin{bmatrix} c_{11} & c_{12} & c_{12} & 0 & 0 & 0 \\ c_{12} & c_{11} & c_{12} & 0 & 0 & 0 \\ c_{12} & c_{12} & c_{11} & 0 & 0 & 0 \\ 0 & 0 & 0 & c_{44} & 0 & 0 \\ 0 & 0 & 0 & 0 & c_{44} & 0 \\ 0 & 0 & 0 & 0 & 0 & c_{44} \end{bmatrix}. \tag{5.21}$$

Note that the coordinate axes x_i are assumed aligned with the axes of the face-centered cubic cells that make up the diamond structure.

The lattice constant, density and elastic moduli for silicon, germanium and diamond are shown in Table 5.1; values are from Landolt–Börnstein [50].

Zincblende Structure. Gallium arsenide (GaAs), indium arsenide (InAs), and silicon carbide (SiC) are binary crystalline compounds. The gallium and indium compounds are very commonly used in high-frequency electronics due

Table 5.1. Crystals with the diamond structure. Note that 1 GPa = 1×10^9 N/m^2; values from Landolt–Börnstein.

Material	a_0 (Å)	ρ (g/cm^3)	c_{11} (GPa)	c_{12} (GPa)	c_{44} (GPa)
Silicon (Si)	5.4307	2.330	165	64	79.2
Germanium (Ge)	5.62	5.323	129	48	67.1
Diamond (C)	3.567	3.515	1040	170	550

to the high carrier mobilities found in these materials. They all have the zincblende structure, which is the same lattice structure as for diamond, except that each of the two interpenetrating fcc crystals are filled by a different atomic species (for GaAs, for instance, one fcc lattice is filled by Ga atoms, the other by As atoms). The symmetry for the zincblende structure is T_d in Schönflies notation, and $\bar{4}3m$ in the Hermann–Mauguin system. The form for the elastic constant matrix for zincblende is the same as for the diamond structure, given by (5.21); again, the coordinate axes x_i are assumed aligned with the cubic cells that make up the zincblende structure.

The lattice constant, density and elastic stiffness for GaAs, InAs and SiC are shown in Table 5.2 [50].

Table 5.2. Crystals with the zincblende structure. Note that 1 GPa = 1×10^9 N/m^2.

Material	a_0 (Å)	ρ (g/cm^3)	c_{11} (GPa)	c_{12} (GPa)	c_{44} (GPa)
GaAs	5.635	5.318	118	53.5	59.4
InAs	6.036	5.667	84.4	46.4	39.6
SiC	4.348	3.166	352	233	140

Wurtzite Structure. Gallium nitride (GaN), aluminum nitride (AlN), indium nitride (InN) and zinc oxide (ZnO) are binary crystalline compounds, having what is commonly known as the zinc oxide or wurtzite structure, with C_{6v} or $6mm$ symmetry. Much attention has recently been paid to the nitride compounds, due to the discovery that blue-light lasers may be fabricated using these materials [51].

The wurtzite structure is formed from two interpenetrating hexagonal close-packed lattices, each sublattice filled with one type of atom. The paired sublattices are stacked in an $ABAB$... sequence along the c axis. The form for the elastic constant matrix is given by (5.22), where the coordinate axes are chosen so that the axis with six-fold symmetry, the c axis, is along x_3.

$$c = \begin{bmatrix} c_{11} & c_{12} & c_{13} & 0 & 0 & 0 \\ c_{12} & c_{11} & c_{13} & 0 & 0 & 0 \\ c_{13} & c_{13} & c_{33} & 0 & 0 & 0 \\ 0 & 0 & 0 & c_{44} & 0 & 0 \\ 0 & 0 & 0 & 0 & c_{44} & 0 \\ 0 & 0 & 0 & 0 & 0 & \dfrac{c_{11}-c_{12}}{2} \end{bmatrix} . \tag{5.22}$$

The lattice constants, density and elastic stiffness for GaN, AlN, InN and ZnO are shown in Table 5.3 [50, 52].

Table 5.3. Crystals with the wurtzite structure. Note that 1 GPa $= 1 \times 10^9$ N/m^2.

Material	a_0 (Å)	c_0 (Å)	ρ (g/cm^3)	c_{11} (GPa)	c_{33} (GPa)	c_{44} (GPa)	c_{12} (GPa)	c_{13} (GPa)
GaN	3.189	5.185	6.095	374	379	101	106	70
AlN	3.112	4.982	3.255	345	395	118	125	120
InN	3.54	5.705	6.880	190	182	10	104	121
ZnO	3.2495	5.207	5.675	209	218	44.1	120	104

Trigonal Crystals. Quartz (SiO_2) and lithium niobate ($LiNbO_3$) are both in the trigonal crystal class. Quartz has D_3 or 32 class symmetry, and lithium niobate has C_{3v} or $3m$ symmetry. For quartz the coordinate axes are chosen with x_3 along the axis of 3-fold symmetry, for lithium niobate the axis of two-fold symmetry is along x_1. The elastic constants have the form

$$c = \begin{bmatrix} c_{11} & c_{12} & c_{13} & c_{14} & 0 & 0 \\ c_{12} & c_{11} & c_{13} & -c_{14} & 0 & 0 \\ c_{13} & c_{13} & c_{33} & 0 & 0 & 0 \\ c_{14} & -c_{14} & 0 & c_{44} & 0 & 0 \\ 0 & 0 & 0 & 0 & c_{44} & c_{14} \\ 0 & 0 & 0 & 0 & c_{14} & \dfrac{c_{11}-c_{12}}{2} \end{bmatrix} . \tag{5.23}$$

The material constants for SiO_2 and $LiNbO_3$ are given in Table 5.4 [53].

Table 5.4. Crystals with the trigonal structure. Note that 1 GPa $= 1 \times 10^9$ N/m^2.

Material	a_0 (Å)	c_0 (Å)	ρ (g/cm^3)	c_{11} (GPa)	c_{33} (GPa)	c_{44} (GPa)	c_{12} (GPa)	c_{13} (GPa)	c_{14}
SiO_2	4.9138	5.4052	2.649	86.6	106.1	57.8	6.7	12.6	−17.8
$LiNbO_3$	5.148	13.863	4.644	202	242	60.1	55	71	8.3

5.5 Strain–Stress Relations

We have discussed the relations that give stress in terms of strain, using the elastic stiffness coefficients c_{ij}. There is of course an inverse relation that gives strain in terms of stress. The inverse relation involves a different set of elastic moduli s_{ij}, which is referred to as the *elastic compliance* of the material. The relation between stress and strain using the compliance matrix is given by

$$\epsilon_i = \sum_{j=1}^{6} s_{ij}\ \tau_j. \tag{5.24}$$

The matrix s is simply the matrix inverse of stiffness coefficients c, $\mathsf{s} = \mathsf{c}^{-1}$. The metric units of the elements s_{ij} are m^2/N. Each element s_{ij} has the value given by the usual formula for the matrix inverse,

$$s_{ij} = (-1)^{i+j}\frac{\Delta_{ij}}{\Delta}, \tag{5.25}$$

where Δ is the determinant of the matrix c and Δ_{ij} is the minor of the matrix c, that is, the determinant of the 2×2 matrix formed by eliminating the ith row and jth column of c. The form for the compliance matrix, in terms of the locations of the zeroes in the matrix s for a material with a particular symmetry, turns out to be identical to the form for the stiffness matrix; see e.g. Mason [49].

The elastic properties of materials are more commonly reported in terms of the stiffness coefficients c_{ij} than the compliances s_{ij}. One important exception to this rule is for isotropic materials, where one frequently finds values for the *Young's modulus* E (also written as Y) and *Poisson's ratio* ν. The Young's modulus has units of $\mathrm{N/m}^2$, and Poisson's ratio is dimensionless. In addition the *shear modulus* G is often used for convenience, although it can be written in terms of E and ρ. These constants are given in terms of the Lamé constants λ and μ by the relations

$$\left.\begin{aligned} E &= \frac{\mu(3\lambda + 2\mu)}{\lambda + \mu}, \\ \nu &= \frac{\lambda}{2\lambda + 2\mu)}, \\ G &= 1/\mu. \end{aligned}\right\} \tag{5.26}$$

The inverse relations, giving the Lamé constants in terms of E and ν, are

$$\left.\begin{aligned} \lambda &= \frac{E\nu}{(1+\nu)(1-2\nu)}, \\ \mu &= \frac{E}{2(1+\nu)}. \end{aligned}\right\} \tag{5.27}$$

The stiffness matrix for an isotropic material is given in terms of these constants by the relation

$$\mathsf{s} = \begin{bmatrix} 1/E & -\nu/E & -\nu/E & 0 & 0 & 0 \\ -\nu/E & 1/E & -\nu/E & 0 & 0 & 0 \\ -\nu/E & -\nu/E & 1/E & 0 & 0 & 0 \\ 0 & 0 & 0 & G & 0 & 0 \\ 0 & 0 & 0 & 0 & G & 0 \\ 0 & 0 & 0 & 0 & 0 & G \end{bmatrix}. \quad (5.28)$$

Example 5.4: Strain for a Bar under Pure Tension. Consider a bar of length $\ell = 1$ μm under pure tension, with a force $F = 10$ nN ($F = 10^{-8}$ N) applied to each end of area $A = 100 \times 100$ nm^2, as shown in Fig. 5.8. The bar is isotropic, with $E = 100$ GPa and $\nu = 1/3$. The stress six-vector is given by $\boldsymbol{\tau} = (F/A, 0, 0, 0, 0, 0)^T = (1 \text{ MPa}, 0, 0, 0, 0, 0)$. Using (5.28), we can immediately write down the stress six-vector, $\boldsymbol{\epsilon} = (1, -\nu, -\nu, 0, 0, 0)\ (F/EA) = 10^{-3}(1, -1/3, -1/3, 0, 0, 0,)$. Hence the bar extends by 1 nm along its length, while each of the transverse dimensions shrinks by 1/30 of a nanometer.

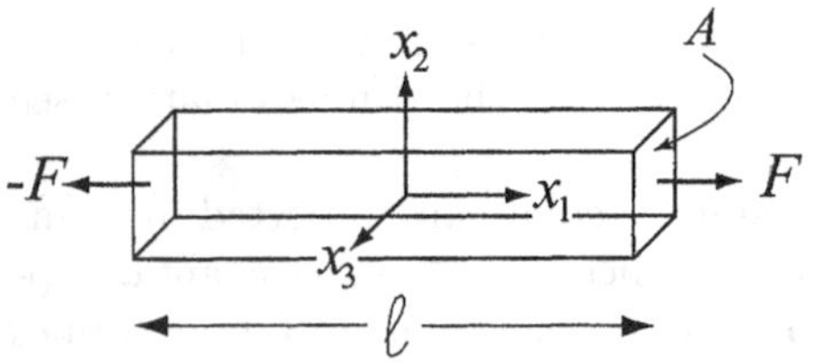

Fig. 5.8. Example 5.4.

5.6 Polycrystalline Materials

Many thin-film materials are polycrystalline in nature, with grain sizes ranging from a few nanometers in diameter to a few micrometers or larger. Within any grain these behave as single crystals, with crystal axes determined by the grain orientation. For volumes containing small numbers of grains, one must model the system as several different materials, interconnected at the grain boundaries. For large volumes, one can average the single crystal values over orientation to obtain what may be an isotropic material, if the grains have no preferred orientation. Many thin-film materials are grown or deposited with preferred grain orientations, possibly with common x_3-axis alignment perpendicular to the film surface, or with flattened grains in the film plane. These materials will have different elastic characteristics in the plane compared to perpendicular to it, and the material constants will not necessarily match the values for the bulk material.

We list in Table 5.5 the properties of some polycrystalline materials commonly used in micro- and nanofabrication. The assumption is that the material volumes are large enough to contain many grains, and that the characteristic are sufficiently close to those of an isotropic material for that approximation to hold.

Table 5.5. Elastic constants for some commonly used polycrystalline materials in micro- and nanofabrication. Note that 1 GPa = 1×10^9 N/m^2.

Material	ρ (g/cm^3)	E (GPa)	ν	Reference
poly Si	2.330	169	0.22	[54]
poly SiO_2	2.20	69 – 92	0.17	[55]
poly Si_3N_4	3.100	357	0.25	[56], [57]
Ag (amorphous)	10.49	83	0.37	
Al (amorphous)	2.70	70	0.35	
Au (amorphous)	19.32	78	0.44	
Cu (amorphous)	8.92	130	0.34	
In (amorphous)	7.31	11	0.3	
Nb (amorphous)	8.57	105	0.40	
Ni (amorphous)	8.91	200	0.31	
Ti (amorphous)	4.51	116	0.32	

5.7 Conclusion

We have described the theory for linear elastic materials, introducing the elastic stiffness and compliance matrices. We have given a few examples of how these are used to relate stress and strain in static problems. Finally, we have given the general form and the numerical values for the elastic moduli for a few common materials.

Exercises

5.1 Find the stress six-vector for an isotropic material with Lamé constants $\lambda = \mu = 100$ GPa, which is strained as described by the relative displacement vector $\boldsymbol{u} = (0.01x + 0.001y, 0.01y, 0)^T$.

5.2 A cube of material with Lamé constants $\lambda = 100$ GPa and $\mu = 75$ GPa is subjected to forces such that the cube sides reduce uniformly by 1%. If the original cube side is 0.1 cm, find the forces on each cube face and their directions.

5.3 Find the form of the elastic stiffness matrix for a material with mirror symmetry about the $x_1 - x_2$ and $x_1 - x_3$ planes.

5.4 Find the coefficients for the stiffness matrix for a material with Young's modulus $E = 100$ GPa and Poisson's ratio $\nu = 0.3$.

5.5 Show that for a solid with linear, isotropic elastic response, the principal axes for strain align with the principal axes for stress.

5.6 Derive the relations (5.26) and (5.27).

5.7 A material with elastic constants $E = 150$ GPa and $\nu = 0.25$ is subjected to a uniform hydrostatic pressure of 100 MPa. Find the resulting stress six-vector.

5.8 A rod of silicon, aligned with its length along the x_1 crystal axis, is strained along its length by 1%, with no strain in any other direction. Find the stress required to achieve this result.

5.9 Find the coordinate transformation rules for the coefficients c_{ij}, from the relations with the linear coefficients α_{ijkl} and their tensor transformation relations.

5.10 Show that the form for the elastic moduli c_{ij} for an isotropic material is given by 5.13, given the form for the elastic moduli for an orthotropic material.

6. Static Deformations of Solids

We now have enough machinery to begin to find solutions to problems involving the static deformation of elastic solids. In Chap. 7 we will discuss the dynamic behavior of solids. Most of our discussion will be restricted to the discussion of linear, isotropic solids, characterized by two material constants (we will mostly use the Young's modulus E and Poisson's ratio ν, but one could just as well use the two Lamé constants λ and μ). Extensions to the treatment of general linear elastic solids, using the 6×6 stiffness matrix c or the compliance matrix s, are possible, but involve such intricate manipulations of complex differential equations that such problems are usually relegated to numerical solutions. We will illustrate some of the differences between isotropic and non-isotropic solids in the examples.

6.1 System of Equations for a Static Deformable Solid

We begin by summarizing the various relations that we have discussed for stress, strain, and their interplay, for a static solid. We shall use both the tensor and six-vector notation for the stress and strain.

The relation between the relative displacement vector $\boldsymbol{u}(\boldsymbol{r})$ and the 3×3 strain tensor $\mathsf{S}(\boldsymbol{r})$ is given by

$$S_{ij}(\boldsymbol{r}) = \frac{1}{2}\left(\frac{\partial u_i}{\partial x_j} + \frac{\partial u_j}{\partial x_i}\right) \tag{6.1}$$

(see (4.7)). Note that the strain tensor must satisfy the compatibility relations, discussed in Appendix 2.

The 3×3 stress tensor $\mathsf{T}(\boldsymbol{r})$ is related to the body force density $\boldsymbol{f}(\boldsymbol{r})$ by

$$\nabla \cdot \mathsf{T}(\boldsymbol{r}) + \boldsymbol{f}(\boldsymbol{r}) = 0, \tag{6.2}$$

(see 4.64), or

$$\sum_{j=1}^{3} \frac{\partial T_{ij}}{\partial x_j} + f_i(\boldsymbol{r}) = 0 \qquad (i = 1 \text{ to } 3). \tag{6.3}$$

For a surface with outward-pointing normal $\hat{\boldsymbol{n}}$ and vector stress (surface force per unit area) $\boldsymbol{t}$, the stress tensor on the surface satisfies

$$\boldsymbol{t} = \mathsf{T} \cdot \hat{\boldsymbol{n}}. \tag{6.4}$$

For a stress-free boundary, then, we have $\mathsf{T} \cdot \hat{\boldsymbol{n}} = 0$.

The stress and strain are assumed to have a linear relationship, and are related by the stiffness or compliance matrices, c or s. As mentioned above, we will mostly restrict our discussion to fully isotropic solids, with a Young's modulus E and Poisson ratio ν. The stress-strain relations, given in terms of the tensor elements, are given by

$$\left.\begin{aligned}
S_{11} &= \frac{1}{E}\left[T_{11} - \nu\left(T_{22} + T_{33}\right)\right], \\
S_{22} &= \frac{1}{E}\left[T_{22} - \nu\left(T_{11} + T_{33}\right)\right], \\
S_{33} &= \frac{1}{E}\left[T_{33} - \nu\left(T_{11} + T_{22}\right)\right], \\
S_{12} &= \frac{1+\nu}{E} T_{12}, \\
S_{13} &= \frac{1+\nu}{E} T_{13}, \\
S_{23} &= \frac{1+\nu}{E} T_{23}.
\end{aligned}\right\} \tag{6.5}$$

Alternatively these can be stated in terms of the Lamé constants as

$$\left.\begin{aligned}
T_{11} &= (\lambda + 2\mu) S_{11} + \lambda S_{22} + \lambda S_{33}, \\
T_{22} &= \lambda S_{11} + (\lambda + 2\mu) S_{22} + \lambda S_{33}, \\
T_{33} &= \lambda S_{11} + \lambda S_{22} + (\lambda + 2\mu) S_{33}, \\
T_{12} &= 2\mu S_{12}, \\
T_{13} &= 2\mu S_{13}, \\
T_{23} &= 2\mu S_{23}.
\end{aligned}\right\} \tag{6.6}$$

For an isotropic solid, we can combine (6.1) and (6.5) to yield the expression (6.7):

$$
\left.
\begin{aligned}
\frac{\partial u_1}{\partial x_1} &= \frac{1}{E}\left(T_{11} - \nu(T_{22} + T_{33})\right), \\
\frac{\partial u_2}{\partial x_2} &= \frac{1}{E}\left(T_{22} - \nu(T_{11} + T_{33})\right), \\
\frac{\partial u_3}{\partial x_3} &= \frac{1}{E}\left(T_{33} - \nu(T_{11} + T_{22})\right), \\
\frac{\partial u_1}{\partial x_2} + \frac{\partial u_2}{\partial x_1} &= 2\frac{1+\nu}{E}T_{12}, \\
\frac{\partial u_2}{\partial x_3} + \frac{\partial u_3}{\partial x_2} &= 2\frac{1+\nu}{E}T_{23}, \\
\frac{\partial u_1}{\partial x_3} + \frac{\partial u_3}{\partial x_1} &= 2\frac{1+\nu}{E}T_{13}.
\end{aligned}
\right\} \tag{6.7}
$$

An alternative form of these equations, involving only the displacement $\boldsymbol{u}$ and the body forces $\boldsymbol{f}$, can be arrived at by combining (6.1), (6.2), and (6.6). These are known as the *Navier equations*,

$$
\mu\nabla^2 u_i + (\lambda + \mu)\frac{\partial}{\partial x_i}(\nabla \cdot \boldsymbol{u}) = -f_i, \qquad (i = 1 \text{ to } 3) \tag{6.8}
$$

where the second term includes the *dilatation* Δ of the displacement,

$$
\begin{aligned}
\Delta &= \nabla \cdot \boldsymbol{u} \\
&= \frac{\partial u_1}{\partial x_1} + \frac{\partial u_2}{\partial x_2} + \frac{\partial u_3}{\partial x_3}.
\end{aligned} \tag{6.9}
$$

6.1.1 Boundary Conditions

The set of differential equations (6.2) through (6.7), or in more compact form (6.8), comprise the complete description of the problem of the static, isotropic linear elastic solid. Many well-defined problems do not include boundary conditions other than those specifying the forces on the surfaces, through (6.4). Some however involve fixed values for the displacement $\boldsymbol{u}$, or its derivatives, on the solid surface. In these problems, therefore, we must add to these equations boundary conditions for the relative displacements, in the simplest cases restricting a surface, or a point on the surface, to zero displacement.

A well-defined problem exists only if the correct number of boundary values are specified; too many boundary conditions, and the problem is overdefined, and may or may not have a solution. Too few conditions does not allow for determination of the various undetermined constants appearing from the integrated solutions of the differential equations.

The reader may be familiar with the types of boundary conditions found in problems involving Laplace's equation,

$$
\nabla^2 \phi = \rho(\boldsymbol{r}), \tag{6.10}
$$

where one encounters situations where the field values $\phi(\boldsymbol{r})$ are given on a closed surface enclosing a volume (Dirichlet boundary conditions), from which one can solve for the field $\phi(r)$ in the interior of the volume. Alternatively, the normal derivative of the field $\partial\phi/\partial\boldsymbol{n}$ might be specified on the surface, in what are known as Neumann boundary conditions. In either case, Laplace's equation (6.10) can then be solved uniquely for $\phi(\boldsymbol{r})$, given the source term $\rho(\boldsymbol{r})$. The question of what consists a well-defined problem for a linear, elastic solid is a significantly more complex question; some discussion of this issue can be found in the text by Sokolnikoff [44] and Morse and Feshbach [4], and references therein.

In the examples that follow we will see how the boundary conditions are combined with the linear solid equations to obtain a full solution to a problem, but will not attempt a formal discussion of the general problem. It is rare to find a problem for which a complete analytic solution exists, and quite often even formulating an approximate approach is very challenging. We will follow the usual approach, where we formulate solutions to a few simple problems, to illustrate the approach, but in general precise solutions of all but the simplest problems require numerical approaches.

6.2 Extensional Forces

A very common problem encountered in the subject of elastic solids is the deformation of solid elements known as *beams*, which typically have one dimension (the length) much larger than the other two, and in the simplest cases have a circular or rectangular cross-section. The ends of the beam are typically plane surfaces perpendicular to the length. In the remainder of this chapter we will mostly focus on this type of structure, as the simplified formalism which has been developed for these elements is instructive and applicable to a wide range of situations. We will orient the x_1 axis along the length of the unstrained beam, with the origin of the $x_2 - x_3$ plane typically at the *centroid* of the cross-section, that is, at the point where the first moments vanish, $\int_A x_2 \mathrm{d}A = \int_A x_3 \mathrm{d}A = 0$, integrating over the beam's cross-sectional area. For rectangular and circular cross-sections, the centroid is at the geometric midpoint.

We begin by discussing the simplest type of problem, namely that of a beam subjected to extensional forces applied along the length (an identical treatment applies for compressional forces, with a corresponding change in sign for all the stresses). Following this example, we treat a beam with torques applied to the end faces, with the torque vector perpendicular to the length. This discussion will lead us to the simplified theory of Euler and Bernoulli, for treating the deflection of (long) beams subjected to lateral forces.

6.2.1 Extension of a Beam from End Forces

We begin by reviewing a problem that we have solved previously, in Chap. 5. This is the problem of a fully isotropic beam of length ℓ and cross-sectional area A, subjected to a uniform axial stress, applied by uniformly distributed forces F and $-F$ on the end faces, as shown in Fig. 6.1. The coordinate system's origin is located at the center of the beam.

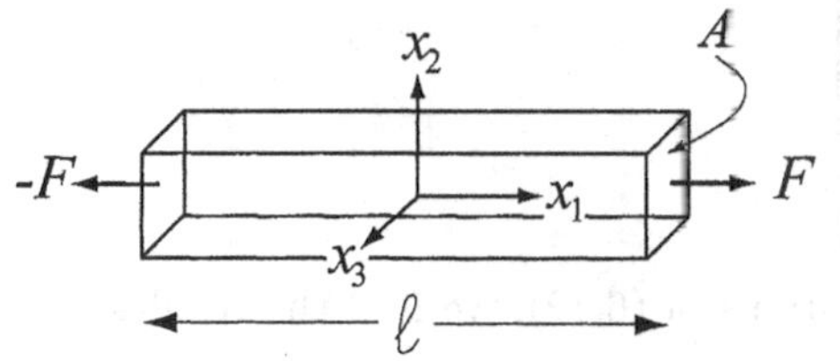

Fig. 6.1. Beam subjected to uniformly applied end forces F and $-F$.

The beam is assumed to be at rest, and as there is no net force, the situation is static. We first have to find an expression for the stress tensor T. The symmetry of the problem implies that there is zero relative displacement at the origin, $\boldsymbol{u}(0) = 0$. Once we have solved for the stress tensor, we can solve for the relative displacement $\boldsymbol{u}(\boldsymbol{r})$ through (6.7) and the application of this boundary condition.

The solution to this problem is obtained by what is commonly called the "inverse approach", where the form for the stress tensor is guessed, and then checked against the stated boundary conditions. From (6.4) and the beam geometry, the normal surface force density on the ends of the beam is simply the element T_{11} of the stress tensor, and as the force F is assumed to be applied uniformly over the cross-section A, we must have $T_{11} = F/A$. There are no other surface forces, so all the other components of the stress tensor must be zero. Furthermore, there are no body forces, so by (6.2) the stresses are constant in the beam interior. We can therefore, as in the previous discussions of this problem, take the stresses in this problem as

$$T_{11} = F/A,\ T_{22} = T_{33} = T_{12} = T_{13} = T_{23} = 0, \tag{6.11}$$

through the volume of the beam. This form clearly satisfies the boundary condition on the stress, so we have already solved half of the problem.

From the stress–strain relations, we can write equations for the relative displacement:

$$\left.\begin{aligned}\frac{\partial u_1}{\partial x_1} &= F/EA, \\ \frac{\partial u_2}{\partial x_2} &= -\nu(F/EA), \\ \frac{\partial u_3}{\partial x_3} &= -\nu(F/EA), \\ \frac{\partial u_1}{\partial x_2} + \frac{\partial u_2}{\partial x_1} &= 0, \\ \frac{\partial u_2}{\partial x_3} + \frac{\partial u_3}{\partial x_2} &= 0, \\ \frac{\partial u_1}{\partial x_3} + \frac{\partial u_3}{\partial x_1} &= 0.\end{aligned}\right\} \tag{6.12}$$

By integrating the first three equations of (6.12), we find the results

$$u_1(x_1, x_2, x_3) = (F/EA)x_1 + f_1(x_2, x_3), \tag{6.13}$$
$$u_2(x_1, x_2, x_3) = -\nu(F/EA)x_2 + f_2(x_1, x_3), \tag{6.14}$$
$$u_3(x_1, x_2, x_3) = -\nu(F/EA)x_3 + f_3(x_1, x_2), \tag{6.15}$$

where the functions f_1, f_2 and f_3 are undetermined; the second set of three equations in (6.12) will impose constraints on their behavior.

We can simplify the situation by using the symmetries of the problem; rotation by 90° about the x_1 axis, inversion with $x_3 \to -x_3$, and inversion with $x_2 \to -x_2$, should all yield the same result. Applying the symmetries as well as the second set of equations gives the result that f_1, f_2 and f_3 are all constant, and with the origin at the center of the beam fixed by the displacement boundary condition $\boldsymbol{u}(0) = 0$, the constants must all be zero. Hence we have the solution

$$u_1(x_1, x_2, x_3) = (F/EA)x_1, \tag{6.16}$$
$$u_2(x_1, x_2, x_3) = -\nu(F/EA)x_2, \tag{6.17}$$
$$u_3(x_1, x_2, x_3) = -\nu(F/EA)x_3. \tag{6.18}$$

The same result could have been achieved without applying the symmetries, and integrating the differential equations, but the use of symmetry greatly reduces the required effort.

6.2.2 Extension of a Beam from a Body Force

We now consider the same beam, but subject it to a uniform body force $\boldsymbol{f} = f\hat{\boldsymbol{x}}_1$ (where f has units of N/m^3, and $\hat{\boldsymbol{x}}_1$ is the unit vector along x_1). This force could arise if the body were uniformly charged with a charge density ρ_Q and subjected to a constant electric field $\boldsymbol{E}$, in which case $\boldsymbol{f} = \rho_Q \boldsymbol{E}$. The force could also come from the action of gravity $\boldsymbol{g}$ on a beam's uniform mass density ρ, in which case $\boldsymbol{f} = \rho \boldsymbol{g}$. The center point of one end of the beam will be clamped, so that point does not displace; we will also prevent the end surface surrounding that point from bending. We place the origin of the coordinate system at the fixed point; see Fig. 6.2.

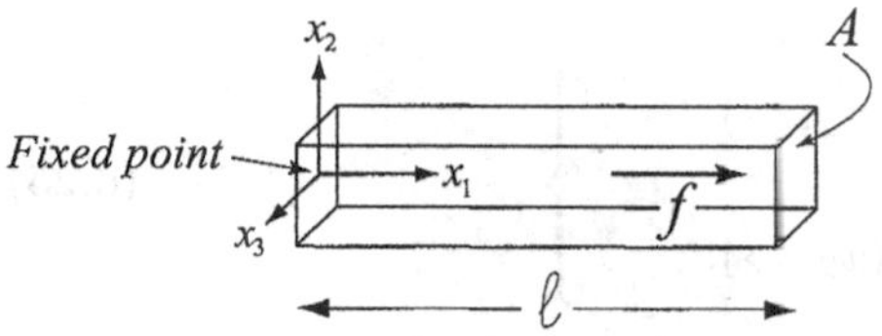

Fig. 6.2. Beam of length ℓ and cross-sectional area A, fixed at one end, subjected to uniform body force f.

The boundary conditions are expressed as

$$\left.\begin{aligned} \boldsymbol{u}(0) &= 0, \\ \frac{\partial u_2}{\partial x_1}(0) &= \frac{\partial u_3}{\partial x_1}(0) = \frac{\partial u_2}{\partial x_3}(0) = 0, \end{aligned}\right\} \tag{6.19}$$

from the stated conditions of zero relative displacement and no bending or turning at the origin.

The total body force is $\boldsymbol{F} = \ell A \boldsymbol{f}$, and to maintain static equilibrium this must be the total force exerted by the clamp at the fixed end of the beam. From the previous example and (6.4), we know that this requires that the stress at $x_1 = 0$ be given by $T_{11} = f\ell$, with all other stress components identically zero (we will assume for simplicity that the restoring stress is uniformly applied over the entire end surface, even though we only restrict the displacement and bending at the center of that surface). The stress in the remainder of the beam can then be calculated using (6.2), which yields

$$T_{11} = f\ell(1 - x_1/\ell), \quad T_{22} = T_{33} = T_{12} = T_{13} = T_{23} = 0. \tag{6.20}$$

We now use the relations (6.7) to calculate the relative displacement:

$$\left.\begin{aligned} \frac{\partial u_1}{\partial x_1} &= \frac{f(\ell - x_1)}{E}, \\ \frac{\partial u_2}{\partial x_2} &= \frac{\partial u_3}{\partial x_3} = -\nu \frac{f(\ell - x_1)}{E}, \\ \frac{\partial u_1}{\partial x_2} + \frac{\partial u_2}{\partial x_1} &= 0, \\ \frac{\partial u_1}{\partial x_3} + \frac{\partial u_3}{\partial x_1} &= 0, \\ \frac{\partial u_2}{\partial x_3} + \frac{\partial u_3}{\partial x_2} &= 0. \end{aligned}\right\} \tag{6.21}$$

Integrating the first relation gives

$$u_1(x_1,x_2,x_3) = \frac{f}{E}(\ell x_1 - x_1^2/2) + p(x_2,x_3). \tag{6.22}$$

Using this (partial) solution in the fourth and fifth relations of (6.21) yields the results

$$\left.\begin{aligned} u_2(x_1,x_2,x_3) &= -x_1\frac{\partial p}{\partial x_2} + q(x_2,x_3), \\ u_3(x_1,x_2,x_3) &= -x_1\frac{\partial p}{\partial x_3} + r(x_2,x_3). \end{aligned}\right\} \tag{6.23}$$

Inserting these results in the remaining equations of (6.21), one can show that the functions p, q and r have the form

$$\left.\begin{aligned} p(x_2,x_3) &= \nu\frac{f}{2E}(x_2^2+x_3^2) + p_2x_2 + p_3x_3 + p_0, \\ q(x_2,x_3) &= q_3x_3 + q_0, \\ r(x_2,x_3) &= r_1x_2 + r_0, \end{aligned}\right\} \tag{6.24}$$

with $r_1 = -q_3$. The details of this calculation are left to Exercise 6-2. Application of the boundary conditions (6.19) at the clamped end of the beam yields

$$\left.\begin{aligned} u_1 &= \frac{f}{2E}(2\ell x_1 - x_1^2 + \nu x_2^2 + \nu x_3^2), \\ u_2 &= -\nu\frac{f}{E}x_1x_2, \\ u_3 &= -\nu\frac{f}{E}x_1x_3. \end{aligned}\right\} \tag{6.25}$$

We see from these relations that the principle effect of the body force is to extend the fractional length of the beam by $f\ell/2E$ (the relative displacement along x_1, at the free end $x_1 = \ell$, is given by $f\ell^2/2E$) . In addition, the two transverse dimensions, along x_2 and x_3, are reduced by Poisson's ratio times this fraction. We also find that the surface of the clamped end of the beam, which without the stress is the plane surface $x_1 = 0$, is now warped towards $+x_1$ due to the forces acting on the beam. If we had clamped the entire right end of the beam, rather than just the center, this warping of the end surface would not have occurred, but the resulting problem is more difficult to solve.

6.3 Flexure of Beams

We now turn from beam extension to beam flexure. Flexure, or bending, can be caused by forces applied transverse to the length of the beam, or, in the simplest case, by pure torque applied to the end faces, with the torque vector perpendicular to the length. We treat the simplest case first, and following this we discuss problems involving more general forces.

6.3.1 Bending by Pure Torque

Consider the beam shown in Fig. 6.3, with length ℓ, width w and depth d. The left end is clamped, so the center of that end does not move, rotate or twist, and the right end is subjected to a torque $\boldsymbol{M} = -M_0\hat{\boldsymbol{x}}_3$ (with $M_0 > 0$). The beam is static, so the clamp at the left end must apply a equal and opposite torque $\boldsymbol{M} = M_0\hat{\boldsymbol{x}}_3$. We specify below the precise distribution of surface forces applied to the ends to achieve these torques.

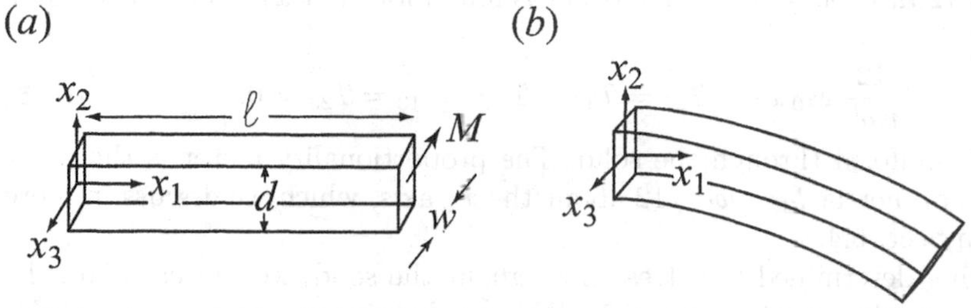

Fig. 6.3. (**a**) Beam with left end fixed under pure flexural torque M, and (**b**) resulting distortion of beam.

The pair of torques makes the beam bend about the x_3 axis; the bending occurs due the extension of the length of the beam along its top surface, perpendicular to the x_2 axis, and a compression along the bottom surface. In the language of strain, the bending results from a positive strain S_{11} on the top surface, and a negative strain along the bottom surface, with a continuous variation of S_{11} between the two surfaces (see Sect. 5.3).

In this type of bending, the line passing through the center of the beam (the center line), parallel to the x_1 axis with $x_2 = x_3 = 0$, is known as the *neutral axis*, as to first order in the displacement it does not undergo any longitudinal strain. Planes perpendicular to this line remain approximately perpendicular to this line in the bent beam, as discussed in Sect. 4.3. Furthermore, the strain S_{11} at a point P, located at (p_1, p_2, p_3) in the undistorted beam, is approximately proportional to p_2, the distance along x_2 from the neutral axis. The stress T_{11} is also proportional to this distance, as discussed in Sect. 5.3. To simplify the problem, we specify that the torque applied to the ends of the undistorted beam is from a surface stress vector $\boldsymbol{t}$ applied parallel to $\hat{\boldsymbol{x}}_3$, proportional to the distance along x_2, so

$$\boldsymbol{t}(\ell, x_2, x_3) = -\boldsymbol{t}(0, x_2, x_3) = t_0 x_2\,\hat{\boldsymbol{x}}_3, \tag{6.26}$$

where we determine the amplitude t_0 below. Note that as in the previous example, we assume the force at the clamped end is distributed over the entire end surface, although the boundary conditions are only applied at one point (the origin).

From (6.26), there is no net force on either surface. The net torque, due only to the force on the surface at $x_1 = \ell$, is given by

$$\begin{aligned} \boldsymbol{M}(x_1 = \ell) &= \int_{-d/2}^{d/2} \int_{-w/2}^{w/2} \boldsymbol{r} \times \boldsymbol{t}(\ell, x_2, x_3)\, \mathrm{d}x_3\, \mathrm{d}x_2 \\ &= -\frac{1}{12} w d^3 t_0\, \hat{\boldsymbol{x}}_3 \end{aligned} \tag{6.27}$$

(this is the torque about the origin). In the problem statement, the net torque was specified as $-M_0 \hat{\boldsymbol{x}}_3$, so from (6.27) the stress amplitude must be given by $t_0 = 12M_0/(wd^3)$. The same result is found for t_0 at $x_1 = 0$. The stress T is thus

$$T_{11} = \frac{12}{wd^3} M_0\, x_2, \quad T_{22} = T_{33} = T_{12} = T_{13} = T_{23} = 0, \tag{6.28}$$

which is uniform through the solid. The proportionality factor is the polar moment of inertia $I_2 = wd^3/12$ about the x_2 axis, which we discuss in more detail in Sect. 6.4.

Having determined the stress throughout the solid, we can calculate the resulting displacement, using (6.7). Writing these equations out, the displacement $\boldsymbol{u}$ must satisfy

$$\left.\begin{aligned} \frac{\partial u_1}{\partial x_1} &= \frac{M_0}{EI_2} x_2, \quad \frac{\partial u_2}{\partial x_2} = \frac{\partial u_3}{\partial x_3} = -\nu \frac{M_0}{EI_2} x_2, \\ \frac{\partial u_1}{\partial x_2} + \frac{\partial u_2}{\partial x_1} &= 0, \quad \frac{\partial u_1}{\partial x_3} + \frac{\partial u_3}{\partial x_1} = 0, \quad \frac{\partial u_2}{\partial x_3} + \frac{\partial u_3}{\partial x_2} = 0. \end{aligned}\right\} \tag{6.29}$$

We now proceed in the same fashion as we did in Sec. 6.2.2. Integrating the first equation in (6.29), and inserting that solution in the fourth and fifth equations, yields the partial solution

$$\left.\begin{aligned} u_1(x_1, x_2, x_3) &= \frac{M_0}{EI_2} x_1 x_2 + p(x_2, x_3), \\ u_2(x_1, x_2, x_3) &= -\frac{M_0}{2EI_2} x_1^2 - \frac{\partial p}{\partial x_2} x_1 + q(x_2, x_3), \\ u_3(x_1, x_2, x_3) &= -\frac{\partial p}{\partial x_3} x_1 + r(x_2, x_3), \end{aligned}\right\} \tag{6.30}$$

where p, q and r are undetermined functions. By further manipulation of these expressions, using all six of the relations in (6.29), one can eventually show (see Exercise 6.3) that

$$\left.\begin{aligned} u_1 &= \frac{M_0}{EI_2} x_1 x_2 + a_1 x_2 + b_1 x_3 + c_1, \\ u_2 &= -\frac{M_0}{2EI_2}(x_1^2 + \nu x_2^2 - \nu x_3^2) + a_2 x_1 + b_2 x_3 + c_2, \\ u_3 &= -\nu \frac{M_0}{EI_2} x_2 x_3 + a_3 x_1 + b_3 x_2 + c_3, \end{aligned}\right\} \tag{6.31}$$

where $a_2 = -a_1$, $a_3 = -b_1$ and $b_3 = -b_2$.

Our boundary conditions state that the left end of the beam is fixed, so that $\boldsymbol{u}(0) = \mathbf{0}$. We also prevent the beam from bending by setting $\partial u_2/\partial x_1 = \partial u_3/\partial x_1 = 0$, and prevent rotation by setting $\partial u_2/\partial x_3 = 0$, all these derivatives evaluated at the origin. These conditions yield $c_1 = c_2 = c_3 = 0$, and $a_1 = a_2 = a_3 = b_1 = b_2 = b_3 = 0$. Hence the final result is

$$\left.\begin{aligned} u_1 &= \frac{M_0}{EI_2} x_1 x_2, \\ u_2 &= -\frac{M_0}{2EI_2}(x_1^2 + \nu x_2^2 - \nu x_3^2), \\ u_3 &= -\nu \frac{M_0}{EI_2} x_2 x_3. \end{aligned}\right\} \quad (6.32)$$

In Fig. 6.4 we display the deformation of the beam in the $x - y$ plane.

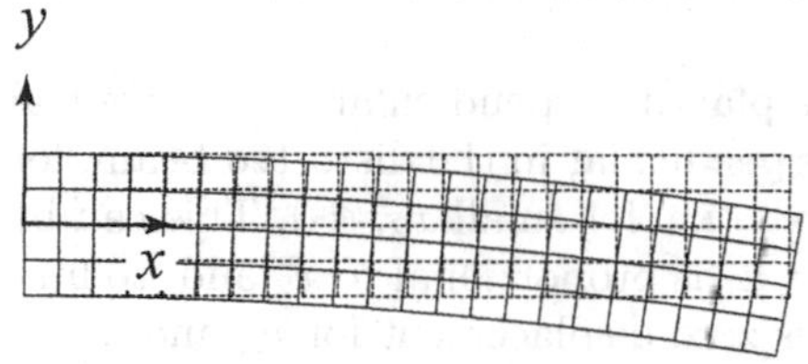

Fig. 6.4. Beam deformed under the displacement (6.32) (*solid lines*), with the undeformed beam shown by the dotted lines.

6.3.2 Displacement of the Neutral Axis

From the relations (6.32), we see that the central axis of the beam, initially at $x_2 = x_3 = 0$, undergoes no compression or extension, i.e. its displacement u_1 is zero. This is as expected for the *neutral axis* of the beam, as mentioned above. However, the neutral axis undergoes a transverse displacement, along x_2, given by $u_2(x_1) = -(M_0/2EI_2)\, x_1^2$, with a displacement at the free end $u_2(\ell) = -(M_0\ell^2/2EI_2)$.

The parabolic shape for the neutral axis is closely approximated (for small total displacement $u_2(\ell)$) by the shape of a circle, as shown in Fig. 6.5. The radius of the circle that best approximates the parabola is given by

$$R = \frac{EI_2}{M_0}, \quad (6.33)$$

i.e. the radius of curvature is inversely proportional to the applied torque M_0, and directly proportional to Young's modulus E and the bending moment I_2.

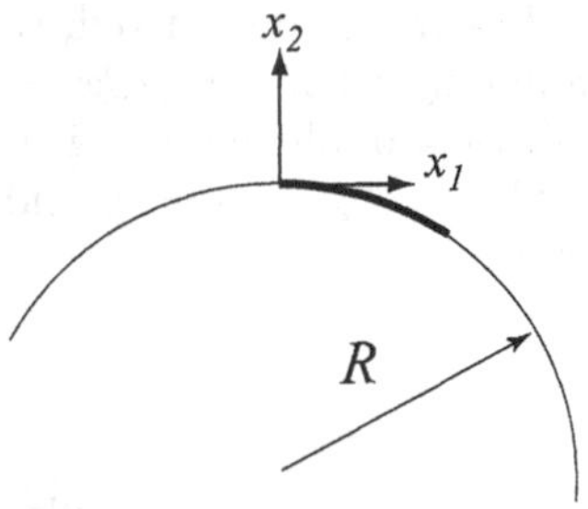

Fig. 6.5. Geometry defining radius of curvature R for a flexural beam; the circle closely approximates the parabolic shape of the beam for a beam length l much less than the radius R

This approximation is quite good for beams with length $\ell \ll R$. The error is only 1% for the vertical displacement of a beam with $\ell = 0.35R$, and the error is 10% for a beam with $\ell = 1.1R$.

The solution (6.32) also shows that planes perpendicular to the neutral axis in the undistorted beam remain perpendicular in the distorted beam, for small displacements from the neutral axis (i.e. for small x_2, x_3). This can be seen from the fact that the displacement u_1 is proportional to x_2 and, to first order in the transverse distance, there is zero displacement for u_2 and u_3.

These two results, the formula for the radius of curvature R given by (6.33), and the preservation of the perpendicular planes under bending, form the essence of the *Euler–Bernoulli theory* of beam bending, which is most accurate for long, thin beams with a large radius of curvature R. In the next section we shall discuss the use of this theory for other problems, as the exact theory becomes rather cumbersome for problems involving loads other than pure torque.

6.3.3 Saint–Venant's Principle

The two problems we have just solved, a beam under uniform, axial tensile stress, and a beam under pure torque, are fairly straightforward. This is because the applied external forces generate uniform and fairly understandable stresses. Frequently, however, the applied forces generate non-uniform or non-intuitive stress distributions, which are difficult to determine: The "inverse approach" becomes difficult to apply.

In such cases, we can use an approximation known as *Saint–Venant's principle* [44]. This principle, published by B. de Saint–Venant in 1855, states that any distribution of forces and torques, applied over a discrete region of a solid, can be replaced by a different distribution of forces and torques that give the same resultant force and torque, and the change in the behavior of the solid will be negligible over a distance that is "sufficiently far removed" from the region in question. The usual rule is that "sufficiently far" is a distance roughly five times the diameter of the area over which the force distribution is applied. Saint–Venant's principle has been rigorously proven for forces applied normal to the surface of a semi-infinite solid, but it has also been shown to be violated in some situations (see [58]). The principle must therefore be applied with some care.

Saint Venant's principle can be used to replace the forces and torques applied over a cross-section of a beam by a single force and a single torque, applied at the center of the cross-section. We will frequently resort to the use of this principle in situations where it is clearly applicable, but in cases of ambiguity, or where precise results are required, one must usually resort to numerical solutions.

6.4 Euler–Bernoulli Theory of Beams

The need for a predictive theory for the bending of beams led to the parallel development of such a model by J. Bernoulli and by L. Euler. Their simple, scalar theory is most accurate for long thin beams, with small total displacements, and can predict the bending of a beam subject to more-or-less arbitrary forces. The theory is based on the fact that the primary effect of a transverse force is to generate a torque transverse to both the force and the beam axis, and the bending induced by that torque is given by the result (6.33). The separate effects of shear, which were not involved in the examples discussed above, are typically very small, and can usually be ignored. If one would like to include shear displacements as well as bending, the solutions are found separately and then superimposed; this is discussed below.

In this theory, plane cross-sections of the beam, initially perpendicular to the long (neutral) axis of the beam, are assumed to remain perpendicular to the neutral axis of the strained beam. This assumption must be dropped when treating shear deflections. Loads that vary spatially along the length of the beam are assumed to cause local bending. The results obtained are usually quite accurate when compared with the full elastic theory, with discrepancies appearing for shorter beams or for larger deflections.

The neutral axis by definition does not undergo any axial strain, and clearly does undergo strain if the problem includes forces directed along the neutral axis (as in the first examples in Sect. 6.2), or problems involving shear; in such problems the effects of the axial forces would be solved for separately, and the two parts of the problem then superimposed. The neutral axis is assumed to pass through the centroid of the beam, and we will assume that the beam's neutral axis is straight in the unstrained beam. This of course need not be the case, except for prismatic beams or for solids of revolution. As usual, we align the $\hat{\boldsymbol{x}}_1$ unit vector along the neutral axis. We will assume that the applied torque $\boldsymbol{M}$ is along $\hat{\boldsymbol{x}}_3$, so that the deflection of the neutral axis is along $\hat{\boldsymbol{x}}_2$, as is shown in Fig. 6.3.

In order to apply (6.33) locally, we need to define the local *polar moment of inertia* I_i, which is defined for each transverse axis x_i as the integral over the cross-sectional area of the coordinate x_i squared, as measured from the neutral axis:

$$I_i = \int_A x_i^2 \, \mathrm{d}A. \tag{6.34}$$

For a rectangular beam with the neutral axis passing through its center, with beam width w along x_3 and thickness t along x_2, the polar moments are $I_2 = wt^3/12$ and $I_3 = tw^3/12$ (see Fig. 6.6). A beam with cylindrical cross-section of radius r has polar moments $I_2 = I_3 = \pi r^4/4$. A beam with elliptical cross-section, with its cross-sectional surface defined by

$$\frac{x_2^2}{r_2^2} + \frac{x_3^2}{r_3^2} = 1, \tag{6.35}$$

has moments $I_2 = (\pi/4) r_2^3 r_3$ and $I_3 = (\pi/4) r_2 r_3^3$. Because the neutral axis is assumed aligned with x_1, the polar moment I_1 is not needed.

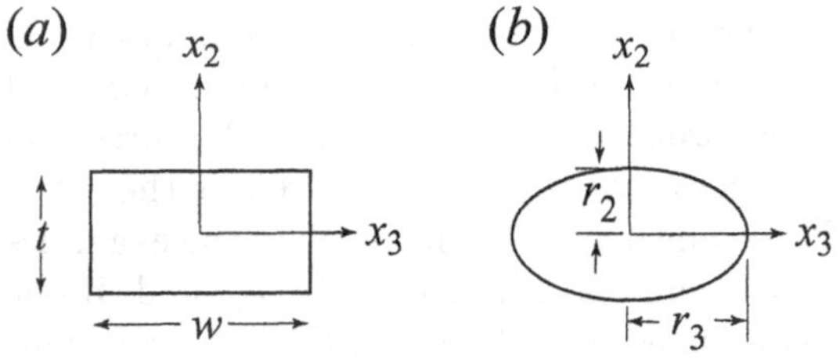

Fig. 6.6. Geometry for calculating the polar moments I_2 and I_3, (**a**) for a rectangular cross-section, and (**b**) for an elliptical cross-section.

The Euler–Bernoulli theory states that the *local* radius of curvature $R(p)$, at the point p along the x_1 axis, can be written in terms of the local moment of inertia and the local torque:

$$R(p) = \frac{E I_2(p)}{M_3(p)}, \tag{6.36}$$

a straightforward application of (6.33). The calculation of the local torque $M_3(p)$, for a problem involving transverse forces as well as end torques, is discussed in Sect. 6.4.1 below.

The deflection $u_2(p)$ of the neutral axis is directly related to the curvature R. We define the bending angle $\theta(p)$ as the angle the local tangent to the displaced neutral axis makes with the original beam axis, as shown in Fig. 6.7.

An increment along an arc length $\mathrm{d}s$ of the displaced neutral axis will change the angle θ by the inverse of the local radius of curvature of the axis at that point, which can in turn be related to the local torque and polar moment:

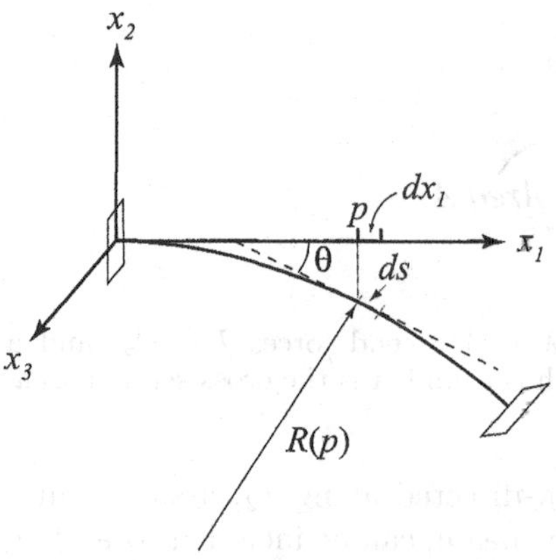

Fig. 6.7. Definition of the angle $\theta(p)$ as the angle the local tangent at p makes with the x_1 axis.

$$\frac{\mathrm{d}\theta}{\mathrm{d}s}(p) = \frac{1}{R(p)} = \frac{M_3(p)}{EI_2(p)}. \tag{6.37}$$

For small angles θ, the arc length $\mathrm{d}s$ is very nearly equal to the displacement $\mathrm{d}x_1$ along the x_1 axis; furthermore, we can approximate $\mathrm{d}u_2/\mathrm{d}x_1(p) = \tan\theta(p) \cong \theta(p)$, so we find an equation very similar to the exact formula for a beam subjected to pure torque:

$$\frac{\mathrm{d}^2 u_2}{\mathrm{d}x_1^2}(p) = \frac{M_3(p)}{EI_2(p)}. \tag{6.38}$$

Within the approximations stated above, the equation (6.38) can handle local variations in the torque and in the moment of inertia.

6.4.1 Torque and Shear due to Local Forces

We now discuss how the local torque and shear are related to the applied forces. Consider the force system sketched in Fig. 6.8. The forces as drawn are all along the x_2 axis, and the torques along x_3. The forces are assumed to act in the $x_1 - x_2$ plane, so they do not generate any torque components about the neutral axis. All resultant torques will therefore be along x_3, and any displacements of the neutral axis will be along x_2. We have included a force and torque at each end of the beam (labelled with subscripts A and B), as well as the distributed force $f(x)$.

We have not explicitly included lateral surface forces or point forces. A point force F_P, located on the neutral axis at a point $x_1 = p$, can be included through the use of the δ-function, by adding to the distributed force $f(x_1)$ the function $F_P/A\,\delta(x_1 - p)$; the force is divided by the cross-sectional area A to give the same resultant when the body force is integrated over volume. Note that the δ-function has units of $(\text{length})^{-1}$.

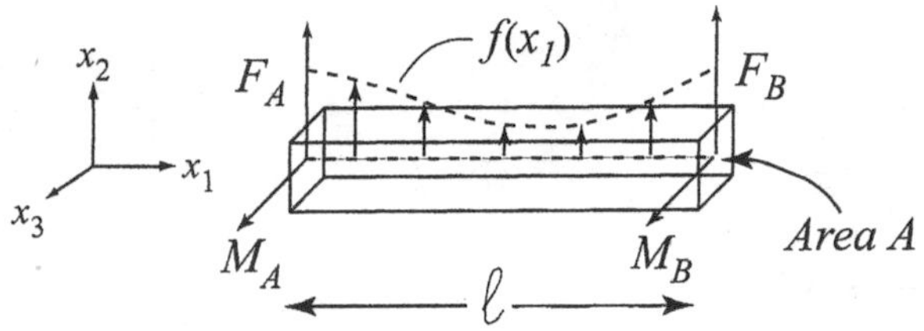

Fig. 6.8. Beam subjected to end torques M_A, M_B, end forces F_A, F_B, and a distributed body force $f(x_1)$. The beam length is ℓ, and A is the cross-section area.

Similarly, a surface vector stress $\boldsymbol{t}(x_1)$, directed along x_2, assumed uniformly distributed across the width w of the beam, can be included by adding to the body force f the force $t(x_1)/w$. In either case, as for the body force, we assume these forces act in the $x_1 - x_2$ plane, generating no torque about x_1, and the effects of shear can be included as discussed in Sect. 6.4.2 below.

Problems that involve forces and torques along directions other than those considered here can be solved by superposition. A problem that involves, e.g., off-axis forces not in the $x_1 - x_2$ plane, can be solved (approximately) by applying Saint–Venant's principle, and replacing each off-axis force along x_2 by an equivalent on-axis force along x_2 and a torque along x_1. One would then solve separately for the displacements and torsion caused by the torque, and add the solutions vectorially.

The beam shown in Fig. 6.8 is assumed to be in static balance, so the forces and torques must add separately to zero, as in Sect. 4.7. Given the distribution of forces as shown, the force balance requirement is that

$$F_A + \int_0^\ell f(x_1) A \, \mathrm{d}x_1 + F_B = 0, \tag{6.39}$$

with all the forces directed along $\hat{\boldsymbol{x}}_2$; A is the cross-sectional area of the beam, over which the force density f is assumed uniform. For the torque balance, we assume the origin for the torque is at the left end of the beam:

$$M_A + \int_0^\ell x_1 f(x_1) A \, \mathrm{d}x_1 + \ell F_B + M_B = 0. \tag{6.40}$$

Note that all the torques are directed along $\hat{\boldsymbol{x}}_3$.

In addition to these overall balance equations, we may divide the beam in two at any point, and equilibrium then requires that each side of the divided beam be individually stable, with the inclusion of the force and torque exerted at the interface by the other part of the beam. The force diagram for the beam divided at x is shown in Fig. 6.9, including the force $F_2(x)$ and torque $M_3(x)$ exerted at the exposed end at x; the actual force and torque distributions are replaced by a single force and torque acting at the center of the surface, using Saint–Venant's principle. Static equilibrium requires that the shear force $F_2(x)$ must be given by

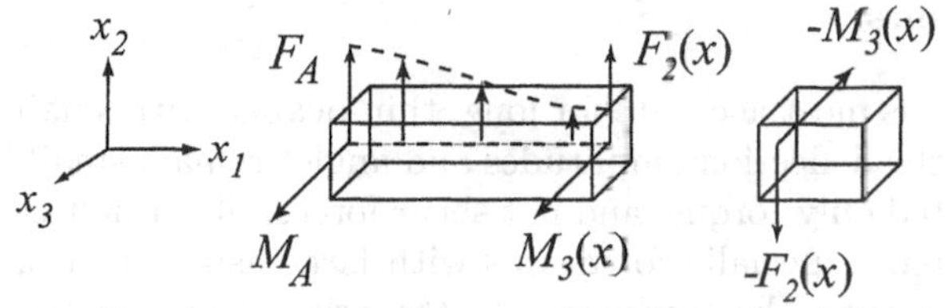

Fig. 6.9. Force diagram for left end of beam divided at $x_1 = x$; the diagram includes the force $F_2(x)$ and torque $M_3(x)$ exerted by the right portion of the beam.

$$F_2(x) = -F_A - \int_0^x f(x_1) A \, \mathrm{d}x_1. \tag{6.41}$$

The torque $M_3(x)$ must similarly be given by

$$M_3(x) = -M_A - \int_0^x x_1 f(x_1) A \, \mathrm{d}x_1 - xF(x). \tag{6.42}$$

Consider now moving along x_1 from x to $x + \mathrm{d}x$. We find that the incremental shear force and torque must satisfy the differential relations

$$\frac{\mathrm{d}F_2}{\mathrm{d}x}(x) = -f(x)A, \tag{6.43}$$

and

$$\begin{aligned} \frac{\mathrm{d}M_3}{\mathrm{d}x}(x) &= -xf(x)A - F_2(x) - x\frac{\mathrm{d}F_2}{\mathrm{d}x}(x) \\ &= -F_2(x). \end{aligned} \tag{6.44}$$

If we differentiate (6.44) again, and use (6.43), we find

$$\frac{\mathrm{d}^2 M_3}{\mathrm{d}x^2} = f(x)A. \tag{6.45}$$

We can use (6.43) and (6.45), in conjunction with (6.38), to write a fourth-order equation for the deflection of the neutral axis of a beam, known as the *beam bending formula*:

$$\frac{\mathrm{d}^4 u_2}{\mathrm{d}x^4} = \frac{f(x)A}{EI_2} \tag{6.46}$$

where of course x is the coordinate along the beam axis, aligned with $\hat{x}_1$. If the beam is rectangular in cross-section, with width w along x_3 and thickness t along x_2, this can be written as

$$\frac{\mathrm{d}^4 u_2}{\mathrm{d}x_1^4} = \frac{12}{Et^2} f(x_1). \tag{6.47}$$

This relation does not, curiously, involve the width of the beam: the resistance to bending to a body force depends only on its thickness. Equivalently, the stiffness of a rectangular beam increases with width at the same rate as the force per unit length.

6.4.2 Shear Displacements

The Euler–Bernoulli theory is most accurate for long, thin beams, with small torques and forces so that the deflection amplitudes and angles remain small. Furthermore, we have treated only torque, and not shear forces. Shear deflection can normally be ignored, especially for beams with large aspect ratios, that is, length much larger than the thickness. As the aspect ratio of the beam becomes smaller,with the ratio of length to thickness decreasing, however, shear becomes more important. The effects of shear are included by superimposing the displacements it induces with those from the applied torque.

A shear force F_2 along $\hat{\boldsymbol{x}}_2$ causes an elemental volume to deform in the fashion shown by Fig. 6.10. The differential element, of length $\mathrm{d}x_1$, shears through a distance $\mathrm{d}u_2$. The stress, for a cross-sectional area A perpendicular to $\hat{\boldsymbol{x}}_1$, is given in six-vector form by $\boldsymbol{\tau} = (0,0,0,0,0,F_2/A)$, so the strain six-vector is given by $\boldsymbol{\epsilon} = (0,0,0,0,0,F_2/(\mu A))$ (see Sect. 5.5). This corresponds to a shear deflection

$$\frac{\mathrm{d}u_2}{\mathrm{d}x_1} = \frac{F_2}{\mu A} \quad \text{(shear only).} \tag{6.48}$$

This deflection is local and additive; if the shear force F_2 varies along the length of the beam, the local shear deflection reflects this.

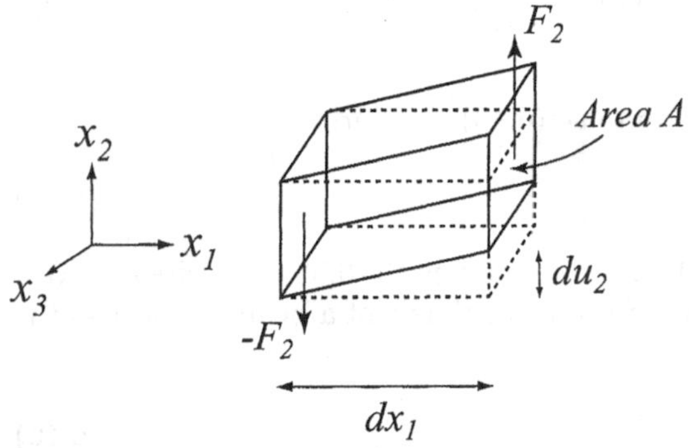

Fig. 6.10. Deflection geometry for shear force F_2.

The displacement due to shear is added to the displacement due to flexure, (6.38). A load involving a combination of a torque M_3 and a shear force F_2 therefore gives a deflection u_2 satisfying

$$\frac{\mathrm{d}^2u_2}{\mathrm{d}x_1^2} = \frac{M_3}{EI_2} + \frac{\mathrm{d}}{\mathrm{d}x_1}\frac{F_2}{\mu A}. \tag{6.49}$$

In this equation, the torque M_3, polar moment I_2, shear force F_2, and area A can all in principle be functions of position. We can relate the displacement to the applied forces by differentiating twice with respect to x_1:

$$\frac{\mathrm{d}^4 u_2}{\mathrm{d}x_1^4} = \frac{\mathrm{d}^2}{\mathrm{d}x_1^2}\left(\frac{M_3(x_1)}{EI_2(x_1)} + \frac{\mathrm{d}}{\mathrm{d}x_1}\frac{F_2}{\mu A}\right). \tag{6.50}$$

For a beam where the moment of inertia I_2 and area A are constant, we obtain

$$\frac{\mathrm{d}^4 u_2}{\mathrm{d}x_1^4} = \frac{A}{EI_2} f(x_1) - \frac{1}{\mu}\frac{\mathrm{d}^2}{\mathrm{d}x_1^2} f(x_1). \tag{6.51}$$

The first term is due to torque, and the second due to shear. With μ and E of the same order, the shear term causes significant deflection only when the body force f varies over a distance of order of, or smaller than, $\sqrt{I_2/A}$. When f varies much more slowly than this, the beam bending formula (6.46) is recovered.

6.4.3 Boundary Conditions

We note that (6.46), or the equivalent equation (6.51) including shear, are linear, fourth-order equations in the displacement u_2. In general one then requires four independent boundary conditions to find a unique solution.

Consider a doubly clamped beam, where each end of the beam is prevented from moving and from bending. The four boundary conditions are then

$$u_2(0) = u_2(\ell) = 0, \quad \frac{\mathrm{d}u_2}{\mathrm{d}x_1}(0) = \frac{\mathrm{d}u_2}{\mathrm{d}x_1}(\ell) = 0. \tag{6.52}$$

These will generate a unique solution to (6.46) or (6.51), given a force distribution $f(x_1)$.

For a doubly pinned beam, the beam ends are prevented from moving but not from bending. The support that holds the pinned end does not apply a local torque to the pinned point, so from (6.38) the second derivative of u_2 is zero. Hence for this type of problem the boundary conditions are

$$u_2(0) = u_2(\ell) = 0, \quad \frac{\mathrm{d}^2 u_2}{\mathrm{d}x_1^2}(0) = \frac{\mathrm{d}^2 u_2}{\mathrm{d}x_1^2}(\ell) = 0. \tag{6.53}$$

Similarly, a cantilevered beam, with one end clamped and the other end free, has the displacement u_2 and its derivative equal to zero at the clamped end, but at the free end the conditions are zero torque, and furthermore, because there is no net force at the free end, the rate of change of torque $\mathrm{d}M/\mathrm{d}x_1$ must be zero, so the third derivative of the displacement is zero:

$$u_2(0) = \frac{\mathrm{d}u_2}{\mathrm{d}x_1}(0) = 0, \quad \frac{\mathrm{d}^2 u_2}{\mathrm{d}x_1^2}(\ell) = \frac{\mathrm{d}^3 u_2}{\mathrm{d}x_1^3}(\ell) = 0. \tag{6.54}$$

By constructing such relations, a variety of problems may be solved; we give a few examples in Sect. 6.4.5.

6.4.4 Point Forces

We now briefly discuss the question of localized, point forces, which can be included in the body force distribution through the use of δ-functions. We invoke Saint–Venant's principle, and assume that the force may be treated as a simple body force for points that are sufficiently far from the point of application, as was discussed in Sect. 6.3.3.

The force F_P, located at $x_1 = p$, adds to the shear force through (6.43),

$$\frac{dF_2}{dx_1}(x_1) = -f(x_1)A - F_P\delta(x_1 - p), \tag{6.55}$$

and adds to the torque through (6.45),

$$\frac{d^2M_3}{dx_1^2}(x_1) = f(x_1)A + F_P\delta(x_1 - p). \tag{6.56}$$

Integrating each of these equations over a differential length centered on p, the contribution from the body force f is negligible, but that from the point force adds a shear force

$$\begin{aligned} \Delta F_2(p) &= F_2(p + \delta/2) - F_2(p - \delta/2) \\ &= -F_P, \end{aligned} \tag{6.57}$$

and adds a torque

$$\begin{aligned} \Delta M_3(p) &= M_3(p + \delta/2) - M_3(p - \delta/2) \\ &= F_P(\delta/2). \end{aligned} \tag{6.58}$$

Hence the shear force increases by a step of height $-F_P$, and the torque by this force acting with a moment arm given by the distance from the point force. If we ignore the effects of shear, the point force adds to the displacement u_2 a term proportional to the cube of the distance from the point p.

6.4.5 Results for Common Bending Problems

We restrict the discussion now to problems in which shear can be ignored, and give the results to a few common problems of general interest: cantilevered and doubly clamped beams, with different load distributions. We consider fully clamped ends, where the beam is both fixed in position and prevented from bending at the clamped point. Our loads range from a point load at a given point, including for the cantilevered beam a point load at the free end, to a few different distributed loads. The results are given in tabular form for ease of reference, in Table 6.1 and Table 6.2. We refer the reader to the text by Roarke and Young [59] for a wider variety of load distributions and clamping conditions.

Table 6.1. Cantilevered beam shapes and deflections for a variety of load distributions. The beam is assumed to have length ℓ and polar moment I_2 along $\hat{x}_2$, and is made of an isotropic material with Young's modulus E and Poisson's ratio ν. The point of maximum deflection is always at $x_1 = \ell$.

Load distribution	Load	Deflection $u_2(x_1)$	Maximum deflection $u_2(\ell)$
Point load at free end	F_P	$\frac{F_P}{EI_2}(\frac{\ell x_1^2}{2} - \frac{x_1^3}{6})$	$\frac{F_P\ell^3}{3EI_2}$
Point load at a	F_P	$\frac{F_P}{EI_2}(\frac{ax_1^2}{2} - \frac{x_1^3}{6})$ for $0 < x_1 < a$	$\frac{F_P}{6EI_2}a^2(3\ell - a)$
		$\frac{F_P}{EI_2}(\frac{a^2x_1}{2} - \frac{a^3}{6})$ for $a < x_1 < \ell$	
Uniform load	f	$\frac{f}{EI_2}(\frac{\ell^2x_1^2}{4} - \frac{\ell x_1^3}{6} + \frac{x_1^4}{24})$	$\frac{f\ell^4}{8EI_2}$
Torque at free end	M_3	$\frac{M_3x_1^2}{2EI_2}$	$\frac{M_3\ell^2}{2EI_2}$

6.5 Torsional Beam

We now examine how beams are affected by torques applied to the ends of the beam, with the torque vector along the beam's long axis. We first examine a circular cylinder, as shown in Fig. 6.11. The cylinder is isotropic, with Young's modulus E and Poisson's ratio ν. The cylinder, of length ℓ and radius R, has a torque M and $-M$ exerted uniformly on each end, generating a twist in the cylinder. We put the origin of our coordinate system at the cylinder's center. We will assume the cylinder twists but does not otherwise deform, so that a plane, perpendicular to the axis of the cylinder and intersecting it at x_1, remains plane under the deformation.

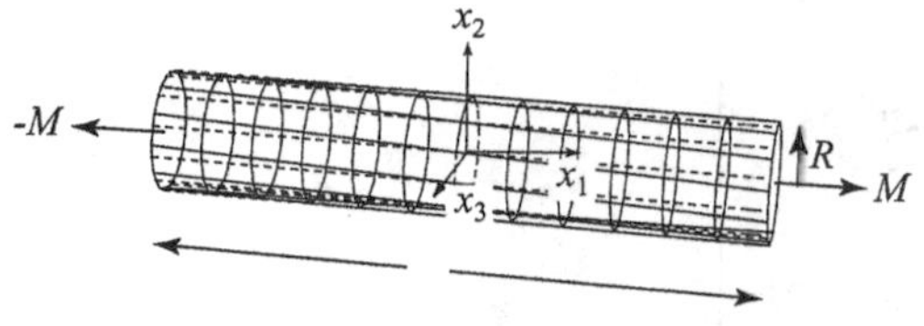

Fig. 6.11. Cylinder undergoing torsion. The undeformed cylinder is shown by the dotted lines, while the deformed cylinder by solid lines.

We assume that each point in the plane rotates about the cylinder axis by an angle $\theta(x_1)$, and the twist angle θ is linear in position along the cylinder axis, $\theta(x_1) = \alpha x_1$. The relative displacement $\boldsymbol{u}(x_1, x_2, x_3)$ is then given by a simple rotation transformation,

Table 6.2. Doubly-clamped beam shapes and deflections for a variety of load distributions. The beams are assumed to have length ℓ and polar moment along $\hat{x}_2$ of I_2, and are made of an isotropic material with Young's modulus E and Poisson's ratio ν.

Load distribution	Load	Deflection $u_2(x_1)$	Point of max. deflection x_1	Maximum deflection u_2
Point load at center	F_P	$\frac{F_P}{EI_2}(\frac{\ell x_1^2}{16} - \frac{x_1^3}{12})$ for $0 < x_1 < \ell/2$	$\ell/2$	$\frac{F_P \ell^3}{192}$
		$\frac{F_P}{EI_2}(\frac{\ell x_1^2}{16} - \frac{x_1^3}{12} + \frac{(x_1 - \ell/2)^3}{6})$ for $\ell/2 < x_1 < \ell$		
Point load at a	F_P	$\frac{F_P}{EI_2}(\frac{a(\ell-a)^2 x_1^2}{2\ell^2} - \frac{(\ell-a)^2(\ell+2a)x_1^3}{6\ell^3})$ for $0 < x_1 < \ell/2$	$\frac{2a\ell}{\ell+2a}$	$\frac{2F_P(\ell-a)^2 a^3}{3EI_2(\ell+2a)^2}$
		$\frac{F_P}{EI_2}(\frac{a(\ell-a)^2 x_1^2}{2\ell^2} - \frac{(\ell-a)^2(\ell+2a)x_1^3}{6\ell^3} - \frac{(x_1-a)^3}{6})$ for $\ell/2 < x_1 < \ell$		
Uniform load	f	$\frac{f}{EI_2}(\frac{\ell^2 x_1^2}{24} - \frac{\ell x_1^3}{12} + x_1^4 24)$	$\ell/2$	$\frac{f\ell^4}{384EI_2}$

$$u = (0, x_2(\cos\theta - 1) - x_3\sin\theta, x_2\sin\theta + x_3(\cos\theta - 1)). \tag{6.59}$$

Restricting the twist to small angles θ, the displacement may be approximated as $u = (0, -\alpha x_1 x_3, \alpha x_1 x_2)$. The strain tensor is then given by (6.1),

$$\mathsf{S} = \begin{bmatrix} 0 & -\alpha x_3/2 & \alpha x_2/2 \\ -\alpha x_3/2 & 0 & 0 \\ \alpha x_2/2 & 0 & 0 \end{bmatrix}. \tag{6.60}$$

We can calculate the stress using (5.17). This yields (translating between the tensor and six-vector notation, being careful with factors of two)

$$\mathsf{T} = \begin{bmatrix} 0 & -\mu\alpha x_3 & \mu\alpha x_2 \\ -\mu\alpha x_3 & 0 & 0 \\ \mu\alpha x_2 & 0 & 0 \end{bmatrix}, \tag{6.61}$$

in terms of the Lamé constant μ. Using (5.27) this may also be written in terms of the Young's modulus and Poisson's ratio, through the relation $\mu = E/2(1+\nu)$. Note that the stress (6.61) does not depend on the position x_1 along the length of the cylinder, only the transverse position given by x_2 and x_3.

From the stress tensor we can calculate the surface stress t, using (6.4). The surface stress along the length of the cylinder is zero, but is non-zero at either end. At the right end of the cylinder, at $x_1 = \ell/2$ with unit normal $\hat{n} = (1, 0, 0)$, the surface stress is given by

$$t = (0, -\mu\alpha x_3, \mu\alpha x_2). \tag{6.62}$$

The surface stress at the other end of the cylinder is equal and opposite. The surface stress is in the plane perpendicular to $\hat{x}_1$, and increases linearly with the radial distance from the axis of the cylinder, as sketched in Fig. 6.12.

The surface torque m (torque per unit area) can be calculated from the surface stress, $m = r \times t = \mu\alpha(x_2^2 + x_3^2, -x_1x_2, -x_1x_3)$. Integrating over the end of the cylinder to get the total torque M yields

$$M = \int_A m(x)\,\mathrm{d}A = \mu\alpha\frac{\pi R^4}{2}\hat{x}_1, \tag{6.63}$$

pointing along the cylinder axis. Hence the torque M required to give a twist angle per unit length of α, i.e. a twist angle $\mathrm{d}\theta/\mathrm{d}x$, is given by

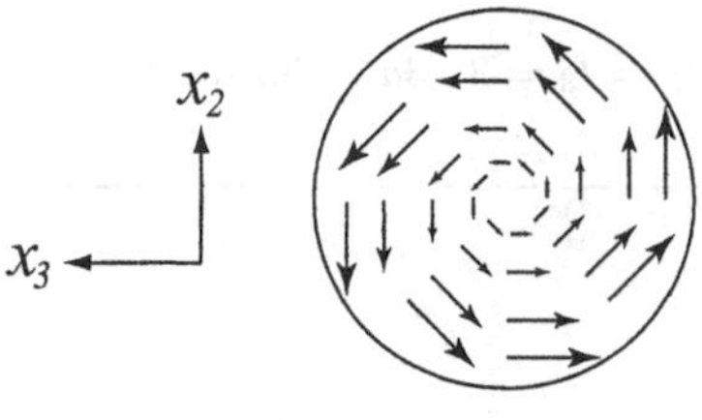

Fig. 6.12. Local stress on end of cylinder undergoing uniform torsion.

$$M = \mu I_1 \alpha = \mu I_1 \frac{d\theta}{dx}(x), \tag{6.64}$$

where $I_1 = \pi R^4/2$ is the rotational moment of inertia about x_1. The factor μI_1 is known as the *torsional rigidity* of the rod, as is written as Γ,

$$\Gamma = \frac{\pi}{2} \mu R^4. \tag{6.65}$$

6.5.1 Torsion of Non-Circular Cylinders

The method we used to treat the static torsion of a circular cylinder does not work for more general shapes, principally because the assumption that plane cross-sections remain plane does not hold. By allowing some out-of-plane displacement, however, a similar calculation may be performed, although the extra degree of freedom complicates the calculation significantly. We will not develop the mathematical approach used in these more general problems, and refer the reader to the text by Sokolnikoff [44].

The primary results of the analytic solutions are given in Table 6.3, giving the torsional rigidity Γ for various cross-sectional geometries, as well as the polar moments of inertia, and the maximum stress $\tau_{\max}$. The torque M giving a twist angle α is then $M = \Gamma\alpha = \Gamma \frac{d\theta}{dx}(x)$.

Table 6.3. Torsional properties of cylinders of different cross-sectional geometries, assuming a twist angle α per unit length. The constant μ is the Lamé constant, related to Young's modulus and Poisson's ratio by $\mu = E/(2(1+\nu))$. Torsion is about the x_1 axis, so the relevant moment are I_2 and I_3.

Cross-section	Dimensions	Torsional rigidity Γ	Polar moments of inertia I_i	Maximum stress $\tau_{\max}$
Circular	Radius R	$(\pi/2)\mu R^4$	$I_2 = I_3 = \pi R^4/2$	$\mu\alpha R$
Ellipsoidal	Semiaxes $x_2 = a$	$\pi\mu \frac{a^3 b^3}{a^2+b^2}$	$I_2 = \pi a b^3/4$	$2\mu\alpha \frac{a^2 b}{a^2+b^2}$
	$x_3 = b$		$I_3 = \pi b a^3/4$	
Square	Side a	$0.1415\mu a^{4*}$	$I_2 = I_3 = a^4/12$	$0.667\mu\alpha a^*$
Rectangular	Sides $x_2 = a$	$0.1415\mu a^{4*}$	$I_2 = I_3 = a^4/12$	$0.667\mu\alpha a^*$
	$x_3 = b$		$I_3 = \pi b a^3/4$	
Equilateral triangle	Side a	$\frac{9\sqrt{3}}{5}\mu a^4$	$I_2 = I_3 = 3\sqrt{3}a^4$	$3\mu\alpha a/2$

* Numerical prefix is approximate.

6.6 Two-Dimensional Problems

We began this chapter with a summary of the relevant equations for fully isotropic linear solids, applicable to any problem. We then specialized to the treatment of beams, and specialized further in the discussion of the essentially one-dimensional treatment of Euler–Bernoulli theory and the treatment of axial torques. In this and the next (final) sections, we discuss briefly the formalism for two- and three-dimensional problems.

The class of two-dimensional problems can be divided into two categories, those of *plane strain* and those of *plane stress*.

6.6.1 Plane Strain

An object is said to be in a state of plane strain, parallel to the x_1-x_2 plane, if the x_3 component of the displacement vector $\boldsymbol{u}$ is zero, and if the components u_1 and u_2 do not depend on x_3. We assume that all body forces and surface forces are likewise independent of x_3 and do not have x_3 components. In such a situation, the strain component S_{33} vanishes, as do the components S_{13} and S_{23}. By symmetry, then, there are only four non-zero components in the strain tensor. Plane strain typically occurs for objects with one long dimension (such as a beam), with no axial loading, and no variation in the cross-section along the object's length.

Using these properties of the strain tensor in (6.6), we can show that for an isotropic material the stress tensor elements T_{13} and T_{23} vanish, while $T_{33} = \nu(T_{11} + T_{22})$ (see Exercise 6.7). Furthermore, the stress tensor cannot depend on x_3.

From the equations relating the stress to the body force density $\boldsymbol{f}$, we see that the body force density is independent of x_3, and furthermore $f_3 = 0$; similar constraints apply to the surface stress, as can be seen through the boundary relations (6.4). The body force equations are then

$$\left.\begin{aligned} \frac{\partial T_{11}}{\partial x_1} + \frac{\partial T_{12}}{\partial x_2} &= -f_1, \\ \frac{\partial T_{21}}{\partial x_1} + \frac{\partial T_{22}}{\partial x_2} &= -f_2, \\ \frac{\partial T_{33}}{\partial x_3} &= 0. \end{aligned}\right\} \tag{6.66}$$

Using the relations (6.6), these can be transformed into the *Navier equations*,

$$\left.\begin{aligned} \mu\nabla^2 u_1 + (\lambda+\mu)\frac{\partial \Delta_2}{\partial x_1} &= -f_1, \\ \mu\nabla^2 u_2 + (\lambda+\mu)\frac{\partial \Delta_2}{\partial x_2} &= -f_2, \end{aligned}\right\} \tag{6.67}$$

where we have used the two-dimensional dilatation Δ_2,

$$\Delta_2 = \frac{\partial u_1}{\partial x_1} + \frac{\partial u_2}{\partial x_2}. \tag{6.68}$$

In addition to these relations, we must satisfy the equations of compatibility, discussed in Appendix 2. In a situation of plane strain, these equations reduce to the single relation

$$\frac{\partial^2 S_{11}}{\partial x_2^2} + \frac{\partial^2 S_{22}}{\partial x_1^2} - 2\frac{\partial^2 S_{12}}{\partial x_1 \partial x_2} = 0. \tag{6.69}$$

With some effort, using the stress–strain relation (6.5), this can be written in terms of the stress as

$$\left(\frac{\partial^2}{\partial x_1^2} + \frac{\partial^2}{\partial x_2^2}\right)(T_{11} + T_{22}) = \frac{1}{1-\nu}\left(\frac{\partial f_1}{\partial x_1} + \frac{\partial f_2}{\partial x_2}\right). \tag{6.70}$$

Let us now proceed by assuming that the body force $\boldsymbol{f}$ is a conservative force, so that it can be written as the gradient of a potential function V, $\boldsymbol{f} = -\nabla V$. This is not a very strong constraint, as many forces are conservative. With this assumption, the right side of (6.70) is simply the two-dimensional gradient operator on the potential function V, as is the left side of that equation. Hence we can write

$$\nabla_2^2 (T_{11} + T_{22}) = \frac{1}{1-\nu}\nabla^2 V, \tag{6.71}$$

(with $\nabla_2^2 = \partial^2/\partial x_1^2 + \partial^2/\partial x_2^2$).

The equations (6.66) and (6.70) determine the behavior of the stress tensor in the volume of the solid. The boundary conditions are determined by the surface stress $\boldsymbol{t}$, through (6.4). This set of two-dimensional equations, with what amount to Neumann boundary conditions, will have a well-defined solution if the problem is well-posed. Indeed, the relations (6.71) will always have a solution of the form

$$\left.\begin{aligned} T_{11} &= V + \frac{\partial^2 \Phi}{\partial x_2^2}, \\ T_{22} &= V + \frac{\partial^2 \Phi}{\partial x_1^2}, \\ T_{12} &= -\frac{\partial^2 \Phi}{\partial x_1 \partial x_2}, \end{aligned}\right\} \tag{6.72}$$

where Φ is an *Airy function.* Inserting these relations into (6.71), one finds that the Airy function must satisfy the equation

$$\nabla_2^4 \Phi = -\frac{1}{1-\nu}\nabla^2 V, \tag{6.73}$$

where the term ∇_2^4 is known as the two-dimensional *biharmonic operator*, and is given by

$$\nabla_2^4 = \frac{\partial^4}{\partial x_1^4} + 2\frac{\partial^4}{\partial x_1^2 \partial x_2^2} + \frac{\partial^4}{\partial x_2^4}. \tag{6.74}$$

If there are no body forces present, so that $\nabla V = 0$, then the Airy function satisfies what is known as the *biharmonic equation*,

$$\nabla^4 \Phi = 0. \tag{6.75}$$

Finally, we can write the explicit boundary conditions on the Airy function using (6.4),

$$\left.\begin{aligned} t_1 &= \left(V + \frac{\partial^2 \Phi}{\partial x_2^2}\right)(\hat{n}\cdot\hat{x}_1) - \frac{\partial^2 \Phi}{\partial x_1 \partial x_2}(\hat{n}\cdot\hat{x}_2), \\ t_2 &= \left(V + \frac{\partial^2 \Phi}{\partial x_1^2}\right)(\hat{n}\cdot\hat{x}_1) - \frac{\partial^2 \Phi}{\partial x_1 \partial x_2}(\hat{n}\cdot\hat{x}_2), \\ t_3 &= 0, \end{aligned}\right\} \tag{6.76}$$

where $\hat{n}$ is the unit vector pointing out from the solid surface. In the absence of body forces, these boundary conditions simplify to

$$\left.\begin{aligned} t_1 &= \frac{\partial^2 \Phi}{\partial x_2^2}(\hat{n}\cdot\hat{x}_1) - \frac{\partial^2 \Phi}{\partial x_1 \partial x_2}(\hat{n}\cdot\hat{x}_2), \\ t_2 &= \frac{\partial^2 \Phi}{\partial x_1^2}(\hat{n}\cdot\hat{x}_1) - \frac{\partial^2 \Phi}{\partial x_1 \partial x_2}(\hat{n}\cdot\hat{x}_2). \end{aligned}\right\} \tag{6.77}$$

The solutions to this set of equations are not guaranteed to be unique, in that the values of the force potential V and the Airy function Φ will still retain some arbitrary coefficients; however, the stress tensor and the corresponding strain tensor and displacement vector will be completely determined.

6.6.2 Plane Stress

An object is in a state of *plane stress*, parallel to the $x_1 - x_2$ plane, if the components T_{13}, T_{23}, and T_{33} are all equal to zero. Using the relations between stress and strain given by the isotropic relations (6.6), we can then write the relation

$$S_{33} = -\frac{\lambda}{\lambda + 2\mu}(S_{11} + S_{22}), \tag{6.78}$$

in terms of the Lamé constants λ and μ. We can then rewrite the relations between stress and strain as

$$\left.\begin{aligned} T_{11} &= \frac{2\lambda\mu}{\lambda + 2\mu}(S_{11} + S_{22}) + 2\mu S_{11}, \\ T_{22} &= \frac{2\lambda\mu}{\lambda + 2\mu}(S_{11} + S_{22}) + 2\mu S_{22}, \\ T_{12} &= 2\mu S_{12}. \end{aligned}\right\} \tag{6.79}$$

Using the relations (6.3), relating the body force density $\boldsymbol{f}$ to the strain tensor T, and replacing the strain tensor with derivatives of the displacement vector $\boldsymbol{u}$ through equations (6.1), we arrive at a relation between the two components u_1 and u_2 and the body force density in the form of a Navier equation,

$$\left.\begin{aligned} \mu\nabla_2^2 u_1 + \left(\frac{2\lambda\mu}{\lambda+2\mu} + \mu\right)\frac{\partial}{\partial x_1}\Delta_2 &= -f_1, \\ \mu\nabla_2^2 u_2 + \left(\frac{2\lambda\mu}{\lambda+2\mu} + \mu\right)\frac{\partial}{\partial x_2}\Delta_2 &= -f_2, \end{aligned}\right\} \tag{6.80}$$

where we have used the two-dimensional dilation Δ_2 defined by (6.68). If we replace the combination $2\lambda\mu/(\lambda+2\mu)$ by λ', we find the same result as for the problem of plane strain, given by (6.67), so that these two problems are formally identical.

Sophisticated techniques have been developed, using the methods of complex analysis, to find exact solutions to a number of problems in plane strain and plane stress. A discussion of these techniques, with references to the original articles developing these ideas, can be found in [44].

6.7 Three-Dimensional Problems

We now briefly discuss the approaches used to solve three-dimensional problems of stress and strain. Discussions of these techniques can be found in more detail in the treatises by Sokolnikoff [44] and Graff [14].

The central equation for a three-dimensional solid is the Navier equation (6.81), which given the description of the body force density $\boldsymbol{f}$ can in principle be solved to yield the resulting static displacement $\boldsymbol{u}$:

$$\mu\nabla^2 u_i + (\lambda+\mu)\frac{\partial}{\partial x_i}(\nabla\cdot\boldsymbol{u}) = -f_i, \qquad (i = 1 \text{ to } 3), \tag{6.81}$$

where $\nabla\cdot\boldsymbol{u} = \Delta$ is the dilatation.

The displacement $\boldsymbol{u}$ can always be written as the sum of the gradient of a scalar field Φ and the curl of a vector field $\boldsymbol{H}$,

$$\boldsymbol{u} = \nabla\Phi + \nabla\times\boldsymbol{H}. \tag{6.82}$$

Note that some authors will include elastic constants in this definition, so that in those treatments the fields Φ and $\boldsymbol{H}$ have units of stress rather than units of $(\text{length})^2$ as here. The dilatation Δ is given by

$$\Delta = \nabla^2\Phi. \tag{6.83}$$

The three-dimensional Navier equation can then be written as

$$\nabla^2\left(\mu u_i + (\lambda+\mu)\frac{\partial\Phi}{\partial x_i}\right) = -f_i. \qquad (i = 1 \text{ to } 3) \tag{6.84}$$

In the absence of body forces, so that $\boldsymbol{f} = 0$, we can define a new vector field $\boldsymbol{\Psi}$, in component form given by

$$\Psi_i = \mu u_i + (\lambda + \mu)\frac{\partial \Phi}{\partial x_i} \qquad (i = 1 \text{ to } 3). \tag{6.85}$$

Each component of $\boldsymbol{\Psi}$ is then a harmonic function satisfying

$$\nabla^2 \Psi_i = 0, \tag{6.86}$$

Combined with the equation for the dilatation,

$$\Delta = \frac{1}{\lambda + 2\mu}\nabla \cdot \Psi, \tag{6.87}$$

we can then (in principle) calculate the displacement $\boldsymbol{u}$.

Including the body forces means that particular solutions to (6.84) must be added to the force-free solutions of (6.86) and (6.87). Of particular interest are problems involving δ-function loads, applied at the point $\boldsymbol{R}$, of the form

$$\boldsymbol{f}(\boldsymbol{r}) = \boldsymbol{F}\,\delta(\boldsymbol{r} - \boldsymbol{R}), \tag{6.88}$$

where $\boldsymbol{F}$ is a vector with units of force. The solutions are then the Green's functions for the Navier equations in an infinite isotropic solid (see Jackson [60] and Morse and Feshbach [4] for a discussion of Green's functions and their uses).

A solution to this problem is given by Sokolnikoff [44],

$$\begin{aligned} u_i(\boldsymbol{r}) &= \frac{\lambda + 3\mu}{8\pi\mu(\lambda + 2\mu)}\frac{F_i}{|\boldsymbol{r} - \boldsymbol{R}|} \\ &+ \sum_{j=1}^{3}\frac{\lambda + \mu}{8\pi\mu(\lambda + 2\mu)}\frac{(r_i - R_i)(r_j - R_j)}{|\boldsymbol{r} - \boldsymbol{R}|^3}F_j, \end{aligned} \tag{6.89}$$

where Fig. 6.13 indicates the physical relations for the various terms. That these solutions satisfy the Navier equations can be verified by inserting this solution in (6.84).

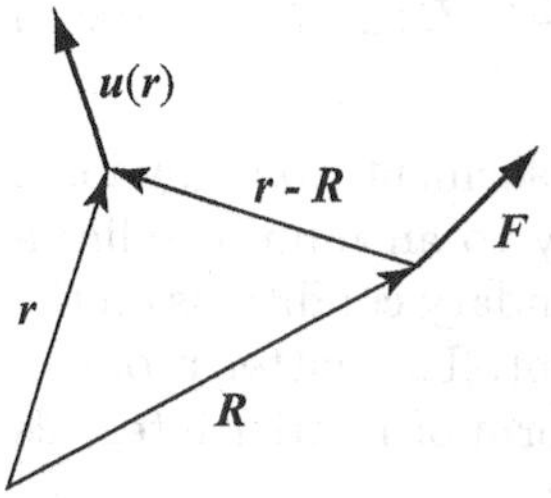

Fig. 6.13. Definition of point of evaluation $\boldsymbol{r}$ for the displacement field $\boldsymbol{u}$, and location of force $\boldsymbol{F}$ at $\boldsymbol{R}$.

If we put the force at the origin, $\boldsymbol{R} = 0$, and align our axes with the force so that $\boldsymbol{F} = F\hat{\boldsymbol{x}}_1$, then the displacement is

$$\left.\begin{aligned} u_1(\boldsymbol{r}) &= \frac{F}{16\pi\mu(1-\nu)}\left(\frac{\lambda+3\mu}{\lambda+\mu}\frac{1}{r} + \frac{r_1^2}{r^3}\right), \\ u_2(\boldsymbol{r}) &= \frac{F}{16\pi\mu(1-\nu)}\frac{r_2 r_1}{r^3}, \\ u_3(\boldsymbol{r}) &= \frac{F}{16\pi\mu(1-\nu)}\frac{r_3 r_1}{r^3}, \end{aligned}\right\} \tag{6.90}$$

using Poisson constant $\nu = \lambda/2(\lambda + \mu)$. The stress tensor components are then

$$\left.\begin{aligned} T_{11}(\boldsymbol{r}) &= -\frac{r_1}{8\pi(1-\nu)r^3}\left(3\left(\frac{r_1}{r}\right)^2 - \frac{\mu}{\lambda+\mu}\right), \\ T_{22}(\boldsymbol{r}) &= -\frac{r_1}{8\pi(1-\nu)r^3}\left(3\left(\frac{r_2}{r}\right)^2 - \frac{\mu}{\lambda+\mu}\right), \\ T_{33}(\boldsymbol{r}) &= -\frac{r_1}{8\pi(1-\nu)r^3}\left(3\left(\frac{r_3}{r}\right)^2 + \frac{\mu}{\lambda+\mu}\right), \\ T_{12}(\boldsymbol{r}) &= -\frac{r_1}{8\pi(1-\nu)r^3}\left(3\left(\frac{r_2}{r}\right)^2 + \frac{\mu}{\lambda+\mu}\right), \\ T_{13}(\boldsymbol{r}) &= -\frac{r_1}{8\pi(1-\nu)r^3}\left(3\left(\frac{r_3}{r}\right)^2 - \frac{\mu}{\lambda+\mu}\right), \\ T_{23}(\boldsymbol{r}) &= -3\frac{r_1}{8\pi(1-\nu)r^3}\frac{r_2 r_3}{r^2}. \end{aligned}\right\} \tag{6.91}$$

More complex problems involving distributed loads can be solved (formally) using these solutions. A general force distribution $\boldsymbol{f}(\boldsymbol{R})$ will cause a displacement given by

$$\begin{aligned} u_i(\boldsymbol{r}) &= \int \frac{\lambda+3\mu}{8\pi\mu(\lambda+2\mu)}\frac{f_i(\boldsymbol{R})}{|\boldsymbol{r}-\boldsymbol{R}|}\,\mathrm{d}^3R \\ &\quad + \sum_{j=1}^{3}\int \frac{\lambda+\mu}{8\pi\mu(\lambda+2\mu)}\frac{(r_i-R_i)(r_j-R_j)}{|\boldsymbol{r}-\boldsymbol{R}|^3}f_j(\boldsymbol{R})\,\mathrm{d}^3R. \end{aligned} \tag{6.92}$$

This formal integral can be used to calculate the displacement from any force distribution. We note that these solutions only apply to an infinite solid; a semi-infinite solid, for example, will have separate boundary conditions on the free surface, and the Green's functions will be different. This latter problem was studied by J. Boussinesq, and a solution in the form of Fourier integrals was found (see Sokolnikoff [44]).

Exercises

6.1 Stress and strain can always be related by a fourth-order elastic stiffness tensor, such that

$$\epsilon_{ij} = \sum_{k,l=1}^{3} c_{ijkl} \tau_{kl}. \tag{6.93}$$

For a linear isotropic solid with Young's modulus E and Poisson's ratio ν, find the non-zero elements c_{ijkl} in the tensor c, and what their values must be.

6.2 Fill in the steps missing in the derivation for a beam subjected to a uniform body force, from (6.21) to (6.25).

6.3 Verify that the solution given in (6.31) is consistent with the displacement relations in (6.29). Fill in the steps missing in the derivation of (6.31).

6.4 Show that the Euler–Bernoulli formula and the exact theory are equivalent for beams with length ℓ much less than the radius of curvature R. Verify the stated errors for beams with lengths approaching and exceeding R.

6.5 Using the beam bending equation (6.47), find the boundary conditions and the relative displacement of a cantilevered beam with one end pinned and the other end free, given a uniform body force $\boldsymbol{f} = f_0 \hat{\boldsymbol{x}}_2$ per unit volume. The beam has length ℓ, width w, thickness t, and the x_1 axis is aligned with the neutral axis.

6.6 As for Exercise 4.5, find the boundary conditions and shape of a beam with the left end clamped and the right end pinned, with a uniform body force $\boldsymbol{f} = f_0 \hat{\boldsymbol{x}}_2$.

6.7 A doubly clamped beam of length ℓ, width w along $\hat{\boldsymbol{x}}_2$, and thickness t along $\hat{\boldsymbol{x}}_3$ is subjected to a point force $\boldsymbol{F}_1 = A\hat{\boldsymbol{x}}_2$ at $x_1 = \ell/3$, and a second point force $\boldsymbol{F}_2 = A\hat{\boldsymbol{x}}_3$ at $x_1 = 2\ell/3$. What are the boundary conditions and relative displacements $u_2(x_1)$ and $u_3(x_1)$ for the beam?

6.8 In Table 6.2, the shape is given for a doubly clamped square beam, of side a and length ℓ, subjected to a uniform body force $\boldsymbol{f} = f_0 \hat{\boldsymbol{x}}_2$. Recalculate this result including the effects of shear, and compare with the tabulated result. How different is the maximum deflection for an aspect ratio $a/\ell = 0.1$? How about for $a/\ell = 1$?

6.9 Verify that for an isotropic material in a state of plane strain, the stress components τ_{13} and τ_{23} vanish, while $\tau_{33} = \nu(\tau_{11} + \tau_{22})$.

7. Dynamical Behavior of Solids

We now turn to the treatment of the dynamics of deformable solids. The basis of the discussion is very similar to that of statics in the previous chapter, as the time-dependence adds only one term to the equations relating stress and strain. The formalism we develop here will allow simple problems to be solved, developing an understanding that can be applied to more complex, but often more realistic, problems; these typically must be solved through numerical techniques.

We begin by describing a one-dimensional version of these equations, that provides a simple approach to describing the longitudinal, torsional, and flexural vibrations of long beams: These are the *normal modes* for long beams, analogous to those discussed in Chaps. 1 and 2. The dynamical equations of motion can be applied to a number of situations, and provide fairly accurate predictions for the frequency and mode shapes of beams with length much larger than their width or thickness, and with wavelengths of the same order as the beam length (and therefore much larger than the cross-sectional dimensions). We also describe, for each case, a simple approach to calculating the stress and strain in the volume of the beams, and indicate some of the resulting inaccuracies in this approximate approach.

Following this discussion, we turn to a description of the full three-dimensional dynamic equations of motion, which will allow us to find the formal description for waves in the infinite solid, and follow with a discussion of plane wave reflections from a plane interface, as well as waves in a plate, and concluding with a discussion of the three-dimensional solution for torsional waves in a cylindrical rod.

7.1 Simple Vibrational Methods: Torsional Vibrations in Beams

The complexity of the full-blown analytic solution to vibrational problems in restricted geometries, such as in beams, is so severe that it is often best left to numerical solutions; this is true even for the simplest problems with isotropic materials. Due to this complexity, a range of alternative, approximate approaches have been developed, which for small amplitudes and low-order modes yield reasonable results for simple geometries. We first discuss the simplest theory, that for longitudinal vibrations in long beams, and we then discuss torsional and flexural motion.

7.1.1 Longitudinal Vibrations

Consider a rod of length ℓ, with cross-sectional area A, with its length oriented along the z axis (see Fig. 7.1). The rod is made of an isotropic material with Young's modulus E, and for simplicity we take Poisson's ratio ν equal to zero. At any point z along the axis, we can consider a differential element of length dz, extending from z to $z+\mathrm{d}z$, with a local longitudinal displacement $U(z)$. Each surface of the element has a force exerted on it by its neighboring elements, with force $-F(z)$ on the surface at z, and $F(z+\mathrm{d}z)$ on the surface at $z+\mathrm{d}z$, as shown in Fig. 7.2. We will assume there are only longitudinal forces.

For static equilibrium, as discussed in Chap. 6, the forces on the two end surfaces must balance, so we would have $F(z) = F(z+\mathrm{d}z)$. Here we interested in the dynamic response of the rod, so this force balance need not hold. We must however specify the *density* ρ of the rod, as this will determine its inertial response; the mass of our differential element is $\mathrm{d}m = \rho A \mathrm{d}z$, and its acceleration is $\partial^2 U/\partial t^2$. We can now write down the dynamic equation of motion for the element:

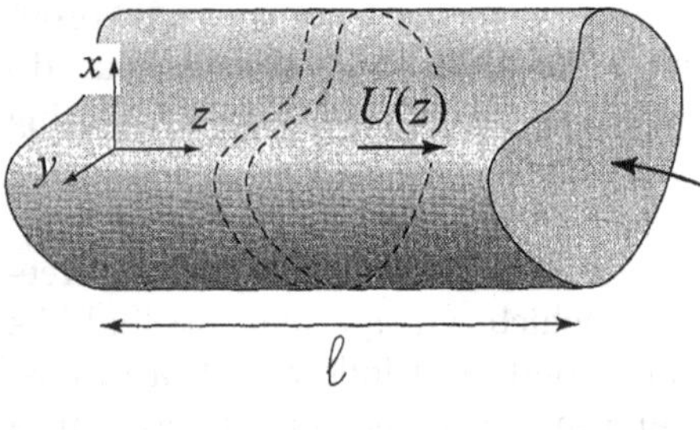

Fig. 7.1. Prismatic bar of arbitrary cross-section, with area A, length ℓ, density ρ, and Young's modulus E. The displacement at each point z is $U(z)$.

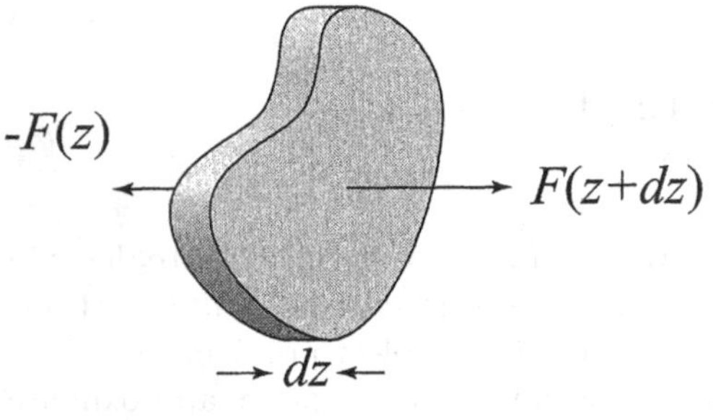

Fig. 7.2. The forces acting on an element of length dz of the prismatic bar.

$$\rho A \,\mathrm{d}z \frac{\partial^2 U}{\partial t^2} = F(z + \mathrm{d}z, t) - F(z, t). \tag{7.1}$$

We can expand the force $F(z + \mathrm{d}z)$ in a Taylor series to first order in z,

$$F(z + \mathrm{d}z) \approx F(z) + \frac{\mathrm{d}F}{\mathrm{d}z}\mathrm{d}z + \ldots, \tag{7.2}$$

and keeping only the first two terms, we can write (7.2) as

$$\rho A \frac{\partial^2 U}{\partial t^2} = \frac{\mathrm{d}F}{\mathrm{d}z}. \tag{7.3}$$

In Chap. 6 we found the relation between the strain and the stress for longitudinal forces, $\partial U/\partial z = F/EA$ (see (6.12)), so that we can write (7.3) as

$$\rho \frac{\partial^2 U}{\partial t^2} = E \frac{\partial^2 U}{\partial z^2}. \tag{7.4}$$

This is a one-dimensional wave equation for the relative longitudinal displacement $U(z, t)$.

We consider for the moment the solutions for the infinitely long rod, so that we can ignore end effects. A general solution to the infinite rod is the d'Alembert solution, which consists of travelling waves of the form $U(z, t) = f(z \pm c_\phi t)$, where $f(s)$ is *any* twice differentiable function of the variable s, and c_ϕ is the phase velocity,

$$c_\phi = (E/\rho)^{1/2}. \tag{7.5}$$

If the argument of f involves the "+" sign, the solutions are waves with waveshape $f(s)$ travelling towards $-z$ at velocity c_ϕ, while those with the "−" sign travel towards $+z$ at the same velocity. Any functional form $f(s)$ will do, and travels without changing its shape, at constant velocity.

A specific form of this type of solution, with a slight change in notation, is formed from the complex exponential of the same argument, of the form

$$U(z, t) \;=\; U_0 \exp i(qz \pm \omega t + \phi). \tag{7.6}$$

Here q is the wavevector and ω the frequency of the travelling wave, U_0 the displacement amplitude, and ϕ the phase. The wavevector q is free to take on any value, as are the amplitude and the phase. In order to satisfy the wave equation (7.4), the frequency ω must be related to the wavevector q by

$$\omega = c_\phi q. \tag{7.7}$$

Hence we find a linear *dispersion relation* for longitudinal vibrations. This is what one would expect from the discussion in Chap. 1, for waves in a chain of atoms: Here we are in the long-wavelength limit for the chain, with wavelengths $2\pi/q$ that are much larger than the interatomic spacing, in which case the atomic chain result reduces to $\omega = c_\phi q = (E/\rho)^{1/2}\, q$.

We note that the solutions (7.6) are complex; we can find an equally general set of solutions by taking the real part of the exponential, to give an alternate *real* form

$$U(z,t) = U_0 \cos(qz \pm \omega t + \phi). \tag{7.8}$$

The set of functions (7.6), or equivalently (7.8), for all wavevectors q and their corresponding frequencies ω, is a *complete set* of functions, in the sense that *any* twice differentiable function $f(z \pm v_\phi t)$ can be written as a linear superposition of functions from this set; this is the basis of the Fourier transform theory discussed in Chap. 2. As the wavevector q in (7.6) can have both positive and negative values, the set of solutions can be restricted to those involving only $-\omega t$ without any loss of generality. This can be seen by taking the real part of the exponential, which is equal to $\cos(qz + \omega t + \phi)$, and recognizing that, because the cosine is an even function, this is equal to $\cos(-qz - \omega t - \phi)$, and can therefore also be described by the exponential with argument $q'z - \omega t + \phi'$, with $q' = -q$ and $\phi' = -\phi$.

The solutions (7.6) apply to the infinite beam. We have not imposed any boundary conditions on the displacement $U(z,t)$. We can now, for example, clamp the ends of the rod, by setting the displacement to zero at $z = -\ell/2$ and $z = \ell/2$:

$$\left.\begin{aligned} U(-\ell/2,t) &= 0, \\ U(\ell/2,t) &= 0. \end{aligned}\right\} \tag{7.9}$$

These boundary conditions will restrict the set of allowed values for the wavevector q in the solutions (7.6) to the set

$$q_n = n\frac{\pi}{\ell}, \qquad (n = 1,2,3\ldots) \tag{7.10}$$

and the corresponding frequencies are restricted to the set

$$\omega_n = nc_\phi\frac{\pi}{\ell}. \qquad (n = 1,2,3\ldots) \tag{7.11}$$

Solutions $U_n(z,t)$, at a given frequency ω_n, are then formed from the linear superposition of the complex solutions with wavevectors $+q_n$ and $-q_n$,

$$U_n(z,t) = (U_{0n}\cos(q_n z) + U'_{0n}\sin(q_n z))\cos(\omega_n t + \phi), \tag{7.12}$$

where we have written these in an explicitly real format. Applying the boundary conditions to (7.12), we see that we must have $U_{0n} = 0$ for n even, and $U'_{0n} = 0$ for n odd. Hence the solutions with these boundary conditions are

$$\left.\begin{aligned} U_n(z,t) &= U_{0n}\cos(n\pi z/\ell)\cos(\omega_n t + \phi) \qquad (n \text{ odd}), \\ U_n(z,t) &= U_{0n}\sin(n\pi z/\ell)\cos(\omega_n t + \phi) \qquad (n \text{ even}). \end{aligned}\right\} \tag{7.13}$$

These solutions are the normal modes for the longitudinal vibrations of a beam of length ℓ; they form the complete description of the longitudinal part of the motion of the beam, and furthermore all of the thermodynamic and energy transport relations worked out in Chapters 1 through 3 can be applied

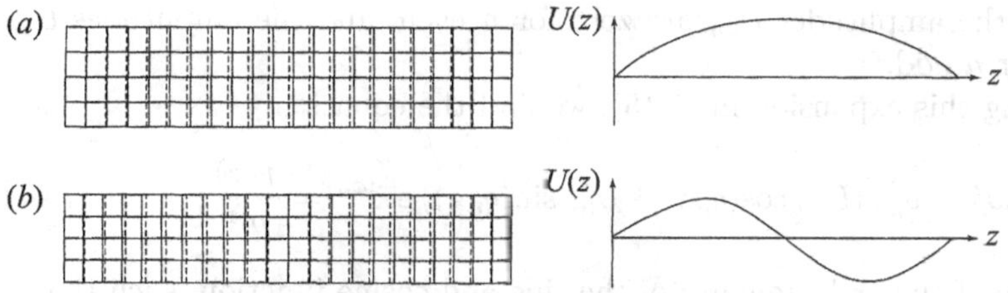

Fig. 7.3. (**a**) Longitudinal displacement for the $n = 1$ mode, with the dotted lines showing the undistorted rod, in cross- section. (**b**) Same geometry, but for the $n = 2$ mode.

to these modes. In Fig. 7.3 we display the first two modes, for $n = 1$ and $n = 2$. These solutions are standing waves, whose maximum displacement point is fixed, and the displacement amplitude oscillates between its positive and negative limits at frequency ω_n.

The equation of motion (7.4) is for an undriven beam. We can very simply add the effect of an external body force, directed along the z axis, by adding to this equation a time-dependent force per unit length $f(z,t)$. The equation of motion for the differential element then includes the force $f(z,t)\mathrm{d}z$ in addition to the forces applied to either end of the element, yielding (after dividing out the common term $\mathrm{d}z$)

$$\rho A \frac{\partial^2 U}{\partial t^2}(z,t) = EA \frac{\partial^2 U}{\partial z^2}(z,t) + f(z,t). \tag{7.14}$$

This equation has solutions consisting of the superposition of the *homogenous* solutions for the undriven problem, and the *particular* solutions associated with the driving force.

We can, without loss of generality, assume that the time dependence of the driving force has the form $f(z,t) = f(z)\cos(\omega_d t)$, oscillating at the drive frequency ω_d; other time dependencies can be written as a superposition of cosines, through the use of the Fourier transform. For ease of notation, we write $f(z,t)$ in complex form, $f(z,t) = f(z)\exp(\mathrm{i}\omega_d t)$, where it is understood that we will take the real part at the end. The particular solution satisfying (7.14) has the same time dependence as $f(z,t)$, $U_p(z,t) = U(z)\exp i(\omega_d t + \phi)$. The z-dependence is determined from (7.14), where the common time dependence has been divided out:

$$-\rho A \omega_d^2 U(z) \mathrm{e}^{\mathrm{i}\phi} = EA\mathrm{e}^{\mathrm{i}\phi} \frac{\partial^2 U}{\partial z^2}(z) + f(z). \tag{7.15}$$

The solution to (7.15) may be found by writing the particular solution as a linear combination of the set of spatial solutions to the undriven problem, which, as mentioned above, form a complete set for functions satisfying the same boundary conditions. We therefore write

$$U(z,t) = \sum_{n=1}^{\infty} \left(U_{0n}\cos(q_n z) + U'_{0n}\sin(q_n z)\right) \mathrm{e}^{\mathrm{i}(\omega_d t - \phi)}. \tag{7.16}$$

Note that the amplitudes U_{0n} are zero for n even, and the amplitudes U'_{0n} are zero for n odd.

Inserting this expansion in (7.15), we find the equation

$$\sum_{n=1}^{\infty} \left(\omega_d^2 - \omega_n^2\right) \left(U_{0n} \cos(q_n z) + U'_{0n} \sin(q_n z)\right) \mathrm{e}^{-\mathrm{i}\phi_n} = \frac{f(z)}{\rho A}. \tag{7.17}$$

Making use of the orthogonality of the sine and cosine function, such that

$$\int_{-\ell/2}^{\ell/2} \cos(q_n z) \cos(q_m z)\, \mathrm{d}z = \frac{1}{2}\delta_{mn}, \tag{7.18}$$

which holds for all $n, m > 0$, with a similar relation for $\sin(q_n z)$ and $\sin(q_m z)$, we can write this as

$$\left\{ \begin{array}{c} U_{0n} \\ U'_{0n} \end{array} \right\} = \frac{2\mathrm{e}^{\mathrm{i}\phi_n}}{\omega_d^2 - \omega_n^2} \int_{-\ell/2}^{\ell/2} \left\{ \begin{array}{c} \cos(q_n z) \\ \sin(q_n z) \end{array} \right\} \frac{f(z)}{\rho A}\, \mathrm{d}z, \quad \left\{ \begin{array}{c} n \text{ odd} \\ n \text{ even} \end{array} \right\}. \tag{7.19}$$

To proceed, we must choose an explicit form for the spatial dependence $f(z)$ of the driving force. If, for example, we choose a constant force $f(z) = f_0$, distributed over the entire length of the rod, then the integrals in (7.19) evaluate to

$$\int_{-\ell/2}^{\ell/2} \left\{ \begin{array}{c} \cos(q_n z) \\ \sin(q_n z) \end{array} \right\} f_0\, \mathrm{d}z = \left\{ \begin{array}{c} 2f_0/q_n \\ 0 \end{array} \right\}, \tag{7.20}$$

so that the amplitudes U_{0n} are given by

$$U_{0n} = \frac{4f_0}{\rho A} \frac{\ell}{n\pi} \frac{1}{\omega_d^2 - \omega_n^2} \mathrm{e}^{\mathrm{i}\phi_n}, \tag{7.21}$$

and the amplitudes U'_{0n} are all identically zero. If f_0 is real, then we can set $\phi_n = 0$, as the phase is chosen to give real amplitudes.

Note that the amplitude for the nth term depends strongly on the relationship between the drive frequency ω_d and the normal mode frequency ω_n, and for $\omega_d \to \omega_n$ the expression (7.21) diverges. This is unphysical, and is due to the fact that we are ignoring the presence of energy loss terms, or dissipation, in the equation of motion. Dissipative terms may be included phenomenologically by adding a term proportional to the velocity $\partial U/\partial t$ in the equation of motion (7.14), which will generate a finite response in place of the divergent response (7.21). We discuss the topic of dissipation, and how it affects the response of resonant mechanical systems, in much more detail in Chap. 8.

If we had chosen a different spatial dependence for the driving term $f(z)$, for instance one with odd parity rather than even (so that $f(-z) = -f(z)$), then the integrals in (7.17) involving the cosine would vanish, so that the terms U_{0n} would be zero, while the terms U'_{0n} would in general be non-zero. The spatial dependence of the driving term will couple better to those modes that have a similar functional dependence on z, while the modes that are effectively orthogonal to the driving force, giving a zero integral in (7.19), will not couple even if the frequency of f is equal to the natural resonance frequency for that mode.

For driving forces $f(z,t)$ that are not pure cosine functions in time, we can write the time dependence of f as a Fourier transform, determining the range of pure frequencies that define the time dependence of f. The frequency components that are close to one of the resonator mode frequencies ω_n, and also have a spatial dependence that couples well to the mode shape (through an integral of the form (7.19)), will generate strong amplitude response in the resonator.

7.1.2 Torsion of Cylinders

We now turn to a discussion of the torsional dynamic motion in a cylinder. Consider a cylindrical rod of length ℓ, made of an isotropic material with Lamé constants λ and μ (see Chap. 5), with polar moment of inertia I_z about the cylinder axis $\hat{z}$ (see Fig. 7.4); the density of the rod is ρ. At any point z along the axis, the rod will have a local torsion angle $\theta(z,t)$. In Sect. 6.5 we discussed the problem for the static torsional response of such a rod, under an externally applied static torque; here we look at the time-dependent response.

As with the longitudinal displacement problem, we begin by considering a differential cross-sectional element of length $\mathrm{d}z$, with torques $-M(z,t)$ and $M(z+\mathrm{d}z,t)$ applied to either face, exerted by the neighboring differential elements. If these torques are not balanced, the element will rotate, at a rate limited by its rotational moment of inertia $\rho I_z\,\mathrm{d}z$. Balancing the inertial response and the elastic torques on the cross-sectional element, we can write the dynamical equation for torque and angular acceleration,

$$\rho I_z \mathrm{d}z \frac{\partial^2 \theta}{\partial t^2} = M(z+\mathrm{d}z,t) - M(z,t). \qquad (7.22)$$

We assume that the torque varies slowly along the beam, and expand the torque in a Taylor series, $M(z+\mathrm{d}z,t) = M(z,t) + (\partial M/\partial z)\,\mathrm{d}z$ to first order

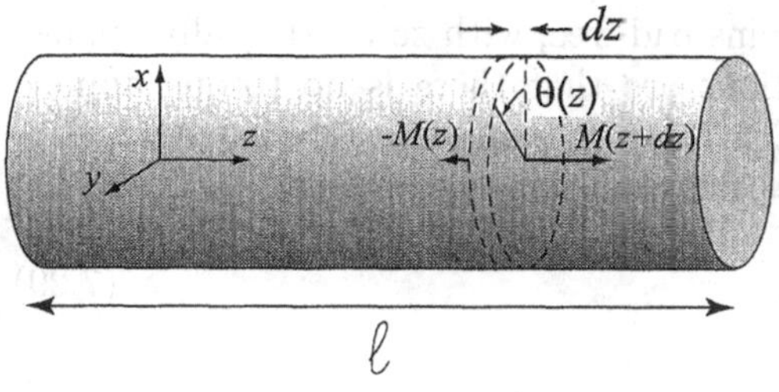

Fig. 7.4. Torsion cylinder geometry, with a cylinder of length ℓ oriented along the x_3 axis.

in z. We also apply the "strength-of-materials" result calculated for static torsion in Chap. 6, (6.64),

$$M(z,t) = \mu I_z \frac{\partial \theta}{\partial z}. \tag{7.23}$$

Combining these relations, we can write (7.22) as

$$\frac{\partial^2 \theta}{\partial t^2} = \frac{\mu}{\rho} \frac{\partial^2 \theta}{\partial z^2}. \tag{7.24}$$

We see that we have generated a one-dimensional wave equation in the torsion angle θ. The solutions to this equation are very similar to those for the longitudinal problem.

For the infinitely long rod, the solutions have the form

$$\theta(z,t) = \theta_0 \mathrm{e}^{\mathrm{i}(qz-\omega t+\phi)}, \tag{7.25}$$

where we anticipate that we need only include the $-\omega$ terms, and of course only the real part of (7.25) is relevant. The dispersion relation between ω and the wavevector q is $\omega = c_\phi\, q$, where now the phase velocity is $c_\phi = (\mu/\rho)^{1/2}$.

Applying clamped boundary conditions at $z = \pm\ell/2$ constrains the wavevectors to the values

$$q_n = n\frac{\pi}{\ell}, \qquad (n = 1, 2, 3 \ldots) \tag{7.26}$$

and the frequencies to

$$\omega_n = n c_\phi \frac{\pi}{\ell}. \qquad (n = 1, 2, 3 \ldots) \tag{7.27}$$

Solutions $\theta_n(z,t)$, at a given frequency ω_n, are then formed from the linear superposition of the solutions with wavevectors $+q_n$ and $-q_n$, with the result

$$\left.\begin{aligned} \theta_n(z,t) &= \theta_{0n} \cos(n\pi z/\ell) \cos(\omega_n t + \phi) \qquad (n \text{ odd}), \\ \theta_n(z,t) &= \theta_{0n} \sin(n\pi z/\ell) \cos(\omega_n t + \phi) \qquad (n \text{ even}). \end{aligned}\right\} \tag{7.28}$$

Our solutions to the problem with both ends fixed are therefore standing waves with alternating cosinusoidal and sinusoidal spatial dependence, depending on the index n, identical to the longitudinal problem. In Fig. 7.5 (a) and (b), we show the first two mode shapes for this choice of boundary conditions.

If the ends of the cylinder were instead free, with zero externally applied torque, then the boundary conditions are that there is no torque applied to either end, $M(-\ell/2,t) = M(\ell/2,t) = 0$. Using the relation $M(z,t) = \mu I_z \partial\theta/\partial z$ from (6.64), this imposes the condition on the angle

$$\left.\frac{\partial \theta}{\partial z}\right|_{z=\pm\ell/2} = 0. \tag{7.29}$$

The allowed solutions are

$$\left.\begin{aligned} \theta_n(z,t) &= \theta_{0n} \sin(n\pi z/\ell) \cos(\omega_n t + \phi) \qquad (n \text{ odd}), \\ \theta_n(z,t) &= \theta_{0n} \cos(n\pi z/\ell) \cos(\omega_n t + \phi) \qquad (n \text{ even}), \end{aligned}\right\} \tag{7.30}$$

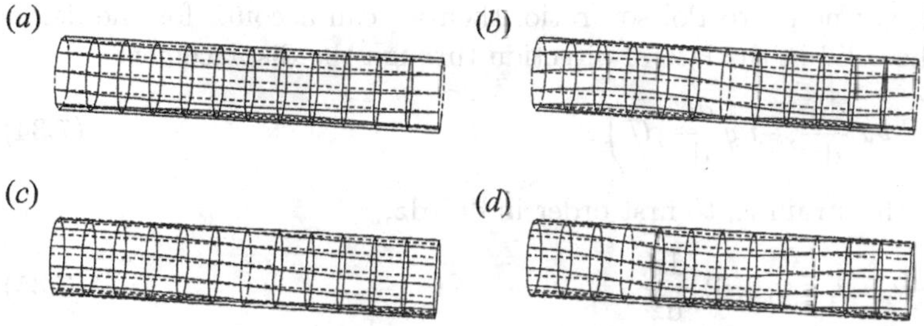

Fig. 7.5. In (**a**) and (**b**) we show the first two modes for torsional vibration for a cylinder fixed at both ends, and in (**c**) and (**d**) those for a cylinder with torsion-free ends.

where again θ_{0n} and ϕ are determined by the initial conditions. In Fig. 7.5 (c) and (d) we show these two mode shapes.

The equation of motion for the torsion angle, (7.24), is for an undriven cylinder, where no external torques are applied. If one wishes to include external torques, with a torque per unit length $m(z,t)$, then the equation of motion would have the form

$$\mu I_z \frac{\partial^2 \theta}{\partial t^2}(z,t) + m(z,t) = \rho I_z \frac{\partial^2 \theta}{\partial z^2}. \tag{7.31}$$

Solutions to this driven equation may be found by following exactly the same prescription as for the longitudinal problem.

7.1.3 Stress and Strain in Longitudinal and Torsional Motion

We briefly discuss the stress and strain in the vibrating longitudinal and torsional rods.

Stress and Strain for Longitudinal Vibrations. The displacement $U(z,t)$ we calculated for the normal resonant modes in the longitudinal rod is an absolute displacement. This absolute displacement is identified with the displacement *vector* $\boldsymbol{u} = (0,0,U)$ (for displacement along z). The strain six-vector (see Sect. 5.2.1) is then given by

$$\epsilon = (0,0,1,0,0,0)\,\frac{\mathrm{d}U}{\mathrm{d}z}. \tag{7.32}$$

The stress six-vector $\boldsymbol{\tau}$ is

$$\boldsymbol{\tau} = (\lambda, \lambda, 2\mu + \lambda, 0, 0, 0)\,\frac{\mathrm{d}U}{\mathrm{d}z}, \tag{7.33}$$

using the Lamé constants λ and μ. The stresses τ_1 and τ_2 appear because the displacement U does not have any contraction associated with the non-zero Poisson ratio $\nu = \lambda/2(\lambda+\mu)$; we assumed at the beginning of this section that we would take $\nu = 0$, or equivalently $\lambda = 0$. With non-zero λ, stress along x and y is required to prevent displacement along x and y.

If we allow a non-zero Poisson ratio, then we can account for the deformation of the solid in the lateral direction through the displacement

$$\boldsymbol{u} = \left(-\nu x \,\frac{\mathrm{d}U}{\mathrm{d}z}, -\nu y \,\frac{\mathrm{d}U}{\mathrm{d}z}, U \right). \tag{7.34}$$

In this case, the strain is, to first order in $\mathrm{d}U/\mathrm{d}z$,

$$\boldsymbol{\epsilon} = (-\nu, -\nu, 1, 0, 0, 0)\,\frac{\mathrm{d}U}{\mathrm{d}z}, \tag{7.35}$$

and the stress is

$$\boldsymbol{\tau} = (0, 0, E, 0, 0, 0)\,\frac{\mathrm{d}U}{\mathrm{d}z}, \tag{7.36}$$

with no out-of-plane stress on the free surfaces. However, the equation of motion (7.1) does not include the inertial effects associated with the lateral motion, perpendicular to the length, that are required by (7.34). Including these effects in the equation of motion is a rather involved. A derivation of an approximate theory was worked out by Love [61], and a somewhat simplified description appears in Graff's book [14]. The resulting equation of motion for U has the form

$$\rho\frac{\partial^2 U}{\partial t^2} = E\,\frac{\partial^2 U}{\partial z^2} + \nu^2 \rho\,\frac{I_3}{A}\,\frac{\partial^4 U}{\partial z^2 \partial t^2}, \tag{7.37}$$

where E is Young's modulus, ρ the density, A the cross-sectional area, and I_3 the polar moment of inertia about z (see Sect. 6.4). Note that this wave equation is *dispersive*, so the relation between wavevector and frequency for a plane wave is non-linear. For a solution of the form $U = U_0 \mathrm{e}^{\mathrm{i}kz - \mathrm{i}\omega t}$, this yields

$$-k^2 + \frac{\nu^2 I_3 \rho}{AE}\omega^2 k^2 + \frac{\rho}{E}\omega^2 = 0. \tag{7.38}$$

For small values of the wavevector k, Love's theory agrees fairly well with experimental and numerical simulations; for very small k, the correction is negligible, and we find the previous linear, non-dispersive result $\omega = c_\phi k$. For large k, however, Love's theory overestimates the correction; it gives a reasonable approximation for $k < 0.3\sqrt{A/I_3}/\nu$. For wavevectors greater than this value, it is reasonable to take the approximate dispersionless equation (7.4), but with the ratio E/ρ replaced by $\approx 0.28E/\rho$.

Stress and Strain for the Torsional Cylinder. We now describe the stress and strain in the cylinder undergoing torsional vibrations. From the discussion in Sect. 6.5, we can write the displacement vector $\boldsymbol{u}$ for torsion by an angle θ as

$$\boldsymbol{u} = (x(\cos\theta - 1) - y\sin\theta, x\sin\theta + y(\cos\theta - 1), 0). \tag{7.39}$$

For small angles this is approximately $\boldsymbol{u} = (-y\theta, x\theta, 0)$. The strain six-vector $\boldsymbol{\epsilon}$ for small angles is then given by

$$\epsilon = \left(0, 0, 0, x\frac{\mathrm{d}\theta}{\mathrm{d}z}, -y\frac{\mathrm{d}\theta}{\mathrm{d}z}, 0\right). \tag{7.40}$$

The stress six-vector $\boldsymbol{\tau}$ is obtained using (5.17),

$$\boldsymbol{\tau} = \left(0, 0, 0, \mu x\frac{\mathrm{d}\theta}{\mathrm{d}z}, -\mu y\frac{\mathrm{d}\theta}{\mathrm{d}z}, 0\right), \tag{7.41}$$

in terms of the Lamé constant μ. The stress is in-plane, proportional to the twist $\mathrm{d}\theta/\mathrm{d}z$. The same stress was found in Sect. 6.5.

7.1.4 Flexural Vibrations

The final one-dimensional problem we will discuss is that of flexural vibrations. There are two levels of complexity in this discussion, the simpler approach where the rotational inertia of the beam is not considered, and the second where it is included (in an approach usually called the Timoshenko beam analysis). We describe each of these approaches in turn.

Simple Approach to Flexural Vibrations. Consider the prismatic beam shown in Fig. 7.6, with length ℓ aligned along the z axis. We will calculate the dynamic behavior of this beam when it flexes, due to displacements $U(z,t)$ in the x-direction.

A differential element of length $\mathrm{d}z$ and cross-sectional area A is subject to forces $F_x(z+\mathrm{d}z)$ and $-F_x(z)$ on each face, directed along x, and torques $M_y(z+\mathrm{d}z)$ and $-M_y(z)$ directed along y (see Fig. 7.7), all from the neighboring elements. Balancing linear forces, we find the equation

$$F_x(z+\mathrm{d}z) - F_x(z) - \rho A\,\mathrm{d}z\frac{\partial^2 U}{\partial t^2} = 0. \tag{7.42}$$

There should be no net torque on the element, so calculating the torque about the left side at z, the torque balance is

$$F_x(z+\mathrm{d}z)\mathrm{d}z + M_y(z+\mathrm{d}z) - M_y(z) = 0. \tag{7.43}$$

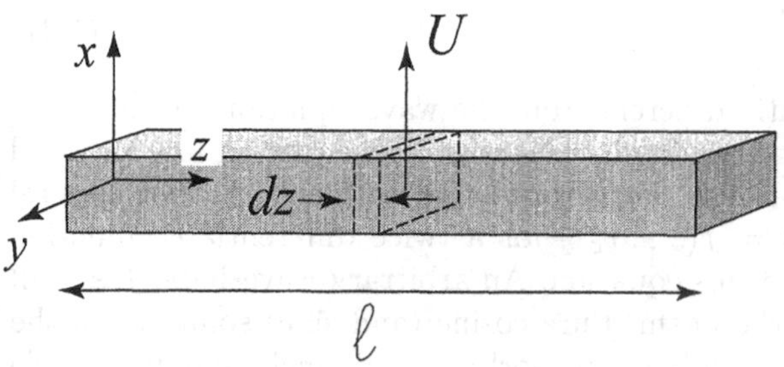

Fig. 7.6. A beam of length ℓ, with transverse displacement $U(z)$ due to forces along x and torques along y.

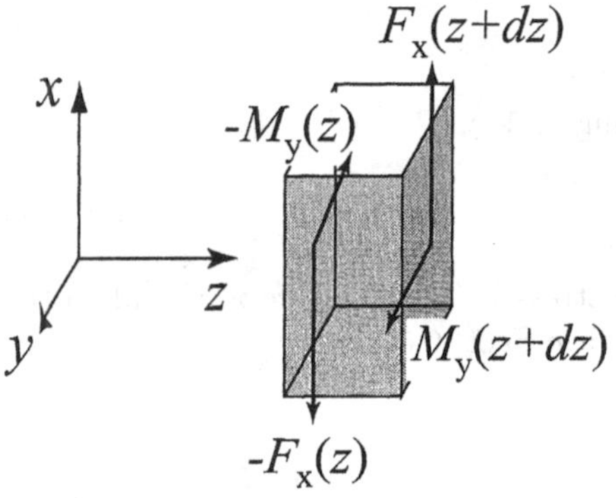

Fig. 7.7. A differential element of length dz is subject to forces $F_x(z + dz)$ and $F_x(z)$, and torques $M_y(z+dz)$ and $-M_y(z)$, which generate the displacement $U(z)$.

Expanding the forces and torques in a Taylor series about the point z, and keeping only first order terms in dz, these equations become

$$\left.\begin{aligned} \frac{\partial F_x}{\partial z} &= \rho A \frac{\partial^2 U}{\partial t^2}, \\ F_x(z) &= -\frac{\partial M_y}{\partial z}. \end{aligned}\right\} \tag{7.44}$$

Combining these equations and the strength of materials result (6.38),

$$M_y = EI_y \frac{\partial^2 U}{\partial z^2}, \tag{7.45}$$

we find the wave equation

$$\frac{\partial^2}{\partial z^2}\left(EI_y \frac{\partial^2 U}{\partial z^2}\right) = -\rho A \frac{\partial^2 U}{\partial t^2}. \tag{7.46}$$

If the beam's bending moment I_y is constant along the length of the beam, which is true for a prismatic beam, then this equation may be written

$$EI_y \frac{\partial^4 U}{\partial z^4} = -\rho A \frac{\partial^2 U}{\partial t^2}. \tag{7.47}$$

This result is significantly different from the wave equations we have described for torsional and longitudinal vibrations. The equation has a second order derivative in time, but a *fourth* order derivative in z. The general d'Alembert solution, $U(z) = f(z \pm v_\phi t)$ for a twice differentiable function f, is no longer a solution to this equation: An arbitrary waveshape $f(s)$ will distort as it travels along the beam. Pure cosine (and sine) solutions of the form $U(z) = A\cos(qz - \omega t)$ will however work, as these include only a single frequency component, and these satisfy (7.47) with the dispersion relation

$$EI_y q^4 = \rho A \omega^2, \tag{7.48}$$

or $\omega = (EI_y/\rho A)^{1/2}q^2$. The nonlinear relation between the wavevector q and the frequency ω means that the waveform phase velocity c_ϕ is now a function of the frequency (or, equivalently, the wavelength). Waveforms that contain more than one frequency will have the different frequency components travelling at different velocities, which is the reason that an arbitrary waveform (one that is not sinusoidal) will change shape as it propagates.

More generally, if we assume a harmonic time dependence for the displacement, $U(z,t) = U(z)\mathrm{e}^{-\mathrm{i}\omega t}$, the spatial dependence must satisfy the differential equation

$$\frac{\mathrm{d}^4 U}{\mathrm{d}z^4}(z) = \left(\frac{\rho A}{EI_y}\right)\omega^2 U(z). \tag{7.49}$$

Defining

$$\beta = (\rho A/EI_y)^{1/4}\omega^{1/2}, \tag{7.50}$$

and assuming a spatial dependence of the form $U(z) = \exp(\kappa z)$, we find the solutions $\kappa = \pm\beta$, $\pm i\beta$. Hence the general solution will have the form

$$U(z) = A\mathrm{e}^{\mathrm{i}\beta z} + B\mathrm{e}^{-\mathrm{i}\beta z} + C\mathrm{e}^{\beta z} + D\mathrm{e}^{-\beta z}, \tag{7.51}$$

or, in terms of real functions only,

$$U(z) = a\cos(\beta z) + b\sin(\beta z) + c\cosh(\beta z) + d\sinh(\beta z). \tag{7.52}$$

These are the general solutions for the infinite beam. The terms involving the hyperbolic sine and cosine however diverge as z goes to large positive and negative values, so in an infinite beam we must set $c = d = 0$, retaining only the cosine and sine terms. In the finite-length beam, however, the hyperbolic terms are important and must be retained.

We can now apply boundary conditions to the ends of the beam, which extends from $z = 0$ to $z = \ell$. Note that, for a given frequency ω, the spatial dependence of the displacement $U(z)$ is given by 4th order equation, so that four boundary conditions are required for a unique solution.

Our first example is for a beam clamped at both ends . Then the displacements $U(0)$ and $U(\ell)$ are zero, as are the slopes $\mathrm{d}U/\mathrm{d}z(0)$ and $\mathrm{d}U/\mathrm{d}z(\ell)$. Using the general solution (7.52), these boundary conditions imply that $a = -c$ and $b = -d$, and the allowed values of β are from a discrete set, satisfying

$$\cos\beta_n\ell\cosh\beta_n\ell - 1 = 0. \tag{7.53}$$

This function is plotted in Fig. 7.8, showing the zero-crossing locations.

The zeroes of this function can be found numerically, with $\beta_n\ell = 0$, 4.73004, 7.8532, 10.9956, 14.1372... The solution $\beta = 0$ is trivial and we discard it. The displacement is then given by

$$U_n = a_n\left(\cos(\beta_n z) - \cosh(\beta_n z)\right) + b_n\left(\sin(\beta_n z) - \sinh(\beta_n z)\right). \tag{7.54}$$

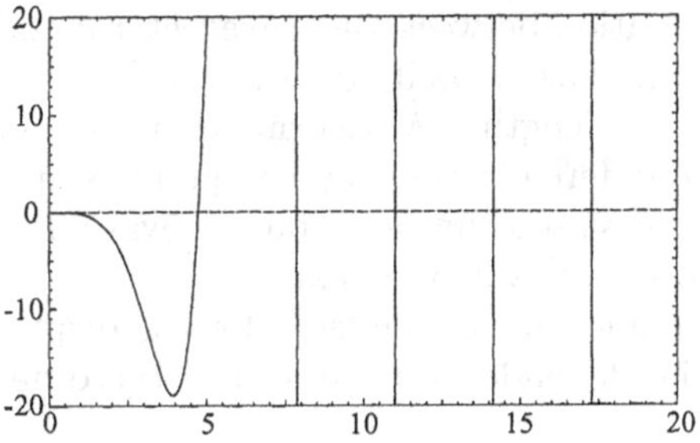

Fig. 7.8. The function $\cos x \cosh x - 1$, whose zero crossings give the mode frequencies for the doubly-clamped beam.

The relative amplitudes for the two terms for the first few modes are $a_n/b_n =$ 1.01781, 0.99923, 1.0000, ..., found from the boundary conditions. The corresponding frequencies are given by

$$\omega_n = \sqrt{\frac{EI_y}{\rho A}}\,\beta_n^2. \tag{7.55}$$

The spatial dependence for the lowest four solutions ($n = 1 - 4$) for the doubly-clamped flexural beam are shown in Fig. 7.9.

We can consider other boundary conditions as well. Another common situation is for a beam clamped at one end, and free at the other (a cantilever geometry). The ends of the cantilevered beam are taken to be at $z = 0$ (fixed end) and $z = \ell$ (free end). The boundary conditions are given by

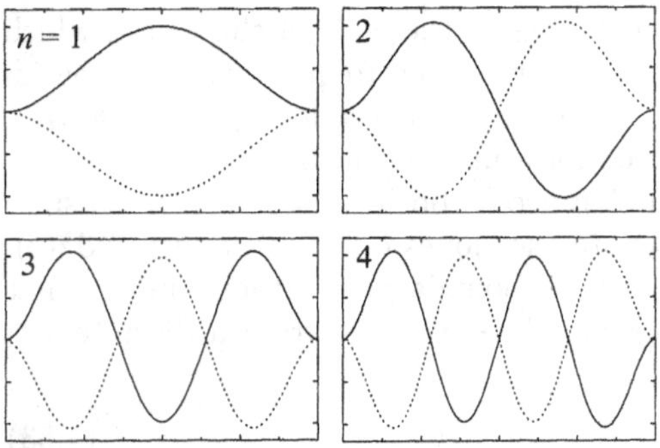

Fig. 7.9. The four lowest frequency modes for the doubly clamped beam, solved using the strength-of-materials approach. The frequencies for these modes are given in the text.

$$\left.\begin{aligned} U(0) &= 0, \\ \frac{\mathrm{d}U}{\mathrm{d}z}(0) &= 0, \\ \frac{\mathrm{d}^2U}{\mathrm{d}z^2}(\ell) &= 0, \\ \frac{\mathrm{d}^3U}{\mathrm{d}z^3}(\ell) &= 0, \end{aligned}\right\} \tag{7.56}$$

where the third and fourth conditions come from the requirement for zero transverse force and zero torque at the free end. Proceeding as before, we find that the frequencies are determined by the equation

$$\cos\beta_n\ell\cosh\beta_n\ell + 1 = 0, \tag{7.57}$$

which has solutions $\beta_n\ell = 1.875, 4.694, 7.855, 10.996\ldots$. The amplitudes are given by $a_n = -c_n$, $b_n = -d_n$, so the functional form is

$$U_n = a_n\left(\cos(\beta_n z) - \cosh(\beta_n z)\right) + b_n\left(\sin(\beta_n z) - \sinh(\beta_n z)\right). \tag{7.58}$$

with a_n/b_n = -1.3622, -0.9819, -1.008, -1.000 .... In Fig. 7.10 we display the first four modes for this problem.

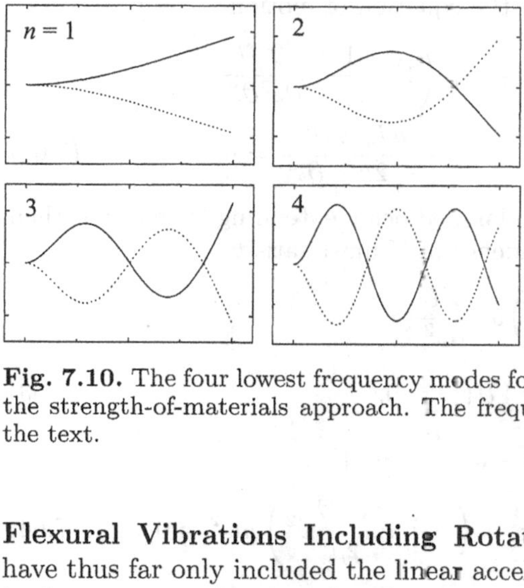

Fig. 7.10. The four lowest frequency modes for the cantilevered beam, solved using the strength-of-materials approach. The frequencies for these modes are given in the text.

Flexural Vibrations Including Rotational Inertia and Shear. We have thus far only included the linear acceleration of the differential element in the equation of motion (7.42). As the element can rotate about the y axis as well as translate along x, we should also include the rotational inertia of each differential element; furthermore, as each element can shear as well as bend, shearing effects need to be included also; we follow the discussion of Timoshenko [62].

The angle between the neutral axis of the beam and the z axis is $\partial U/\partial z$ (for small deflections). The angular velocity component Ω_y corresponding to this angle is $\Omega_y = \partial^2 U/\partial z \partial t$. The torque required to change this component of the angular velocity, for a differential element of thickness $\mathrm{d}z$, is given by

$$\begin{aligned} M_y &= \rho\, I_y\, \mathrm{d}z\, \frac{\partial \Omega_y}{\partial t} \\ &= \rho\, I_y\, \frac{\partial^3 U}{\partial z\, \partial^2 t}. \end{aligned} \tag{7.59}$$

This torque should be entered in the torque balance equation (7.43), which should therefore be replaced by

$$F(z + \mathrm{d}z)\, \mathrm{d}z + M_y(z + \mathrm{d}z) - M_y(z) - \rho\, I_y\, \mathrm{d}z\, \frac{\partial^3 U}{\partial z\, \partial^2 t} = 0. \tag{7.60}$$

Eliminating terms, we find a new equation to replace (7.46),

$$\frac{\partial^2}{\partial z^2}\left(E\, I_y\, \frac{\partial^2 U}{\partial z^2}\right) = -\rho\, A\, \frac{\partial^2 U}{\partial t^2} + \rho\, I_y\, \frac{\partial^4 U}{\partial z^2 \partial t^2}. \tag{7.61}$$

A second important effect we have neglected is shear. The tangent to the neutral axis of the beam deflects both due to rotation of each element of the beam and due to shear deformation. The shear displacement is described by the equation (6.48), and should be added to the displacement due to bending, yielding (after some manipulation) the equation of motion

$$\begin{aligned} \frac{\partial^2}{\partial z^2}\left(E\, I_y\, \frac{\partial^2 U}{\partial z^2}\right) + \rho\, A\, \frac{\partial^2 U}{\partial t^2} - \rho\, I_y\left(1 + \frac{1}{2\mu}\right)\frac{\partial^4 U}{\partial z^2 \partial t^2} & \\ + \frac{\rho\, I_y\, E}{2\mu}\frac{\partial^4 U}{\partial t^4} = 0. & \end{aligned} \tag{7.62}$$

If we consider again the doubly clamped beam extending from 0 to ℓ, then at a frequency ω the spatial dependence of U must satisfy

$$\begin{aligned} E\, I_y\, \frac{\mathrm{d}^4 U}{\mathrm{d}z^4}(z) + \rho\, I_y\left(1 + \frac{1}{2\mu}\right)\omega^2 \frac{\mathrm{d}^2 U}{\mathrm{d}z^2}(z) & \\ + \left(-\omega^2 \rho\, A + \frac{\rho\, I_y\, E}{2\mu}\omega^4\right) U(z) = 0. & \end{aligned} \tag{7.63}$$

Taking exponential solutions $U(z) = \mathrm{e}^{\kappa z}$, we find

$$E\, I_y\, \kappa^4 + \rho\, I_y\left(1 + \frac{1}{2\mu}\right)\omega^2 \kappa^2 + \left(-A + \frac{I_y\, E}{2\mu}\omega^2\right)\rho\, \omega^2 = 0, \tag{7.64}$$

which can be solved to yield

$$\begin{aligned} \kappa^2 = & \frac{1}{2\, E\, I_y}\Big(-\rho I_y (1 + 1/2\mu)\omega^2 \\ & \pm \sqrt{(\rho I_y (1 + 1/2\mu)\omega^2)^2 - 4 E I_y (I_y E \omega^2/2\mu - A)\rho \omega^2}\Big). \end{aligned} \tag{7.65}$$

We can build solutions that satisfy the clamped boundary conditions from these solutions, and we find the same relations for the allowed values of κ as above, but the relation between κ and the frequency ω is more complicated, as given by (7.65). In the limit where rotational and shear effects can be ignored, which applies when $1/2\mu \to 0$, we recover the relations (7.50).

Stress and Strain for Flexural Vibrations. We now work out expressions for the stress and strain in a flexural beam. We return to the simpler Euler–Bernoulli theory for flexure, with the controlling equation of motion (7.47). At rest, the neutral axis is the line $x = y = 0$, i.e. along the z axis. Hence the neutral axis displaces by $\boldsymbol{u}(0,0,z) = (U(z), 0, 0)$. For points off the neutral axis, the displacement is such that planes perpendicular to the neutral axis remain plane and perpendicular to the *local* neutral axis. To first order, a point initially at (x,y,z) is displaced by $\boldsymbol{u}(x,y,z) = (U, 0, -xU')$, where $U' = \mathrm{d}U/\mathrm{d}z$. The strain is then

$$\epsilon = (0, 0, -xU'', 0, 0, 0)\,, \tag{7.66}$$

where $U'' = \mathrm{d}^2U/\mathrm{d}z^2$, and the stress is given by

$$\tau = -\left(\lambda xU'', \lambda xU'', (2\mu + \lambda)xU'', 0, 0, 0\right). \tag{7.67}$$

As for the longitudinal vibrations, we find additional stresses because we are neglecting the additional lateral displacements as the beam flexes. In order to obtain a stress with only $\tau_3 \neq 0$, we posit the additional displacements such that

$$\boldsymbol{u}(x,y,z) = (U + \nu\frac{x^2}{2}U'', \nu xyU'', -xU'), \tag{7.68}$$

with the Poisson ratio $\nu = \lambda/2(\lambda + \mu)$. The strain for this corrected displacement is, dropping terms involving the third derivative U''' as small,

$$\epsilon = (\nu xU'', \nu xU'', -xU'', 0, 0, \nu xU'')\,, \tag{7.69}$$

with stress

$$\tau = (0, 0, -ExU'', 0, 0, \mu\nu xU'')\,. \tag{7.70}$$

We have eliminated the perpendicular stress on the free surfaces, but added an in-plane stress. In Fig. 7.11 we display the shape of lateral displacement and of the cross-section of the beam resulting from the displacement (7.68).

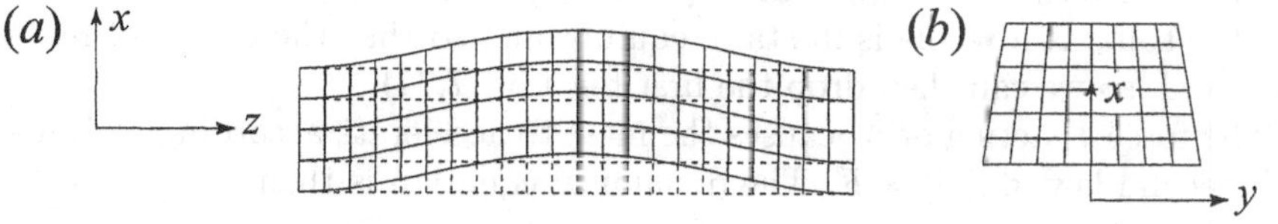

Fig. 7.11. (**a**) Transverse displacement along the length of the doubly clamped beam, showing the stationary beam as dotted lines. (**b**) Cross-section of flexural beam, showing the displacement along y due to a non-zero Poisson ratio.

7.2 Dynamical Equations of Motion in an Isotropic Solid

We will now develop the full equations of motion for stress and strain in a three dimensional solid. We will focus on the description of the isotropic solid, with the elastic modulus E and Poisson ratio ν, or equivalently the Lamé constants λ and μ.

Consider a small volume V within a solid body, centered on a point P. If the solid has mass density ρ, the mass m enclosed by the volume is

$$m = \int_V \rho \, \mathrm{d}V. \tag{7.71}$$

This mass is subject to surface stresses $\boldsymbol{t}$ and body forces $\boldsymbol{f}$. The total force F_i applied to the volume along any one of the coordinate axes x_i is given by

$$\begin{aligned} F_i &= \int_V f_i(\boldsymbol{r}) \, \mathrm{d}V + \int_S t_i \, \mathrm{d}S \\ &= \int_V f_i(\boldsymbol{r}) \, \mathrm{d}V + \int_S (\mathsf{T} \cdot \hat{\boldsymbol{n}})_i \, \mathrm{d}S, \end{aligned} \tag{7.72}$$

where S is the surface enclosing V, $\mathrm{d}S$ is the differential surface element, and $\hat{\boldsymbol{n}}$ the unit vector perpendicular to the surface element $\mathrm{d}S$, pointing outwards from the volume V. We have used the relation (6.4) between the surface stress three-vector $\boldsymbol{t}$ and the local stress tesnor T.

A net force F_i will cause the mass m to accelerate, i.e. its momentum $\boldsymbol{p}$ will change. The momentum is given by

$$\boldsymbol{p} = \int_V \rho \frac{\partial \boldsymbol{r}}{\partial t} \, \mathrm{d}V, \tag{7.73}$$

where $\boldsymbol{r}$ is the position vector from a fixed (inertial) origin to the point P. We write the position as $\boldsymbol{r} = \boldsymbol{R} + \boldsymbol{r}_0 + \boldsymbol{u}$, where $\boldsymbol{R}$ is the position of the center of mass of the solid, $\boldsymbol{r}_0$ the initial position of P with respect to the center of mass, and $\boldsymbol{u}$ the relative displacement of this point. We can then write the momentum as

$$\boldsymbol{p} = \int_V \rho \left(\frac{\mathrm{d}\boldsymbol{R}}{\mathrm{d}t} + \frac{\partial \boldsymbol{u}}{\partial t} \right) \mathrm{d}V, \tag{7.74}$$

as $\boldsymbol{r}_0$ is by definition time-independent.

The first term in the integral of (7.74) is just $m \, \mathrm{d}\boldsymbol{R}/\mathrm{d}t$. We will assume that the solid as a whole is in static equilibrium, so that the center of mass $\boldsymbol{R}$ is fixed, so we can then drop the first term in (7.74).

The force $\boldsymbol{F}$ acting on V causes the mass to accelerate according to Newton's second law, $\mathrm{d}\boldsymbol{p}/\mathrm{d}t = \boldsymbol{F}$. The dynamical equation is then

$$\frac{\mathrm{d}}{\mathrm{d}t} \int_V \rho \frac{\partial \boldsymbol{u}}{\partial t} \, \mathrm{d}V = \int_V \boldsymbol{f}(\boldsymbol{r}) \, \mathrm{d}V + \int_S \mathsf{T} \cdot \mathrm{d}\boldsymbol{S}. \tag{7.75}$$

We apply Green's divergence theorem to the surface integral, which in component form is given by

$$\begin{aligned}\left(\int_S \mathsf{T}\cdot \mathrm{d}\boldsymbol{S}\right)_i &= \int_S \sum_{j=1}^{3} T_{ij}\, \mathrm{d}S_j \\ &= \int_V \sum_{j=1}^{3} \frac{\partial T_{ij}}{\partial x_j}\, \mathrm{d}V, \end{aligned} \tag{7.76}$$

and is given in vector notation by the expression

$$\int_S \mathsf{T}\cdot \mathrm{d}\boldsymbol{S} = \int_V \nabla\cdot \mathsf{T}\, \mathrm{d}V. \tag{7.77}$$

We can now write Newton's law (7.75) as a single integral over volume,

$$\int_V \left(\rho \frac{\partial^2 \boldsymbol{u}}{\partial t^2} - \boldsymbol{f}(\boldsymbol{r}) - \nabla\cdot\mathsf{T}\right) \mathrm{d}V = 0. \tag{7.78}$$

The volume V over which we are integrating was chosen to be small, and is arbitrary; the expression (7.78) must hold for all choices of volume V, and therefore the integrand itself must be zero, i.e.

$$\rho \frac{\partial^2 \boldsymbol{u}}{\partial t^2} - \boldsymbol{f}(\boldsymbol{r}) - \nabla\cdot\mathsf{T} = 0. \tag{7.79}$$

Equation (7.79) is the central equation used for the description of the dynamical behavior of solids. The strain S is related to the relative displacement $\boldsymbol{u}$ by (7.80),

$$S_{ij} = \frac{1}{2}\left(\frac{\partial u_i}{\partial x_j} + \frac{\partial u_j}{\partial x_i}\right), \tag{7.80}$$

and the stress T is related to the strain in a linear solid by the elastic moduli α_{ijkl},

$$T_{ij} = \sum_{k=1}^{3}\sum_{l=1}^{3} \alpha_{ijkl}\, S_{kl}. \tag{7.81}$$

(see Sect. 5.1).

We now specialize to the case with no body forces, so that $\boldsymbol{f}(\boldsymbol{r}) = 0$. Then we can write the dynamical equation of motion (7.79) in component form as

$$\begin{aligned}\rho \frac{\partial^2 u_i}{\partial t^2} &= \sum_{j=1}^{3} \frac{\partial T_{ij}}{\partial x_j} \\ &= \sum_{j=1}^{3}\sum_{\ell=1}^{3}\sum_{m=1}^{3} \alpha_{ij\ell m} \frac{\partial S_{\ell m}}{\partial x_j} \\ &= \sum_{j=1}^{3}\sum_{\ell=1}^{3}\sum_{m=1}^{3} \alpha_{ij\ell m} \frac{\partial^2 u_\ell}{\partial x_j \partial x_m}, \end{aligned} \tag{7.82}$$

using the symmetry of the moduli $\alpha_{ij\ell m}$.

In an isotropic solid, the stress-strain relation (7.81) can be simplified to a relation involving the two Lamé constants λ and μ (see Sect. 5.5),

$$T_{ij} = \lambda(S_{11} + S_{22} + S_{33})\delta_{ij} + 2\mu S_{ij}. \tag{7.83}$$

We then obtain the equation of motion

$$\rho\frac{\partial^2 \boldsymbol{u}}{\partial t^2} = (\lambda + \mu)\nabla(\nabla \cdot \boldsymbol{u}) + \mu\nabla^2 \boldsymbol{u}. \tag{7.84}$$

This equation can be written in indexed form as

$$\rho\frac{\partial^2 u_i}{\partial t^2} = (\lambda + \mu)\frac{\partial}{\partial x_i}\sum_{j=1}^{3}\frac{\partial u_j}{\partial x_j} + \mu\sum_{j=1}^{3}\frac{\partial^2 u_i}{\partial x_j^2}. \tag{7.85}$$

This is the central equation of motion for an isotropic linear solid.

7.3 Waves in Infinite Isotropic Solids

We can now apply the dynamical equation of motion (7.84) to some specific problems. We fist look at the simplest problem, that of waves in an infinite, isotropic solid.

The relations (7.84) are linear, second-order differential equations in both the time and spatial coordinates. We will find solutions for particular types of disturbances, and find that this equation supports plane wave solutions with two different velocities, a longitudinal velocity c_ℓ and a transverse velocity c_t. In an anisotropic solid, there are in general three different velocities, a longitudinal velocity c_ℓ and two transverse velocities, c_{t1} and c_{t2}.

7.3.1 Dilational and Rotational Waves

We first treat the problem of plane waves in an infinite isotropic solid. If we take the divergence of both sides of (7.84), we find

$$\rho\nabla \cdot \frac{\partial^2 \boldsymbol{u}}{\partial t^2} = (\lambda + \mu)(\nabla \cdot \nabla)(\nabla \cdot \boldsymbol{u}) + \mu\nabla \cdot (\nabla^2 \boldsymbol{u}), \tag{7.86}$$

which can be written

$$\rho\frac{\partial^2}{\partial t^2}(\nabla \cdot \boldsymbol{u}) = (\lambda + 2\mu)\nabla^2(\nabla \cdot \boldsymbol{u}). \tag{7.87}$$

We can write (7.87) in terms of the dilatation $\Delta = \nabla \cdot \boldsymbol{u}$, defined in Sect. 6.1, so that

$$\rho\frac{\partial^2 \Delta}{\partial t^2} = (\lambda + 2\mu)\nabla^2 \Delta. \tag{7.88}$$

This is a three-dimensional wave equation for Δ, with solutions of the form

$$\Delta(\boldsymbol{r}, t) = A \exp i(\pm \boldsymbol{q} \cdot \boldsymbol{r} - \omega t) \tag{7.89}$$

with wavevector $\boldsymbol{q}$ and angular frequency ω. Inserting this solution in (7.88), we find the dispersion relation

$$\omega = \sqrt{\frac{\lambda + 2\mu}{\rho}}q, \tag{7.90}$$

in terms of the wavevector amplitude $q = |\boldsymbol{q}|$. This is a linear dispersion relation, giving the dilatational or *longitudinal* wave velocity c_ℓ, given by

$$c_\ell = \sqrt{\frac{\lambda + 2\mu}{\rho}}. \tag{7.91}$$

Dilatational waves, with nonzero compression and rarefaction of the solid, correspond to longitudinal waves, as the displacement $\boldsymbol{u}$ is parallel to the wavevector $\boldsymbol{q}$. This is apparent from the spatial dependence of the dilatation Δ, which requires the displacement $\boldsymbol{u}$ have the dependence

$$\boldsymbol{u}(\boldsymbol{r}) = A\,\hat{\boldsymbol{q}}\,\exp i(\pm \boldsymbol{q}\cdot\boldsymbol{r} - \omega t), \tag{7.92}$$

directed along $\boldsymbol{q}$. In Fig. 7.12 we display a schematic longitudinal wave.

The other type of plane wave solution is found by taking the curl of (7.84), so that

$$\rho\nabla \times \frac{\partial^2 \boldsymbol{u}}{\partial t^2} = (\lambda + \mu)(\nabla \times \nabla)(\nabla \cdot \boldsymbol{u}) + \mu\nabla \times (\nabla^2 \boldsymbol{u}), \tag{7.93}$$

The curl of a divergence is zero, so the first term vanishes. We define the rotation vector $\boldsymbol{\Omega}$ of the displacement field as

$$\boldsymbol{\Omega} = \frac{1}{2}\nabla \times \boldsymbol{u}, \tag{7.94}$$

where this should not be confused with the rotation tensor Ω defined in Chap. 4 (see Sect. 4.1). We can then re-write (7.93) as

$$\rho\frac{\partial^2 \boldsymbol{\Omega}}{\partial t^2} = \mu\nabla^2\boldsymbol{\Omega}. \tag{7.95}$$

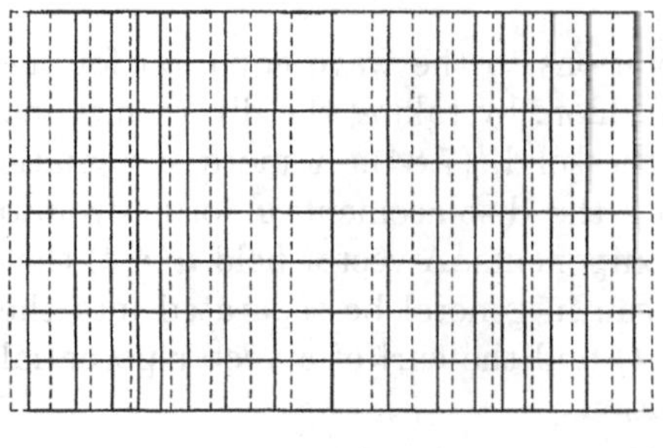

Fig. 7.12. A longitudinal plane wave travelling in the horizontal direction. The dotted lines show the undistorted solid. Note this is identical to the longitudinal wave in a beam, with Poisson ratio set to zero; for the dilatational wave, however, the solution holds for non-zero Poisson ratio.

This is again a wave equation, with the same plane wave solutions as for longitudinal waves. The plane wave solutions have the form $\boldsymbol{\Omega} = \boldsymbol{\Omega}_0 \exp i(\pm\boldsymbol{q}\cdot\boldsymbol{r} - \omega t)$, with vector amplitude $\boldsymbol{\Omega}_0$ perpendicular to $\boldsymbol{q}$. We find the dispersion relation between ω and $\boldsymbol{q}$ given by $\omega = c_t q$, with wave velocity

$$c_t = \left(\frac{\mu}{\rho}\right)^{1/2}. \tag{7.96}$$

This is the velocity of a rotational wave. These waves are also known as *transverse* waves, because the displacement $\boldsymbol{u}$ is perpendicular to the wavevector $\boldsymbol{q}$. The displacement corresponding to the rotation $\boldsymbol{\Omega}$ can be written

$$\boldsymbol{u}(\boldsymbol{r}) = \frac{1}{q^2}(\boldsymbol{\Omega}_0 \times \boldsymbol{q}) \exp i(\pm\boldsymbol{q}\cdot\boldsymbol{r} - \omega t), \tag{7.97}$$

where the displacement amplitude is perpendicular to both $\boldsymbol{\Omega}$ and $\boldsymbol{q}$. In Fig. 7.13 we display the characteristic displacement amplitude as a function of position for a transverse or rotational wave.

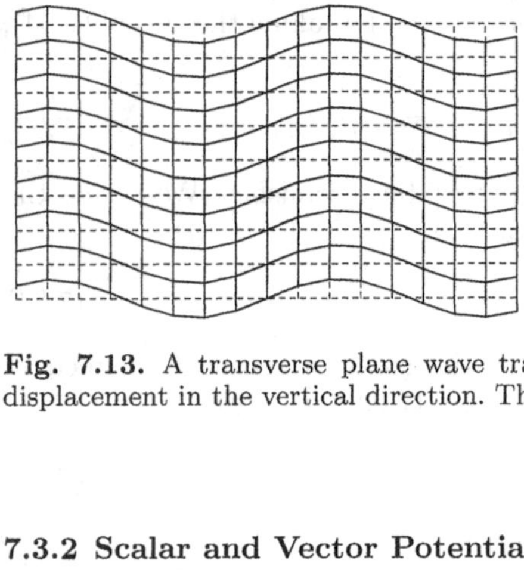

Fig. 7.13. A transverse plane wave travelling in the horizontal direction, with displacement in the vertical direction. The dotted lines show the undistorted solid.

7.3.2 Scalar and Vector Potentials $\boldsymbol{\Phi}$ and $\boldsymbol{H}$

We found two linearly independent solutions to the wave equation for the isotropic solid, in a somewhat ad hoc fashion, by taking the divergence and curl of the full dynamic equation for the solid, (7.84). A more systematic approach can be developed by splitting the displacement $\boldsymbol{u}$ formally into separate curl-free and divergence-free components. A vector field $\boldsymbol{u}(\boldsymbol{r},t)$ that is continuous and twice differentiable, can in general be separated into the gradient of a scalar potential field $\Phi(\boldsymbol{r},t)$ and the curl of a vector potential field $\boldsymbol{H}(\boldsymbol{r},t)$, through the defining equation

$$\boldsymbol{u}(\boldsymbol{r},t) = \nabla\Phi(\boldsymbol{r},t) + \nabla \times \boldsymbol{H}(\boldsymbol{r},t). \tag{7.98}$$

This separation is unique.

By taking the divergence of (7.98), we find

$$\begin{aligned} \nabla^2 \Phi &= \nabla \cdot \boldsymbol{u} \\ &= \Delta, \end{aligned} \tag{7.99}$$

giving the dilatation Δ, and by taking the curl,

$$\begin{aligned} \nabla \times \nabla \times \boldsymbol{H} &= \nabla \times \boldsymbol{u} \\ &= 2\boldsymbol{\Omega}, \end{aligned} \tag{7.100}$$

giving the rotation vector $\boldsymbol{\Omega}$. These equations uniquely define the scalar potential Φ to within a Laplacian, and the vector potential $\boldsymbol{H}$ to within a divergence. It is common to define $\boldsymbol{H}$ such that its divergence is zero, $\nabla \cdot \boldsymbol{H} = 0$, and to force Φ to vanish at infinity. These additional constraints completely define Φ and $\boldsymbol{H}$.

We can insert the defining equation (7.98) into the wave equation (7.84), yielding

$$\begin{aligned} (\lambda + \mu)\nabla\nabla \cdot (\nabla\Phi + \nabla \times \boldsymbol{H}) + \mu\nabla^2(\nabla\Phi + \nabla \times \boldsymbol{H}) \\ = \rho\frac{\partial^2}{\partial t^2}(\nabla\Phi + \nabla \times \boldsymbol{H}). \end{aligned} \tag{7.101}$$

Through the use of the vector identities $\nabla \cdot \nabla\Phi = \nabla^2\Phi$, $\nabla^2(\nabla\Phi) = \nabla(\nabla^2\Phi)$, and $\nabla \cdot (\nabla \times \boldsymbol{H}) = 0$, this can be written as

$$\nabla\left((\lambda + 2\mu)\nabla^2\Phi - \rho\frac{\partial^2\Phi}{\partial t^2}\right) + \nabla \times \left(\mu\nabla^2\boldsymbol{H} - \rho\frac{\partial^2\boldsymbol{H}}{\partial t^2}\right) = 0. \tag{7.102}$$

The individual terms in (7.102) may be individually set to zero, without loss of generality. We then find two separate wave equations for Φ and $\boldsymbol{H}$,

$$\rho\frac{\partial^2\Phi}{\partial t^2} = (\lambda + 2\mu)\nabla^2\Phi, \tag{7.103}$$

and

$$\rho\frac{\partial^2\boldsymbol{H}}{\partial t^2} = \mu\nabla^2\boldsymbol{H}. \tag{7.104}$$

One form of solution to the equation in Φ is a plane wave, with

$$\Phi = \Phi_0 e^{i\boldsymbol{q}\cdot\boldsymbol{r} - i\omega t}, \tag{7.105}$$

with longitudinal wave velocity $c_\ell = ((\lambda + 2\mu)/\rho)^{1/2}$, and dispersion relation $\omega = c_\ell q$. The displacement corresponding to (7.105) is

$$\boldsymbol{u}(\boldsymbol{r}, t) = \Phi_0\, \boldsymbol{q}\, e^{i\boldsymbol{q}\cdot\boldsymbol{r} - i\omega t}, \tag{7.106}$$

corresponding to a longitudinal wave with displacement amplitude $\boldsymbol{A} = \Phi_0\boldsymbol{q}$, directed along $\boldsymbol{q}$. These are the same as the dilatational waves in (7.92). Such waves are also known as irrotational or *primary waves*, and in shorthand referred to as *P*-waves.

A plane wave solution to the vector wave equation (7.104) is

$$\boldsymbol{H} = \boldsymbol{H}_0\, \mathrm{e}^{\mathrm{i}\boldsymbol{q}\cdot\boldsymbol{r}-\mathrm{i}\omega t}, \tag{7.107}$$

with $\boldsymbol{H}_0 \cdot \boldsymbol{q} = 0$ (so that $\boldsymbol{H}$ has zero divergence), and dispersion relation $\omega = c_t\, q$, with transverse wave velocity $c_t = (\mu/\rho)^{1/2}$. The displacement corresponding to (7.107) is

$$\boldsymbol{u}(\boldsymbol{r},t) = \boldsymbol{q} \times \boldsymbol{H}_0\, \mathrm{e}^{\mathrm{i}\boldsymbol{q}\cdot\boldsymbol{r}-\mathrm{i}\omega t}, \tag{7.108}$$

corresponding to a plane wave perpendicular to $\boldsymbol{q}$ and $\boldsymbol{H}_0$, and amplitude qH_0. These waves are the transverse waves found in (7.97), and are also called secondary, or S-waves.

Another way to look at these types of solutions is to consider a general plane wave solution for the displacement $\boldsymbol{u}$,

$$\boldsymbol{u}(\boldsymbol{r},t) = \boldsymbol{A}\mathrm{e}^{\mathrm{i}(\pm\boldsymbol{q}\cdot\boldsymbol{r}-\omega t)}, \tag{7.109}$$

for some arbitrary vector amplitude $\boldsymbol{A}$. Inserting this *ansatz* into the wave equation (7.84), we find the vector equation

$$(\lambda+\mu)\boldsymbol{q}(\boldsymbol{q}\cdot\boldsymbol{u}(\boldsymbol{r},t)) + \mu q^2\boldsymbol{u}(\boldsymbol{r},t) = \rho\omega^2\boldsymbol{u}(\boldsymbol{r},t). \tag{7.110}$$

Cancelling out the common terms, we find

$$(\lambda+\mu)(\boldsymbol{q}\cdot\boldsymbol{A})\boldsymbol{q} + \mu q^2\boldsymbol{A} = \rho\omega^2\boldsymbol{A}. \tag{7.111}$$

We break up the displacement vector $\boldsymbol{A}$ into the component $A_{||}$ along $\boldsymbol{q}$, and the vector $\boldsymbol{A}_\perp$ perpendicular to $\boldsymbol{q}$. We then find a dispersion relation for each of these components,

$$(\lambda+2\mu)q^2 = \rho\omega^2 \qquad \text{(parallel component)}, \tag{7.112}$$

and

$$\mu q^2 = \rho\omega^2 \qquad \text{(perpendicular component)}. \tag{7.113}$$

Hence we find that transverse plane waves, with displacement vector $\boldsymbol{A}$ perpendicular to the direction of motion $\boldsymbol{q}$, travel with velocity $c_t = (\mu/\rho)^{1/2}$, while longitudinal plane waves, with displacement parallel to $\boldsymbol{q}$, travel with velocity $c_\ell = ((\lambda+2\mu)/\rho)^{1/2}$. These different approaches therefore all give the same result.

7.3.3 Time-Dependent Body Forces

We now re-introduce the body force terms in the equations of motion. For a linear isotropic solid we can write the equivalent of the wave equation (7.84), now including body forces, as

$$\rho\frac{\partial^2\boldsymbol{u}}{\partial t^2} = (\lambda+\mu)\nabla(\nabla\cdot\boldsymbol{u}) + \mu\nabla^2\boldsymbol{u} + \rho\boldsymbol{f}. \tag{7.114}$$

The formal separation of $\boldsymbol{u}$ into the gradient of the scalar field Φ and the curl of the vector field $\boldsymbol{H}$ can be applied to the body force vector field $\boldsymbol{f}$ as well, with a new scalar field ψ and a vector field $\boldsymbol{A}$, such that

$$\boldsymbol{f} = \nabla\psi + \nabla \times \boldsymbol{A}. \tag{7.115}$$

With some manipulation of vector identities, the equation equivalent to (7.101) can be shown to have the form

$$\begin{aligned}\nabla & \left((\lambda + 2\mu)\nabla^2\Phi + \rho\psi - \rho\frac{\partial^2\Phi}{\partial t^2} \right) \\ & + \nabla \times \left(\mu\nabla^2\boldsymbol{H} + \rho\boldsymbol{A} - \rho\frac{\partial^2\boldsymbol{H}}{\partial t^2} \right) = 0.\end{aligned} \tag{7.116}$$

Setting each term separately to zero yields two driven wave equations in Φ and $\boldsymbol{H}$,

$$(\lambda + 2\mu)\nabla^2\Phi + \rho\psi = \rho\frac{\partial^2\Phi}{\partial t^2}, \tag{7.117}$$

and

$$\mu\nabla^2\boldsymbol{H} + \rho\boldsymbol{A} = \rho\frac{\partial^2\boldsymbol{H}}{\partial t^2}. \tag{7.118}$$

Solutions to these equations may be found by using either Fourier analysis, or using Green's functions techniques, which correspond to the solutions for an impulse force located at a single point. A discussion of the Green's function approach may be found in the text by Graff [14]. We develop the Fourier approach below.

Fourier Transform Approach. We first assume a harmonic time dependence for the driving force, and consider the same frequency component in the displacement, with the body force of the form $\boldsymbol{f}(\boldsymbol{r},t) = \boldsymbol{f}(\boldsymbol{r})\exp(-\mathrm{i}\omega t)$, and the displacement of the form $\boldsymbol{u}(\boldsymbol{r},t) = \boldsymbol{u}(\boldsymbol{r})\exp(-\mathrm{i}\omega t)$. In this case, the central equation (7.114) has the form

$$(\lambda + \mu)\nabla(\nabla \cdot \boldsymbol{u}(\boldsymbol{r})) + \mu\nabla^2\boldsymbol{u}(\boldsymbol{r}) + \rho\boldsymbol{f}(\boldsymbol{r}) = -\rho\omega^2\boldsymbol{u}(\boldsymbol{r}). \tag{7.119}$$

Next we assume the body force is driven with a plane wave spatial dependence, so that $\boldsymbol{f}(\boldsymbol{r}) = \boldsymbol{f}_0\exp(i\boldsymbol{q}\cdot\boldsymbol{r})$. The displacement then has similar spatial dependence, $\boldsymbol{u}(\boldsymbol{r}) = \boldsymbol{u}_0\exp(i\boldsymbol{q}\cdot\boldsymbol{r})$. Inserting this in (7.119) yields the result

$$(\lambda + \mu)(\boldsymbol{q}\cdot\boldsymbol{u}_0)\boldsymbol{q} + \mu q^2\boldsymbol{u}_0 + \rho\boldsymbol{f}_0 = \rho\omega^2\boldsymbol{u}_0. \tag{7.120}$$

We separate the problem into one involving a transverse component, with $\boldsymbol{q}$ perpendicular to the displacement vector $\boldsymbol{u}_0$, and one with a longitudinal component, with $\boldsymbol{q}$ parallel to $\boldsymbol{u}_0$. In the former case we have

$$\boldsymbol{u}_0 = \frac{1}{c_t^2\, q^2 - \omega^2}\boldsymbol{f}_0 \qquad \text{(transverse waves)}, \tag{7.121}$$

and in the latter case we have

$$u_0 = \frac{1}{c_\ell^2 q^2 - \omega^2} f_0 \qquad \text{(longitudinal waves).} \tag{7.122}$$

In either case, if the body force frequency ω and spatial wave vector $\boldsymbol{q}$ match the dispersion relation for the wave in question, i.e. if $\omega = c_\ell q$ for the longitudinal case, or if $\omega = c_t q$ for the transverse case, the displacement amplitude u diverges (in a real system, the inclusion of dissipation will keep the response finite; see Chap. 8).

If the specified body force vector $\boldsymbol{f}_0$ is not purely parallel or purely perpendicular to the spatial wave vector $\boldsymbol{q}$, separate solutions are found for the parallel and perpendicular components, and the solutions for $\boldsymbol{u}$ can then be added together vectorially.

The general solution to an arbitrary body force time- and space-dependence may now be constructed from the specific results given here. We demonstrate how this is done for a pure transverse body force, i.e. one that has an arbitrary time and spatial dependence, but with the restriction that the body force spatial wavevector $\boldsymbol{q}$ is always perpendicular to the body force vector $\boldsymbol{f}$. In this case the displacement vector $\boldsymbol{u}$ is related to the body force through (7.121). The completely general case is left to the reader, who can follow the prescription given here.

We take an arbitrary time and spatial dependence $\boldsymbol{f}(\boldsymbol{r}, t)$, subject to the condition that the response consists solely of transverse waves, and construct the Fourier transform $\boldsymbol{F}(\boldsymbol{q}, \omega)$ of $\boldsymbol{f}(\boldsymbol{r}, t)$ through the equation

$$\boldsymbol{F}(\boldsymbol{q}, \omega) = \frac{1}{(2\pi)^4} \int_{-\infty}^{\infty} \int_V \boldsymbol{f}(\boldsymbol{r}, t) \mathrm{e}^{-\mathrm{i}(\boldsymbol{q}\cdot\boldsymbol{r} + \omega t)} \, \mathrm{d}\boldsymbol{r} \, \mathrm{d}t. \tag{7.123}$$

The spatial integral is over the whole volume V of the solid, and the time integral over all time (or equivalently over the time for which the force $\boldsymbol{f}$ is applied).

The displacement $\boldsymbol{u}(\boldsymbol{r}, t)$ can be expressed by its Fourier transform $\boldsymbol{U}(\boldsymbol{q}, \omega)$ by a relation identical to (7.123),

$$\boldsymbol{U}(\boldsymbol{q}, \omega) = \frac{1}{(2\pi)^4} \int_{-\infty}^{\infty} \int_V \boldsymbol{u}(\boldsymbol{r}, t) \mathrm{e}^{-\mathrm{i}(\boldsymbol{q}\cdot\boldsymbol{r} + \omega t)} \, \mathrm{d}\boldsymbol{r} \, \mathrm{d}t. \tag{7.124}$$

The Fourier components $\boldsymbol{U}$ and $\boldsymbol{F}$ are then directly related by the transverse relation (7.121),

$$\boldsymbol{U}(\boldsymbol{q}, \omega) = \frac{\rho}{\mu q^2 - \rho\omega^2} \boldsymbol{F}(\boldsymbol{q}, \omega). \tag{7.125}$$

We can then apply the inverse Fourier transform to obtain the formal relation between the displacement $\boldsymbol{u}$ and the body force transform $\boldsymbol{F}$,

$$\boldsymbol{u}(\boldsymbol{r}, t) = \int_0^{\infty} \int \frac{\rho}{\mu q^2 - \rho\omega^2} \boldsymbol{F}(\boldsymbol{q}, \omega) \mathrm{e}^{\mathrm{i}(\boldsymbol{q}\cdot\boldsymbol{r} - \omega t)} \, \mathrm{d}\boldsymbol{q} \, \mathrm{d}\omega, \tag{7.126}$$

where the integral over $\boldsymbol{q}$ is over all wavevector space, and the integral over ω over positive frequencies only. This can further be manipulated to generate a relation between $\boldsymbol{u}$ and $\boldsymbol{f}$, but the result is not particularly illuminating.

7.4 Waves in Infinite Crystalline Solids

We now turn to a discussion of crystalline solids. We return to the general wave equation (7.82), which applies to any crystalline solid. We assume a displacement solution of the form

$$\boldsymbol{u}(\boldsymbol{r}) = u_0 \boldsymbol{e} \exp i(\pm \boldsymbol{q} \cdot \boldsymbol{r} - \omega t), \tag{7.127}$$

where u_0 is the (scalar) amplitude and $\boldsymbol{e}$ is the polarization vector assumed to be of unit length. Inserting this solution in (7.82) we find the relation

$$\rho \omega^2 e_i = \sum_{j=1}^{3} \sum_{\ell=1}^{3} \sum_{m=1}^{3} \alpha_{ij\ell m} q_j q_m e_\ell . \tag{7.128}$$

We now define the Christoffel tensor D, corresponding to the wavevector $\boldsymbol{q}$, through the relation

$$D_{ij} = \frac{1}{\rho} \sum_{j=1}^{3} \sum_{m=1}^{3} \alpha_{ij\ell m} \hat{q}_j \hat{q}_m, \tag{7.129}$$

in terms of the vector components $\hat{q}_i$ of the unit wavevector $\hat{\boldsymbol{q}} = \boldsymbol{q}/q$. The Christoffel tensor is a function of the direction of the wavevector $\boldsymbol{q}$, so it must be re-evaluated for each wavevector direction in the crystal; it does not however change as the wavevector amplitude q, or equivalently the wavelength $\lambda = 2\pi/q$, of the plane wave is varied.

The definition of the Christoffel tensor allows us to rewrite the wave equation (7.128) in the form of an eigenvector-eigenvalue equation $(\mathsf{D} - v^2 \mathsf{1}) \cdot \boldsymbol{e} = 0$ or, in component form,

$$\left(D_{i\ell} - v^2 \delta_{i\ell}\right) e_\ell = 0. \tag{7.130}$$

The three eigenvalues v of (7.130) correspond to the three phase velocities of the plane wave solutions. In an isotropic solid, the eigenvalues would be the longitudinal and transverse wave velocities c_ℓ and c_t, the latter appearing twice as it is a degenerate eigenvalue. The eigenvectors $\boldsymbol{e}$ of (7.130) are the corresponding polarization vectors, which for the isotropic solid would be the longitudinal wavevector $\hat{q}$ and the two degenerate transverse unit vectors perpendicular to $\hat{q}$.

The general eigenvalues v can be determined by solving the characteristic equation,

$$\begin{vmatrix} D_{11} - v^2 & D_{12} & D_{13} \\ D_{21} & D_{22} - v^2 & D_{23} \\ D_{31} & D_{32} & D_{33} - v^2 \end{vmatrix} = 0. \tag{7.131}$$

This equation will in general have three distinct solutions v_i, corresponding to a quasi-longitudinal and two quasi-transverse velocities. The velocity solutions v_i depend on the direction of the wavevector $\boldsymbol{q}$ through the $\boldsymbol{q}$-dependence of the Christoffel tensor; however, the velocities are independent of the wavevector magnitude q, so the continuum approach is dispersionless (see Sect. 1.3.2).

In a cubic crystal, the expression for the Christoffel tensor D is significantly simplified, as is discussed in the beautiful monograph by Wolfe [63]. The fourth-rank, $3 \times 3 \times 3 \times 3$ tensor $\alpha_{ij\ell m}$ is related to the 6×6 elastic moduli c_{pq} by mapping the index pairs $ij \to p$, and $\ell m \to q$, according to the rule

$$\left.\begin{array}{ccccccc} ij \text{ or } \ell m & 11 & 22 & 33 & 23 & 13 & 12 \\ p \text{ or } q & 1 & 2 & 3 & 4 & 5 & 6 \end{array}\right\} \tag{7.132}$$

(see Sect. 2.8). In a cubic crystal the c_{ij} have the form

$$\mathsf{c} = \begin{bmatrix} c_{11} & c_{12} & c_{12} & 0 & 0 & 0 \\ c_{12} & c_{11} & c_{12} & 0 & 0 & 0 \\ c_{12} & c_{12} & c_{11} & 0 & 0 & 0 \\ 0 & 0 & 0 & c_{44} & 0 & 0 \\ 0 & 0 & 0 & 0 & c_{44} & 0 \\ 0 & 0 & 0 & 0 & 0 & c_{44} \end{bmatrix}, \tag{7.133}$$

(see Chap. 5). Using this notation, the Christoffel tensor may be written as

$$\left.\begin{aligned} \rho D_{ii} &= c_{11}\hat{q}_i^2 + c_{44}(1 - \hat{q}_i^2) \\ \rho D_{ij} &= (c_{12} + c_{44})\hat{q}_i\hat{q}_j, \quad (i \neq j) \end{aligned}\right\} \tag{7.134}$$

Using the elastic constants for e.g. GaAs, we can now calculate the phase velocities v_i of the three phonon modes, as a function of wavevector direction

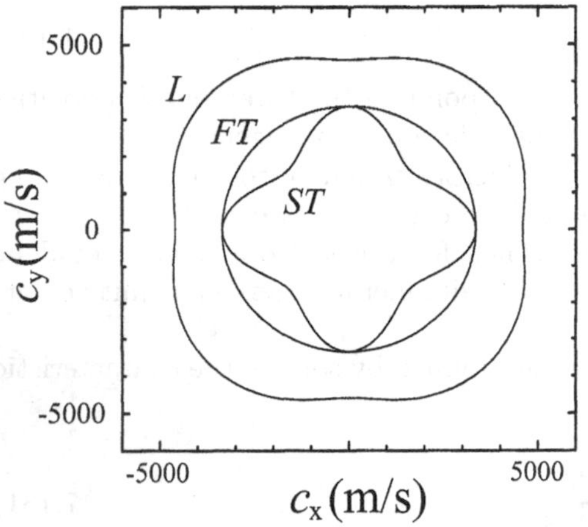

Fig. 7.14. Phase velocity calculated for GaAs in the $x_1 - x_2$ plane, with the GaAs crystal axes taken aligned with the coordinate axes. The phase velocity is plotted for the three polarizations, longitudinal (L), slow transverse (ST) and fast transverse (FT). The distance from the origin gives the magnitude, and the x_1 and x_2 coordinates the directions of the corresponding wavevector $\boldsymbol{q}$.

$\hat{q}$. These are displayed in Fig. 7.14 for wavevectors in the $x_1 - x_2$ plane, where we have taken the crystallographic axes as the coordinate axes.

7.4.1 Energy Flow and the Poynting Vector in Crystalline Solids

From Fig. 7.14, we see that the phase velocity depends on the direction of the wavevector in the crystal. This spatial dependence has a very significant impact on energy transport in a crystalline solid: The direction in which energy moves through a crystal is not necessarily in the direction of the wavevector. This rather non-intuitive result is similar to that found for the transmission velocity of wave trains through dispersive media, where the frequency is a nonlinear function of wavevector. In a dispersive medium, the phase velocity $c_\phi = \omega/q$ determines the wavefront velocity of a monochromatic wave, but is a function of the wavevector q. A wave train, such as a Gaussian-shaped pulse, built from the superposition of many single-frequency waves, moves at the group velocity $c_g = \mathrm{d}\omega/\mathrm{d}q$ (see Sect. 1.3.2).

In a non-isotropic solid, the vector phase velocity (the solution to the characteristic equation (7.131)) is given by $\boldsymbol{c}_\phi = (\omega/q^2)\boldsymbol{q}$, but the group velocity is $\boldsymbol{c}_g = \nabla_{\boldsymbol{q}}\omega = \mathrm{d}\omega/\mathrm{d}\boldsymbol{q}$. The group velocity may be expressed in terms of the elastic constants of the solid, through the relation (see [63])

$$
\begin{aligned}
\alpha_{g,\,i} &= \frac{\partial \omega}{\partial q_i} \\
&= \frac{1}{\rho\omega}\sum_{j\,\ell\,m} \alpha_{j\ell mi}\, e_j\, k_\ell\, e_m.
\end{aligned}
\tag{7.135}
$$

In classical mechanics, a force $\boldsymbol{F}$ acting on a mass m moving at velocity $\boldsymbol{v}$ delivers mechanical energy at a rate $P = \boldsymbol{v} \cdot \boldsymbol{F}$. For a solid surface, the stress T is the equivalent of force per unit area, so the *vector* power flow density (per unit area) delivered to the surface is $\boldsymbol{v}_\phi \cdot \mathsf{T}$.

We define the *Poynting vector* $\boldsymbol{P}$, analogous to that in electromagnetism, as the average vector power flow density, through the relation

$$
\boldsymbol{P} = -\left\langle \boldsymbol{v}_\phi \cdot \mathsf{T} \right\rangle, \tag{7.136}
$$

where the brackets $\langle\,\rangle$ designate a time average. The minus sign represents the flow out from the surface. For a monochromatic wave with displacement $\boldsymbol{u} = u_0 \boldsymbol{e} \exp i(\boldsymbol{q} \cdot \boldsymbol{r} - \omega t)$, we can write the Poynting vector as

$$
\begin{aligned}
\boldsymbol{P} &= \left\langle (\boldsymbol{e} \cdot \mathsf{c} \cdot \boldsymbol{q}\boldsymbol{e})\omega\, u_0^2 \sin^2(\boldsymbol{q} \cdot \boldsymbol{r} - \omega t) \right\rangle \\
&= \frac{1}{2}(\boldsymbol{e} \cdot \mathsf{c} \cdot \boldsymbol{q}\boldsymbol{e})\omega\, u_0^2,
\end{aligned}
\tag{7.137}
$$

or in component form

$$
P_m = \frac{\omega u_0^2}{2} \sum_{ij\ell} \alpha_{ijlm} e_i q_j e_\ell. \tag{7.138}
$$

We see that we can write the Poynting vector in terms of the vector group velocity $\boldsymbol{c}_g$ as

$$\boldsymbol{P} = \frac{1}{2}\rho\omega^2 u_0^2 \boldsymbol{c}_g. \tag{7.139}$$

The direction of power flow is thus the same direction as the group velocity, but not necessarily the same as the phase velocity. Hence even for a *monochromatic* phonon wave, power does not flow along the wavevector direction, but along the group velocity direction.

Geometrically, the group velocity is the three-dimensional gradient of the frequency $\omega(\boldsymbol{q})$, and is therefore normal to the surfaces of constant ω. Because the frequency is related to the wavevector through the phase velocity, $\omega = c_\phi q$, the surfaces of constant ω are the same as those of the phase velocity, as plotted in Fig. 7.14. The local normal to this surface gives the direction of the group velocity, which is plotted for the slow transverse mode in GaAs in Fig. 7.15.

Another way to display the group velocity's dependence on direction is shown in Fig. 7.16, where for each q-vector in the $x_1 - x_2$ plane, we show the corresponding group velocity vector. The *phonon focusing* effect, due to accumulation of velocity vectors in certain directions, is quite clear. For an isotropic source of plane waves, emanating equally in all directions $\hat{\boldsymbol{q}}$, power will flow preferentially in the directions of flow concentration shown in the figure.

In an isotropic solid, the figure equivalent to Fig. 7.15 would be a circle. The rather strong distortion of the group velocity shown in the figure has

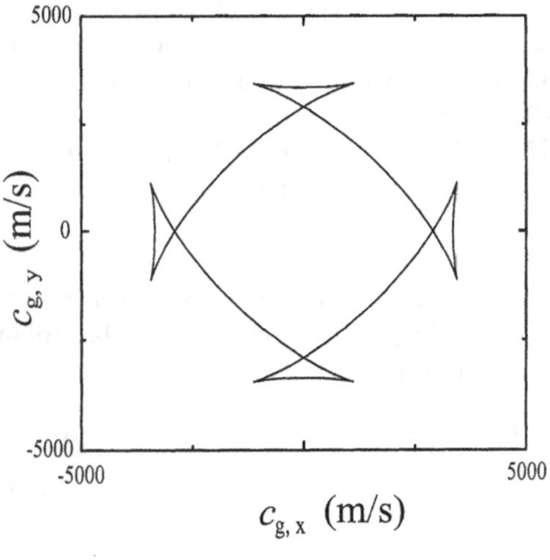

Fig. 7.15. Slow transverse group velocity in the $x_1 - x_2$ plane for GaAs. The group velocity depends on the direction of the wavevector, but does not necessarily point in the same direction as the wavevector. The cusps in the group velocity appear when the curvature of the frequency surface goes to zero.

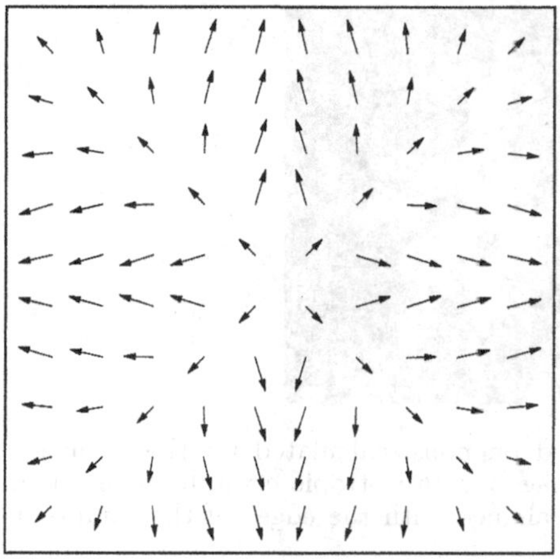

Fig. 7.16. Slow transverse group velocity vector field in the $x_1 - x_2$ plane for GaAs. The base of each vector is located at a q-vector point, with the origin at the center of the figure, and the direction and length of the vectors in the field correspond to the direction and relative magnitude of the corresponding group velocity.

important implications for the flow of energy in a solid. In a series of experiments with many beautiful results from Wolfe's group (see e.g. [64], [65], and the monograph by Wolfe [63]), the dramatic redirection of energy due to crystalline focusing effects has been imaged both experimentally and theoretically.

These experiments were designed to measure the distribution of phonons emitted from a point source on one side of a crystal striking the opposite side, as shown in Fig. 7.17. In actuality the detector was at a fixed point, and

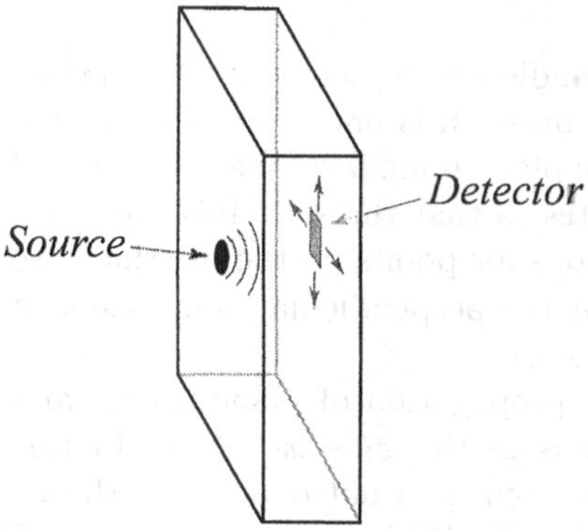

Fig. 7.17. Schematic experiment used to measure the effects of phonon focusing in a crystal. Phonons emitted from a point source on one side of the crystal are detected on the opposing side, and a map of phonon intensity thereby generated.

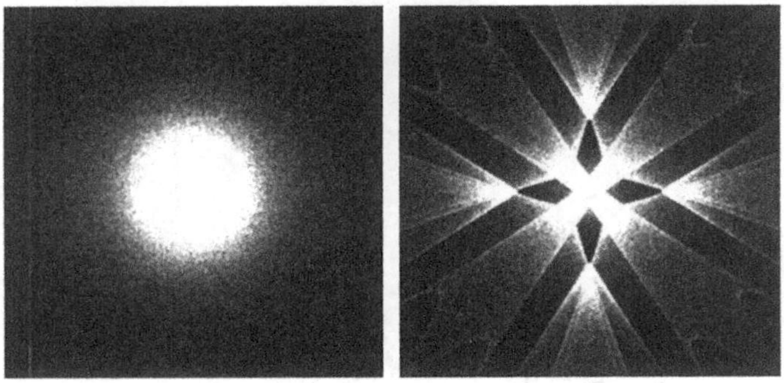

Fig. 7.18. Intensity maps of transmitted phonons, calculated for the geometry shown in Fig. 7.17. On the left is the image for an isotropic crystal, on the right the image for GaAs, with its ⟨100⟩ axes aligned with the edges of the sample in Fig. 7.17.

the source scanned over the opposite face, but the results are equivalent, and the schematic shown in the figure is somewhat easier to understand.

In Fig. 7.18 we display the numerically-simulated results of such an experiment for an isotropic solid and for a crystal of GaAs, oriented with its ⟨100⟩ crystal axes aligned with the coordinate axes. The dramatic redirection of phonons due to the crystalline anisotropy is quite striking.

The significant difference between isotropic and crystalline solids leads to complications in understanding and predicting many effects due to phonon transport in real solids. In the remainder of this chapter we will concentrate on understanding simple aspects of isotropic solids.

7.5 Waves in Semi-infinite Isotropic Solids

We now turn to a discussion of how to handle the presence of an *interface* in a dynamic problem. The simplest such problem is one where the solid is semi-infinite, filling space up to one infinite plane boundary, the other side of which is vacuum. We choose our coordinates so that the solid fills the space $x_3 > 0$. We must specify boundary conditions for points on the interface; we choose to take a stress-free surface, so that the perpendicular component of the surface stress $t_3 = (\mathsf{T} \cdot \hat{\boldsymbol{n}})_3$ is equal to zero.

We shall restrict our discussion to the propagation of plane waves, and orient our axes so that the wavevector $\boldsymbol{q}$ is in the $x_1 - x_3$ plane. In this case, the spatial dependence of the displacement will not depend on the x_2 coordinate, i.e. $\boldsymbol{u} = \boldsymbol{u}(x_1, x_3, t)$; note that the displacement vector $\boldsymbol{u}$ can still have a non-zero component along x_2. We can then write $\boldsymbol{u}$ in terms of the scalar and vector potentials Φ and $\boldsymbol{H}$,

$$\left.\begin{aligned} u_1 &= \frac{\partial \Phi}{\partial x_1} - \frac{\partial H_2}{\partial x_3}, \\ u_2 &= \frac{\partial H_1}{\partial x_3} - \frac{\partial H_3}{\partial x_1}, \\ u_3 &= \frac{\partial \Phi}{\partial x_3} + \frac{\partial H_2}{\partial x_1}, \end{aligned}\right\} \tag{7.140}$$

where all derivatives with respect to x_2 are zero. The stress-strain relations (7.83) and (7.80) are

$$\left.\begin{aligned} T_{11} &= \lambda\left(\frac{\partial u_1}{\partial x_1} + \frac{\partial u_3}{\partial x_3}\right) + 2\mu\frac{\partial u_1}{\partial x_1}, \\ T_{22} &= \lambda\left(\frac{\partial u_1}{\partial x_1} + \frac{\partial u_3}{\partial x_3}\right), \\ T_{33} &= \lambda\left(\frac{\partial u_1}{\partial x_1} + \frac{\partial u_3}{\partial x_3}\right) + 2\mu\frac{\partial u_3}{\partial x_3}, \\ T_{23} &= \mu\frac{\partial u_2}{\partial x_3}, \\ T_{13} &= \mu\left(\frac{\partial u_3}{\partial x_1} + \frac{\partial u_1}{\partial x_3}\right), \\ T_{12} &= \mu\frac{\partial u_2}{\partial x_1}. \end{aligned}\right\} \tag{7.141}$$

In terms of the scalar and vector potentials, this system of equations may be written

$$\left.\begin{aligned} T_{11} &= \lambda\left(\frac{\partial^2 \Phi}{\partial x_1^2} + \frac{\partial^2 \Phi}{\partial x_3^2}\right) + 2\mu\left(\frac{\partial^2 \Phi}{\partial x_1^2} - \frac{\partial^2 H_2}{\partial x_1 \partial x_3}\right), \\ T_{22} &= \lambda\left(\frac{\partial^2 \Phi}{\partial x_1^2} + \frac{\partial^2 \Phi}{\partial x_3^2}\right), \\ T_{33} &= \lambda\left(\frac{\partial^2 \Phi}{\partial x_1^2} + \frac{\partial^2 \Phi}{\partial x_3^2}\right) + 2\mu\left(\frac{\partial^2 \Phi}{\partial x_3^2} + \frac{\partial^2 H_2}{\partial x_1 \partial x_3}\right), \\ T_{23} &= \lambda\left(\frac{\partial^2 H_1}{\partial x_3^2} - \frac{\partial^2 H_3}{\partial x_1 \partial x_3}\right), \\ T_{13} &= \mu\left(2\frac{\partial^2 \Phi}{\partial x_1 \partial x_3} - \frac{\partial^2 H_2}{\partial x_3^2} + \frac{\partial^2 H_2}{\partial x_1^2}\right), \\ T_{12} &= \mu\left(\frac{\partial^2 H_1}{\partial x_1 \partial x_3} - \frac{\partial^2 H_3}{\partial x_1^2}\right). \end{aligned}\right\} \tag{7.142}$$

As usual we take the gauge for $\boldsymbol{H}$ so that $\nabla \cdot \boldsymbol{H} = 0$, or

$$\frac{\partial H_1}{\partial x_1} + \frac{\partial H_3}{\partial x_3} = 0. \tag{7.143}$$

The boundary condition on the plane surface, $\mathsf{T} \cdot \hat{\boldsymbol{x}}_3 = 0$, translates to $T_{13} = T_{23} = T_{33} = 0$ at $x_3 = 0$.

Finally, we have that the scalar potential, and each component of the vector potential, satisfies their respective wave equations, as given by (7.103) and (7.104):

$$\left.\begin{aligned} \nabla^2 \Phi - \frac{1}{c_\ell^2}\frac{\partial^2 \Phi}{\partial t^2} &= 0, \\ \nabla^2 H_i - \frac{1}{c_t^2}\frac{\partial^2 H_i}{\partial t^2} &= 0, \quad (i = 1 \text{ to } 3). \end{aligned}\right\} \tag{7.144}$$

We are now prepared to apply these equations to specific situations in the semi-infinite solid. We discuss three situations of interest: Waves incident on the surface at normal incidence (i.e. with $\boldsymbol{q} = q\,\hat{\boldsymbol{x}}_3$), waves at oblique incidence, so that $\boldsymbol{q}$ has both an x_1 and an x_3 component, and *surface waves* or *Rayleigh waves*, which are waves that are transmitted along the plane surface of the interface, without transmission of energy into the bulk.

7.5.1 Waves Incident at Normal Incidence

A wave in the solid $x_3 > 0$, incident perpendicular to the plane $x_3 = 0$, has wavevector $\boldsymbol{q} = q\,\hat{\boldsymbol{x}}_3$, so that the displacement $\boldsymbol{u}$ has no x_1 or x_2 dependence. From our gauge choice $\nabla \cdot \boldsymbol{H} = 0$, we can set $H_3 = 0$. We have two possible wavevectors $\boldsymbol{q}$ for a given frequency ω, namely $\boldsymbol{q}_\ell = q_\ell \hat{\boldsymbol{x}}_3$ with $q_\ell = \omega/c_\ell$ for longitudinal polarization, and $\boldsymbol{q}_t = q_t \hat{\boldsymbol{x}}_3$ with $q_t = \omega/c_t$ for transverse polarization. The scalar potential Φ and vector potential $\boldsymbol{H}$ thus have the form

$$\left.\begin{aligned} \Phi(x_3, t) &= \Phi_\pm e^{i(\pm q_\ell x_3 - \omega t)}, \\ H_1(x_3, t) &= H_{1\pm} e^{i(\pm q_t x_3 - \omega t)}, \\ H_2(x_3, t) &= H_{2\pm} e^{i(\pm q_t x_3 - \omega t)}, \\ H_3(x_3, t) &= 0, \end{aligned}\right\} \tag{7.145}$$

where the "+" sign is chosen for a wave travelling in the direction $-x_3$ (towards the interface) and the "−" sign for those travelling away from the interface. A complete solution includes waves travelling in both directions, with different amplitudes and possibly different phases.

Inserting these equations in the expressions for the stress, we find the relations

$$\left.\begin{aligned} T_{11} &= T_{22} = -\lambda\, q_\ell^2\, \Phi(x_3, t), \\ T_{33} &= -(\lambda + 2\mu)\, q_\ell^2\, \Phi(x_3, t), \\ T_{23} &= -\mu\, q_t^2\, H_1(x_3, t), \quad T_{13} = \mu\, q_t^2\, H_2(x_3, t), \quad T_{12} = 0. \end{aligned}\right\} \tag{7.146}$$

In order to satisfy the boundary conditions, we require $\Phi(0,t) = H_1(0,t) = H_2(0,t) = 0$. This can be satisfied if the "+" and "−" amplitudes are equal but opposite in sign,

$$\left.\begin{aligned} \Phi(x_3,t) &= \Phi_0 \left(e^{i(q_\ell x_3-\omega t)} - e^{i(-q_\ell x_3-\omega t)}\right), \\ H_{1,2}(x_3,t) &= H^0_{1,2} \left(e^{i(q_t x_3-\omega t)} - e^{i(-q_t x_3-\omega t)}\right). \end{aligned}\right\} \quad (7.147)$$

The corresponding displacement amplitudes can be calculated from (7.140), giving

$$\left.\begin{aligned} u_1 &= iq_t H^0_2 \left(e^{i(q_t x_3-\omega t)} + e^{i(-q_t x_3-\omega t)}\right), \\ u_2 &= iq_t H^0_1 \left(e^{i(q_t x_3-\omega t)} + e^{i(-q_t x_3-\omega t)}\right), \\ u_3 &= iq_\ell \Phi_0 \left(e^{i(q_\ell x_3-\omega t)} + e^{i(-q_\ell x_3-\omega t)}\right). \end{aligned}\right\} \quad (7.148)$$

With an unimportant change of phase, these can be written

$$\left.\begin{aligned} u_1 &= A_1 \cos(q_t x_3 - \omega t), \\ u_2 &= A_2 \cos(q_t x_3 - \omega t), \\ u_3 &= A_3 \cos(q_\ell x_3 - \omega t), \end{aligned}\right\} \quad (7.149)$$

with independent amplitudes A_1, A_2 and A_3.

We see that regardless of the transverse (non-zero u_1 or u_2) or longitudinal (non-zero u_3) nature of the transverse wave, the form of the solution is the same. A plane interface reflects both longitudinal and transverse normally-incident waves, preserving their polarization and amplitude. Note further that while the boundary conditions impose zero stress at the interface, the displacement is not necessarily zero.

7.5.2 Waves Incident at Arbitrary Angles

We now consider waves incident at an arbitrary angle. We distinguish two cases, one where the polarization of the wave, i.e. the displacement vector, is perpendicular to both the wavevector $\boldsymbol{q}$ and to the plane normal $\hat{x}_3$, a wave known as a *shear horizontal wave*, or SH wave, and the other where the polarization is in the plane formed by the wavevector and the plane normal, which are either dilatational waves with $\boldsymbol{H} = 0$, or are shear vertical (SV) waves with $\Phi = 0$ (or some combination thereof). We shall see that for the latter case, *mode conversion* occurs at the interface between the Φ and $\boldsymbol{H}$ waves, while in the bulk these waves do not interact.

Dilatational and Shear Vertical Waves. If there is no displacement out of the plane formed by the wavevector and the plane normal, then given our geometry, there is no displacement along $\hat{x}_2$. Hence we have $u_2 = 0$, and all derivatives with respect to x_2 vanish, since as discussed above we align our axes so that $q_2 = 0$. We then immediately have the stresses $T_{12} = T_{23} = 0$ in the interior of the solid. Furthermore, from the expression for u_2 in terms of H_1 and H_3, and the divergenceless of $\boldsymbol{H}$, we may set $H_1 = H_3 = 0$. We therefore only have to solve for Φ and H_2, each of which satisfies the wave equation (7.144).

We can build the solution from a superposition of waves Φ_+, Φ_-, H_{2+} and H_{2-}. The solutions have the form

$$\Phi(x_1, x_3, t) = \Phi_+ \, \mathrm{e}^{\mathrm{i}(\boldsymbol{\alpha}\cdot\boldsymbol{x}-\omega t)} + \Phi_- \, \mathrm{e}^{\mathrm{i}(\boldsymbol{\beta}\cdot\boldsymbol{x}-\omega t)}, \tag{7.150}$$

$$H_2(x_1, x_3, t) = H_{2+} \, \mathrm{e}^{\mathrm{i}(\boldsymbol{\gamma}\cdot\boldsymbol{x}-\omega t)} + H_{2-} \, \mathrm{e}^{\mathrm{i}(\boldsymbol{\delta}\cdot\boldsymbol{x}-\omega t)}, \tag{7.151}$$

with wavevectors $\boldsymbol{\alpha}$, $\boldsymbol{\beta}$, $\boldsymbol{\gamma}$ and $\boldsymbol{\delta}$. The wave equations impose the dispersion relations $\omega = c_\ell \alpha = c_\ell \beta$ and $\omega = c_t \gamma = c_t \delta$, where $\alpha = |\boldsymbol{\alpha}|$, etc. The components α_2, β_2, γ_2 and δ_2 all vanish. Because $\boldsymbol{\alpha}$ is associated with a "+" wave, α_3 is positive, $\boldsymbol{\beta}$ is associated with a "−" wave so β_3 is negative, and so on.

The stress-free nature of the plane boundary imposes restrictions on the wavevectors and amplitudes. These conditions are given, for T_{33}, by

$$\begin{aligned}\left(\lambda\alpha^2 + 2\mu\alpha_3^2\right)\Phi_+ \mathrm{e}^{\mathrm{i}\alpha_1 x_1} &+ \left(\lambda\beta^2 + 2\mu\beta_3^2\right)\Phi_- \mathrm{e}^{\mathrm{i}\beta_1 x_1} \\ +2\mu\gamma_1\gamma_3 H_{2+} \mathrm{e}^{\mathrm{i}\gamma_1 x_1} &+ 2\mu\delta_1\delta_3 H_{2-} \mathrm{e}^{\mathrm{i}\delta_1 x_1} = 0,\end{aligned} \tag{7.152}$$

and for T_{13} by

$$\begin{aligned}2\mu\alpha_1\alpha_3\Phi_+ \mathrm{e}^{\mathrm{i}\alpha_1 x_1} &+ 2\mu\beta_1\beta_3\Phi_- \mathrm{e}^{\mathrm{i}\beta_1 x_1} + \mu\left(-\gamma_3^2 + \gamma_1^2\right) H_{2+} \mathrm{e}^{\mathrm{i}\gamma_1 x_1} \\ &+ \mu\left(-\delta_3^2 + \delta_1^2\right) H_{2-} \mathrm{e}^{\mathrm{i}\delta_1 x_1} = 0.\end{aligned} \tag{7.153}$$

Note that these expressions are evaluated at $x_3 = 0$, with the common time dependence divided out.

The relations (7.152) and (7.153) must hold for all x_1; the only way this can occur is if the coefficients of x_1 are all equal, i.e. if $\alpha_1 = \beta_1 = \gamma_1 = \delta_1$. Hence the x_1-components of the various wavevectors must all be equal.

Another way to state this result is in the language of optics, which is illustrated in Fig. 7.19, where we introduce the incidence angle θ_1 for Φ_-, the reflection angle θ_2 for Φ_+, ζ_1 for H_{2+}, and ζ_2 for H_{2-}. The angles θ_i and ζ_i are given by the relative magnitudes of the respective wavevectors, such that $\tan(\theta_1) = \alpha_1/\alpha_3 = -\beta_1/\beta_3$, and so on. The incoming wave Φ_- reflects into the wave Φ_+, with angle of incidence equal to angle of reflection, $\theta_1 = \theta_2 = \theta$, and the accompanying wave $\boldsymbol{H}_-$ with incoming angle ζ_1, reflects into the outgoing wave $\boldsymbol{H}_+$ with equal angle $\zeta_2 = \zeta_1 = \zeta$. Furthermore, the angles of incidence and reflection of the Φ and $\boldsymbol{H}$ waves are related, in that

$$c_t \sin\, \theta_1 = c_\ell \, \sin\zeta_1. \tag{7.154}$$

We can then divide out the spatial dependence of (7.152) and (7.153), leaving the relations

$$\left.\begin{aligned}(2\mu\sin^2\theta - \lambda - 2\mu)(\Phi_+ + \Phi_-) + \frac{c_\ell^2}{c_t^2}\,\mu\,\sin 2\zeta(H_{2+} - H_{2-}) &= 0,\\ \mu\sin 2\theta(\Phi_+ - \Phi_-) - \frac{c_\ell^2}{c_t^2}\,\mu\,\cos 2\zeta(H_{2+} + H_{2-}) &= 0.\end{aligned}\right\} \tag{7.155}$$

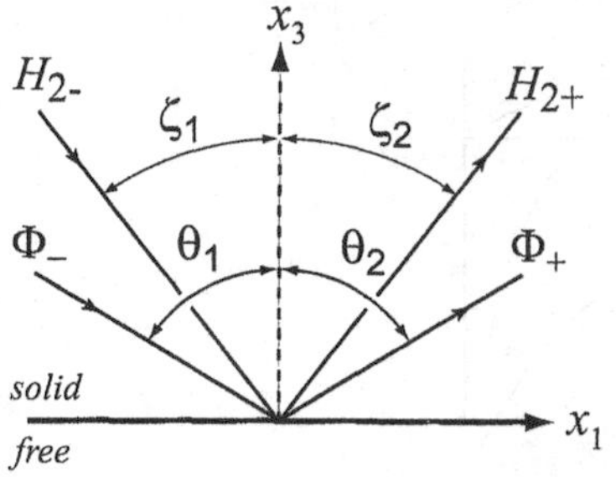

Fig. 7.19. A plane interface normal to $\hat{x}_3$ reflects P-waves described by Ψ with incidence angle θ_1 equal to reflection angle θ_2, and S-waves described by H_2 with incidence angle ζ_1 equal to reflection angle ζ_2. These waves are coupled at the interface so that θ_1 and ζ_1 are also related, as described in the text.

Given the incident amplitudes Ψ_- and H_{2-}, these two relations give the corresponding reflected amplitudes Φ_+ and H_{2+}.

Consider for example a pure longitudinal P wave, incident on the interface, so that $H_{2-} = 0$. Solving for Φ_+ and H_{2+} in terms of the incoming wave amplitude Φ_-, we find

$$\left.\begin{aligned} \Phi_+ &= \frac{\mu \sin 2\theta \sin 2\zeta - (\lambda + 2\mu)\cos^2 2\zeta}{\mu \sin 2\theta \sin 2\zeta + (\lambda + 2\mu)\cos^2 2\zeta}\,\Phi_-, \\ H_{2+} &= \frac{2\mu \sin 2\theta \sin 2\zeta}{\mu \sin 2\theta \sin 2\zeta + (\lambda + 2\mu)\cos^2 2\zeta}\,\Phi_-. \end{aligned}\right\} \tag{7.156}$$

We note that a common factor of μ may be divided out, leaving the ratio $(\lambda + 2\mu)/\mu = (2 - 2\nu)/(1 - 2\nu)$, so that the amplitude depends on Poisson's ratio ν only. The ratio of wave velocities c_ℓ/c_t, as can be seen from (7.91) and (7.96), can be written $c_\ell/c_t = ((2 - 2\nu)/(1 - 2\nu))^{1/2}$. We can then plot the relative amplitudes Φ_+/Φ_- and H_{2+}/Φ_- as a function of incidence angle θ. This is shown in Fig. 7.20 for $\nu = 1/4$ and $\nu = 1/3$.

We see that for a pure longitudinal wave incident on the interface at an angle θ, the reflected wave in general includes both a longitudinal component Φ_+, reflected at the angle θ, and a non-zero transverse H_{2+} component reflected at the angle ζ, given by (7.154), with $\zeta \le \theta$; the reflection is pure longitudinal for normal incidence ($\theta = 0°$), as well as for grazing incidence ($\theta = 90°$). For small ν, however, there are also two *critical angles* where the transverse reflected amplitude goes to zero. As shown in Fig. 7.20, for $\nu = 1/4$ this occurs at the incidence angles $\theta = 60°$ and $\theta = 77.5°$. The more general case, with the creation of a transverse component from a pure longitudinal incident wave, is an example of *mode conversion.*

We can also examine the reflection of a pure secondary, or transverse wave, with $\Psi_- = 0$ and H_{2-} non-zero. Solving (7.155) in that case yields the relations

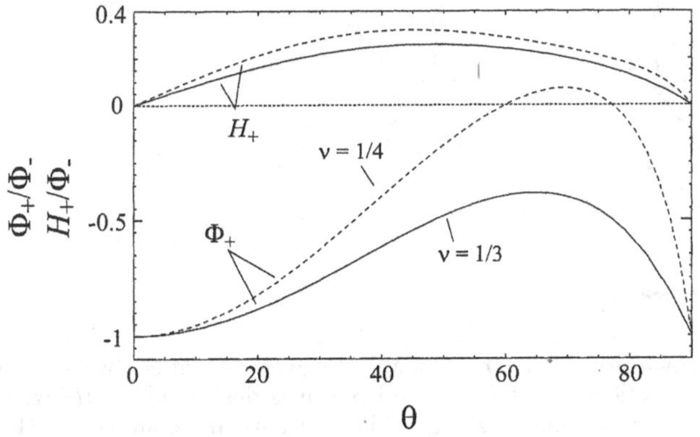

Fig. 7.20. Relative amplitude of the reflected longitudinal wave Ψ_+ and the transverse wave H_{2+} for Poisson's ratio $\nu = 1/4$ and $\nu = 1/3$, for an incident longitudinal wave Φ_-. Figure following Graff [14].

$$\left.\begin{aligned} \Psi_+ &= \frac{\mu \sin 2\theta \sin 2\zeta - (\lambda + 2\mu)\cos^2 2\zeta}{\mu \sin 2\theta \sin 2\zeta + (\lambda + 2\mu)\cos^2 2\zeta} H_{2-}, \\ H_{2+} &= -\frac{2(\lambda + 2\mu)\sin 2\theta \sin 2\zeta}{\mu \sin 2\theta \sin 2\zeta + (\lambda + 2\mu)\cos^2 2\zeta} H_{2-}. \end{aligned}\right\} \qquad (7.157)$$

Again we find mode conversion, with the incident transverse wave generating a reflected transverse wave with reflection angle ζ equal to the incidence angle ζ, and a reflected longitudinal wave with angle θ satisfying (7.154), with $\theta \geq \zeta$. This is shown in Fig. 7.21 for $\nu = 1/3$ and $\nu = 1/4$.

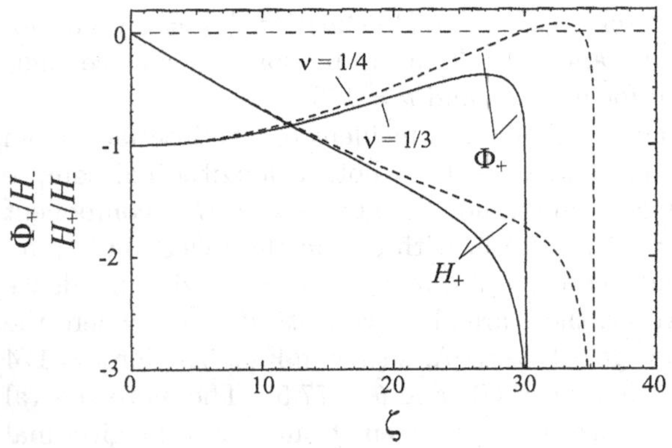

Fig. 7.21. Relative amplitude of the reflected longitudinal wave Ψ_+ and the reflected transverse wave H_{2+} for Poisson's ratio $\nu = 1/4$ and $\nu = 1/3$, for an incident S, or transverse wave H_{2-}. Figure following Graff [14].

Shear Horizontal Waves. We now turn briefly to a discussion of the reflection of pure *shear horizontal* (SH) waves, transverse waves with a polarization perpendicular to both the plane interface normal and to the plane of reflection of the wave; such a wave has displacement parallel to the interface, along x_2. We use the same geometry as in the previous section.

A SH wave has a displacement with only u_2 non-zero. From (7.140), we see that we must specify both of the components H_1 and H_3 in order to find u_2, and we can set Φ and H_2 equal to zero. Each component of $\boldsymbol{H}$ must satisfy the wave equation (7.144), and we assume plane waves solutions for both components, of the form

$$\left.\begin{aligned} H_1(x_1,x_3,t) &= H_{1+}\,\mathrm{e}^{\mathrm{i}(\boldsymbol{\alpha}\cdot\boldsymbol{x}-\omega t)} + H_{1-}\,\mathrm{e}^{\mathrm{i}(\boldsymbol{\beta}\cdot\boldsymbol{x}-\omega t)}, \\ H_3(x_1,x_3,t) &= H_{3+}\,\mathrm{e}^{\mathrm{i}(\boldsymbol{\gamma}\cdot\boldsymbol{x}-\omega t)} + H_{3-}\,\mathrm{e}^{\mathrm{i}(\boldsymbol{\delta}\cdot\boldsymbol{x}-\omega t)}. \end{aligned}\right\} \qquad (7.158)$$

The "−" components move towards the interface $x_3 = 0$, so that β_3 and δ_3 are negative, while the "+" components move away from the interface, so that α_3 and γ_3 are positive. The wavevectors $\boldsymbol{\alpha}$, $\boldsymbol{\beta}$, γ and $\boldsymbol{\delta}$ in this equation all have zero $\hat{\boldsymbol{x}}_2$ component, because we assume no x_2 dependence in the displacement. Also, because the frequency ω is the same for all the components, and each component satisfies the wave equation with the same speed c_t, the lengths of the wavevectors are all equal, so that $\alpha_1^2 + \alpha_3^2 = \beta_1^2 + \beta_2^2 = \ldots$. Finally, the zero divergence condition for $\boldsymbol{H}$ gives the relation

$$\alpha_1 H_{1+}\mathrm{e}^{\mathrm{i}\boldsymbol{\alpha}\cdot\boldsymbol{x}} + \beta_1 H_{1-}\mathrm{e}^{\mathrm{i}\boldsymbol{\beta}\cdot\boldsymbol{x}} + \gamma_3 H_{3+}\mathrm{e}^{\mathrm{i}\boldsymbol{\gamma}\cdot\boldsymbol{x}} + \delta_3 H_{3-}\mathrm{e}^{\mathrm{i}\boldsymbol{\delta}\cdot\boldsymbol{x}} = 0, \qquad (7.159)$$

where the time dependence has been divided out. This relation holds for all points in the semi-infinite solid, so the wavevectors' x_1 components must all be equal. From the equality of the wavevector lengths, this in turn implies that the magnitudes of the x_3 components are all equal, except for the sign. Dividing out the common x_1 dependence, we find

$$\left.\begin{aligned} \alpha_1 H_{1+} &= \gamma_3 H_{3+}, \\ \beta_1 H_{1-} &= \delta_3 H_{3-}. \end{aligned}\right\} \qquad (7.160)$$

The interface is stress-free, with $T_{i3} = 0$; examining (7.142), this gives one additional constraint on the vector potential, namely that from T_{23},

$$\alpha_3^2(H_{1+} + H_{1-}) + \alpha_1^2(H_{1+} + H_{1-}) = 0. \qquad (7.161)$$

This final relation gives the general result for reflection of a shear horizontal wave, namely that $H_{1+} = -H_{1-}$, and equivalently $H_{3+} = -H_{3-} = (\alpha_1/\alpha_3)H_{1+}$. Hence a shear horizontal wave reflects with a 180° change in phase in each of its components, with reflection angle equal to the incidence angle. There is no mode conversion at the interface.

7.5.3 Rayleigh Waves

In addition to the bulk waves we have discussed so far, the equations of motion for the semi-infinite solid also admit solutions restricted to the surface of the interface. These surface waves are known as *Rayleigh waves*, as they were first described by Lord Rayleigh in the late 19th century [66].

Consider a plane wave travelling along the $x_3 = 0$ surface of the solid, in the $\hat{\boldsymbol{x}}_1$ direction; we assume as before that everything is independent of x_2. The displacement is mixed longitudinal and shear vertical, with $u_2 = 0$, and therefore we include the scalar potential Φ and the H_2 component of the vector potential, setting $H_1 = H_3 = 0$ (the former to give zero divergence for $\boldsymbol{H}$, the latter to have $u_2 = 0$). We then take a functional dependence for Φ and H_2 that corresponds to a plane wave along x_1, but decaying exponentially with distance x_3 into the solid:

$$\left.\begin{aligned} \Phi(x_1, x_3, t) &= \Phi_0 \mathrm{e}^{-\kappa x_3} \mathrm{e}^{\mathrm{i}(q x_1 - \omega t)}, \\ H_2(x_1, x_3, t) &= H_2^0 \mathrm{e}^{-\sigma x_3} \mathrm{e}^{\mathrm{i}(q x_1 - \omega t)}. \end{aligned}\right\} \tag{7.162}$$

The frequency and wavevector for the two components are the same (note this is never true for the bulk waves). As the two potentials must satisfy wave equations with different wave velocities c_ℓ and c_t, the decay coefficients κ and σ must necessarily differ. From the wave equations (7.144), we find these are given by

$$\left.\begin{aligned} \kappa^2 &= q^2 - \omega^2/c_\ell^2, \\ \sigma^2 &= q^2 - \omega^2/c_t^2. \end{aligned}\right\} \tag{7.163}$$

In order that κ and σ be real, so that the potentials decay with distance from the surface rather than oscillating, the frequency and in-plane wavevector must satisfy $|q| > \omega/c_\ell, \omega/c_t$.

The one remaining criterion is that the boundary condition on the exposed surface must be stress-free, so that $T_{i3} = 0$. Using (7.142), we see that the condition on T_{23} is satisfied immediately, but the conditions on T_{13} and T_{33} impose the constraints

$$\left.\begin{aligned} \lambda(-q^2 + \kappa^2)\Phi_0 + 2\mu(\kappa^2\Phi_0 - \mathrm{i}qsH_2^0) &= 0, \\ -2\mathrm{i}q\kappa\Phi_0 - (\sigma^2 + q^2)H_2^0 &= 0, \end{aligned}\right\} \tag{7.164}$$

where the common dependence on x_1 and t has been divided out, and the potentials are evaluated at $x_3 = 0$. These equations may be used to find the relationship between the amplitudes,

$$H_2^0 = -\frac{2\mathrm{i}q\kappa}{\sigma^2 + q^2}\Phi_0. \tag{7.165}$$

We can also find the characteristic equation for the wavevector k, or equivalently solve for the travelling *Rayleigh wave velocity* $c_R = \omega/q$. After some manipulation, the latter may be expressed as

$$\gamma^6 - 8\gamma^4 + (24 - 16\frac{c_t^2}{c_\ell^2})\gamma^2 + 16\frac{c_t^2}{c_\ell^2} - 16 = 0, \tag{7.166}$$

where we define $\gamma = c_R/c_t$ as the ratio of the Rayleigh wave velocity to the transverse wave velocity. The equation (7.166) is known as the *Rayleigh–Lamb equation.* The coefficients involve the ratio of the transverse to the longitudinal wave velocities, which may be expressed in terms of the Poisson ratio as

$$\frac{c_t}{c_\ell} = \sqrt{\frac{1-2\nu}{2-2\nu}}. \tag{7.167}$$

The decay wavelengths κ and σ may be expressed in terms of q, γ and the velocity ratio as

$$\left.\begin{aligned} \kappa^2 &= (1-(c_t^2/c_\ell^2)\gamma^2)q^2, \\ \sigma^2 &= (1-\gamma^2)q^2. \end{aligned}\right\} \tag{7.168}$$

In order that κ and σ be real, we must have $\gamma < 1$, so for Rayleigh waves to exist the roots of (7.166) must be between 0 and 1. It can be shown that there is one real root that satisfies this condition for $0 < \nu < 1/2$ [67], and a plot of the corresponding Rayleigh wave velocity c_R as a function of ν is shown in Fig. 7.22.

The displacement amplitudes for a Rayleigh wave are then given by (7.140),

$$\left.\begin{aligned} u_1(x_1,x_3,t) &= (\mathrm{e}^{-\kappa x_3} - \frac{2\kappa\sigma}{\sigma^2+q^2}\mathrm{e}^{-\sigma x_3})\mathrm{i}q\Phi_0\mathrm{e}^{\mathrm{i}(qx-\omega t)}, \\ u_2(x_1,x_3,t) &= 0, \\ u_3(x_1,x_3,t) &= (-\mathrm{e}^{-\kappa x_3} + \frac{2q^2}{\sigma^2+\kappa^2}\mathrm{e}^{-\sigma x_3})\kappa\Phi_0\mathrm{e}^{\mathrm{i}(qx-\omega t)}. \end{aligned}\right\} \tag{7.169}$$

We note that the displacements along x_1 and x_3 are both travelling waves with wavevector q, and decay in the solid's interior in a length of order $1/\sigma$.

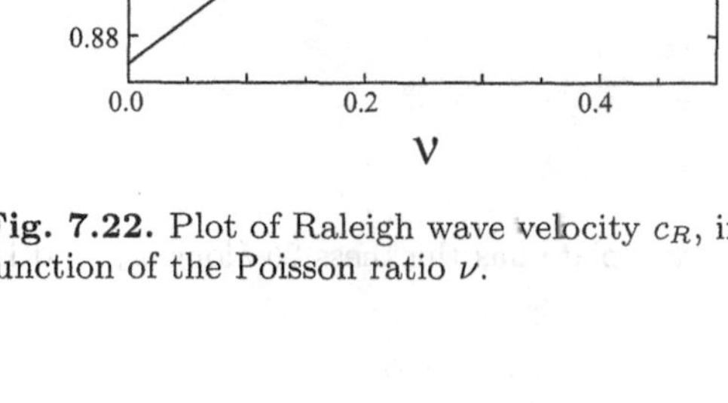

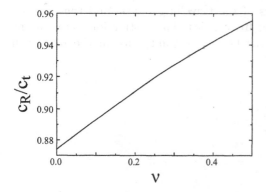

Fig. 7.22. Plot of Raleigh wave velocity c_R, in units of transverse velocity c_t, as a function of the Poisson ratio ν.

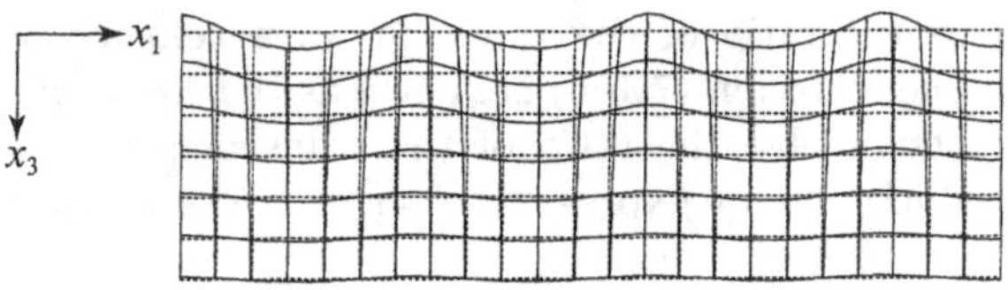

Fig. 7.23. Rayleigh wave projected onto the $x_1 - x_3$ plane; solid lines show the displacements from the Rayleigh wave, dotted lines the undeformed semi-infinite solid.

The displacements are however out of phase by 90°, so that the wave is circularly polarized; both polarizations are transmitted with the same velocity. In Fig. 7.23 we display the form for the displacement.

7.6 Waves in Plates

We now begin a discussion of waves in an infinite plate. We take a solid slab of thickness $2b$ in the x_3 direction, and infinite in extent along the x_1 and x_2 directions. The origin of the coordinate system is at the center of the slab, so the slab extends from $-b$ to $+b$ along x_3; the geometry is shown in Fig. 7.24. The surfaces on both sides of the slab are stress-free, so that $T_{i3} = 0$, and we assume the slab is an isotropic solid. Even though the solid is then defined by only two material parameters, the slab will display a very complex behavior due to the restricted geometries. Finally, we assume no dependence of the displacement (or any of the associated stress and strain) on x_2, although the displacement u_2 along x_2 can be non-zero. In general, therefore, the displacement $\boldsymbol{u}$ has the functional dependence $\boldsymbol{u}(x_1, x_3, t)$.

In the semi-infinite solid, we found we needed certain combinations of P and S waves to keep the surface stress-free. There were however no restrictions on the allowed wavevectors of the travelling waves, and therefore no restrictions on the corresponding frequencies. The distinction between the interface and the plate is that the boundary conditions on the plate will

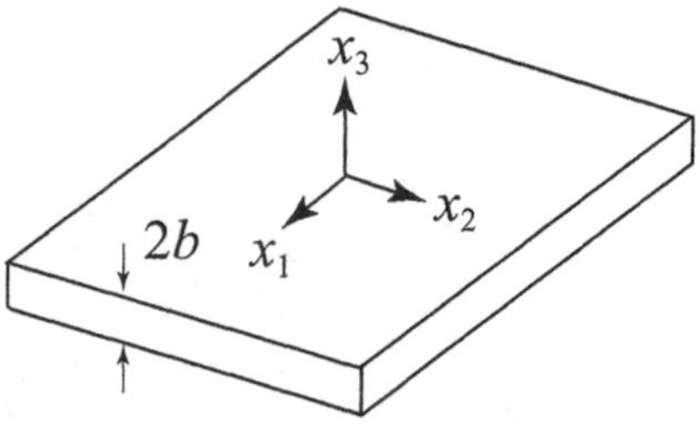

Fig. 7.24. Geometry for waves in a plate; plate has thickness $2b$ along x_3, and is infinite in extent along x_1 and x_2.

impose strong constraints on the allowed wavevectors, and correspondingly the allowed frequencies, of the travelling modes. The solutions will thus be transformed from a continuous distribution of wavevectors to a discrete set.

7.6.1 Shear Horizontal Waves in Plates

We begin by treating plane waves with displacement only in the x_2 direction, propagating along x_1. These are the shear horizontal waves discussed in Sect. 7.5.2 above, and as we shall see, are the simplest type of waves in this geometry. Problems involving the polarization perpendicular to the slab surface, or along the direction of propagation, are much more complicated.

The displacement function u_2 is given by (7.140), and involves the H_1 and H_3 components of the vector potential. These components each satisfy the wave equation (7.144), with wave velocity c_t. With propagation along x_1 only, with the solid restricted along x_3, the solutions have the form

$$\begin{aligned} H_1(x_1,x_3,t) &= H_{1+}\mathrm{e}^{\mathrm{i}(q_1x_1+q_3x_3-\omega t)} + H_{1-}\mathrm{e}^{\mathrm{i}(q_1x_1-q_3x_3-\omega t)}, \\ H_3(x_1,x_3,t) &= H_{3+}\mathrm{e}^{\mathrm{i}(q_1x_1+q_3x_3-\omega t)} + H_{3-}\mathrm{e}^{\mathrm{i}(q_1x_1-q_3x_3-\omega t)}, \end{aligned} \qquad (7.170)$$

with the wavevector components q_1 and q_3 satisfying the dispersion relation $\omega^2 = c_t^2(q_1^2+q_3^2)$.

The vector potential must be divergenceless, which translates to the relations

$$\left.\begin{aligned} q_1H_{1+} &= -q_3H_{3+}, \\ q_1H_{1-} &= +q_3H_{3-}. \end{aligned}\right\} \qquad (7.171)$$

We now apply the stress-free boundary conditions on the two surfaces $x_3 = \pm b$. The relevant relation involving H_1 and H_3 is $T_{23} = 0$, which gives

$$\begin{aligned} q_3^2H_{1+}\mathrm{e}^{\pm\mathrm{i}q_3b} &+ q_3^2H_{1-}\mathrm{e}^{\mp\mathrm{i}q_3b} \\ &- q_1q_3H_{3+}\mathrm{e}^{\pm\mathrm{i}q_3b} + q_1q_3H_{3-}\mathrm{e}^{\mp\mathrm{i}q_3b} = 0. \end{aligned} \qquad (7.172)$$

Using (7.171), this can be written as

$$\left.\begin{aligned} H_{1+}\mathrm{e}^{\mathrm{i}q_3b} + H_{1-}\mathrm{e}^{-\mathrm{i}q_3b} &= 0, \\ H_{1+}\mathrm{e}^{-\mathrm{i}q_3b} + H_{1-}\mathrm{e}^{\mathrm{i}q_3b} &= 0. \end{aligned}\right\} \qquad (7.173)$$

Forming sums and differences of these expressions, we find

$$\left.\begin{aligned} H_{1+}\cos q_3b + H_{1-}\cos q_3b &= 0, \\ H_{1+}\sin q_3b - H_{1-}\sin q_3b &= 0, \end{aligned}\right\} \qquad (7.174)$$

and this in turn yields the final result

$$\left.\begin{aligned} \cos q_3b\,\sin q_3b &= 0, \\ H_{1+} &= \pm H_{1-}. \end{aligned}\right\} \qquad (7.175)$$

Hence the solutions to the slab with shear horizontal waves are composed of symmetric ($H_{1-} = H_{1+}$) modes that satisfy $\cos q_3 b = 0$, and antisymmetric ($H_{1+} = -H_{1-}$) modes with $\sin q_3 b = 0$. The allowed wavevectors are constrained along the x_3 axis to $q_3 = n\pi/2b$ with $n \geq 0$ an integer, n odd corresponding to symmetric solutions and n even corresponding to antisymmetric ones. There is no constraint on the wavevector component q_1 along the x_1 direction, so that the full set of allowed modes in this system have wavevectors of the form $\boldsymbol{q_n} = (q_1, 0, n\pi/2b)$, with the values of q_1 unconstrained. The corresponding allowed frequencies have the form $\omega^2 = c_t(q_1^2 + q_3^2)$. In Fig. 7.25 we display the dispersion relation for the first four modes.

The displacement functions u_2 corresponding to these modes may be found from the expression (7.140),

$$\left.\begin{aligned} u_2(x_1, x_3, t) &= A_n \cos\frac{n\pi x_3}{2b} \mathrm{e}^{\mathrm{i}q_1 x_1 - \omega t} \quad (n \text{ even}), \\ u_2(x_1, x_3, t) &= A_n \sin\frac{n\pi x_3}{2b} \mathrm{e}^{\mathrm{i}q_1 x_1 - \omega t} \quad (n \text{ odd}). \end{aligned}\right\} \tag{7.176}$$

Note that these expressions are complex, only the real part of these expressions has any physical meaning. The displacement for even n corresponds to symmetric displacements about the plane of symmetry $x_3 = 0$, with maximum displacement at the surface, and odd n odd corresponds to antisymmetric displacements, but again with maximum displacement at the surface. In Fig. 7.26 we display the first few mode shapes.

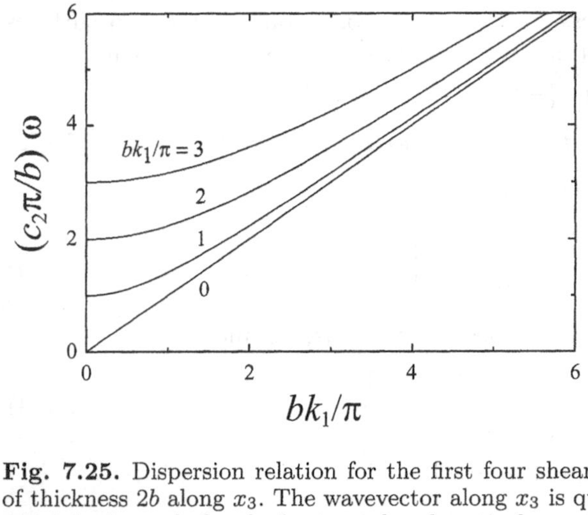

Fig. 7.25. Dispersion relation for the first four shear horizontal modes in a slab of thickness $2b$ along x_3. The wavevector along x_3 is quantized in multiples of π/b, which is the scale for the horizontal and vertical axes in the plot.

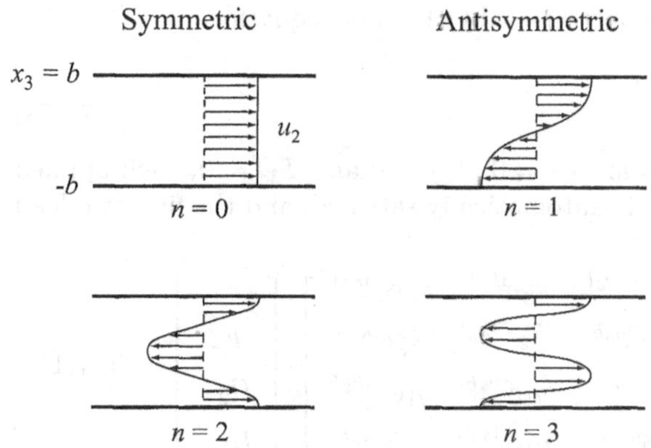

Fig. 7.26. Displacement amplitudes u_2 as a function of position x_3 along the thickness of the slab. The first four shear horizontal modes, two symmetric and two antisymmetric, are shown.

7.6.2 The General Problem of Waves in a Plate

We now turn to the significantly more complex problem of waves with displacement in the $x_1 - x_3$ plane, with the same geometry as described above: An isotropic plate with extent $-b \leq x_3 \leq b$. A realistic choice for the boundary conditions would be to have zero stress on either surface. However, the resulting problem is quite intricate, and working through the detailed solution not very illuminating. We therefore consider a simpler, albeit less realistic, boundary condition, with zero stress along the x_1 and x_2 directions, but with zero displacement (rather than zero stress) along x_3. We restrict the spatial dependence of the displacement to x_1 and x_3 only; the more general case with dependence on x_2 involves more complicated notation but the approach is the same. A major result that we will find is the *Rayleigh–Lamb equation for the plate.*

Given that $\boldsymbol{u}$ is in the $x_1 - x_3$ plane, and $\boldsymbol{H}$ has zero divergence, we only need to consider the H_2 component of the vector potential, as well as the scalar potential Φ. These each satisfy the wave equations (7.144), with wave velocities c_t and c_ℓ respectively, so the solutions have the form

$$\left.\begin{aligned} \Phi(x_1, x_3, t) &= \Phi_+ e^{i(q_1 x_1 + q_3 x_3 - \omega t)} \\ &+ \Phi_- e^{i(q_1 x_1 - q_3 x_3 - \omega t)}, \\ H_2(x_1, x_3, t) &= H_{2+} e^{i(k_1 x_1 + k_3 x_3 - \omega t)} \\ &+ H_{2-} e^{i(k_1 x_1 - k_3 x_3 - \omega t)}. \end{aligned}\right\} \tag{7.177}$$

The boundary conditions relating H_2 and Φ must hold for all x_1, so the wavevector components along x_1 must be equal, $k_1 = q_1$. The remaining wavevector components are constrained by the wave equations,

$$\begin{aligned} \omega^2 &= c_\ell^2(q_1^2+q_3^2) \\ &= c_t^2(q_1^2+k_3^2). \end{aligned} \tag{7.178}$$

The boundary conditions are $u_3 = 0$, $T_{13} = 0$ and $T_{23} = 0$, each applied at $x_3 = \pm b$. The last of these is automatically satisfied, and the first two lead to the set of equations

$$\begin{bmatrix} -2q_1q_3\mathrm{e}^{\mathrm{i}q_3b} & 2q_1q_3\mathrm{e}^{-\mathrm{i}q_3b} & k_3^2\mathrm{e}^{\mathrm{i}k_3b} & k_3^2\mathrm{e}^{-\mathrm{i}k_3b} \\ -2q_1q_3\mathrm{e}^{-\mathrm{i}q_3b} & 2q_1q_3\mathrm{e}^{\mathrm{i}q_3b} & k_3^2\mathrm{e}^{-\mathrm{i}k_3b} & k_3^2\mathrm{e}^{\mathrm{i}k_3b} \\ q_3\mathrm{e}^{\mathrm{i}q_3b} & -q_3\mathrm{e}^{-\mathrm{i}q_3b} & q_1\mathrm{e}^{\mathrm{i}k_3b} & q_1\mathrm{e}^{-\mathrm{i}k_3b} \\ q_3\mathrm{e}^{-\mathrm{i}q_3b} & -q_3\mathrm{e}^{\mathrm{i}q_3b} & q_1\mathrm{e}^{-\mathrm{i}k_3b} & q_1\mathrm{e}^{\mathrm{i}k_3b} \end{bmatrix} \cdot \begin{bmatrix} \Phi_+ \\ \Phi_- \\ H_{2+} \\ H_{2-} \end{bmatrix} = 0. \tag{7.179}$$

This system of equations has a non-trivial solution if the determinant of the coefficients vanishes, i.e. when

$$(\mathrm{e}^{2iq_3b} - \mathrm{e}^{-2iq_3b})(\mathrm{e}^{2ik_3b} - \mathrm{e}^{-2ik_3b}) = 0, \tag{7.180}$$

which can also be written as

$$\sin(2q_3b)\sin(2k_3b) = 0. \tag{7.181}$$

The four non-trivial solutions to this equation are given by

$$\left.\begin{aligned} \sin(q_3b) &= 0, \\ \cos(q_3b) &= 0, \\ \sin(k_3b) &= 0, \\ \cos(k_3b) &= 0. \end{aligned}\right\} \tag{7.182}$$

To each of these solutions, there corresponds a particular eigenvector, that is, a set of values for the amplitudes $\Phi_\pm$ and $H_{2\pm}$. For example, for the first solution in (7.182), we find the wavevectors $q_{3n} = n\pi/2b$, $n = 0, 2, 4\ldots$, with amplitudes $\Phi_+ = \Phi_-$ and $H_{2+} = H_{2-} = 0$. These are symmetric dilatational waves, or P-waves, with a displacement vector given by

$$\left.\begin{aligned} u_{1n} &= 2\mathrm{i}q_1\Phi_+\cos(n\pi x_3/2b)\mathrm{e}^{\mathrm{i}q_1x_1}, \\ u_{2n} &= 0, \\ u_{3n} &= -2(n\pi/2b)\Phi_+\sin(n\pi x_3/2b)\mathrm{e}^{\mathrm{i}q_1x_1}, \end{aligned}\right\} \tag{7.183}$$

for $n = 0, 2, 4, \ldots$.

Similarly, for the second solution, we find the wavevectors $q_{3n} = n\pi/2b$, $n = 1, 3, 5, \ldots$, and the amplitudes $\Phi_+ = -\Phi_-$, $H_{2+} = H_{2-} = 0$. The corresponding displacement amplitude is

$$\left.\begin{aligned} u_{1n} &= -2q_1\Phi_+\sin(n\pi x_3/2b)\mathrm{e}^{\mathrm{i}q_1x_1}, \\ u_{2n} &= 0, \\ u_{3n} &= 2i(n\pi/2b)\Phi_+\cos(n\pi x_3/2b)\mathrm{e}^{\mathrm{i}q_1x_1}, \end{aligned}\right\} \tag{7.184}$$

for $n = 1, 3, 5, \ldots$. This solution forms antisymmetric P-waves.

The other two solutions may be found in a similar fashion, and also form symmetric and antisymmetric sets, but consist of shear modes rather than dilatational ones, and are a result of the superposition of SV waves. For the third solution we have $k_{3n} = n\pi/2b$, $n = 0, 2, 4 \ldots$, $H_+ = -H_-$ and $\Phi_+ = \Phi_- = 0$, and displacement

$$\left.\begin{aligned} u_{1n} &= -2i(n\pi/2b)H_+ \cos(n\pi x_3/2b)\mathrm{e}^{\mathrm{i}q_1 x_1}, \\ u_{2n} &= 0, \\ u_{3n} &= -2q_1 H_+ \sin(n\pi x_3/2b)\mathrm{e}^{\mathrm{i}q_1 x_1}, \end{aligned}\right\} \qquad (7.185)$$

representing a symmetric SV wave. For the fourth solution $k_{3n} = n\pi/2b$, $n = 1, 3, 5 \ldots$, $H_+ = H_-$ and $\Phi_+ = \Phi_- = 0$, and

$$\left.\begin{aligned} u_{1n} &= -2i(n\pi/2b)H_+ \sin(n\pi x_3/2b)\mathrm{e}^{\mathrm{i}q_1 x_1}, \\ u_{2n} &= 0, \\ u_{3n} &= 2q_1 H_+ \cos(n\pi x_3/2b)\mathrm{e}^{\mathrm{i}q_1 x_1}, \end{aligned}\right\} \qquad (7.186)$$

an antisymmetric SV wave.

A set of selected modes are shown in Fig. 7.27.

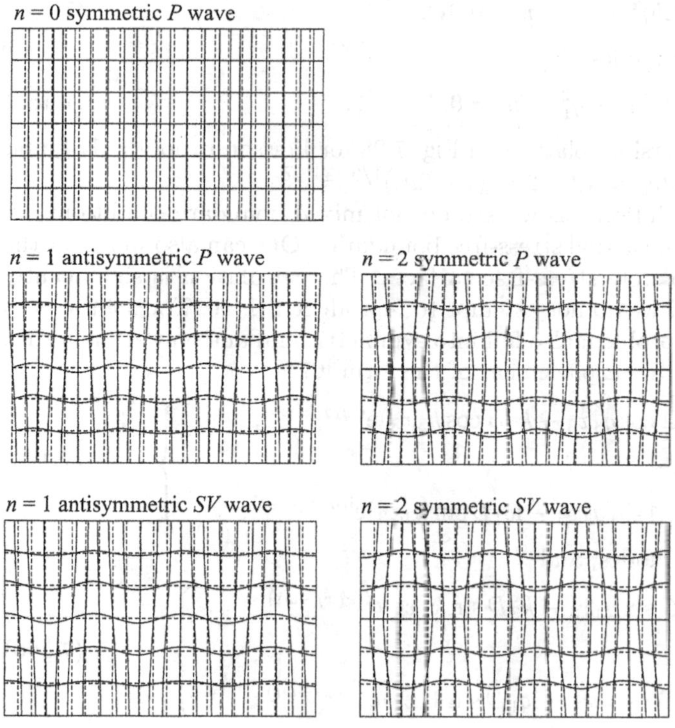

Fig. 7.27. Five of the lowest plate modes, solved for using mixed boundary conditions.

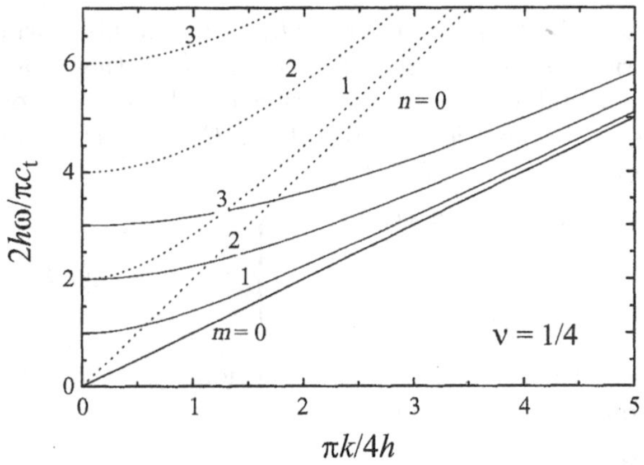

Fig. 7.28. Dispersion relations for dilational and shear modes in an infinite plate with mixed boundary conditions. The dotted lines are for the dilational modes, and the solid lines for the shear modes; mode numbers are as marked.

The dispersion relations for the various modes may be obtained from (7.178), which for the dilatational modes are

$$\omega_n^2 = c_\ell^2(n\pi/2b)^2 + q_1^2, \quad n = 0, 1, 2, \ldots, \tag{7.187}$$

while for the shear modes,

$$\omega_m^2 = c_t^2(m\pi/2b)^2 + q_1^2, \quad m = 0, 1, 2, \ldots. \tag{7.188}$$

We plot these dispersion relations in Fig. 7.28 for Poisson ratio $\nu = 1/4$, and correspondingly $c_\ell/c_t = ((2-2\nu)/(1-2\nu))^{1/2} = \sqrt{3}$.

This detailed solution was worked out for mixed boundary conditions, an approximation to the actual stress-free boundaries. One can also work out the solutions for completely stress-free surfaces. The calculation is significantly more complex, and we do not attempt to reproduce it here. The solutions to this more realistic problem also fall into symmetric and antisymmetric forms, with solutions for the displacement of the form [67]

$$\left.\begin{aligned} u_1 &= (\mathrm{i}q_1 A\cos q_3x_3 + k_3 B\cos k_3x_3)\mathrm{e}^{\mathrm{i}(q_1x_1-\omega t)}, \\ u_2 &= 0, \\ u_3 &= (-q_3A\sin q_3x_3 - \mathrm{i}q_1 B\sin k_3x_3)\mathrm{e}^{\mathrm{i}(q_1x_1-\omega t)}, \end{aligned}\right\} \tag{7.189}$$

for the symmetric modes, and

$$\left.\begin{aligned} u_1 &= (\mathrm{i}q_1 C\sin q_3x_3 - k_3 D\sin k_3x_3)\mathrm{e}^{\mathrm{i}(q_1x_1-\omega t)}, \\ u_2 &= 0, \\ u_3 &= (q_3C\cos q_3x_3 - \mathrm{i}q_1 D\sin k_3x_3)\mathrm{e}^{\mathrm{i}(q_1x_1-\omega t)}, \end{aligned}\right\} \tag{7.190}$$

for the antisymmetric modes. The boundary conditions applied to these solutions yield the dispersion relations, which for the symmetric modes are

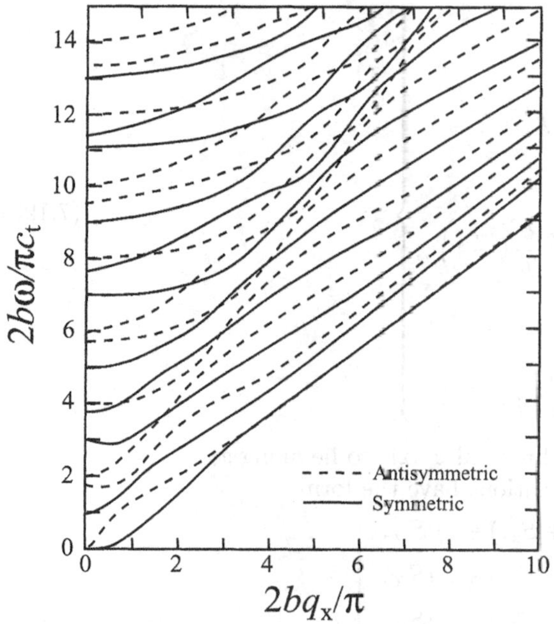

Fig. 7.29. Dispersion relations from the Rayleigh–Lamb equation. Adapted from Graff [14].

$$\frac{\tan k_3 b}{\tan q_3 b} = -\frac{4q_1^2 q_3 k_3}{(k_3^2 - q_1^2)^2}, \tag{7.191}$$

and for the antisymmetric modes are given by

$$\frac{\tan k_3 b}{\tan q_3 b} = -\frac{(k_3^2 - q_1^2)^2}{4q_1^2 q_3 k_3}. \tag{7.192}$$

These sets of equations are known as the *Rayleigh–Lamb* equations for the plate, and generate a very complex dispersion relation, similar but more complex than that displayed in Fig. 7.28. We show the dispersion relations calculated for a Poisson ratio of $\nu = 0.3$, with the axes scaled by the plate thickness b, in Fig. 7.29. A thorough discussion of these equations and their implications may be found in the text by Graff [14].

7.7 Waves in Rods

We turn now to the discussion of waves in systems with boundaries in two dimensions, and infinite extent in the third. We will discuss only one example of such a system, a cylindrical rod of radius R, with its length oriented along x_3; such rods transmit waves known as *Pochhammer waves.* The stress S is related to the displacement $\boldsymbol{u} = (u_r, u_\theta, u_z)$ in cylindrical coordinates by

$$\left.\begin{aligned}
S_{rr} &= \frac{\partial u_r}{\partial r}, \\
S_{\theta\theta} &= \frac{1}{r}\left(\frac{\partial u_\theta}{\partial \theta} + u_r\right), \\
S_{zz} &= \frac{\partial u_z}{\partial z}, \\
S_{r\theta} &= \frac{1}{2r}\left(\frac{\partial u_r}{\partial \theta} + r^2\frac{\partial}{\partial r}(\frac{u_r}{r})\right), \\
S_{z\theta} &= \frac{1}{r}\frac{\partial u_z}{\partial \theta} + \frac{\partial u_\theta}{\partial z}, \\
S_{rz} &= \frac{1}{2}\left(\frac{\partial u_r}{\partial z} + \frac{\partial u_z}{\partial r}\right),
\end{aligned}\right\} \tag{7.193}$$

where we have chosen the cylindrical z axis to lie along x_3.

The linear stress-strain relations have the form

$$\left.\begin{aligned}
T_{rr} &= \lambda(S_{rr} + S_{\theta\theta} + S_{zz}) + 2\mu S_{rr}, \\
T_{\theta\theta} &= \lambda(S_{rr} + S_{\theta\theta} + S_{zz}) + 2\mu S_{\theta\theta}, \\
T_{zz} &= \lambda(S_{rr} + S_{\theta\theta} + S_{zz}) + 2\mu S_{zz}, \\
T_{r\theta} &= 2\mu S_{r\theta}, \\
T_{z\theta} &= 2\mu S_{z\theta}, \\
T_{rz} &= 2\mu S_{rz}.
\end{aligned}\right\} \tag{7.194}$$

There are three classes of waves in a cylinder, torsional, longitudinal, and flexural. Each has a specific symmetry, and we treat each in turn.

7.7.1 Torsional Waves

Torsional waves in a cylindrical rod have the simplest functional dependence, with no axial (z) or azimuthal (θ) dependence, and no displacements along the radial or axial directions. We can therefore set $u_r = u_z = 0$, and all derivatives with respect to θ are equal to zero. We take solutions that are plane waves along the cylinder axis, so that $u_\theta(r, z, t) = f(r) \exp i(kqz - \omega t)$. The radial dependence that satisfies the stress-strain relations can be shown to be either of the form $f(r) = r$ or $f(r) = J_1(r)$, where J_1 is the first-order Bessel function.

There are no forces on the surface of the cylinder, so that the stress on the surface must satisfy $T_{r\theta}(r = R) = 0$ (the other surface forces automatically vanish from the conditions given in the previous paragraph, as the reader can verify). This imposes a condition on the frequency given by the derivatives of J_1,

$$\eta_t R J_2(\eta_t R) = 0, \tag{7.195}$$

where J_2 is a second-order Bessel function, and η_t is the renormalized wavevector given by

$$\eta_t^2 = (\omega/c_t)^2 - q^2. \tag{7.196}$$

The solutions to (7.195) consist of the solutions $\eta_0 = 0$, and the roots ζ_n of the Bessel function, $J_1(\zeta_n) = 0$, which are given by $\eta_m R = \zeta_n = 5.136, 8.417, 11.620\ldots$.

The corresponding displacement u_θ is given by

$$u_{\theta,0} = U_0\, r\, \mathrm{e}^{\mathrm{i}q_0 z - \mathrm{i}\omega t}, \tag{7.197}$$

where $q_0 = \omega/c_t$, and the set

$$u_{\theta,n} = U_n\, J_1(r)\, \mathrm{e}^{\mathrm{i}q_n z - \mathrm{i}\omega t}, \tag{7.198}$$

where $q_n^2 = (\omega/c_t)^2 - \zeta_n^2/R^2$. The amplitudes U_i are determined by the initial conditions.

7.7.2 Longitudinal Waves

For longitudinal waves we have axial symmetry with no θ dependence, allowing displacements along r and z with dependence on the same variables. The boundary conditions require that T_{rr} and T_{rz} vanish at $r = R$. Solutions are assumed to have the form $u_r, u_z = f(r)\exp(\mathrm{i}qz - \mathrm{i}\omega t))$, with wavevector q and frequency ω. The solutions to the stress-strain relations, and the corresponding boundary conditions, are significantly more complicated than for torsional waves, and can be found in [67]. We merely state the frequency equations, which are

$$(q^2 - \eta_t^2)^2 \frac{\eta_\ell R J_0(\eta_\ell R)}{J_1(\eta_\ell R)} + 4q^2\eta_t^2 \frac{\eta_t R J_0(\eta_t R)}{J_1(\eta_t R)} = 2\eta_\ell^2(\eta_t^2 + q^2), \tag{7.199}$$

where $\eta_\ell^2 = (\omega/c_\ell)^2 - q^2$ and $\eta_t^2 = (\omega/c_t)^2 - q^2$. These may be simplified for small wavevector (large wavelength) and low frequency to extract the usual expression for transverse waves,

$$\omega = \sqrt{\frac{E}{\mu}}\, q = c_t q. \tag{7.200}$$

For the more general result we must resort to numerical solutions.

7.7.3 Flexural Waves

Flexural waves are the most complicated, involving displacements along all three coordinates. As usual, we take plane waves along the axial direction, and assume the functional dependence on r and θ is separable, so that the displacements have the form

$$\left.\begin{aligned} u_r(r,\theta,z) &= \Gamma_r(r)\Theta_r(\theta)\mathrm{e}^{\mathrm{i}(qz-\omega t)}, \\ u_\theta(r,\theta,z) &= \Gamma_\theta(r)\Theta_\theta(\theta)\mathrm{e}^{\mathrm{i}(qz-\omega t)}, \\ u_r(r,\theta,z) &= \Gamma_z(r)\Theta_z(\theta)\mathrm{e}^{\mathrm{i}(qz-\omega t)}. \end{aligned}\right\} \tag{7.201}$$

The stress-strain equations, applied to this *ansatz*, allow solutions of the form

$$\left.\begin{aligned} \Gamma_i(r) &= J_n(\eta r) + \mathrm{i}Y_n(\eta r), \\ \Theta_i(\theta) &= \mathrm{e}^{\pm in\theta}, \end{aligned}\right\} \tag{7.202}$$

where J_n and Y_n are the Bessel functions of the first and second kind, and $\eta^2 = (\omega/c_{t,l})^2 - q^2$ is the reduced wave vector. The solutions are fairly complicated, involving waves with both the transverse and longitudinal wave velocities; to simplify, we retain only the first order ($n = 1$) dependence on θ. Applying the stress-free surface boundary condition at $r = R$ yields the horrendous frequency equation

$$\begin{aligned} J_1(\tilde{\eta_\ell})J_1^2(\tilde{\eta_t}) \times \Big[2(\tilde{\eta_t}^2 - \tilde{q}^2)^2 \left(\frac{\tilde{\eta_t}J_0(\tilde{\eta_t})}{J_1(\tilde{\eta_t})}\right)^2 \\ +2\tilde{\eta_t}^2(5\tilde{q}^2 + \tilde{\eta_t}^2)\frac{\tilde{\eta_t}J_0(\tilde{\eta_t})}{J_1(\tilde{\eta_t})}\frac{\tilde{\eta_\ell}J_0(\tilde{\eta_\ell})}{J_1(\tilde{\eta_\ell}} \\ +2\tilde{\eta_t}^2\left(2\tilde{\eta_t}^2\tilde{q}^2 - \tilde{\eta_t}^2 - 9\tilde{q}^2\right)\frac{\tilde{\eta_\ell}J_1(\tilde{\eta_\ell})}{J_0(\tilde{\eta_\ell})} \\ \left(\tilde{\eta_t}^6 - 10\tilde{\eta_t}^4 - 2\tilde{\eta_t}^4\tilde{q}^2 + 2\tilde{\eta_t}^2\tilde{q}^2 + \tilde{\eta_t}^2\tilde{q}^4 - 4\tilde{q}^4\right)\frac{\tilde{\eta_\ell}J_0(\tilde{\eta_\ell})}{J_1(\tilde{\eta_\ell})} \\ \tilde{\eta_t}^2\left(-\tilde{\eta_t}^4 + 8\tilde{\eta_t}^2 - 2\tilde{\eta_t}^2\tilde{q}^2 + 8\tilde{q}^2 - \tilde{q}^4\right)\Big] \\ = 0, \end{aligned} \tag{7.203}$$

where $\tilde{\eta_{\ell,t}} = \eta_{\ell,t}R$ and $\tilde{q} = qR$. Solutions to this equation can be found numerically for a given value of Poisson's ratio, fixing the relationship between η_ℓ and η_t. The lowest branch of solutions may be approximated for small q by

$$\frac{\omega R}{c_t} \cong \left[\frac{1+\nu}{2}\right]^{1/2}(qR)^2. \tag{7.204}$$

Note that the torsional and longitudinal waves have linear dispersion relations, while the flexural wave has a quadratic relation, $\omega \propto q^2$.

7.8 Summary

In this chapter we have covered a number of approaches to understanding the dynamics of solids. We began with the Euler–Bernoulli and Timoshenko analyses, where we discussed the solutions of the problems of flexural, longitudinal and torsional motion of beams. We then turned to the full stress-strain tensor analysis of time-dependent forces, and discussed applications of these ranging from the simplest situation of the infinite isotropic solid, to gradually more complex problems, the complexity increasing as the number of boundaries increases. We now have a clear understanding of the description of classical waves in a solid, in a number of geometries ranging from the infinite solid to the nearly one-dimensional beam. In the next chapter we turn to a description of the dissipation of energy in solids.

Exercises

7.1 Find frequency dependence for the undamped response of a longitudinal beam to a force per unit length $f(z,t) = f_0 \cos(3\pi z/\ell)\cos(\omega_d t)$, directed along z.

7.2 Find the first four terms $n = 1-4$ for the undamped response of a longitudinal beam to a force per unit length $f(z,t) = f_0 z \cos(\omega_d t)$, directed along z.

7.3 Taking a phenomenological damping term $\gamma\partial\theta/\partial t$ in the dynamic equation of motion for the driven torsional rod, find an expression for the response to a uniform driving torque $m(z,t) = m_0 \cos(\omega_d t)$, for the $n = 1$ mode.

7.4 Work out the frequency dependence for the undamped response of a flexural beam to a uniformly distributed force per unit length $f(z,t) = f_0 e^{i\omega t}$.

7.5 Find the four lowest flexural frequencies for a doubly clamped beam made of a material with $E = 100\ \mathrm{N/m^2}$, with dimensions $L \times w \times t = 10 \times 0.1 \times 0.1\ \mu\mathrm{m}^3$. Find the fractional correction from the Timoshenko theory for each of these modes.

7.6 Consider placing a lumped mass at the end of a circular rod. The rod has a torsional moment about its length of I_3; the mass has a torsional moment of inertia I about the same axis. Assume the rod has length L, Lamé constant μ, and mass ρ. Find the lowest torsional resonance frequency for the combined rod and mass.

7.7 Using the Love theory for longitudinal waves in a rod, find the correction to the speed of sound of longitudinal waves as a function of wavevector.

7.8 Find an expression for the dilatation δ for a plane harmonic dilatational wave travelling in an isotropic medium.

7.9 Consider a sphere of radius R. Using spherical coordinates, find the equation that governs the frequency relations for the spherically symmetric, radial vibration modes of the sphere.

7.10 Find expressions for the amplitudes of the reflected P and S waves from a P (longitudinal) wave, incident at an angle θ from $x_3 > 0$ on a completely fixed surface $x_3 = 0$.

7.11 As in Ex. 7.8, find expressions for the amplitudes of the reflected P and S waves from a SV wave, incident at an angle ζ from $x_3 > 0$ on a completely fixed surface $x_3 = 0$.

7.12 As in Ex. 7.8, find expressions for the amplitudes of the reflected P and S waves from a SH wave, incident at an angle ζ from $x_3 > 0$ on a completely fixed surface $x_3 = 0$.

7.13 Consider a plane interface at $x_3 = 0$ between two media with different material properties. Find an expression for the reflection and transmission of an SH wave incident at normal incidence on the interface.

7.14 As in Ex. 7.8, consider a plane interface at $x_3 = 0$ between two media with different material properties. Find an expression for the reflection and transmission of an SH wave incident at an angle ζ on the interface.

7.15 Plot the group velocity as a function of frequency for the SH waves in a plate.

8. Dissipation and Noise in Mechanical Systems

The dynamical theory of solids that we have described in Chaps. 4 through 7 is fundamentally based on the *harmonic approximation* to the atomic Hamiltonian. This approximation, as discussed in Chap. 2, leads to the linear relation between the continuum stress and strain. When we worked out the vibrational properties of beams in Chap. 7, we found a mechanical response that had perfectly defined resonance frequencies, with no natural width, and that gave an infinite response for forces applied at one of the resonance frequencies. Furthermore, the waves in infinite and semi-infinite solids were found to travel without any attenuation, travelling without transfer of energy from the wavevector and frequency of interest to other mechanical modes in the solid. This theory therefore does not describe the dissipation of energy from the mechanical modes. The dissipation and attenuation of energy in a mechanical structure is however thermodynamically required: In order that a mechanical normal mode achieve thermal equilibrium with its environment, it must be able to exchange energy with that environment. In a mechanical system, the environment comprises all the other mechanical normal modes of the solid, as well as any electronic degrees of freedom or other types of excitations that might exist.

There are a number of fundamental processes that lead to the transfer of energy between the different modes in a solid, including interactions between the particular normal mode of interest with all the other mechanical normal modes of the system, interactions with electrons and phonons, as well as other effects. These were discussed in Chap. 3. There we outlined the quantum mechanical description of the primary terms that couple phonons, that is, the mechanical normal modes, to one another, and to the electrons in a solid. There are also non-intrinsic effects that lead to the dissipation of energy, such as the motion of defects or ions in the solid due to the imposed strain, and interactions with surface contaminants. Here we will describe the phenomenological theory that allows these interactions to be included in a simple but quantitative way in the continuum theory.

The exchange of energy between the normal modes and the other degrees of freedom in the solid occurs through random and irreversible processes. There is therefore, due to the fluctuation-dissipation theorem , *mechanical noise* associated with the damping in the system. Below we outline the fundamental concepts involved in the classical fluctuation-dissipation theorem as it applies to a single normal mode resonance interacting with a complex environment, quantified by the Langevin equation. For the particular case of a flexural beam resonator described by the Euler–Bernoulli theory, we describe the thermally-driven noise, and its impact on measurements of the resonator frequency.

8.1 Langevin Equation

The one-dimensional simple harmonic oscillator is frequently used as a prototype for resonant systems, as it captures the principal characteristics for resonance, while allowing for a simple phenomenological description in terms of only two parameters (the resonance frequency and the damping coefficient). We therefore begin by describing the lossy harmonic oscillator.

Consider a system comprising an inertial mass m that interacts with its environment through a conservative potential $U(x)$, and, in addition, through a complex interaction term that is characterized by friction as well as noise. A common example of the latter is the interaction of a small particle with the molecules in a fluid, which can be described by an average energy dissipation quantified by Stokes' law, and also includes noise-induced Brownian motion. The connection between friction and Brownian motion was first made by Einstein in 1905.

We take the conservative potential to have a local energy minimum at $x = 0$, so that we can expand the potential $U(x)$ to second order in the displacement x, $U(x) = U(0) + U''(0)x^2/2 + \ldots$. The first-order term that involves the first derivative $U'(0)$ vanishes, and the second derivative $U''(0)$ is positive, because of the stipulation that we are at a local minimum of U. Defining the frequency ω_0 through the relation

$$m\omega_0^2 = U''(0), \tag{8.1}$$

we can write the dynamic equation in the absence of friction, for small displacements x, as

$$m\,\frac{\mathrm{d}^2x}{\mathrm{d}t^2} + m\,\omega_0^2\,x = 0. \tag{8.2}$$

The solution to this lossless equation can be written in complex notation as

$$x(t) = x_0\,\exp(-\mathrm{i}\omega_0 t + \mathrm{i}\phi). \tag{8.3}$$

In this equation, the minus sign on the time dependence is taken by convention, and the phase angle ϕ and amplitude x_0 are determined by the initial conditions; at the end of the calculation we of course will take the real part of this solution.

We now include the friction and noise in the system, which is by its nature defined in statistical terms. These effects, which are due to interactions of the mass m with a large number of degrees of freedom in the environment, can be included in (8.2) by adding a time-dependent environmental force term $F_{\mathrm{env}}(t)$,

$$m\frac{\mathrm{d}^2x}{\mathrm{d}t^2}+m\,\omega_0^2\,x=F_{\mathrm{env}}(t). \tag{8.4}$$

In many dissipative systems, the environmental force can be separated into a *dissipation* (or loss) term, which typically is proportional to the *ensemble average* velocity $\langle \mathrm{d}x/\mathrm{d}t\rangle$, averaged over an ensemble of identical systems (see a text on statistical mechanics for a discussion of ensemble averages), and a *noise* term due to a random force $F_N(t)$. We thus rewrite (8.4) as

$$m\frac{\mathrm{d}^2x}{\mathrm{d}t^2}+m\,\omega_0^2\,x=-m\gamma\frac{\mathrm{d}x}{\mathrm{d}t}+F_N(t), \tag{8.5}$$

where dissipative constant γ relates to the rate of dissipation to the average velocity. Equations of the form (8.5) are known as *Langevin equations*.

The dissipative term in (8.5) causes energy to be transferred from the harmonic oscillator to the environment. This term appears because the motion of the oscillator generates statistical correlations in the environmental force term F_{env} in (8.4), which are proportional to the average velocity $\dot{x}$, and last long enough to do non-zero work on the oscillator. This *first moment* of the environmental force has been extracted in the Langevin equation (8.5), giving the definition for the dissipative constant γ in terms of the correlations in the environmental force, and leaving the random force F_N defined so that it has zero statistical average value, $\langle F_N(t)\rangle = 0$. Thermal equilibrium in a system controlled by (8.5) is then achieved through the second moment of the noise force, which must satisfy

$$2mk_{\mathrm{B}}T\gamma=\int_{-\infty}^{\infty}\langle F_N(0)F_N(t')\rangle\,\mathrm{d}t'. \tag{8.6}$$

In this equation, the angle brackets $\langle\ldots\rangle$ indicate that an ensemble average of the product of the forces at time $t=0$ and t' must be taken. If we assume that the noise force is uncorrelated for time scales over which the oscillator responds, we can simplify (8.6) to read

$$\langle F_N(t)F_N(t')\rangle = 2m\gamma k_{\mathrm{B}}T\delta(t-t'), \tag{8.7}$$

where $\delta(t)$ is the Dirac delta function (see Appendix A).

We can also define the spectral density of the noise force, as the Fourier transform of (8.7). For a time-dependent force $F_N(t)$, the Fourier transform $\mathcal{F}_N(\omega)$ of the average force $\langle F_N(t)\rangle$ is given by

$$\mathcal{F}_N(\omega)=\frac{1}{2\pi}\int_{-\infty}^{\infty}\langle F_N(t)\rangle\,\mathrm{e}^{\mathrm{i}\omega t}\mathrm{d}t. \tag{8.8}$$

Because the ensemble average of the force F_N is zero, this translates to $\mathcal{F}_N(\omega) = 0$. The spectral density of the force is defined as the Fourier transform of the *correlation function* $K(s)$ of the force. The latter is defined as

$$K(s) = \langle F_N(t) F_N(t+s) \rangle , \tag{8.9}$$

the ensemble average of the product of forces at time t and time $t+s$, assumed to be independent of time t and depending only on the separation in time s. This assumption about the statistical nature of the force, that its correlation function does not depend on the absolute time t, is because we assume this is a stationary random process.

For a general correlation function (8.9), the spectral density $S(\omega)$ is defined as

$$S(\omega) = \frac{1}{2\pi} \int_{-\infty}^{\infty} K(s) e^{i\omega s} ds. \tag{8.10}$$

From (8.7), for a completely uncorrelated force noise, we can write the correlation function as

$$K(s) = 2m\gamma k_B T \delta(s). \tag{8.11}$$

For the particular correlation function (8.11), we find the spectral density

$$S(\omega) = \frac{m\gamma k_B T}{2\pi}, \tag{8.12}$$

independent of frequency. This is known as a *white* spectral density, and is characteristic of time-uncorrelated functions.

The mean energy of the harmonic oscillator, $\langle E \rangle$, is given by

$$\langle E \rangle = \frac{1}{2} m \langle \dot{x}^2 \rangle + \frac{1}{2} k \langle x^2 \rangle . \tag{8.13}$$

The energy of an undriven oscillator, described by (8.5), will equilibrate to the energy of the environment by losing any initial non-equilibrium energy through the velocity-dependent dissipation term, and then gaining and losing energy stochastically through the noise term $F_N(t)$. The noise force (8.7) will produce this equilibrium. The proof involves a detailed calculation, that may be found e.g. in Reif [68]. The mean energy $\langle E \rangle$ equilibrates to the temperature T, so that

$$\langle E \rangle = k_B T, \tag{8.14}$$

with the individual equilibration of the kinetic and potential energies,

$$\frac{1}{2} m \langle \dot{x}^2 \rangle = \frac{1}{2} k \langle x^2 \rangle = \frac{1}{2} k_B T. \tag{8.15}$$

The relation (8.7) provides a fundamental and important relation between environmental noise, dissipation, and temperature. It was first discovered by Einstein in 1905 in relation to Brownian motion in liquids, but has applications to most systems in which the average damping of an object is proportional to its velocity. Due to the ubiquity and the simplicity of these relations, many systems in which the dissipative terms are more completely understood in terms of the microscopic interactions are nevertheless approximated by Langevin equations of the form (8.5). We note that the quantity x appearing in (8.4) is not necessarily a linear displacement; x may represent a torsional angle for a torsional resonator, the dilatation Δ for a longitudinal resonator, or some more complex quantity related to a mechanical disturbance. The same discussion applies equally well to e.g. electronic circuits, where a parallel combination of an inductor, a resistor and a capacitor is described by exactly the same dynamical equations for a harmonic oscillator, but where x represents either the charge on the capacitor or the current through the inductor. The mass m is replaced by a term involving the capacitance or the inductance, and the frictional term would involve the value of the resistance. The noise force $F_N(t)$ is then directly related to the Johnson–Nyquist noise in the resistor.

8.1.1 Dissipation and Quality Factor

In the absence of the noise term $F_N(t)$, the solution of (8.5) is given by $x(t) = x_0 \exp(-\mathrm{i}\omega t + \phi)$, where the complex-valued frequency is given by

$$\omega^2 + \mathrm{i}\,\gamma\,\omega - \omega_0^2 = 0. \tag{8.16}$$

The frequency ω has both real and imaginary parts, $\omega = \omega_R + i\,\omega_I$, which from solving (8.16) are given by

$$\left.\begin{aligned} \omega_R &= \sqrt{\omega_0^2 - \gamma^2/4}, \\ \omega_I &= -\gamma/2. \end{aligned}\right\} \tag{8.17}$$

The imaginary part of ω gives damping, as we can rewrite the time dependence as

$$x(t) = x_0 \mathrm{e}^{\omega_I t} \mathrm{e}^{-\mathrm{i}\omega_R t} = x_0 \mathrm{e}^{-\gamma t/2} \mathrm{e}^{-\mathrm{i}\omega_R t}. \tag{8.18}$$

This represents an oscillation at frequency ω_R, damped to $1/e$ the initial amplitude in a time $\tau = 2/\gamma$. The energy of the oscillation, proportional to x^2, damps to $1/e$ its initial value in the time $\tau_E = 1/\gamma$.

We define the quality factor Q through the equation

$$Q^{-1} \equiv 2\left|\frac{\omega_I}{\omega_R}\right| = \frac{\gamma}{\sqrt{\omega_0^2 - \gamma^2/4}}. \tag{8.19}$$

In the limit of small damping, $\gamma \ll \omega_0$, we can approximate the quality factor by $Q \cong \omega_0/\gamma = \omega\tau/2$. The time dependence may then be written in the form $x(t) = x_0 \exp(-\omega_0 t/2Q)\exp(-\mathrm{i}\omega_R t)$, so that the energy damps with a characteristic time $\omega_0 \tau_E \cong 1/Q$.

The damping in a harmonic oscillator is therefore captured by the dimensionless quality factor Q, or equivalently the dissipation $D = 1/Q$. We can rewrite the Brownian motion result (8.7) in the form $\langle F_N^2 \rangle = 2m\omega_0 k_\mathrm{B} T/Q$.

8.1.2 Attenuation

The discussion above has been for a single harmonic oscillator, parameterized by a mass m, a resonance frequency ω, and dissipation $1/Q$. A related type of problem is where a single frequency strain wave is transmitted through a solid, and its attenuation as a function of distance is measured. If the strain wave travels along x, and is described by the relative displacement $\boldsymbol{u}(x,t)$, then by solving equations analogous to the Langevin equation (8.5), the travelling wave solutions are found to have the form

$$\begin{aligned} \boldsymbol{u}(x,t) &= \boldsymbol{u}_0 \; \mathrm{e}^{\mathrm{i}(qx-\omega t)} \\ &= \boldsymbol{u}_0 \; \mathrm{e}^{-Ax} \; \mathrm{e}^{\mathrm{i}(q_R x-\omega t)}, \end{aligned} \tag{8.20}$$

where $\boldsymbol{u}_0$ is the initial displacement, q and ω the wavevector and frequency of the displacement, where q is complex-valued. In the second line of (8.20) q has been separated into a real and imaginary part, $q = q_R + \mathrm{i}A$. The attenuation A, also called the logarithmic decrement, describes the rate at which the displacement decays as the disturbance passes through the solid.

Attenuation and dissipation are intimately related, as they represent two characterizations of the same physical process. A wave travelling through a medium with dissipation $1/Q$ for Q large decays in time as $\exp(-\omega_0 t/2Q)$. As it travels with phase velocity $v_\phi = \omega_0/q_R$, this is equivalent to a decay in position $\exp(-Ax) = \exp(-Av_\phi t)$, giving the identity

$$A = \frac{\omega_0}{2Qv_\phi}. \tag{8.21}$$

Mechanical dissipation is usually measured by monitoring the behavior of a mechanical resonator as a function of time; many such measurements were made on macroscopic resonators with resonance frequencies in the audio range. Attenuation measurements have also been made on waves generated using a variety of ultrasonic transducers, and cover a range of frequencies from a few MHz to few tens of GHz. We now turn to a discussion of a variety of intrinsic and extrinsic sources of damping in mechanical systems, some of which are well understood in terms of microscopic processes and some of which are more phenomenological.

8.2 Zener's Model of an Anelastic Solid

Zener was the first to develop a simple model allowing inelastic mechanical processes to be included in the theory of deformable solids, through a straightforward generalization of Hooke's law [69]. In Zener's model, the Hooke's stress-strain relation in one dimension, $\tau = E\epsilon$, relating the stress τ to the strain ϵ, is generalized to allow for mechanical relaxation in the solid, that is, an explicit dependence on the *rate* at which processes occur:

$$\tau + \tau_\epsilon \frac{d\tau}{dt} = E_R \left(\epsilon + \tau_\tau \frac{d\epsilon}{dt} \right), \tag{8.22}$$

where E_R is the relaxed value of Young's modulus. Strains applied slowly in time generate responses with the relaxed modulus, while rapidly varying strains involve a different value for the modulus. Note that τ_ϵ and τ_τ are time constants, not to be confused with the stress τ.

We consider stress and strain variations that are harmonic in time, so that $\tau(t) = \tau_0 \, e^{-i\omega t}$ and $\epsilon(t) = \epsilon_0 \, e^{-i\omega t}$. At low frequencies, such that $\omega \ll 1/\tau_\epsilon, 1/\tau_\tau$, the relation between the amplitudes τ_0 and ϵ_0 becomes the standard Hooke's law relation with a real elastic modulus $E = E_R$, with $\tau_0 = E_R \epsilon_0$. At high frequencies, with $\omega \gg 1/\tau_\epsilon, 1/\tau_\tau$, this relation is again real, but with $E = E_U$, the *unrelaxed* Young's modulus, given by

$$E_U = \frac{\tau_\tau}{\tau_\epsilon} E_R. \tag{8.23}$$

For intermediate frequencies, the Young's modulus $E(\omega)$, with $\tau_0 = E(\omega)\epsilon_0$, becomes complex-valued of the form

$$\begin{aligned} E(\omega) &= \left(\frac{1+\omega^2 \bar{\tau}^2}{1+\omega^2 \tau_\epsilon^2} - \frac{i\omega\bar{\tau}}{1+\omega^2\tau_\epsilon^2} \Delta \right) E_R \\ &= E_{\text{eff}}(\omega) \left(1 - \frac{i\omega\bar{\tau}}{1+\omega^2\bar{\tau}^2} \Delta \right), \end{aligned} \tag{8.24}$$

where we have defined the mean relaxation time $\bar{\tau} = \sqrt{\tau_\epsilon \tau_\tau}$, the fractional modulus difference $\Delta = (E_U - E_R)/E_R$, and the effective Young's modulus

$$E_{\text{eff}} = \frac{1+\omega^2\bar{\tau}^2}{1+\omega^2\tau_\epsilon^2} E_R. \tag{8.25}$$

A generalization of these expressions to include the 6×6 form for the relaxed and unrelaxed elastic moduli c_{ij}^R and c_{ij}^U is straightforward.

The imaginary part of the Young's modulus in (8.24) implies that the stress τ will include a component that is 90° out of phase with the strain ϵ. If the strain is $\epsilon_0 e^{-i\omega t}$, the stress will be $(\tau_R + i\tau_I)e^{-i\omega t}$ with $\tau_R = E_{\text{eff}}(\omega)\epsilon_0$ and $\tau_I = -\left[E_{\text{eff}}(\omega) \frac{\omega\bar{\tau}}{1+\omega^2\bar{\tau}^2} \Delta \right] \epsilon_0$. The work done by the stress is proportional to $\tau \cdot d\epsilon/dt$, where only the real parts of the time-dependent stress and strain are used. When averaged over time, the only non-zero part of the work is proportional to $\tau_I d\epsilon/dt$, representing energy lost from the oscillation. Hence the imaginary part of $E(\omega)$ determines the loss.

For small Δ, we can define a quality factor Q as the amplitude of the ratio of the imaginary to the real part of E:

$$Q^{-1} = \frac{\omega\bar{\tau}}{1+\omega^2\bar{\tau}^2} \Delta. \tag{8.26}$$

This expression shows that the dissipation $1/Q$ is frequency dependent, becoming small for high and low frequency excitations, where the effective

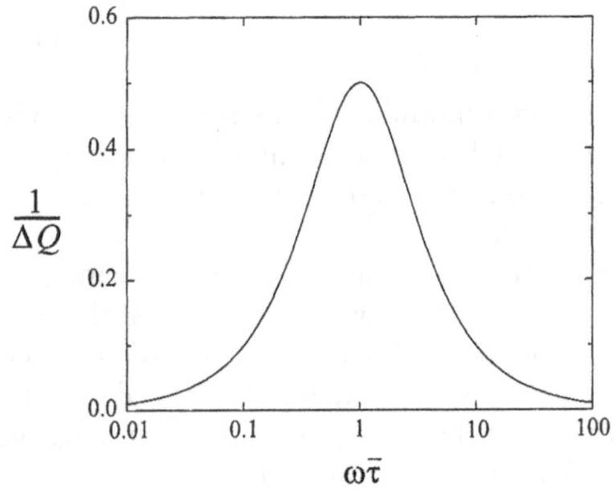

Fig. 8.1. Dependence of the dissipation $1/Q$ on the excitation frequency ω, from Zener's model for an anelastic solid.

elastic constant becomes real, and reaching its maximum value at $\omega = 1/\bar{\tau}$, where $(1/Q)_{max} = \Delta/2$; this dependence is shown in Fig. 8.1.

We have not discussed the nature of the relaxation times $\tau_{\tau,\epsilon}$; their inclusion in Zener's model through (8.22) is purely phenomenological, and detailed identification in terms of physical processes is done on a case-by-case basis. Solids have many possible sources of anelastic relaxation, including such diverse effects as ion motion, grain boundary rearrangements, electronic and magnetic excitations, and fundamental processes such as thermoelastic relaxation or phonon-phonon interactions. Many of these processes are thermally-activated, and therefore temperature dependent.

8.2.1 Two-Level Systems

As a phenomenological example, we consider a generic two-level system, which might describe two states available for an ionic defect in a crystalline solid. If the two states differ by an energy $\Delta\mathcal{E}$, then the occupation probabilities of the upper and lower energy states are related by the thermodynamic relation,

$$\frac{P_{\text{upper}}}{P_{\text{lower}}} = \mathrm{e}^{-\Delta\mathcal{E}/k_B T}. \tag{8.27}$$

If the system makes transitions from the upper to the lower state at a rate Γ_{down}, due to the emission of energy $\Delta\mathcal{E}$, then in order to maintain the relative occupations, upward transitions must occur at a rate $\Gamma_{\text{up}} = \exp(-\Delta\mathcal{E}/k_B T)\Gamma_{\text{down}}$. The upwards rate is controlled by the probability that the environment can provide the energy $\Delta\mathcal{E}$, which for example might come from absorption of phonons of frequency $\mathcal{W} = \Delta\mathcal{E}/\hbar$. In the presence of an acoustic strain field of frequency ω', some of the transition energy may be provided by the strain field. In the usual limit where the strain field frequency

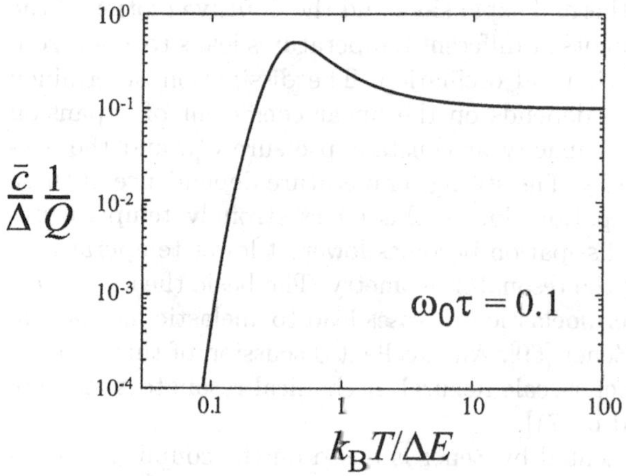

Fig. 8.2. Temperature-dependent dissipation $1/Q$ due to a two-level system, with high-temperature relaxation rate $\omega\tau_0 = 0.1$.

$\omega' \ll \omega$, the strain field causes a slow variation of the energy difference $\Delta\mathcal{E}$, and transitions between the two states absorb some of the strain energy, but at a rate controlled by the thermal phonon population about the much higher frequency ω. In this case, our two-level system will absorb energy from the strain field at a temperature-dependent rate $1/\bar{\tau}$ with

$$\frac{1}{\bar{\tau}} = \frac{1}{\tau_0} e^{-\Delta\mathcal{E}/k_B T}. \tag{8.28}$$

A generic two-level system will therefore display a temperature-dependent quality factor given by (8.26), with a temperature-dependent $\bar{\tau}$ given by (8.28). In Fig. 8.2 we display a typical dependence, for a strain field with $\omega\tau_0 = 0.1$. Such a temperature dependence for the quality factor has been observed in both metallic and insulating high-Q mechanical resonators (see e.g. [70]). The characteristic peak shown in Fig. 8.2, which is usually accompanied by other peaks from other microscopic interactions, is known as a *Debye peak*. Particular microscopic processes are given names as well, with point-defect relaxation peaks termed *Snoek peaks*, dislocation relaxation termed *Bordoni peaks*, and so on. A thorough discussion of these issues is given by Nowick and Berry [25].

8.3 Thermoelastic Relaxation

Thermoelastic relaxation is the term given to the generation of (high-frequency) thermal phonons from the strain field present in a mechanical resonator. Strain generated by the low-frequency mechanical motion generates differences in the temperature at different locations in the resonator, due to a non-zero coefficient of thermal expansion, and the diffusive motion of the thermal phonons between points at different temperatures leads to a non-zero dissipation $1/Q$ for the mechanical oscillation. The dissipation determined by this fundamental process depends on the linear coefficient of expansion $\alpha = (1/L)\mathrm{d}L/\mathrm{d}T$, the heat capacity at constant pressure C_p, and the average resonator temperature T_0. The strong temperature dependence of these parameters makes the dissipation due to this effect strongly temperature-dependent. In general, the dissipation becomes lower at lower temperatures, depending on the details of the resonator geometry. The basic theoretical arguments describing how thermoelastic processes lead to anelastic mechanical response are described by Zener [69]. An excellent discussion of this process, with applications to nanometer-scale flexural mechanical resonators made of Si and GaAs, may be found in [71].

The basic argument presented by Zener is based on the coupling of temperature and strain. Consider a solid with a (scalar) strain $\epsilon(\boldsymbol{r}, t)$, and a position- and time-dependent temperature $T(\boldsymbol{r}, t) = T_0 + \Theta(\boldsymbol{r}, t)$, where Θ is the local temperature variation from the average T_0. Variations of temperature cause thermal expansion of the solid, so the strain is related to both the stress τ and the local temperature variation Θ, with the approximate linear expression

$$\epsilon = E_{\mathrm{iso}}^{-1}\,\tau + \alpha\,\Theta, \tag{8.29}$$

where E_{iso} is the isothermal Young's modulus, and α is the constant-stress coefficient of linear expansion. The temperature variations in the solid are determined by both the diffusion of heat within the solid, and by generation of heat from mechanical processes. The latter is assumed to be proportional to the *rate of strain* $\mathrm{d}\epsilon/\mathrm{d}t$ in the solid, so that the local temperature satisfies

$$c_V \frac{\partial \Theta}{\partial t} = -\kappa \nabla^2 \Theta - \gamma \frac{\partial \epsilon}{\partial t}, \tag{8.30}$$

where c_V is the constant volume heat capacity and κ is the thermal conductivity. Landau [72] identifies the coefficient γ as $\gamma = E_{\mathrm{iso}}\alpha T_0/(1-2\nu)$, in terms of Poisson's ratio ν.

A strain plane wave $\epsilon(r,t) = \epsilon_0 \exp(\mathrm{i}qr - \mathrm{i}\omega t)$ at frequency ω with wave-vector q will therefore generate a temperature disturbance of the form

$$\Theta(r,\, t) = -\frac{\mathrm{i}\omega\gamma}{\mathrm{i}\omega c_V + \kappa q^2}\,\epsilon_0\, \mathrm{e}^{\mathrm{i}qr - \mathrm{i}\omega t}. \tag{8.31}$$

Taking the stress to have the same plane wave form as the strain, $\tau(r,t) = \tau_0 \exp(\mathrm{i}qr - \mathrm{i}\omega t)$, inserting (8.31) in (8.29), we find the relation between stress and strain of the form

$$\epsilon_0 = E_{\mathrm{iso}}^{-1}\,\frac{\kappa q^2 + \mathrm{i}\omega c_V}{\kappa q^2 + \mathrm{i}\omega(c_V + \alpha\gamma)}\tau_0, \tag{8.32}$$

Clearly (8.32) has the form of the response of the standard linear solid (8.24), so the discussion of Sect. 8.2 applies to thermoelastic solids.

The series of equations (8.29-8.32) provides the conceptual framework for the full discussion of mechanical dissipation due to thermoelasticity. Of course the full equations for deformable solids must be employed to obtain accurate results for any given problem; an excellent derivation for the problem of dissipation in flexural, doubly-clamped resonating beams appears in [71]. The authors show that the thermoelastic dissipation $1/Q_t$ for a thin beam under transverse flexure is approximately given by a form related to (8.32),

$$\frac{1}{Q_t} = \Delta_t \frac{\omega\tau_t}{1+\omega^2\tau_t^2}, \tag{8.33}$$

where the characteristic time τ_t is given by

$$\tau_t = \frac{t^2}{\pi^2 D}, \tag{8.34}$$

where t is the beam thickness, and D is the thermal diffusivity of the solid, $D = \kappa/c_V$. The time τ_t is thus the mean time for a thermal phonon to diffuse across the thickness of the beam. Zener's result (8.32) is thus a simple Lorentzian similar to that found for a two-level system, although with a different temperature dependence.

One interesting aspect of the result (8.33) is that for a given material, the highest value of the dissipation is given by $1/Q_{t,\,\max} = \Delta_t$, and is achieved when the resonator frequency ω is equal to the inverse characteristic time $1/\tau_t$. This is independent of the beam geometry, but is a function of temperature. In Fig. 8.3, following [71], we display this maximum value as a function of temperature for single-crystal Si and GaAs, which is not atypical for materials presently used for high-Q resonators.

Zener's result (8.33) is only an approximate result, quite accurate for thin beams with large diffusivity but less accurate for thicker beams or smaller values of diffusivity. The exact result is given by [71]

$$\frac{1}{Q} = \Delta_Z \frac{6}{\xi^2}\left(1 - \frac{\sinh(\xi)+\sin(\xi)}{\cosh(\xi)+\cos(\xi)}\right), \tag{8.35}$$

with the dimensionless parameter $\xi = t(\omega/2D)^{1/2}$. Errors in the limit of large ξ, where Zener's result is less accurate, are of order 20%. However, the more exact solution (8.35) does not have the obvious Lorentzian dependence of Zener's result, and is not as easy to evaluate.

Zener's theory applies to transverse flexure of thin beams, and a fairly different result is found for beams undergoing longitudinal motion. The calculations for longitudinal waves are somewhat more involved, with solutions presented by Landau and Lifshitz [72], and by Chadwick (see [73]).

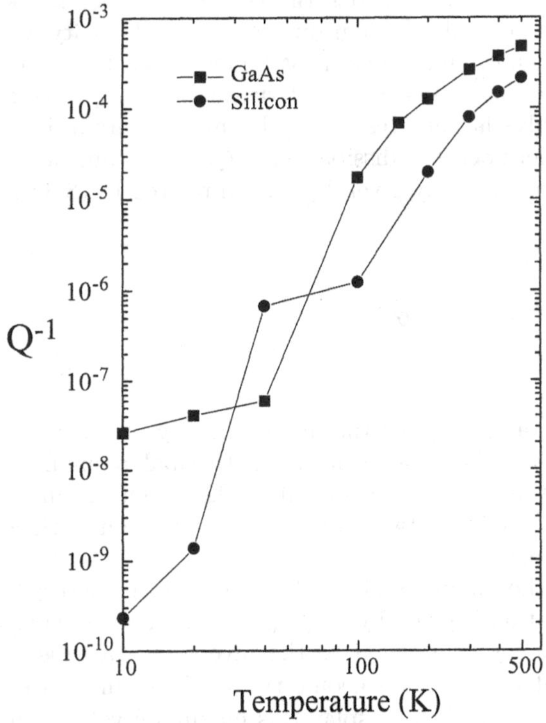

Fig. 8.3. Maximum dissipation achievable due to thermoelastic dissipation for a Si and a GaAs flexural resonator as a function of temperature. By permission of the authors a[71].

8.4 Phonon–Phonon Interactions

Another fundamental source of dissipation in mechanical systems comes from the interaction of the strain field with thermal phonons that are already present in the solid, in contrast to the process of thermoelastic relaxation, where the thermal phonons are generated by the strain field. The theory of dissipation due to phonon-phonon interactions was first worked out by Akhiezer [74], Bömmel and Dransfeld [75], and Woodruff and Ehrenreich [76]. A slightly different version of a theory for this process is presented by Mason [27]. The theory developed by these authors applies in the limit where the thermal phonons are assumed to always be in local thermal equilibrium with the instantaneous state of induced strain. This strain periodically modulates the frequencies of the thermal phonons, changing the local temperature, and the resulting spatial variation in temperature causes heat to flow between points of differing strain. The resulting increase in entropy removes energy from the strain field.

This theory applies in the limit where the thermal relaxation time for the phonon gas, τ_{th}, is much less than the period $2\pi/\omega$ of the strain field. In the opposite limit where the relaxation time is much greater than the strain field period, a quantum approach rather than a thermodynamic approach must be used, where the rate of thermal phonon emission from the lower-frequency strain field must be calculated from second and third-order terms in the Hamiltonian for the phonon gas. A summary of the theory in this limit is given by Klemens [28], and was discussed in Chap. 3.

As discussed in Chap. 3, the relaxation time constant for the strain-generated temperature variation is given by

$$\tau_{\text{th}} = \frac{3\kappa}{c_V \bar{v}_{\text{D}}^2}, \tag{8.36}$$

where κ is the thermal conductivity, c_V the specific heat, and $\bar{v}_{\text{D}}$ the Debye velocity.

This process causes inelastic relaxation as described by Zener's theory. A standing low-frequency wave of frequency ω will therefore have dissipation $1/Q$ given by

$$\frac{1}{Q_{\text{ph-ph}}} = \frac{\Delta c}{\bar{c}} \frac{\omega\tau_{\text{th}}}{1+\omega^2\tau_{\text{th}}^2}, \tag{8.37}$$

where Δc is the difference between the relaxed and the unrelaxed elastic modulus, caused by allowing the temperature variation. This is given approximately by ([27])

$$\Delta c = \gamma^2 c_V T. \tag{8.38}$$

where γ is Grüneisen's constant. The dissipation may therefore be written using the phase velocity relation $v_\phi^2 = \bar{c}/\rho$,

$$\frac{1}{Q_{\text{ph-ph}}} = \frac{\gamma^2 c_V T}{\rho v_\phi^2} \frac{\omega\tau_{\text{th}}}{1+\omega^2\tau_{\text{th}}^2}, \tag{8.39}$$

The corresponding expression for the attenuation $A_{\text{ph-ph}}$ of a propagating sound wave of frequency ω is given by

$$\begin{aligned} A_{\text{ph-ph}} &= \frac{\omega}{2Q_{\text{ph-ph}} v_\phi} \\ &= \frac{\gamma^2 c_V T}{2\rho v_\phi^3} \frac{\omega^2\tau_{\text{th}}}{1+\omega^2\tau_{\text{th}}^2}. \end{aligned} \tag{8.40}$$

Each of the expressions (8.39) and (8.40) apply in the limit $\omega\tau_{\text{th}} < 1$, where the relaxation of the thermal phonons is faster than the oscillation frequency of the strain wave.

8.4.1 Attenuation in Metals

Low frequency acoustic waves in metals interact strongly with the electrons in the metal. One method for calculating the strength of the interaction is to use the semiclassical approach of Pippard [77]. The ion motion from an acoustic wave generates electric and magnetic fields due to space charge and displacement currents, and the electron gas is assumed to move so as to maintain zero net charge imbalance. This in turn generates Joule heating due to the nonzero electrical resistance of the electrons, and causes damping of the acoustic wave. A Boltzmann transport theory approach ultimately yields the logarithmic decrement for both transverse and longitudinal waves, A_ℓ and A_t, respectively, parameterized by the product of phonon wavevector q and electron mean free path Λ, $a = q\Lambda$ [78]:

$$A_\ell = \frac{n_e m_e}{\rho c_\ell \tau}\left(\frac{a^2 \arctan a}{3(a - \arctan a)} - 1\right) \tag{8.41}$$

for longitudinal waves, and

$$A_t = \frac{n_e m_e}{\rho c_t \tau}\left(\frac{2}{3}\frac{a^3}{(1 + a^2)\arctan a - a} - 1\right) \tag{8.42}$$

for transverse waves.

At low frequencies, or in very dirty metals, the parameter $a = q\Lambda \ll 1$ indicates that the phonon wavelength is much less than the relaxation length for the electrons; in this limit these expressions simplify to

$$\left.\begin{aligned} A_\ell &= \frac{4}{15}\frac{n_e m_e v_\mathrm{F}}{\rho c_\ell^3}\omega^2 \Lambda, \\ A_t &= \frac{1}{5}\frac{n_e m_e v_\mathrm{F}}{\rho c_\ell^3}\omega^2 \Lambda, \end{aligned}\right\} \quad (q\Lambda \ll 1) \tag{8.43}$$

with attenuation increasing with the square of the phonon frequency. At high frequencies, or for very pure metals, the opposite limit where the phonon wavelength is much smaller than the electron mean free path, and we find attenuation scaling linearly with phonon frequency, but independent of electron mean free path:

$$\left.\begin{aligned} A_\ell &= \frac{\pi}{6}\frac{n_e m_e v_\mathrm{F}}{\rho c_\ell^2}\omega, \\ A_t &= \frac{4}{3\pi}\frac{n_e m_e v_\mathrm{F}}{\rho c_t}\omega. \end{aligned}\right\} \quad (q\Lambda \geq 1) \tag{8.44}$$

The two limits given by (8.43) and (8.44) can be achieved experimentally, for example in a piece of copper. With electron mean free paths of order 10^{-6} cm at room temperature, corresponding to a scattering time $\tau \sim 10^{-14}$ s, the limit (8.43) will hold even for phonons up to a few tens of GHz. At low temperatures, however, mean free paths as long as 0.3 cm can be achieved, so that (8.43) only holds for phonon frequencies below a few MHz; at higher frequencies, the expression (8.44) will apply.

For transverse acoustic waves, Pippard's derivation gives the dissipation $1/Q$ as

$$\frac{1}{Q_{\mathrm{el-ph},t}} = \frac{2}{3}\frac{nm_e v_\mathrm{F}}{\rho\omega\Lambda}\left(\frac{(q\Lambda)^3}{(q^2\Lambda^2+1)\tan^{-1}(q\Lambda) - q\Lambda} - \frac{3}{2}\right), \tag{8.45}$$

In the limit where the mean free path is short compared to the phonon wavelength, $q\Lambda \ll 1$, this approaches the limit

$$\frac{1}{Q_{\mathrm{el-ph},t}} \cong \frac{nm_e v_\mathrm{F}}{5\rho\omega} q^2\Lambda \qquad (q\Lambda \ll 1) \tag{8.46}$$

and in the limit where the mean free path is long,

$$\frac{1}{Q_{\mathrm{el-ph},t}} \cong \frac{nm_e v_\mathrm{F}}{\rho\omega}\frac{4q}{3\pi}. \qquad (q\Lambda \geq 1) \tag{8.47}$$

For longitudinal waves the dissipation is given by a somewhat different expression,

$$\frac{1}{Q_{\mathrm{el-ph},\ell}} = \frac{nm_e v_\mathrm{F}}{\rho\omega\Lambda}\left[\frac{(q\Lambda)^3}{3(q\Lambda - \tan^{-1}(q\Lambda))} - \left(1 + \frac{q^2\Lambda^2}{3}\right)\right]. \tag{8.48}$$

In the dirty ($q\ell \ll 1$) and clean ($q\ell \gg 1$) limits this expression can be simplified to

$$\frac{1}{Q_{\mathrm{el-ph},\ell}} \cong \frac{4}{15} nm_e v_\mathrm{F}\rho\omega q^2\ell, \qquad (q\ell \ll 1) \tag{8.49}$$

and

$$\frac{1}{Q_{\mathrm{el-ph},\ell}} \cong \frac{\pi}{6}\frac{nm_e v_\mathrm{F}}{\rho\omega} q. \qquad (q\ell \gg 1) \tag{8.50}$$

The limiting forms for both the longitudinal and transverse waves are very similar, with only a slight difference in the numerical factors.

The interaction of acoustic waves with electrons in metals have proved useful as a tool in understanding and mapping the Fermi surface in real metals. The theory for the interaction of acoustic waves and electrons with non-spherical Fermi surfaces has been worked out, and is summarized in the review article by Rayne and Jones [78]; a discussion of experimental results is also provided. Magnetoacoustic effects, where the motion of the electrons is modified by the presence of a strong magnetic field, thus modulating their interaction with acoustic waves, have also proved a useful tool in mapping out Fermi surfaces. A review of these effects is given by Peverley [79].

8.5 Dissipation in Nanoscale Mechanical Resonators

The Zener formalism for the anelastic solid may be used with the continuum theory of solids to include dissipation in, for example, the description of the vibrations of torsional, longitudinal and flexural beams. Here we apply it to determine the equation of motion for a doubly-clamped flexural beam. We will then calculate the thermal noise induced by the finite value of the quality factor for the beam, and calculate how this affects the precision with which the frequency of the resonator is known. We will work out formulas for the amplitude noise, phase noise, frequency noise, and Allan variance, all of which are different representations of the same quantity. The method we describe can be applied to calculate the effects of other noise sources on resonator behavior.

Our model resonator is shown in Fig. 8.4, comprising a beam of length L, width w and thickness t, with the long axis along $\hat{\boldsymbol{z}}$. The equation of motion for the beam is given by the Euler–Bernoulli theory, with

$$\rho\frac{\partial^2 U}{\partial t^2}(z,t) + E\frac{\partial^4 U}{\partial z^4}(z,t) = 0, \tag{8.50B}$$

for a displacement $U(z,t)$ of the beam neutral axis at a point z along the beam, with elastic modulus E and density ρ.

The Euler–Bernoulli formula, evaluated for harmonic motion at frequency ω, so that the displacement amplitude has the form $U(z,t) = U(z)\mathrm{e}^{-\mathrm{i}\omega t}$, using the Zener formula for the effective elastic modulus (8.24), becomes

$$\omega^2 \rho\, A\, U(z) = E(\omega)\, I_y \left(1 - \frac{\mathrm{i}}{Q}\right) \frac{\partial^4 U}{\partial z^4}(z). \tag{8.51}$$

The spatial solutions $U(z)$ are the same as for (7.47). For a doubly-clamped beam, these have the form (7.54),

$$U_n(z) = a_n\left(\cos(\beta_n z) - \cosh(\beta_n z)\right) + b_n\left(\sin(\beta_n z) - \sinh(\beta_n z)\right), \tag{8.52}$$

with $a_n/b_n = 1.01781, 0.99923, 1.0000, \ldots$ and n indicates the mode number. Our beam extends from $z = 0$ to $z = L$, and we choose to normalize the eigenfunctions so that

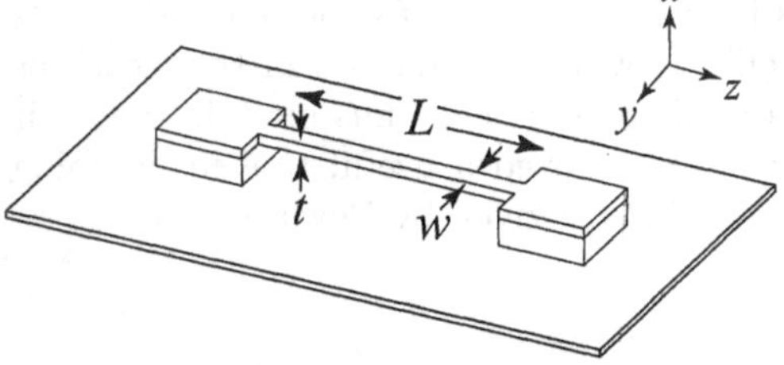

Fig. 8.4. Doubly-clamped beam with length L, width w and thickness t. The end supports are assumed infinitely rigid.

$$\int_0^L U_n(z)U_m(z)\,\mathrm{d}z = L^3\delta_{mn}, \tag{8.53}$$

where for $m \neq n$ the integral is zero, because the eigenfunctions are orthogonal, and for $m = n$ the amplitude is chosen so that the integral yields L^3. This normalization fixes $a_n = L$, with b_n given by the ratios above.

In the presence of dissipation, the time-dependent amplitudes have the form $U(z,t) = U(z)\mathrm{e}^{-\mathrm{i}\omega' t}$, where the damped eigenfrequencies ω'_n are given by

$$\begin{aligned} \omega'_n &= \left(1+\frac{\mathrm{i}}{2Q}\right)\sqrt{\frac{EI_y}{\rho A}}\,\beta_n^2 \\ &= \left(1+\frac{\mathrm{i}}{2Q}\right)\omega_n, \end{aligned} \tag{8.54}$$

in the limit of small dissipation Q^{-1}. The imaginary part of ω'_n indicates that the nth eigenmode will decay in amplitude as $\exp(-\omega_n t/2Q)$, similar to the behavior of the damped simple harmonic oscillator discussed earlier.

8.5.1 Driven Damped Beams

We now add to (8.50B) a harmonic driving force $F(z,t) = f(z)\exp(-\mathrm{i}\omega_c t)$, where $f(z)$ is the position-dependent force per unit length, and ω_c is the drive frequency, often also called the carrier frequency. We assume the force is uniform across the beam cross-section and directed along the x-axis; we further assume that the carrier frequency ω_c is close to the fundamental beam frequency ω_1. The equation of motion for U is now given by [62]

$$\rho A\frac{\partial^2 U}{\partial t^2} + E A\frac{\partial^4 U}{\partial z^4} = f(z)\,\mathrm{e}^{-\mathrm{i}\omega_c t}. \tag{8.55}$$

We solve this equation for times long in comparison to the damping time for the beam, $\omega_c t/Q \gg 1$. Any transient at the natural resonance frequency then damps out, so the time dependence $U(z,t)$ has the form $U(z,t) = U(z)\mathrm{e}^{-\mathrm{i}\omega_c t}$, with motion only at the carrier frequency ω_c. The amplitude $U(z)$ may be complex, so that the motion is not necessarily in phase with the force F. The displacement can be expanded in terms of the eigenfunctions $U_n(z)$,

$$U(z) = \sum_{n=1}^{\infty} a_n\,U_n(z)\,\mathrm{e}^{-\mathrm{i}\omega_c t}, \tag{8.56}$$

with amplitude a_n for the nth mode. Inserting this in (8.55) yields the equation

$$-\omega_c^2\rho A\sum_{n=1}^{\infty} a_n U_n(z) + E A\sum_{n=1}^{\infty} a_n\frac{\partial^4 U_n(z)}{\partial z^4} = f(z). \tag{8.57}$$

Using the defining relation for the eigenfunctions, (7.47), the relation for the eigenfrequencies ω_n', (8.54), and the normalization of the eigenfunctions $U_n(z)$, this can be written

$$(\omega_n'^2 - \omega_c^2)a_n = \frac{1}{\rho A L^3}\int_0^L U_n(z)\,f(z)\,\mathrm{d}z, \tag{8.58}$$

for each term n in the expansion. For ω_c close to ω_1, only the $n = 1$ term in (8.58) has a significant amplitude, given by

$$a_1 = \frac{1}{\rho A L^3}\frac{1}{\omega_1^2 - \omega_c^2 - \mathrm{i}\,\omega_1^2/Q}\int_0^L U_1(z)\,f(z)\,\mathrm{d}z, \tag{8.59}$$

in the limit of small dissipation Q^{-1}.

We will for simplicity assume a uniform force distribution, $f(z) = f_0$; for the doubly-clamped beam, the integral in (8.59) is then the first moment η_1 of the eigenfunction $U_1(z)$,

$$\begin{aligned} \eta_1 &= \frac{1}{L^2}\int_0^L U_1(z)\,\mathrm{d}z \\ &= 0.8309. \end{aligned} \tag{8.60}$$

The amplitude can then be written as

$$a_1 = \frac{\eta_1}{\omega_1^2 - \omega_c^2 - \mathrm{i}\,\omega_1^2/Q}\frac{f_0}{M}, \tag{8.61}$$

where $M = \rho A L$ is the mass of the beam.

The position-dependent displacement of the beam for ω_c close to ω_1, is thus given by

$$U(z,t) = \frac{\eta_1}{\omega_1^2 - \omega_c^2 - \mathrm{i}\,\omega_1^2/Q}\frac{f_0}{M}U_1(z)\mathrm{e}^{-\mathrm{i}\omega_c t}. \tag{8.62}$$

If the force distribution $f(z)$ is instead chosen to be proportional to the eigenfunction $U_1(z)$, the integral (8.59) is unity, and the amplitude is then given by

$$a_1 = \frac{1}{\omega_1^2 - \omega_c^2 - \mathrm{i}\,\omega_1^2/Q}\frac{f_0}{M}, \tag{8.63}$$

where $f(z) = f_0(U_1(z)/L)$.

We wish to point out that the response functions, (8.61) or (8.63), while similar to that of a damped, one-dimensional harmonic oscillator, differ in the Q-dependent denominator; for a harmonic oscillator, the equivalent response has the form

$$a = \frac{1}{\omega_0^2 - \omega_c^2 - \mathrm{i}\,\omega_c\omega_0/Q}\frac{F_0}{M}, \tag{8.64}$$

where the driving force is $F_0\,\mathrm{e}^{-\mathrm{i}\omega_c t}$, the time-dependent amplitude is $a\mathrm{e}^{-\mathrm{i}\omega_c t}$, the natural resonance frequency is ω_0, and the mass is M. The difference between (8.61) and (8.64) is only apparent for small values of Q: For values of Q greater than 15, the fractional difference at any frequency is less than 1%.

8.5.2 Dissipation-Induced Amplitude Noise

The displacement of a forced, damped beam driven near its fundamental frequency is given by (8.62). In the absence of noise, this solution represents pure harmonic motion at the carrier frequency ω_c.

The finite value of Q, however, necessitates the presence of noise, from the fluctuation-dissipation theorem. Regardless of the origin of the dissipation mechanism, it acts to thermalize the motion of the resonator, so that in the presence of dissipation only (no driving force), the mean total energy $\langle \mathcal{E}_n \rangle$ for each mode n of the resonator will be given by

$$\langle \mathcal{E}_n \rangle = k_{\mathrm{B}} T, \tag{8.65}$$

where T is the physical temperature of the resonator, or, more precisely, the temperature of the dissipation source. This is because each mode n has two degrees of freedom, the amplitude and velocity of the mode displacement, and in thermal equilibrium each degree of freedom has an energy $\frac{1}{2} k_{\mathrm{B}} T$.

The thermalization occurs due to the presence of a noise force $f_N(z,t)$ per unit length of the beam. Each point on the beam experiences a noise force with the same spectral density, but fluctuating independently from the force at other points; the noise at any two points on the beam is uncorrelated. The noise can then equivalently be written as an expansion in terms of the eigenfunctions $U_n(z)$,

$$f_N(z,t) = \frac{1}{L} \sum_{n=1}^{\infty} f_{N_n}(t) U_n(z), \tag{8.66}$$

where the time-dependent noise force amplitude f_{N_n} associated with each mode n is uncorrelated with the noise for other modes n'; the factor $1/L$ in (8.66) appears because of the normalization of the eigenfunctions $U_n(z)$.

Noise is generally characterized by its mean value (its *first moment*), and by its spectral density, or *second moment*. The spectral density $S_f(\omega)$ for a force noise $f(t)$ is defined as

$$S_f(\omega) = \frac{1}{2\pi} \int_{-\infty}^{\infty} f^2(t) \mathrm{e}^{\mathrm{i}\omega t} \mathrm{d}t. \tag{8.66B}$$

The noise force amplitude $f_{N_n}(t)$ has a white spectral density $S_{f_n}(\omega)$, with zero mean. The magnitude of the spectral density $S_{f_n}(\omega)$ may be evaluated by requiring that it achieve thermal equilibrium, as given by (8.65), for each mode n. The spectral density of the noise-driven amplitude a_n of the nth mode, driven by the noise force $f_{N_n}(t)$, is then given by (8.63),

$$S_{a_n}(\omega) = \frac{1}{(\omega_n^2 - \omega^2)^2 + (\omega_n^2/Q)^2} \frac{S_{f_n}(\omega)}{M^2}. \tag{8.67}$$

The SI units for $S_{f_n}(\omega)$ are $(\mathrm{N/m})^2/(\mathrm{rad/sec}) = \mathrm{kg}^2/(\mathrm{sec}^3\text{-rad})$. Those for $S_{a_n}(\omega)$ are 1/(rad/sec), because a_n is dimensionless.

The kinetic energy in the nth mode associated with the spectral density $S_{a_n}(\omega)$ is given by

$$\begin{aligned} \langle \mathcal{T}_n \rangle &= \frac{1}{2} \int_0^L \rho A \left| \frac{\partial U_n}{\partial t}(z,t) \right|^2 \mathrm{d}z \\ &= \frac{1}{2} \int_0^\infty \int_0^L \rho\, A\, \omega^2\, S_{a_n}(\omega)\, U_n^2(z)\, \mathrm{d}z\, \mathrm{d}\omega \\ &= \frac{1}{2} \int_0^\infty \rho A L^3 \omega^2 S_{a_n}(\omega) d\omega \\ &\approx \frac{1}{4} \frac{Q\, L^2}{\omega_n} \frac{S_{f_n}(\omega)}{M}, \end{aligned} \tag{8.68}$$

where the last equality becomes exact in the limit $Q^{-1} \to 0$. The error in (8.68) for finite Q is less than 1% for $Q > 10$.

In order that this yield thermal equilibrium, so that $\langle \mathcal{T}_n \rangle = \frac{1}{2} k_{\mathrm{B}} T$, the spectral density S_{f_n} must be given by

$$S_{f_n}(\omega) = \frac{2 k_{\mathrm{B}} T M \omega_n}{\pi Q L^2}. \tag{8.69}$$

The term L^2 appears in (8.69) because f_{N_n} is the force per unit length of beam. An equivalent derivation for a one-dimensional simple harmonic oscillator with natural resonance frequency ω_0 yields the force density

$$S_{\mathrm{F}}(\omega) = \frac{2 k_{\mathrm{B}} T M \omega_0}{\pi Q}. \tag{8.70}$$

Given the result (8.69), we can write the spectral density of the thermally-driven amplitude,

$$S_{a_n}(\omega) = \frac{\omega_n}{(\omega_n^2 - \omega^2)^2 + (\omega_n^2/Q)^2} \frac{2 k_{\mathrm{B}} T}{\pi M L^2 Q}. \tag{8.71}$$

When superposed with a driving force with a carrier frequency $\omega_c = \omega_1$, the power consists of a δ-function peak at the carrier superposed with the Lorentzian given by (8.71). This response is sketched in Fig. 8.5.

8.5.3 Dissipation-Induced Phase Noise

The form we have used in (8.71) to write the noise amplitude represents frequency-distributed amplitude noise. Equivalent expressions can be written for the phase noise, fractional frequency noise, or the Allan variance [80], which are useful for applications in time-keeping and narrowband filter applications. The resonator is driven by the carrier signal near its resonance

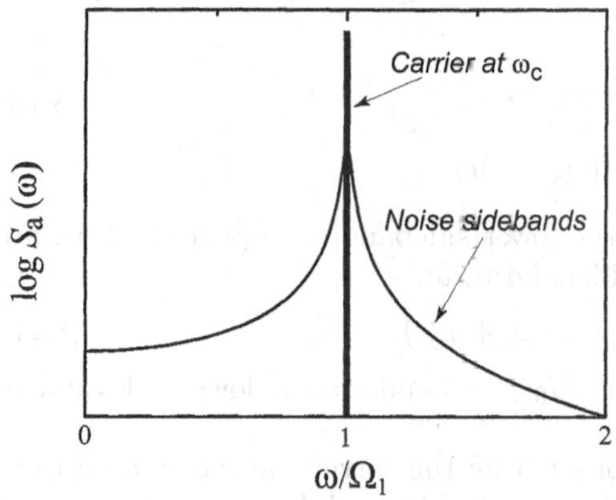

Fig. 8.5. Frequency spectrum of a driven beam in the presence of noise.

frequency ω_1, and in addition is driven by dissipation-induced noise, with spectral density given by (8.69). The time-dependent amplitude can then be written in the form

$$a(t) = a_0 \sin(\omega_c t + \phi(t) + \theta), \tag{8.72}$$

where $\phi(t)$ represents a phase variation from the main carrier signal at frequency $\omega_c \approx \omega_1$; the amplitude a_0 is assumed constant, and θ is a phase offset. Following Robins [81], we pick one frequency component at ω for the phase variation, $\phi(t) = \phi_0 \sin(\omega t)$. Assuming small maximum deviation ϕ_0, the amplitude may be written

$$\begin{aligned} a(t) &= a_0 \sin(\omega_c t + \theta) \\ &\quad + a_0 \frac{\phi_0}{2} \sin\left((\omega_c + \omega)t\right) \\ &\quad - a_0 \frac{\phi_0}{2} \sin\left((\omega_c - \omega)t\right). \end{aligned} \tag{8.73}$$

The phase variation at ω generates sidebands spaced $\pm\omega$ from the carrier, with amplitude $\pm\, a_0\, \phi_0/2$. The lower sideband is phase-coherent with the upper sideband, with the opposite sign; this is characteristic of phase noise. Independent sideband signals can be generated by adding an amplitude noise source $M(t)$ to the phase noise $\phi(t)$, so that the amplitude is written

$$a(t) = a_0\,(1 + M(t)) \sin(\omega_c t + \phi(t) + \theta). \tag{8.74}$$

We consider a single component at ω for both the phase and amplitude modulation, so that

$$\left.\begin{aligned} M(t) &= M_0 \sin(\omega t) \\ \phi(t) &= \phi_0 \sin(\omega t). \end{aligned}\right\} \tag{8.75}$$

Again assuming small variations, this can be written as

$$\begin{aligned} a(t) = &\ a_0 \sin(\omega_c t + \theta) \\ &+ \frac{1}{2} a_0 (M_0 + \phi_0) \sin((\omega_c + \omega)t) \\ &+ \frac{1}{2} a_0 (M_0 - \phi_0) \sin((\omega_c - \omega)t). \end{aligned} \tag{8.76}$$

Setting the amplitude $M_0 = \phi_0$, the lower sideband disappears and we are left with the independent upper sideband term,

$$a(t) = a_0 \sin(\omega_c t + \theta) + a_0 \phi_0 \sin((\omega_c + \omega)t). \tag{8.77}$$

Choosing the opposite sign relation $M_0 = -\phi_0$ allows the lower sideband to be chosen.

A noise signal at a frequency offset from the carrier can be created from the superposition of a phase and an amplitude modulation of the original carrier. In applications where the resonator is to be used as a frequency source or a clock, the amplitude modulation is unimportant: use of a zero-crossing detector, or a perfect limiter, eliminates the effects of the amplitude modulation. We therefore ignore this noise source. This is equivalent, from the arguments leading to (8.73), to limiting the noise to that which is phase coherent between the upper and lower sidebands, with amplitudes such that $a(\omega_c + \omega) = -a(\omega_c - \omega)$. Noise which has the opposite phase relation, with $a(\omega_c + \omega) = +a(\omega_c - \omega)$, is due to amplitude modulation. Noise which is associated with only one sideband consists of the sum or difference of these two "modes".

Dissipation-induced noise, of the form given by (8.71), is intrinsically phase-incoherent on opposite sides of the carrier signal at ω_c. Half of the noise power is therefore associated with amplitude modulation, and half with phase modulation; the phase noise power is therefore half the original, total, noise power.

We can evaluate the dissipation-induced phase noise for a resonator driven at its fundamental resonance frequency. We drive the resonator with an external force f per unit length, distributed uniformly over the beam, at the frequency $\omega_c = \omega_1$. The amplitude for the response at the fundamental frequency is given by (8.61),

$$a_1 = -\mathrm{i} \frac{\eta_1}{\omega_1^2} \frac{Q f}{M}. \tag{8.78}$$

The amplitude lags the force by 90°, and includes the multiplicative factor Q. Dissipation generates incoherent noise, distributed about the carrier with noise power given by (8.71). The phase noise power density $S_\phi(\omega)$, defined in the same way as the force noise spectral density in (8.66B), at frequency ω from the carrier frequency is then given by

$$\begin{aligned} S_\phi(\omega) &= \frac{1}{2} \frac{S_{a_1}(\omega_1 + \omega)}{|a_1|^2} \\ &= \frac{\omega_1}{(2\omega_1 \omega + \omega^2)^2 + (\omega_1^2/Q)^2} \frac{k_B T}{\pi |a_1|^2 L^2 M Q}. \end{aligned} \tag{8.79}$$

For frequencies that are well off the peak resonance, $\omega \gg \omega_1/Q$, but small compared to the resonance frequency, $\omega \ll \omega_1$, we may approximate the denominator in (8.79),

$$\begin{aligned} S_\phi(\omega) &\approx \frac{k_B T}{4\pi\omega_1\omega^2|a_1|^2 L^2 M Q} \\ &\approx \frac{k_B T \omega_1}{8\pi E_c Q \omega^2} \quad (\omega_1/Q \ll \omega \ll \omega_1). \end{aligned} \tag{8.80}$$

Here we define the energy E_c at the carrier frequency, $E_c = M\omega_1^2 L^2|a_1|^2/2$. This can also be written in terms of the power $P_c = \omega_1 E_c/Q$ needed to maintain the carrier amplitude, i.e. that needed to counter the loss due to the non-zero value of $1/Q$:

$$S_\phi(\omega) \approx \frac{k_B T}{8\pi P_c Q^2}\left(\frac{\omega_1}{\omega}\right)^2 \quad (\omega_1/Q \ll \omega \ll \omega_1). \tag{8.81}$$

We can also write this expression in terms of frequency $f = 2\pi\omega$,

$$S_\phi(f) \approx \frac{k_B T}{4 P_c Q^2}\left(\frac{\nu_1}{f}\right)^2 \quad (\nu_1/Q \ll f \ll \nu_1). \tag{8.82}$$

8.5.4 Frequency Noise

The phase fluctuations discussed above can also be viewed as *frequency* fluctuations, where the amplitude $a(t)$ has a time dependence given by

$$a(t) = a_0 \sin\left(\int_{-\infty}^{t} \omega(t')dt' + \theta\right). \tag{8.83}$$

The time-dependent frequency $\omega(t)$ is related to the carrier frequency ω_c and the phase $\phi(t)$ by

$$\omega(t) = \frac{d(\omega_c t + \phi(t))}{dt} = \omega_c + \frac{d\phi}{dt}. \tag{8.84}$$

We can define time-dependent frequency variation $\delta\omega(t)$ as

$$\delta\omega(t) = \omega(t) - \omega_c = \frac{d\phi(t)}{dt}. \tag{8.85}$$

We consider a single phase modulation component, so that $\phi(t) = \phi_0 \sin(\omega t)$. The frequency variation

$$\delta\omega(t) = \delta\omega_0 \cos(\omega t) = \omega\, \phi_0 \cos(\omega t) \tag{8.86}$$

represents a sinusoidal variation of the frequency, with amplitude $\delta\omega_0 = \omega\, \phi_0$, modulated at ω. The time-dependent amplitude in (8.83) can be calculated in the same manner as for the phase variation, and leads to

$$\begin{aligned} a(t) &= a_0 \sin(\omega_c t + (\delta\omega_0/\omega)\sin(\omega t) + \theta) \\ &= a_0 \sin(\omega_c t) + \frac{1}{2} a_0 \frac{\delta\omega_0}{\omega} \sin((\omega_c + \omega)t) \\ &- \frac{1}{2} a_0 \frac{\delta\omega_0}{\omega} \sin((\omega_c - \omega)t), \end{aligned} \tag{8.87}$$

a result very similar to that for phase variations, (8.73).

The arguments leading to the spectral density, (8.81), may be re-worked to yield the equivalent expression for the frequency variation noise density $S_{\delta\omega}$. A more useful quantity is the *fractional* frequency variation, defined as $y = \delta\omega/\omega_c$. The noise density for y is related to that for the phase noise density by

$$\begin{aligned} S_y(\omega) &= \left(\frac{\partial y}{\partial \phi}\right)^2 S_\phi(\omega) \\ &= \left(\frac{\omega}{\omega_c}\right)^2 S_\phi(\omega), \end{aligned} \tag{8.88}$$

where we use the fact that modulation at ω generates sidebands at $\pm\omega$ from the carrier at ω_c. From (8.81) we then have

$$S_y(\omega) \approx \frac{k_B T}{8\pi P_c Q^2} \quad (\omega_c/Q \ll \omega \ll \omega_c). \tag{8.89}$$

In the frequency domain this is

$$S_y(f) \approx \frac{\pi k_B T}{4 P_c Q^2} \quad (\nu_c/Q \ll f \ll \nu_c). \tag{8.90}$$

8.5.5 Allan Variance

A third useful quantity, commonly used to compare frequency standards, is the Allan variance $\sigma_A(\tau_A)$ [82, 83]. The phase and frequency noise are defined in the frequency domain; the Allan variance is defined in the time domain, as the variance over time in the measured frequency of a source, each measurement averaged over a time interval τ_A, with zero dead time between measurement intervals. The defining expression for the square of the Allan variance is

$$\sigma_A^2(\tau_A) = \frac{1}{2f_c^2} \frac{1}{N-1} \sum_{m=2}^{N} (\bar{f}_m - \bar{f}_{m-1})^2, \tag{8.91}$$

where $\bar{f}_m$ is the average frequency measured over the mth time interval, of length $\Delta t = \tau_A$, and f_c is the nominal carrier frequency. The squared Allan variance is related to the phase noise density by [83]

$$\sigma_A^2(\tau_A) = 2\left(\frac{2}{\omega_c \tau_A}\right)^2 \int_0^\infty S_\phi(\omega) \sin^4(\omega\tau_A/2) d\omega, \tag{8.92}$$

where $\omega_c = 2\pi f_c$ and ω is the modulation frequency.

For the approximate form for the phase noise density (8.81), the Allan variance is

$$\sigma_A(\tau_A) = \sqrt{\frac{k_B T}{8 P_c Q^2 \tau_A}}. \tag{8.93}$$

Defining the dimensionless drive energy ε_c as the ratio of drive energy per cycle to the thermal energy, $\varepsilon_c = 2\pi P_c/\omega_c k_B T$, we have

$$\sigma_A(\tau_A) = \frac{1}{Q}\sqrt{\frac{\pi}{4\varepsilon_c \omega_c \tau_A}}. \tag{8.94}$$

We see that the Allan variance falls inversely with the square root of the product $\omega_c \tau_A$, and it is also proportional to the dissipation Q^{-1}. Other things being equal, increasing the resonator frequency ω_c lowers the Allan variance.

Exercises

8.1 Plot the Akhiezer phonon–phonon limited quality factor Q for Si at room temperature as a function of frequency, from 1 MHz to 10 GHz, using numerical values for the various parameters.

8.2 Find the resonance frequency, and estimate the value of the thermoelastic dissipation for a flexural resonator made from Si, with length 1 μm, width and thickness 0.1 μm. Use Fig. 8.3 to calculate the dissipation at room temperature and at 10 K.

8.3 Find numerical values for the dissipation of longitudinal and transverse phonons at 1 MHz and 1 GHz in copper at room temperature.

9. Experimental Nanostructures

In this chapter we turn from a theoretical description of solid mechanics to a description of experimental nanostructures, built to investigate various mechanical properties and applications of nanomechanical devices. The structures that we have chosen to describe here were almost all fabricated using semiconductor processing techniques, developed for integrated circuit fabrication, as this is the most straightforward and controllable method for fabricating mechanical structures at the submicron scale. There are many references in this chapter to particular techniques for fabricating the structures; we have devoted two chapters, following this one, to the detailed description of a number of semiconductor processing techniques and how they can be applied to nanostructure fabrication.

We begin this chapter with an outline of a somewhat generic process sequence for fabricating a suspended mechanical structure. We then follow this description with a number of specific examples of structures that have been described in the literature, giving detailed fabrication parameters and measurement techniques.

9.1 A Simple Fabrication Sequence

The fabrication sequence for a complete mechanical nanostructure, with an integrated metal lead for actuating and mechanical motion measuring the resulting displacement, is shown in Fig. 9.1.

This particular example uses a structural material that includes, from the bottom up, a bulk substrate known as a *handle*, typically with a thickness of a few hundred microns, a sacrificial layer with a thickness of a few tenths of a micron, and a top layer that will form the active mechanical structure, with a thickness that can range from under 0.01 μm to a few microns. For reference, 1 micron $= 10^{-6}$ m is roughly one-fiftieth the diameter of a human hair. The sacrificial layer is designed to have different chemical properties from both the handle and the top layer, so that it can be selectively removed. This type of structural material is known as a heterostructure.

The fabrication begins with the deposition of a photolithographic or electron beam *resist*, on the surface of the structural material. This resist is then

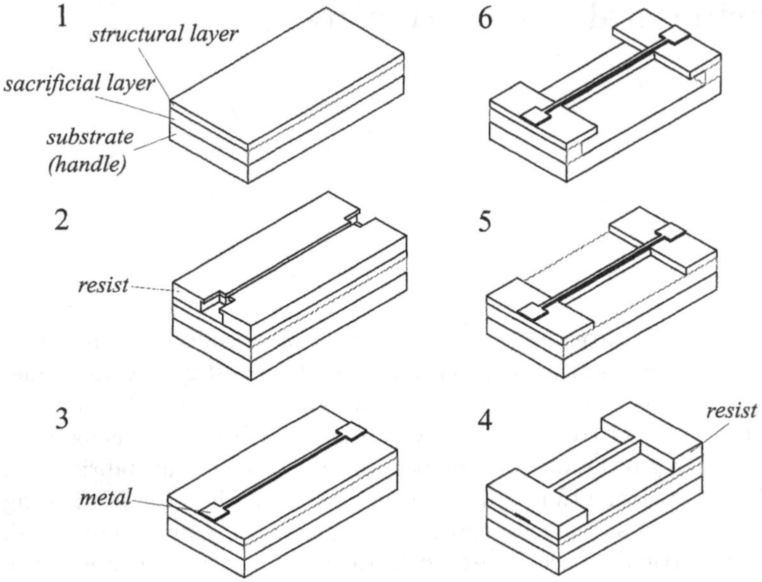

Fig. 9.1. Fabrication sequence for a doubly clamped resonator. Starting at top left, the heterostructure substrate consists of the bulk substrate, a sacrificial layer, and a top structural layer (1). Moving counterclockwise around the figure, photoresist or electron-beam resist is coated on the substrate, and patterned (2), after which metal is deposited and removed from the resist-coated areas (3). A second resist layer is patterned for an etch process (4), the structural layer etched and the resist removed (5), and in the final step a timed etch of the sacrificial layer suspends the structure (6).

selectively exposed to either ultraviolet light or a charged particle beam (typically electrons). The exposed area can then be dissolved away when the whole structure is placed in a developer, generating openings in the resist layer (this is then known as a negative resist process). Some resists however become less soluble after exposure, so the *unexposed* area is washed away and the exposed area remains after developing, in what is known as a positive resist process.

A metal film is then deposited over the patterned resist and substrate, and then *lifted off*, where the resist and the metal overlying the resist are removed in a wet solvent, leaving only the metal that was deposited in the openings in the resist pattern. A second layer of resist is then deposited and patterned, which forms a protective coating (a mask) for etching the top structural layer of the substrate, using either a wet chemical etch or a dry, charged ion-based etch. This opens windows through the top structural layer, that reach down through to the underlying sacrificial layer. In the final step, the sacrificial layer is removed, again using either a dry or a wet timed etch, resulting in a suspended beam with a metal wire along its top surface.

We now describe in detail a number of actual nanostructure examples, many of which followed fabrication processes similar to that just described. We refer the reader to Chaps. 10 and 11 for more information regarding the detailed processing techniques, as well as to texts on semiconductor processing techniques (see e.g. Ghandhi [84] or Brodie and Muray [85]).

9.2 Flexural Resonators

9.2.1 Radiofrequency Flexural Resonators

One of the early demonstrations of a radiofrequency mechanical resonator was fabricated and measured by the author with M.L. Roukes at Caltech [86]. The resonator consists of a doubly-clamped beam of Si, etched from a wafer of single crystal Si using a technique similar to the SCREAM process developed by MacDonald [87]. The SCREAM process is described in detail in Chapter 10. This example also demonstrates the important use of the *magnetomotive* displacement actuation and sensing technique; at present, this technique remains one of the principle methods used for actuating and sensing nanomechanical resonators, so we describe it here in some detail.

The fabrication sequence involves a number of processing steps that are described in Chap. 10; here we list the parameters used for each step, and the reader who is not familiar with these techniques can refer to that chapter for more information. The processing started with a Si wafer, on top of which a 1 μm thick SiO_2 layer was thermally grown, in a 1100°C furnace with water-saturated oxygen. Electron beam lithography was used to define a 100 nm thick Ni film, using a liftoff pattern. The Ni pattern was used as a mask for a reactive ion etch (RIE) step, which cut through the silicon oxide and silicon beneath it, to a depth of about 2 μm. The Ni was then removed using a wet etch, and a thin dry thermal oxide was grown on the sidewalls of the exposed silicon. The oxide on the horizontal surface of the Si was then removed through a second, brief reactive ion etch, leaving only the sidewalls of the Si protected by oxide. Finally, the beams were released using an isotropic RIE process, using NF_3 (nitrogen trifluoride) as the reactive gas, completing the mechanical structure. Gold was then evaporated on top of the beam to provide a metal lead for measurements. Figure 9.2 shows a schematic of these process steps, except for the last gold evaporation, and an electron micrograph of a completed device is shown in Fig. 9.3.

The Au lead on the top surface of the resonator allows the beam to be mechanically driven, and its motion detected, using a combination of a magnetic field and a source of radiofrequency (rf) current. The electrical circuit is shown in Fig. 9.4. The beam was placed in a large magnetic field aligned out of the plane of the Si substrate, along the x axis. A current is passed through the electrode on the top surface of the beam, in the direction of the z axis. The perpendicular arrangement of magnetic field and the direction of

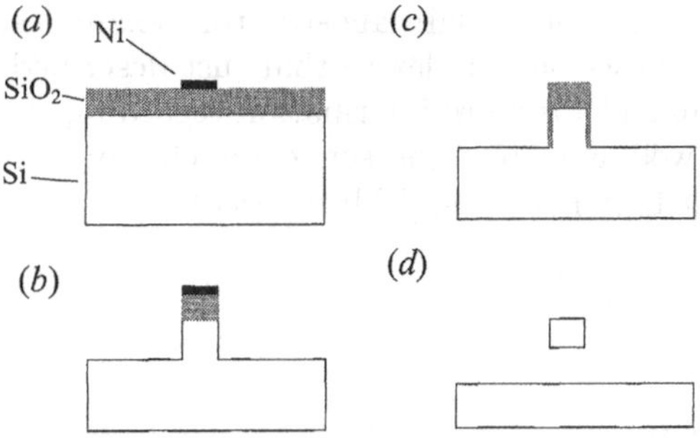

Fig. 9.2. Fabrication sequence for single-crystal Si beam, in cross-section. In (**a**) an oxidized Si wafer has a patterned Ni film deposited on its surface; in (**b**) this pattern is transferred by reactive ion etch into the underlying oxide and silicon wafer. After this step, the Ni is removed and a second, thin oxide layer is grown on the Si wafer. The oxide at the base is removed by anisotropic reactive ion etching, shown in (**c**). Finally, in (**d**) the beam is released from the substrate using an isotropic reactive ion etch, and the oxide layers removed using hydrofluoric acid. Following this step, a Au layer is evaporated to allow electrical measurements (not shown).

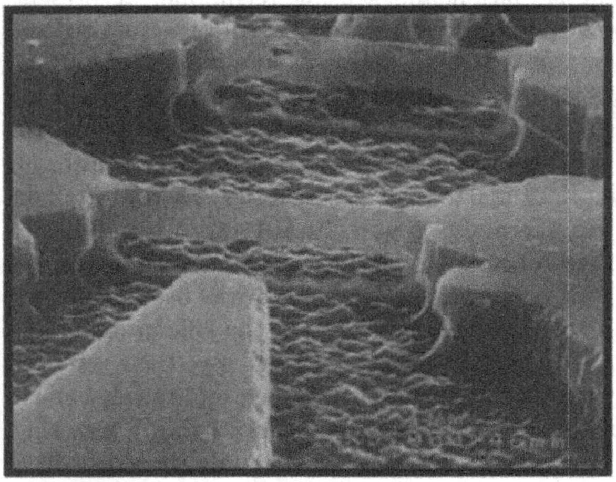

Fig. 9.3. A pair of beams fabricated using the technique described in the text. The beams have a length of about $L = 7.7$ μm, a height $h = 0.8$ μm and a width of about $w = 0.33$ μm.

current flow generates a radiofrequency Lorentz force directed in the plane of the Si wafer, along y, at the same frequency as the rf current. This is the actuation part of the magnetomotive technique.

The resulting displacement of the beam along y, travelling through the magnetic field, generates an electromotive force, or voltage, at the same frequency as the Lorentz force and the rf current. This voltage is sensed, allowing the displacement to the measured

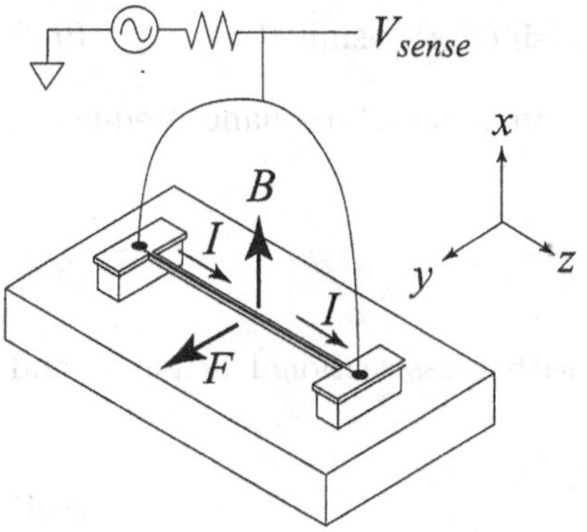

Fig. 9.4. Measurement scheme for driving and detecting the motion of a flexural resonator. An electrical current I is passed through a metal electrode on the top surface of the beam, and the perpendicular magnetic field B generates a Lorentz force F that drives the beam in the plane of the device. The resulting motion generates an electromotive force along the beam, which is detected by the room-temperature electronics (a network analyzer). The beam has length L, and is oriented along the z axis, with motion along y.

Sweeping the frequency of the applied current allows one to trace out the response of the flexural resonance, measuring the amplitude of the voltage as a function of the frequency of the applied current. This measurement technique is extremely simple to implement, works over frequencies from audio frequencies to very high radio frequencies (above 1 GHz), and as a result has been used by a number of researchers to complete measurements on a variety of mechanical structures.

The force per unit length f, for a magnetic field B perpendicular to the direction of current flow I in the metal electrode, is given by $f = IB$. An oscillating current at frequency ω_c, $I = I_0 e^{-i\omega_c t}$, therefore generates a uniform force per unit length, given by the real part of $f(z) = BI_0 e^{-i\omega_c t} = f_0 e^{-i\omega_c t}$. This force can be included to the Euler–Bernoulli theory for a flexural beam, as discussed in Sect. 7.1.4. The response of the flexural beam is worked out in Sect. 8.5.1. The dynamic equation of motion for the beam displacement $U(z,t)$, as a function of position z along the beam, is given by (8.55),

$$\rho A \frac{\partial^2 U}{\partial t^2} - E I_y \frac{\partial^4 U}{\partial z^4} = f_0 \, e^{-i\omega_c t}, \tag{9.1}$$

for a beam of mass density ρ, cross-sectional area A, bending modulus I_y, and Young's modulus E, the latter complex in the Zener model for the anelastic solid. This equation can be solved by expanding $U(z,t)$ in terms of the normal modes $U_n(z)e^{-i\omega_n t}$ of the flexural beam, which depend on the clamping conditions at the ends of the beam. Here we have a doubly clamped beam, with a mode shape $U_n(z)$ for the fundamental resonance $n = 1$ given by (see Sect. 8.5)

$$\frac{U_1}{L}(z) = (\cos(\beta_1 z) - \cosh(\beta_1 z)) + \alpha_1 (\sin(\beta_1 z) - \sinh(\beta_1 z)), \tag{9.2}$$

with $\alpha_1 = 0.9825$ and $\beta_1 = 4.73004/L$. The fundamental resonance frequency, ω_1, is given by

$$\omega_1 = \sqrt{\frac{EI_y}{\rho A}}\,\beta_1^2. \tag{9.3}$$

For the structure shown in Fig. 9.3, with cross-sectional height h and width w, the bending moment I_y is

$$I_y = \frac{hw^3}{12}. \tag{9.4}$$

Treating Si as an isotropic material with elastic modulus $E \approx 1.7 \times 10^{11}$ N/m^2, density $\rho = 2.33$ g/cm^3, we estimate the resonance frequency to be

$$\nu_1 = \frac{\omega_1}{2\pi} = 8.75\,\frac{w}{L^2}\ \text{GHz} - \mu\text{m}, \tag{9.5}$$

where the dimensions w and L are measured in µm. For the structure in Fig. 9.3 with $w = 0.33$ µm and $L = 7.7$ µm, we estimate $\nu_1 = 50$ MHz. The actual frequency for this structure was measured to be 70.7 MHz; the difference may be due to residual strain in the substrate.

For a frequency ω_c near the fundamental resonance frequency ω_1, the displacement of the neutral axis of the beam is very nearly given by $U(z,t) = \mathcal{A}U_1(z)e^{-i\omega_c t}$, with $U_1(z)$ the mode shape (9.2). The dimensionless amplitude of motion $\mathcal{A}$ is given by (8.59),

$$\begin{aligned}\mathcal{A} &= \frac{1}{\rho A L^3}\,\frac{1}{\omega_1^2 - \omega_c^2 - i\omega_1^2/Q}\int_0^L U_1(z)\, f_0\, dz,\\ &= \frac{\eta_1}{\omega_1^2 - \omega_c^2 - i\omega_c\omega_1/Q}\,\frac{f_0}{M},\end{aligned} \tag{9.6}$$

where $\eta_1 = 0.8309$ is the integral of $U_1(z)$ over the beam length, and M the total mass of the beam. In this linear theory, the amplitude $\mathcal{A}$ is proportional to the force amplitude f_0, and therefore to the current I_0 driving the beam. This holds for small displacements, but as we shall see below, the linear response regime is easily exceeded.

The beam's motion through the magnetic field generates an electromotive force (EMF) V, which can be monitored to determine the motion of the beam. The EMF is given by the real part of the rate of change of the enclosed magnetic flux Φ,

$$\begin{aligned}V(\omega) &= \frac{d\Phi}{dt} = B\int_0^L \frac{\partial U}{\partial t}(z)\, dz\\ &= -i\omega_c\, B\, \mathcal{A}\, L\, e^{-i\omega_c t}\\ &= -i\frac{\eta_1}{\omega_1^2 - \omega_c^2 - i\omega_c\omega_1/Q}\,\frac{I_0 B^2 L}{M}\, e^{-i\omega_c t}.\end{aligned} \tag{9.7}$$

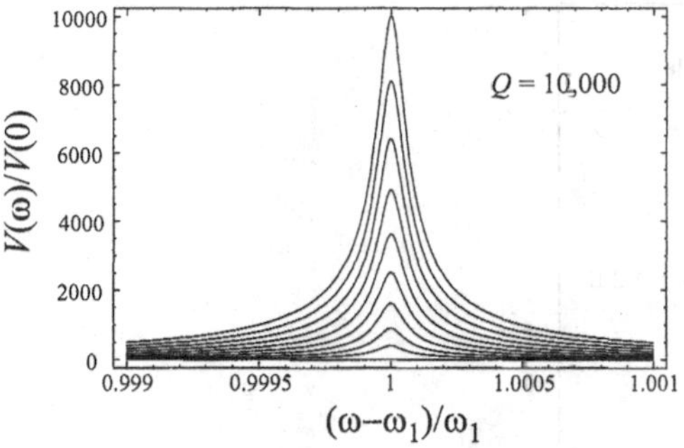

Fig. 9.5. Voltage calculated from (9.7), for a quality factor $Q = 10^4$, for fixed current amplitude and magnetic field ranging from 0 to 1 in steps of 0.1 (arbitrary units). The vertical axis is in units of the response at $\omega = 0$ with $B = 1$, and the horizontal axis is in units of the fundamental frequency ω_1.

The EMF depends quadratically on the magnetic field B and linearly on the current amplitude I_0. For a large quality factor Q, the resonance is at $\omega_c \approx \omega_1$. We display the form for (9.7) in Fig. 9.5, for $Q = 10^4$, and for a stepped sequence of magnetic field strengths B. Electrically, the magnetomotive response is almost identical to that of a parallel combination of an inductor L, a capacitor C, and a resistor R, such that the resonance frequency is $\omega_1 = 1/\sqrt{LC}$, and the quality factor is $Q = \sqrt{L/C}/R$.

The measured EMF as a function of drive current frequency ω_c, for fixed current amplitude I_0 and different magnetic field strengths B, is shown in Fig. 9.6 for the actual device shown in Fig. 9.3. Note that while the response closely resembles that shown in Fig. 9.5, there are non-idealities due to the fact that the beam in not impedance-matched to the cable, giving an overall background response (which is not magnetic-field dependent), and the response at larger magnetic fields is not purely harmonic, indicative of the nonlinearities encountered at large displacement amplitudes that we discuss below.

Other Materials. The resonator design in the previous section was based on the use of single-crystal silicon. GaAs and SiN have been used by other researchers; two other materials that have recently been demonstrated are silicon carbide (SiC) and aluminum nitride (AlN).

Silicon carbide can be grown epitaxially on $\langle 100 \rangle$ oriented silicon, using atmospheric-pressure chemical vapor deposition (APCVD) [88]. SiC is a very light, very stiff material, with a longitudinal sound speed $c_\ell = \sqrt{E/\rho} = 15.7$ km/s, roughly twice that of Si. It is chemically resistant, so that processing this material is not entirely trivial. Roukes and co-workers [89] developed a method by which SiC could be patterned using a combination of electron-beam lithography and electron cyclotron resonance (ECR)-based etching. Resonance frequencies as high as 134 MHz were measured, and the quality

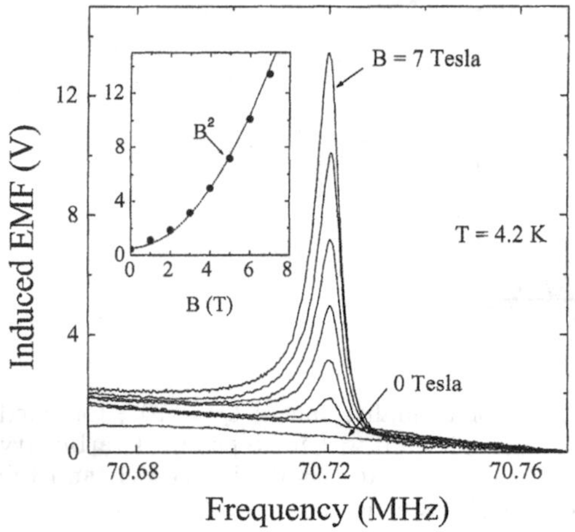

Fig. 9.6. Induced EMF as a function of drive frequency. Magnetic field strength ranged from 0 to 7 T at a temperature of 4.2 K. Drive amplitude was 10 mV. Inset: Induced EMF at resonance as a function of field strength. The vertical axis has same units as the main plot.

factor for a beam with a resonance frequency of 70 MHz was measured to be about 4000.

The authors started with a single-crystal Si substrate, on top of which a 259 nm thick 3C-SiC layer was grown by APCVD. A Cr mask (30-60 nm thick) was then deposited using an electron-beam lithography patterned liftoff mask, and used to protect the underlying SiC for anisotropic RIE using a gas mixture of NF_3, Ar and O_2. The underlying Si was then removed using an isotropic reactive ion etch using NF_3 and Ar, which did not damage the SiC. The Cr mask was then removed using a wet etchant, and a second layer of lithography used to pattern Au electrodes on the top surface of the already-suspended structures. In Fig. 9.7 we show a set of beams fabricated in this manner. The structures were then measured in vacuum at 4.2 K using the magnetomotive technique described above.

Fig. 9.7. Left: Top view of a family of 150 nm wide beams, having lengths from 2 to 8 μm. Right: Side view of a family of 600 nm wide beams, with lengths ranging from 8 to 17 μm. Figure by permission of the authors [89].

Aluminum nitride (AlN) is another interesting material to which these techniques have been applied. This material can be grown in single-crystal form on ⟨111⟩ oriented Si, and then patterned and etched. The growth is by MOCVD, using triethyl aluminum (TEA) and ammonia (NH_3) as the precursors. Films up to several microns thick can be grown, with c-axis alignment in the growth direction. The author has demonstrated the fabrication and measurement of flexural beams made from this material [90]. A 0.17 μm thick, single-crystal AlN film was grown on bulk Si, and Au electrodes with a Ti adhesion layer (35 nm/5 nm) were patterned using electron beam lithography. The electrode material was protected with a 60 nm thick film of Ni, all of which were deposited using the same liftoff mask pattern. An anisotropic reactive ion etch process, using Cl_2 gas, was used to etch through the AlN layer, stopping on the Si. The Si was then wet-etched using an isotropic silicon etchant, a mixture of ammonium fluoride, nitric acid and water. The suspended AlN is not affected by this etch, and was sufficiently strong to withstand the surface tension effects of submerging the structure in the etchant as well as rinsing and drying it. Critical point drying, a common technique for releasing very delicate structures from wet solutions, was not needed.

The completed resonators were measured using the magnetomotive technique at a temperature of 4.2 K in vacuum. An example of a set of resonance curves measured for different magnetic field strengths is shown in Fig. 9.8.

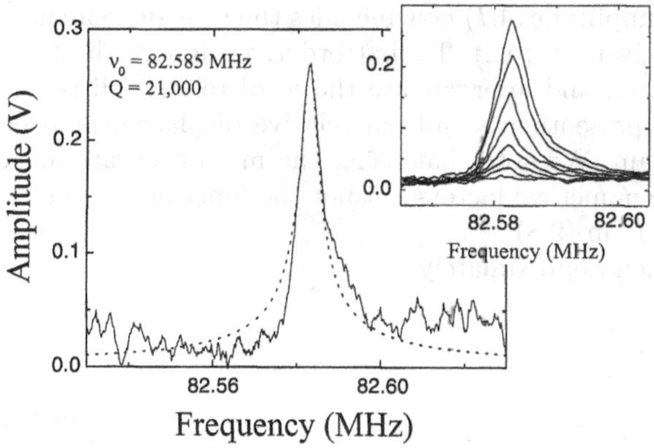

Fig. 9.8. Measured resonance of a 3.9 μm long beam of AlN, measured at 4.2 K in a transverse magnetic field of 8 T. Applied rf power was -85 dBm. *Inset:* Measured resonance for a constant rf power of -75 dBm, while varying the magnetic field through integer values from 1 T (smallest peak) to 8 T (largest peak). From [90].

9.2.2 Nonlinear Resonators

The doubly-clamped flexural resonator, at large displacement amplitudes, displays a striking nonlinearity in its response. This comes about because the flexure causes the beam to lengthen, which at large amplitudes adds a significant correction to the overall elastic response of the beam. We calculate this effect for a beam of length L, width w and thickness t, oriented with its long axis along z.

The total energy of the beam can be written in terms of its potential energy PE and kinetic energy T,

$$PE = \frac{1}{2}\int_V \boldsymbol{\epsilon}^T \cdot \mathsf{c} \cdot \boldsymbol{\epsilon}\, \mathrm{d}\boldsymbol{r}, \tag{9.8}$$

and

$$T = \frac{\rho}{2}\int_V \left(\frac{\partial \boldsymbol{u}}{\partial t}\right)^2 \mathrm{d}\boldsymbol{r}, \tag{9.9}$$

the integrals over the volume of the beam; $\boldsymbol{\epsilon}$ is the six-vector form for the strain. To evaluate these integrals we need the form for the displacement vector $\boldsymbol{u}$, which will give the kinetic energy T directly, and can be used to evaluate the strain, which then gives the potential energy PE. The correction will appear due to a change in the total beam volume due to lengthening.

The displacement of the neutral axis, in its fundamental mode, was discussed in Sect. 7.1.4. Here we write the displacement as $U(z,t) = \mathcal{A}(t)U_1(z)$, where the (dimensionless) amplitude $\mathcal{A}(t)$ now includes the time dependence. The mode shape $U_1(z)$ is given by (9.2). To first order, we ignore the lateral displacement of the beam, and approximate the local relative displacement $\boldsymbol{u} \simeq (U(z,t),0,0)$, representing a uniform relative displacement over the cross-section of the beam. We note that using the more accurate form given by (7.69) changes the numerical factors but not the functional form for the strain potential energy U in (9.8).

The kinetic energy is then approximately

$$\begin{aligned} T &\approx \frac{\rho}{2} A\dot{\mathcal{A}}^2 \int_0^L U_1(z)^2 \mathrm{d}z \\ &\approx \frac{1}{2} M\, L^2 \dot{\mathcal{A}}^2, \end{aligned} \tag{9.10}$$

where $M = \rho A L$ is the beam mass and $\dot{\mathcal{A}}(t) = \mathrm{d}\mathcal{A}/\mathrm{d}t$ the dimensionless velocity. To evaluate the potential energy, we approximate the strain $\boldsymbol{\epsilon}$ by (7.66),

$$\boldsymbol{\epsilon} \approx (0,0,-xU''(z,t),0,0,0). \tag{9.11}$$

We set all but the diagonal elements of c to zero, so the strain energy density is $E\epsilon^2/2$ in terms of Young's modulus E. The total strain energy is then

$$
\begin{aligned}
PE &= \frac{Ew\mathcal{A}^2}{2}\int_{-t/2}^{t/2}\int_0^L x^2\left(\frac{\mathrm{d}^2U_1}{\mathrm{d}z^2}\right)^2 \mathrm{d}z\mathrm{d}x \\
&= \frac{E}{2}\frac{wt^3}{12}\beta_1^4 L^3\mathcal{A}^2 = \frac{M}{2}\omega_1^2\, L^2\, \mathcal{A}^2.
\end{aligned}
\tag{9.12}
$$

Following the Lagrangian formalism, we can now write the equation of motion for the dimensionless amplitude $\mathcal{A}(t)$. We have $\mathcal{L} = T - PE$, with the equation of motion for our one degree of freedom given by

$$
\frac{\partial \mathcal{L}}{\partial \mathcal{A}} - \frac{\mathrm{d}}{\mathrm{d}t}\frac{\partial \mathcal{L}}{\partial \dot{\mathcal{A}}} = 0,
\tag{9.13}
$$

so that we find

$$
M\frac{\mathrm{d}^2\mathcal{A}}{\mathrm{d}t^2} + M\omega_1^2\mathcal{A} = 0.
\tag{9.14}
$$

This returns the expected (undamped) natural oscillatory response $\mathcal{A}(t) = \mathcal{A}_0 \mathrm{e}^{-\mathrm{i}\omega_1 t}$.

We can add the effect of the external driving force by adding to the potential energy the work done by the force per unit length f,

$$
-\int_0^L f\,U(z,t)\,\mathrm{d}z = -\eta_1 L^2 f\mathcal{A},
\tag{9.15}
$$

and we add the damping to yield the damped equation of motion for the amplitude $\mathcal{A}(t)$,

$$
M\frac{\mathrm{d}^2\mathcal{A}}{\mathrm{d}t^2} + M\gamma\frac{\mathrm{d}\mathcal{A}}{\mathrm{d}t} + M\omega_1^2\mathcal{A} = \eta_1 f,
\tag{9.16}
$$

where γ is the damping per unit mass. This is the same equation of motion that we found in Sect. 8.5.1.

We can now look at the correction to this equation due to the lengthening of the beam under displacement. The integrals in z along the beam length L for the total potential and kinetic energies should account for that fact that due to the displacement amplitude $\mathcal{A}$, the beam is slightly lengthened, as shown (greatly exaggerated) in Fig. 9.9. The integral should actually be along the differential length segment $\mathrm{d}\ell$ rather than $\mathrm{d}z$, with $\mathrm{d}\ell = \sqrt{\mathrm{d}z^2 + \mathrm{d}U^2} = \sqrt{1 + (\mathrm{d}U/\mathrm{d}z)^2}\,\mathrm{d}z$. For small displacements U, we can approximate $\mathrm{d}\ell$ by

$$
\mathrm{d}\ell \approx \mathrm{d}z + \frac{1}{2}\left(\frac{\mathrm{d}U}{\mathrm{d}z}\right)^2 \mathrm{d}z.
\tag{9.17}
$$

The integral therefore comprises two terms. The first term will yield the same potential and kinetic energies PE and T as we found earlier. The second term gives rise to corrections ΔPE and ΔT, proportional to $\mathcal{A}^2$. The correction to the potential energy is given by

$$
\begin{aligned}
\Delta PE &= \frac{Ewt^3}{24}\int_0^L \left(\frac{\mathrm{d}U_1}{\mathrm{d}z}\right)^2\left(\frac{\mathrm{d}^2U_1}{\mathrm{d}z^2}\right)^2 \mathrm{d}z\,\mathcal{A}^4 \\
&\approx 0.27\,\frac{M}{2}\,\omega_1^2\,\beta_1^2\,L^4\,\mathcal{A}^4.
\end{aligned}
\tag{9.18}
$$

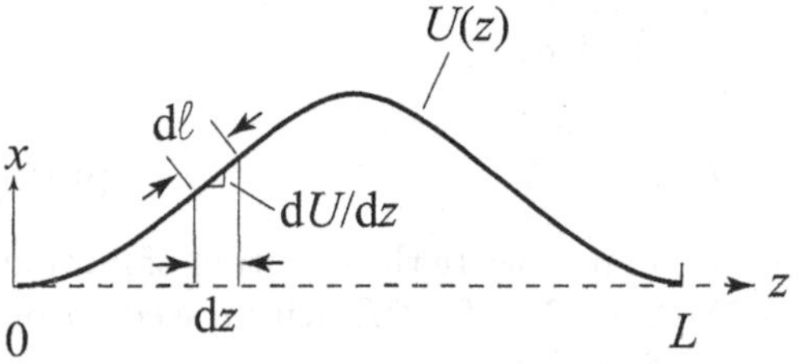

Fig. 9.9. Illustration of beam lengthening, used to calculate first-order correction to the strain energy.

Similarly, the correction to the kinetic energy is

$$\begin{aligned} \Delta T &= \frac{\rho}{2} A \int_0^L U_1^2 \left(\frac{\mathrm{d}U_1}{\mathrm{d}z}\right)^2 \mathrm{d}z \mathcal{A}^2 \dot{\mathcal{A}}^2 \\ &\approx 0.45 \frac{M}{2} \beta_1^2 L^4 \mathcal{A}^2 \dot{\mathcal{A}}^2. \end{aligned} \tag{9.19}$$

Hence the Lagrangian has the form

$$\begin{aligned} \mathcal{L} &= T + \Delta T - PE - \Delta PE \\ &= \frac{M}{2} L^2 \dot{\mathcal{A}}^2 \left(1 + K_1 \mathcal{A}^2\right) \\ &\quad - \frac{M\omega_1^2}{2} L^2 \mathcal{A}^2 \left(1 + K_2 \mathcal{A}^2\right), \end{aligned} \tag{9.20}$$

where $K_1 = 0.45\,\beta_1^2 L^2 \approx 10.1$ and $K_2 = 0.27\,\beta_1^2 L^2 \approx 6.04$. The Lagrange equation of motion then yields

$$\left(1 + K_1 \mathcal{A}^2\right) \frac{\mathrm{d}^2 \mathcal{A}}{\mathrm{d}t^2} + 2K_1 \mathcal{A} \left(\frac{\mathrm{d}\mathcal{A}}{\mathrm{d}t}\right)^2 + \omega_1^2 \left(1 + 2K_2 \mathcal{A}^2\right) \mathcal{A} = 0. \tag{9.21}$$

This is a rather challenging equation to solve; we approximate by assuming that the frequency of motion ω is close to the resonance frequency ω_1, so

$$\frac{\mathrm{d}^2 \mathcal{A}}{\mathrm{d}t^2} + \omega_1^2 \mathcal{A} + \omega_1^2 K_3 \mathcal{A}^3 = 0, \tag{9.22}$$

where $K_3 \approx K_1 + 2K_2 \approx 22.2$. The damping and the driving force terms can be added in through the Lagrangian formalism as earlier, which yields the equation

$$\frac{\mathrm{d}^2 \mathcal{A}}{\mathrm{d}t^2} + \gamma \frac{\mathrm{d}\mathcal{A}}{\mathrm{d}t} + \omega_1^2 \mathcal{A} + \omega_1^2 K_3 \mathcal{A}^3 = \eta_1 f, \tag{9.23}$$

an equation known as the driven Duffing equation. For small $\mathcal{A}$, the response is the same as for the linear damped equation. For larger amplitudes, the cubic term generates a positive correction term, as $K_3 > 0$, which makes the effective spring constant stiffer, increasing the resonance frequency as the amplitude increases. For $K_3 < 0$, the frequency would decrease with increasing amplitude. The nonlinearity becomes significant when $K_3\mathcal{A}^2 \geq 10^{-1}$, or $\mathcal{A} \geq 0.05$.

We can analyze the steady-state response to a harmonic force $f(t) = f_0 \cos\omega_c t$. We follow the analysis of Yurke et al. [91], and take a solution of the form

$$\mathcal{A}(t) = \frac{1}{2}\left(\mathcal{A}_0 e^{i\omega_c t} + \mathcal{A}_0^* e^{-i\omega_c t}\right), \tag{9.24}$$

where the amplitude $\mathcal{A}_0$ is complex; note that $\mathcal{A}(t)$ in this equation is real. Inserting this in (9.23), dropping terms of order $e^{\pm 3i\omega_c t}$ as these will be damped out, and matching the terms in $e^{i\omega_c t}$, we find

$$-\omega_c^2 \mathcal{A}_0 + i\omega_c\gamma\mathcal{A}_0 + \omega_1^2\mathcal{A}_0 + \frac{3}{4}\omega_1^2 K_3|\mathcal{A}_0|^2\mathcal{A}_0 = \eta_1 f_0, \tag{9.25}$$

where the common time dependence has been divided out. The complex conjugate of this expression is found for the terms in $e^{-i\omega_c t}$. Writing $\mathcal{A}_0 = a_0 e^{i\phi_0}$, where a_0 is strictly real, this yields

$$\left(\omega_1^2 - \omega_c^2 + \frac{3\omega_1^2 K_3}{4} a_0^2 + i\omega_c\gamma\right) a_0 = \eta_1 f_0 e^{-i\phi}. \tag{9.26}$$

We approximate this expression for drive frequencies close to the fundamental resonance, so $\omega_c = \omega_1(1+\delta)$, with $|\delta| \ll 1$, and drop terms of order δ^2. The energy of the oscillator is proportional to $E = a_0^2$, for which we can write

$$\left(-2\omega_1^2\delta + \frac{3\omega_1^2 K_3}{4} E\right)^2 E + \omega_1^2\gamma^2 E = \eta_1^2 f_0^2, \tag{9.27}$$

or, with $Q = \omega_1/\gamma$ and $\kappa = 3K_3/4$,

$$E^3 - \frac{4\delta}{\kappa}E^2 + \left(\frac{1}{Q^2\kappa^2} + \frac{4}{\kappa^2}\delta^2\right)E - \frac{\eta_1^2 f_0^2}{\kappa^2} = 0. \tag{9.28}$$

Similarly we derive an expression for the phase ϕ,

$$\begin{aligned}\tan\phi &= \frac{\gamma(1+\delta)}{2\omega_1\delta - 3\omega_1 K_3 E/4} \\ &= \frac{1}{Q}\frac{1+\delta}{2\delta - \kappa E}.\end{aligned} \tag{9.29}$$

In Fig. 9.10, we show the solutions for the energy E and the phase ϕ as a function of the detuning δ, for an oscillator with $Q = 100$, $\kappa = 0.1$ and a range of driving amplitudes. The response curve at large amplitudes tilts towards higher frequencies. For a drive amplitude above a certain critical value f_c, the tilt is sufficient that we find three possible values of E and ϕ, rather than the single valued solution for the linear regime. This multiple-valued solution is typical of the Duffing equation. Which value of energy and phase the system will actually respond with depends on the history and the particular bias point. In the absence of active feedback, the stable solution is as shown in Fig. 9.11.

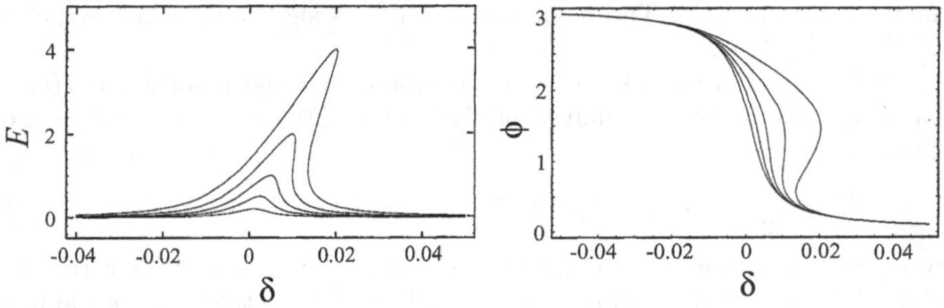

Fig. 9.10. Energy and phase for an oscillator with $Q = 100$, with $\eta_1^2 f_0^2/\kappa^2$ ranging from 2×10^{-5} to 4×10^{-3}.

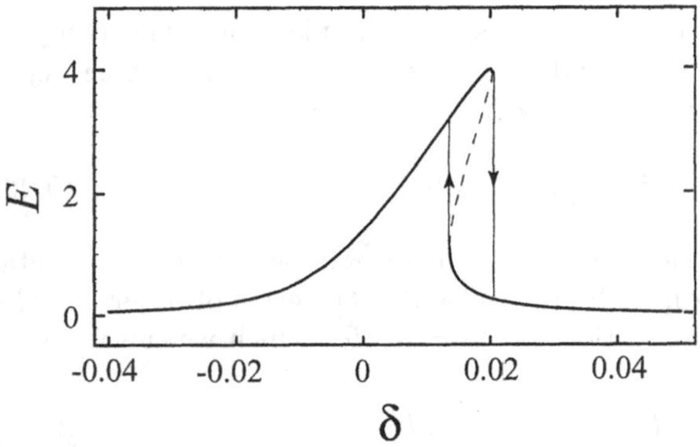

Fig. 9.11. Stable sections of the energy curve; as the frequency increases from below ω_1, the system follows the upper curve until the curve tips over, at which point the system jumps discontinuously to the lower curve. On reducing the frequency, the discontinuity occurs at the point where the curve tilts up, and the system jumps to the upper curve. The dashed region of the response curve is not stable.

The critical value of drive amplitude at which the response curve becomes multiple valued is where the derivative $\mathrm{d}E/\mathrm{d}\delta$ goes to infinity. This can be solved for using (9.27), and we find the critical point is at [91]

$$\left.\begin{aligned} \delta_c &= \frac{\sqrt{3}\omega_1}{Q}, \\ E_c &= \frac{2\delta_c}{3\omega_1\kappa} = \frac{2}{\sqrt{3}Q}, \\ \eta_1 f_c &= \left(\frac{8\sqrt{3}\omega_1^2}{9\kappa Q^3}\right)^{1/2}. \end{aligned}\right\} \tag{9.30}$$

The critical point in energy is also a critical point in the phase as a function of frequency.

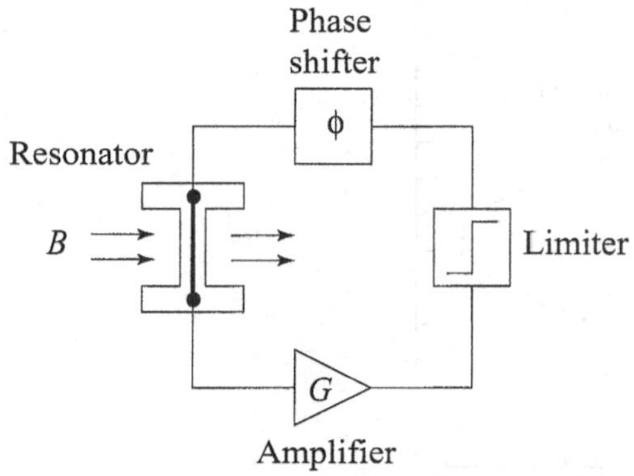

Fig. 9.12. Phase-locked loop used to control the relative phase of a resonator, which due to the closed loop can drive itself if the amplifier gain G is sufficiently large. The limited is used to fix the amplitude of the amplifier output, and the phase shifter has an externally-controlled phase that acts as the independent variable in the loop.

The cubic nonlinearity in the response of a doubly clamped beam has been used in a striking manner by D. Greywall and co-workers from Bell Laboratories. These authors fabricated a doubly-clamped flexural resonator from single crystal Si, with dimensions $3600 \times 127 \times 26\ \mu m^3$, and placed the resonator on the mixing stage of a dilution refrigerator, operating at a temperature of 100 mK. The resonance was actuated and detected using the magnetomotive technique, described in the previous section; a transverse magnetic field of 1 T was used for the measurements. In the simple version of this technique described earlier, the resonator is driven by a current source whose frequency is swept to trace out the amplitude-frequency and phase-frequency response. An alternative approach, used by these authors, was to drive the resonator using a phase-locked loop, in which the resonator motion is detected, amplified, a constant phase shift added to the resulting signal, and the resonator then driven by its own phase-shifted signal (see Fig. 9.12). The resonator was thereby allowed to *self-resonate*, and the phase difference between the driving force and the resonator motion set externally. The phase-frequency curves, with the phase shift the independent variable, were measured for different values of the drive amplitude, are shown in Fig. 9.13.

By operating the resonator at the inflection point in phase versus frequency, the authors were able to achieve a substantially lower phase noise than could be achieved in the linear response regime [92, 91]. A reduction in the phase noise by a factor of ten was achieved in this manner, as shown in Fig. 9.14. This figure displays the measured frequency fluctuations as a function of averaging time (proportional to the Allan variance, as discussed in Sect. 8.5.5). At short times the noise floor limits the measured fluctuations; at longer averaging times, the linear oscillator is limited by $1/f$ noise,

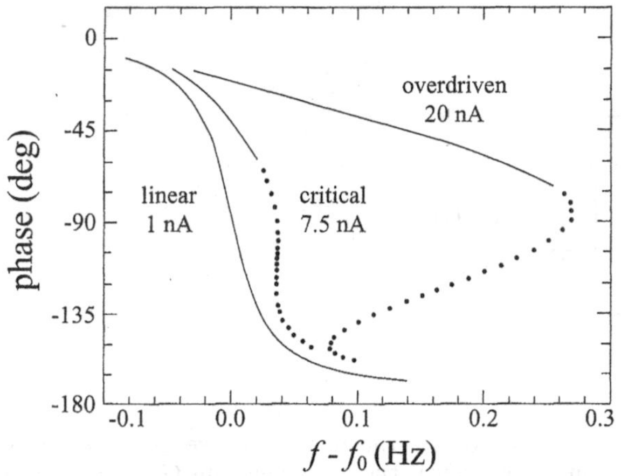

Fig. 9.13. Measurement of the phase-frequency relation for a self-excited resonator, with the phase difference between the drive and response controlled externally (points), or with the frequency slowly ramped (*solid lines*). By permission of the authors [92].

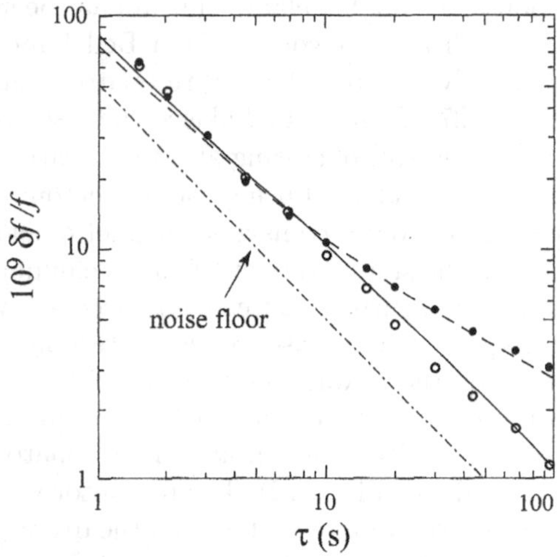

Fig. 9.14. Measurement of the frequency fluctuations in a linear and a nonlinear oscillator, showing clear improvement at averaging times of 100 seconds and longer. The measurements were made at a temperature of 100 mK. By permission of the authors [92].

while a clear improvement is visible for the nonlinear oscillator operated at the inflection point.

Other applications for this highly phase-stable regime could include force sensing and time- and frequency-keeping applications. Whether the measured performance can be achieved at more practical temperatures, such as at room temperature, remains to be seen.

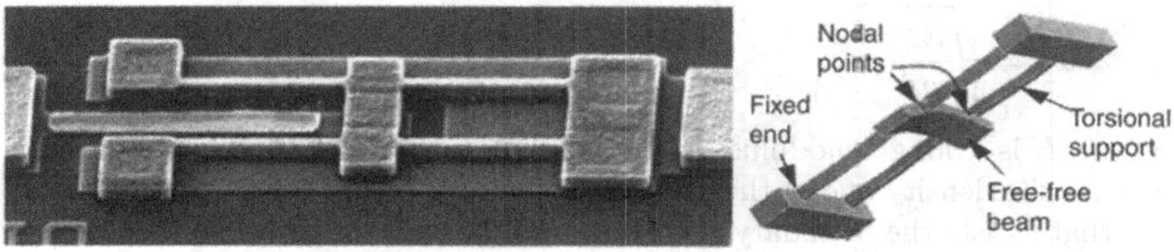

Fig. 9.15. Micrograph of free–free resonator, fabricated using a polysilicon-based MEMS process. Also shown is a schematic view of the flexure of the beam, and the corresponding flexure of the torsional support rods. From [93], by permission of the authors.

9.2.3 Free–Free Resonators

The nonlinearity of the doubly-clamped beam limits the amplitude to which it can be driven without inducing nonlinearity and frequency shifts in the response. One way to avoid this limitation is to use a cantilevered beam, with only one end clamped. Even for fairly large amplitude motion, this type of resonator does not display any significant nonlinearity.

A variant on a cantilevered beam, a *free-free* mechanical resonator, was demonstrated by Wang, Wong, and Nguyen [93]. This structure comprises a flexural beam held in place by a set of four carefully designed legs, positioned at the nodal points of the desired flexural mode shape. In Fig. 9.15 we display the structure, along with a schematic drawing of the mechanical design. The flexural beam is designed to resonate at its fundamental frequency, with nodes at the support points.

We orient the coordinate axes so that the beam length L is along the z axis, and the displacement $U(z,t)$ of the beam is along the x axis, with bending therefore about the y axis. The Euler–Bernoulli theory, applied to this beam, yields the solutions found previously in Sect. 7.1.4. The mode shape has the form given by (7.52), with $U(z,t) = U_n(z)\mathrm{e}^{-\mathrm{i}\omega_n t}$ for the nth mode, with

$$U_n(z) = a_n \cos(\beta_n z) + b_n \cosh(\beta_n z) + c_n \sin(\beta_n z) + d_n \sinh(\beta_n z). \tag{9.31}$$

The boundary conditions at the ends $z = 0$ and $z = L$ are fixed by the requirements for no bending or flexure, so that

$$\left.\begin{aligned} \frac{\mathrm{d}^2 U}{\mathrm{d}z^2}(0) = \frac{\mathrm{d}^2 U}{\mathrm{d}z^2}(\ell) &= 0, \\ \frac{\mathrm{d}^3 U}{\mathrm{d}z^3}(0) = \frac{\mathrm{d}^3 U}{\mathrm{d}z^3}(\ell) &= 0. \end{aligned}\right\} \tag{9.32}$$

These fix the prefactors in (9.31), $a_n = b_n$ and $c_n = d_n$, with

$$c_n = \frac{\cosh \beta L - \cos \beta L}{\sinh \beta L - \sin \beta L}\, a_n, \tag{9.33}$$

along with the equation that gives the mode frequency ω_n,

$$\omega_n = \sqrt{\frac{EI_y}{\rho A}}\,\beta_n^2, \tag{9.34}$$

where E is Young's modulus, I_y the bending moment of inertia about the y axis, ρ the density and A the cross-sectional area of the beam. The values of β_n that satisfy the boundary conditions are the same as those for the doubly-clamped beam, with $\beta_n \ell = 4.73004, 7.8532, 10.9956, 14.1372\ldots$ The nodal support points are located where the displacement (9.31) is zero, which for the fundamental mode with $\beta_n = 4.73004/L$ are at $z = (0.5 \pm 0.276)L$.

The design frequencies for the flexural resonators that were demonstrated ranged from 30 to 90 MHz. The elastic modulus for the polysilicon from which the resonators were fabricated was $E = 165$ GPa, and the density was $\rho = 2.25\ \mathrm{g/cm^3}$; Poisson's ratio was measured to be $\nu = 0.226$.

The authors, in addition to using a free-free flexural beam, also designed the structure with the goal of reducing the mechanical coupling between the flexural motion and the rigid substrate. The support rods were connected at the nodes of the flexural motion, so that they would not in principle undergo any translational motion. The flexure does however cause them to torsion, sending torsional waves down their length. Further decoupling was therefore achieved by designing the length of the torsional rods to be one-quarter of the corresponding wavelength for the torsional waves. For this choice of rod length, the base of the support rods, connected rigidly to the substrate, is at a node for the torsional motion. This provides a second level of mechanical decoupling. The support rods had a width w and thickness t, and the polysilicon from which they were fabricated has a torsion modulus $G = E/2$. The torsion wavelength λ_c for a signal at frequency ω_c is

$$\lambda_c = \frac{2\pi}{\omega_c}\sqrt{\frac{G\gamma}{\rho I_s}}, \tag{9.35}$$

where I_s is the torsional moment of inertia, given by

$$I_s = hw\frac{h^2 + w^2}{12}. \tag{9.36}$$

γ is the torsional constant, given for this structure with $h > w$ by

$$\gamma = 0.229hw^3 \tag{9.37}$$

(see [94]).

The resonators were fabricated using a multilayer polysilicon process. A 2 μm thick thermal oxide was grown on a bulk Si substrate, and coated with a 0.2 μm thick, LPCVD-grown Si_3N_4 layer. A 0.3 μm layer of polysilicon layer was then grown by LPCVD and doped by ion implantation, and then patterned to form electrostatic drive electrodes. A sequence of oxide and polysilicon layers were then deposited, patterned and etched to form the mechanical structure, which had a thickness of 2 μm, and was spaced 0.16 μm vertically above the electrodes that were used to drive and detect the motion of the resonator (see below). The flexural resonators had widths ranging from 10 μm for the lower-frequency resonators, to 6 μm for the higher frequency ones, and lengths ranging from 23 to 12 μm, depending on the design frequency.

The authors used electrostatic forces to drive the motion of the resonators, and the resulting displacement-driven capacitance changes were used to sense the motion; see Sect. 9.3 following for a discussion of this actuation and sensing technique. The electric field that generated the motion was applied between polysilicon electrodes that were deposited first on the substrate, forming the bottom half of a capacitor plate, and the resonator structure defined on top of them, which formed the (flexible) upper half of the resonator structure. A static voltage applied to the electrodes pulls the resonator down, making the capacitor spacing smaller; a radiofrequency voltage applied in combination with the static voltage could then be used to drive the resonator near its resonance frequency. An innovation in the structural design was the inclusion of small dimples on the bottom surface of the resonator, onto which the resonator would be pulled using the electrostatic voltage. These dimples, with a design height of 0.12 μm, would prevent the resonator structure from pulling down further as the voltage on the electrodes was increased beyond the "contact" voltage. The dimple height was chosen so that the combination of the static voltage with the oscillating rf voltage was sufficient to drive the resonators into flexure, and also generate a sufficiently large response voltage that a second electrode could be used to sense the motion.

This design, which included a number of other innovations, successfully achieved quite high quality factors at room temperature (in the range of 7000-9000), with resonators operating in the range of 30-90 MHz. The resonators were measured in vacuum to reduce air-induced damping. The resonators do not exhibit the striking nonlinearity present in doubly-clamped beams, and furthermore demonstrated that quite high frequencies could be achieved using all-optical lithography, with no lateral dimensions less than 1 μm.

We note that the same authors have pursued another quite different approach to isolating resonant structures from losses in the supports. Shown in Fig. 9.16 is a disc resonator, supported only by a small column in the center of the disc. External electric fields drive a "breathing" mode that in principle does not generate any strain, and therefore any loss, at the support point.

9.3 Parametric Resonators: The Mathieu Oscillator

The parametric resonator provides a very useful instrument that has been developed by a number of researchers, in part because a parametric resonator can serve as a mechanical amplifier, over a narrow band of frequencies. The simplest parametric resonator is a child's swing, where the child pumps energy into the swing's motion by varying the mass distribution. Parame-

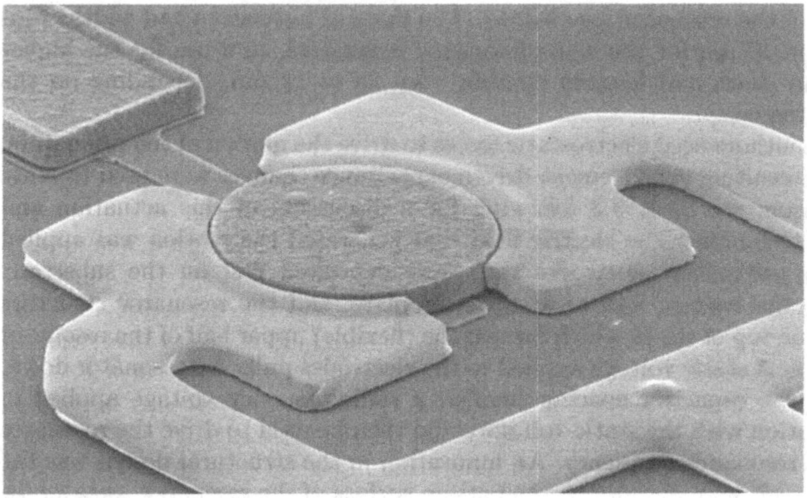

Fig. 9.16. Micrograph of disc resonator. Electrodes on the sides are used for electrostatic actuation and detection. From [95], by permission of the authors.

tric modulation has been studied since the 19th century, especially in the parametrically-driven pendulum.

The basic elements of a mechanical parametric resonator can be understood by considering a simple parallel plate capacitor, one side of which is part of the mechanical resonator, the other side of which is fixed; see Fig. 9.17. The resonator motion changes the spacing between the two plates, so that for a displacement x, the plate spacing becomes $d + x$ (x is positive for displacement to the right). For large area plates with area A, the electrical capacitance C between the plates is given by

$$C = \epsilon_0 \frac{A}{d+x} \approx \epsilon_0 \frac{A}{d} \left(1 - \frac{x}{d} - \frac{x^2}{d^2} + \dots\right), \tag{9.38}$$

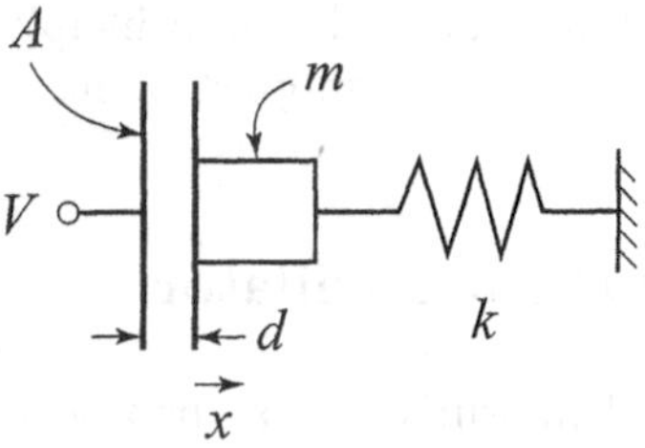

Fig. 9.17. Model for an electrostatically modulated resonator, with mass m, spring constant k, modulated by an electrostatic voltage V applied across a parallel plate capacitor with area A and spacing d.

where we assume the displacement $x \ll d$, and we ignore fringing fields; in the second part of (9.38) we have used the Taylor expansion for small x. If we apply a voltage $V(t)$ across the capacitor plates, the electrostatic energy is given by

$$\mathcal{E}_{\mathrm{el}} = \frac{1}{2} C\, V^2(t). \tag{9.39}$$

The resonator has mass m and spring constant k. The Lagrangian for the system, including the electrostatic energy, is given by

$$\mathcal{L} = \frac{1}{2} m\dot{x}^2 - \frac{1}{2} kx^2 - \mathcal{E}_{\mathrm{el}}, \tag{9.40}$$

where $\dot{x} = \mathrm{d}x/\mathrm{d}t$. Lagrange's equation of motion is then

$$m\ddot{x} + kx = -\frac{\partial \mathcal{E}_{\mathrm{el}}}{\partial x}. \tag{9.41}$$

The dependence of the electrostatic energy $\mathcal{E}_{\mathrm{el}}$ on displacement x is through the capacitance C. For the approximate relation given by (9.38), we have

$$m\ddot{x} + m\gamma\dot{x} + kx = -\epsilon_0 \frac{A}{d^2} V^2(t) \left(1 + 2\frac{x}{d}\right), \tag{9.42}$$

where we have also added the damping term $m\gamma\dot{x}$ (see Sect. 9.2.1, and Chap. 8).

The term on the right will drive the resonator motion, and involves the square of the driving voltage, including both a term that is independent of displacement x, as with a standard harmonic oscillator, as well as a term that is linear in the displacement, coming from the third-order term in the Taylor expansion. Defining $r_1 = \epsilon_0 A/d^2$, and using the quality factor $Q = \omega_0/\gamma$, we can re-write the equation of motion as

$$m\ddot{x} + m\frac{\omega_0}{Q}\dot{x} + kx = -r_1(1 + 2x/d)V^2(t). \tag{9.43}$$

We define the spring modulation $\Delta k(t)$,

$$\Delta k(t) = 2\frac{r_1}{d} V^2(t), \tag{9.44}$$

and the driving force term $F_{\mathrm{D}}(t)$,

$$F_{\mathrm{D}}(t) = -r_1 V^2(t). \tag{9.45}$$

Re-arranging the terms then yields

$$m\ddot{x} + m\frac{\omega_0}{Q}\dot{x} + \left[k + \Delta k(t)\right] x = F_{\mathrm{D}}(t). \tag{9.46}$$

This equation forms the basis for the parametric resonator: It includes both a time-dependent spring constant $\Delta k(t)$, which allows us to modulate one of the parameters of the harmonic oscillator, and an external driving term with which motion can be generated. In this simplistic model, both the parametric modulation and the driving force are proportional to the square of the applied voltage $V^2(t)$; by adding another capacitor plate on the opposite side of the mass m, with an independent driving voltage, we can make a system in which the parametric modulation and the driving force can be modulated independently.

We therefore assume that $\Delta k(t)$ and $F_{\mathrm{D}}(t)$ are independent, and we modulate each harmonically, with the driving term $F_{\mathrm{D}}(t)$ at frequency ω_{D} and the parametric modulation $\Delta k(t)$ at frequency ω_P:

$$\left.\begin{aligned}\Delta k(t) &= \Delta k_0 \sin \omega_P t,\\ F_{\mathrm{D}}(t) &= F_0 \cos(\omega_{\mathrm{D}} t + \phi).\end{aligned}\right\} \tag{9.47}$$

The phase ϕ introduces a phase difference between the two terms.

We then have

$$m\ddot{x} + m\frac{\omega_0}{Q}\dot{x} + (k + \Delta k_0 \sin \omega_P t)x = F_0 \cos(\omega_{\mathrm{D}} t + \phi). \tag{9.48}$$

This equation is known as a damped Mathieu equation [96], including both a standard simple harmonic oscillator driving term, and a *parametric modulation* term, due to the electronically-controllable spring constant.

In the absence of parametric modulation, the resonator has a natural resonance frequency $\omega_0 = \sqrt{k/m}$; for small damping ($Q^{-1} \ll 1$), the resonator will display a strong resonance when the driving term is near the natural resonance frequency, $\omega_{\mathrm{D}} \approx \omega_0$. If we now turn on the parametric modulation, and keep the driving term weak (F_0 small), the equation (9.48) will exhibit additional strong resonances whenever $\omega_P = 2\omega_0/n$ for integers $n = 1, 2, 3 \ldots$: The parametric term resonates for all *submultiples* of the first natural harmonic frequency $2\omega_0$ (including of course the fundamental ω_0, resonant at $n = 2$).

We focus on the response at the harmonic at $n = 1$, so we take $\omega_P = 2\omega_0$, pumping the parameter at twice the natural resonance frequency. We also take the drive frequency equal to the natural frequency, $\omega_{\mathrm{D}} = \omega_0$. We can then work out the solution to (9.48), following the analysis of Rugar and Grütter [97].

We introduce the complex variable a, defined by

$$a = \frac{\mathrm{d}x}{\mathrm{d}t} + \mathrm{i}\Omega^* x, \tag{9.49}$$

with the complex frequency Ω defined as

$$\Omega = \left[\sqrt{1 - \frac{1}{4Q^2}} + \frac{\mathrm{i}}{2Q}\right]\omega_0. \tag{9.50}$$

(9.49) can be inverted to yield x and $\mathrm{d}x/\mathrm{d}t$ in terms of a and a^*:

$$\left.\begin{aligned}x &= -\frac{i\,(a - a^*)}{\Omega + \Omega^*},\\ \frac{\mathrm{d}x}{\mathrm{d}t} &= \frac{\Omega a + \Omega^* a^*}{\Omega + \Omega^*}.\end{aligned}\right\} \tag{9.51}$$

A simplified equation of motion for a then results,

$$\frac{da}{dt} = i\Omega a + i\frac{\Delta k_0}{m}\sin 2\omega_0 t\,\frac{a - a^*.}{\Omega_1 + \Omega_1^*} + \frac{F_0}{m}\cos(\omega_0 t + \phi). \tag{9.52}$$

The solutions have the form $a = Ae^{i\omega_0 t} + Be^{-i\omega_0 t}$, with complex amplitudes A and B. Inserting this form in (9.52), making the approximations $Q \gg 1$, $\Omega + \Omega^* \approx 2\omega_0$, and $\Omega - \omega_0 \approx i\omega_0/2Q$, we find $B \approx 0$ and

$$A = F_0\,\frac{Q\omega_0}{k}\left[\frac{\cos\phi}{1 + Q\Delta k_0/2k} + i\frac{\sin\phi}{1 - Q\Delta k_0/2k}\right]. \tag{9.53}$$

The oscillator motion that corresponds to this solution is

$$x(t) = X_A\cos\omega_0 t + X_B\sin\omega_0, \tag{9.54}$$

with the amplitudes calculable from (9.53) using (9.51).

We can now see how a parametric oscillator can be used as an amplifier: The gain G of the amplifier is given by the ratio of the amplitude $(X_A^2 + X_B^2)^{1/2}$ with and without the parametric modulation Δk_0. It is easily shown than that the gain is

$$G = \left[\frac{\cos^2\phi}{(1 + Q\Delta k_0/2k)^2} + \frac{\sin^2\phi}{(1 - Q\Delta k_0/2k)^{1/2}}\right]^{1/2}. \tag{9.55}$$

In the limit $\Delta k_0 \to 0$, which is equivalent to turning off the parametric modulation, we get the response $A = QF_0(\omega_0/k)e^{i\phi}$, identical to that for a simple harmonic oscillator; the gain in this case is $G = 1$ For non-zero Δk_0, the response depends on the relative phase ϕ, as is typical for a parametrically driven system: For $\phi = 0, \pi, 2\pi, \ldots$, the response is de-amplified, with a smaller amplitude than in the absence of the parametric signal. For these values at the phase, then, G is less than one. For $\phi = \pi/2, 3\pi/2, \ldots$, the response is *larger* than in the absence of the parametric signal, yielding amplification, with $G > 1$. As the amplitude of the parametric signal is increased, to the limit $\Delta k_0 = 2k/Q$, the response diverges, with G increasing to infinity (until, of course, the nonlinearities in the elastic response of the mechanical resonator begin to dominate). In Fig. 9.18 we display the phase and modulation amplitude dependence of the gain G.

If we were to remove the restriction on the parametric frequency ω_P, we would find this behavior occurs at each of the resonances $2\omega_0/n$: For very small modulation amplitude Δk_0, the response is either deamplified or amplified, depending on the relative phase of the drive and the parametric modulation, while for large enough modulation amplitude the response will diverge. The regions of stable and unstable response can be mapped out as a function of parametric modulation frequency ω_P and modulation amplitude Δk_0; for most values, the response is stable, with "tongues" of unstable response about $\omega_P = 2\omega_0/n$ reaching towards the $\Delta k_0 = 0$ axis. In most parametrically driven systems it is very difficult to experimentally observe more than one of the parametric resonances (typically that with $n = 1$).

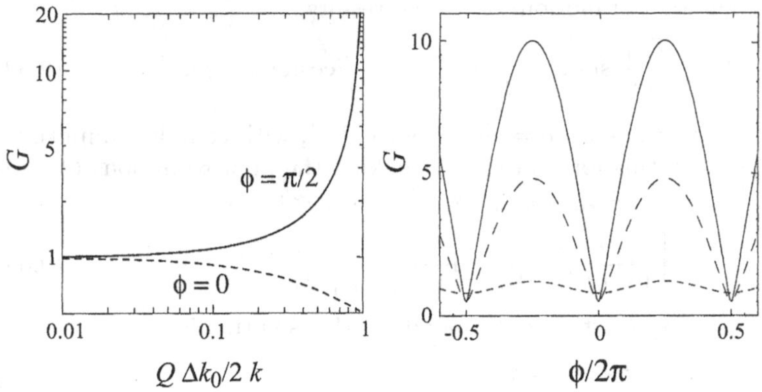

Fig. 9.18. Gain calculated for a parametric amplifier, as a function of modulation amplitude on the left, plotted for $\phi = 0$ and $\phi = \pi/2$, and as a function of phase on the right, for modulation amplitude $Q\Delta k_0/2k = 0.2$, 0.8 and 0.9.

However, using micro- and nanomachined resonators with high Q, high displacement sensitivity, and excellent control over the parametric modulation, Turner and MacDonald proved that it was possible to find several (four) of these parametric resonances [98].

Multiple Parametric Resonances. The system Turner and MacDonald studied [98] was a symmetrically-driven torsional resonator. In this chapter, we have been discussing flexural resonators; the torsional resonator is very similar in form, with the torsion angle θ replacing the displacement x, the moment of inertia I about the torsion axis replacing the mass m, and the torsion spring constant κ replacing the linear spring constant k. The natural resonance frequency is $\omega_0 = (\kappa/I)^{1/2}$. The external force is replaced by a torque τ, which in this case is applied by applying a voltage between the plates of an interdigitated capacitor, one side of which is a cantilevered structure that is part of the torsional resonator (see Fig. 9.19). The electrostatic force between the capacitor plates includes a component that points out of the plane of the interdigitated structure [99], due to the presence of the substrate below the interdigitated structure. This out-of-plane force, applied at a point some distance from the center axis of the torsional rod, generates a torque about the rod. The resonator in this experiment included capacitors on either side of the torsion axis, with voltages applied to the capacitors, so that for an upwards force on one side, there is an accompanying downward force on the other side, so that there is no *net* force, only a torque. The torque however depends on the torsion angle of the resonator: For small torsion angles, the torque is approximately linear in the torsion angle, so that a voltage of the same form as (9.46) gives a torque $\tau = (\tau_0 + \tau_1 \cos \omega_P t)\theta$, providing the parametric response. An additional drive torque τ_D can be generated by applying a voltage between the substrate and the resonator,

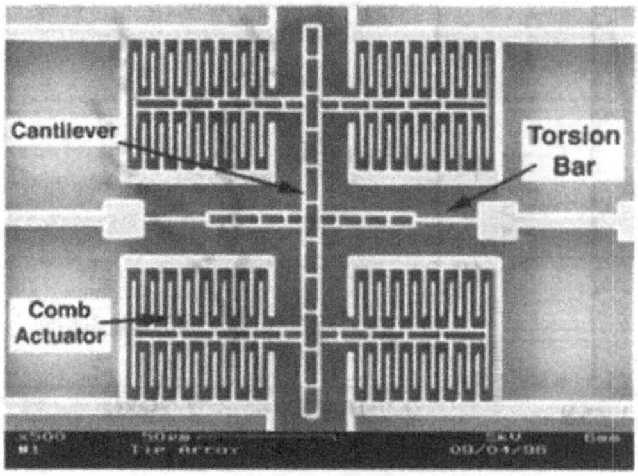

Fig. 9.19. Micrograph of the torsional oscillator used by Turner and MacDonald. The cantilever torsions about the horizontal torsion bar, which has a torsional spring constant $\kappa = 2.75 \times 10^{-8}$ N-m. The cantilever has a torsional moment $I = 2.12 \times 10^{-19}$ kg-m^2, and the quality factor in vacuum (at a pressure of 18 mtorr) was measured to be $Q = 3000$. The device was fabricated using the SCREAM process described in Chap. 10. Figure by permission of the authors [98].

which does not depend on the angle θ for small angles. Taken together, these yield the equation of motion in the torsion angle θ,

$$I\ddot{\theta} + I\frac{\omega_0}{Q}\dot{\theta} + \kappa\theta = (\tau_0 + \tau_1 \cos\omega_P t)\,\theta + \tau_D \cos(\omega_D t + \phi), \tag{9.56}$$

of the same form as (9.48).

These authors were able to map out the stable and unstable regions in the $\omega_P - \tau_1$ plane, for the resonance indices $n = 1 - 4$. The results for frequencies near the $n = 1$ and $n = 3$ resonances are shown in Fig. 9.20. The separation between regions of stable and unstable response are extremely sharp, allowing much higher precision in determining the resonator parameters, as well as their variation due to external effects, than can be done using a simple harmonic oscillator response (by at least an order of magnitude).

Parametric Amplification. The first example of a micromechanical parametric amplifier was developed by Rugar and Grütter [97]. These authors fabricated a silicon cantilever with dimensions 500 µm in length, 10 µm wide and several microns in thickness. The spring constant was about 1 N/m, and the resonance frequency was 33.57 kHz. The quality factor, measured in vacuum at room temperature, was above 10^4. The cantilever was parametrically modulated using an electrostatic force applied through a pair of capacitor plates, as discussed above. The motion was detected using a laser interferometer. This displacement sensing technique, which works very well for structures with minimum dimensions above a few µm, had a displacement sensitivity of about 10^{-4} Å/$\sqrt{\text{Hz}}$. The driving force F_D was applied separately, using a piezoelectric mount for the cantilever, which was driven at ω_D. The parametric modulation was at twice the natural resonance frequency of

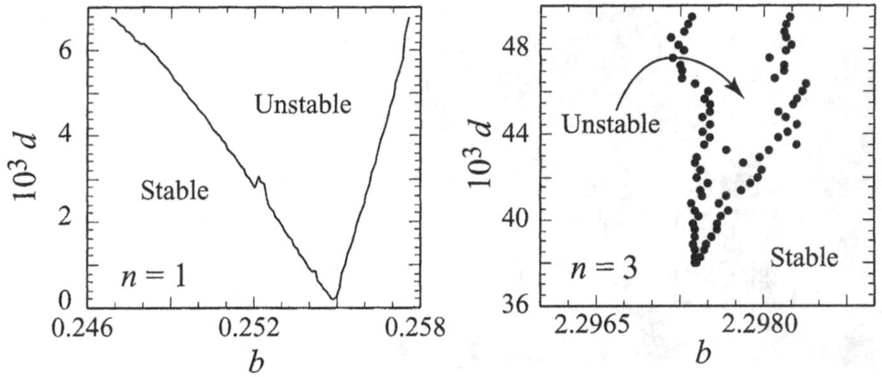

Fig. 9.20. Stability maps for the $n = 1$ and $n = 3$ resonances; the parameters b and d correspond to the frequency $\omega_P^2 = b\omega_0^2$ and amplitude $\Delta k_0/m = d\omega_0^2$ in (9.48). Figure by permission of the authors [98].

the cantilever, and the drive signal was set at the resonance frequency, as described above.

Amplification was observed in the response, defined as the ratio of the motional amplitude with and without parametric modulation. Gains G as high as $\times 100$ were observed, with the correct dependence on the relative phase ϕ. The measurements reflected an almost textbook response, as shown in Fig. 9.18.

A similar system was explored by D. Carr et al. [100]. These authors fabricated a torsional resonator using a SIMOX substrate (see Chap. 10), yielding a resonator $t = 200$ nm thick, with a central element 15×4 μm^2. A thin layer of Cr (5 nm) and Au (10 nm) were deposited on the resonator's surface, to allow for the application of electrostatic voltages. A micrograph of the torsional structure is shown in Fig. 9.21, showing a resonator with dimensions $l \times w = 4 \times 4$ μm^2.

The oscillator was driven by applying a voltage between the substrate (held at ground) and the suspended torsional element. A perfectly symmetric system would experience only a flexural force, but small asymmetries in the design allow the generation of a torque as well. The high quality factor of the resonator then allows selection of a particular mode by frequency control. Motion was detected using a laser reflected from the surface of the resonator; the reflected signal depends strongly on interference between reflections from the resonator and the substrate, yielding good displacement sensitivity. The resonator had a natural resonance frequency of $\omega_0/2\pi = 485$ kHz, measured in vacuum. The authors demonstrated the phase-dependent amplification of a signal at the natural resonance frequency $\omega_D = \omega_0$, using a parametric drive at $\omega_P = 2\omega_0$, i.e. the $n = 1$ resonance of the Mathieu equation.

In Fig. 9.22 we display the measured amplification as a function of the parametric modulation amplitude for phase $\phi = \pi/2$. We also display the

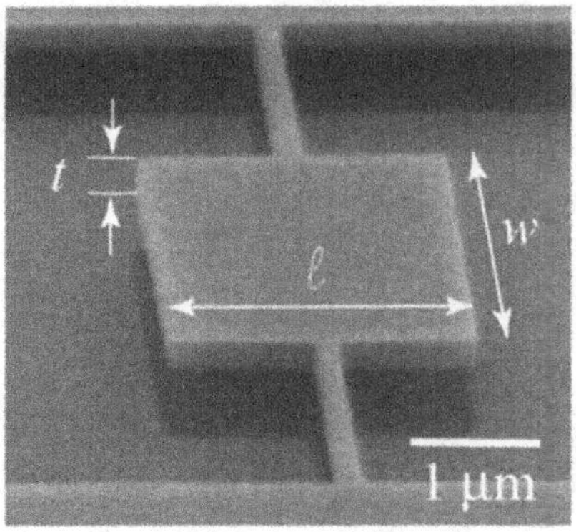

Fig. 9.21. Torsional resonator fabricated using the SIMOX process, a structure the authors used to explore parametric amplification. Figure by permission of the authors [100].

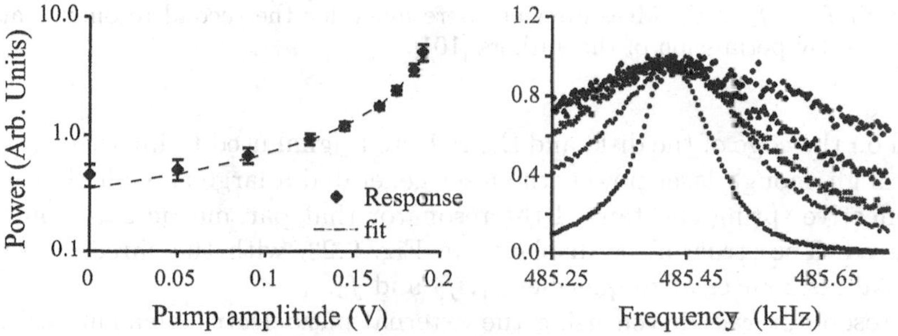

Fig. 9.22. (*Left*) Measured oscillation amplitude as a function of the amplitude of the parametric modulation amplitude, measured in volts. The signal frequency is at the natural resonance frequency of the resonator, and the parametric modulation at twice that frequency. (*Right*) Response curves as a function of drive frequency about the resonator's natural resonance frequency; each curve is for a different modulation amplitude, as in the left figure. By permission of the authors [100].

expected narrowing of the resonance curve, as a function of the drive frequency ω_D, as the pump amplitude is varied. The maximum amplification achieved was about a factor of 10, with noise effects in the electronics and the resonator contributing significantly to the measured response.

9.3.1 Optical Parametric Resonator

In an interesting further development on this approach, Zalatudinov et al. [101], again at Cornell, demonstrated an optically-driven parametric oscillator. The oscillator consists of a disk of silicon, suspended by a pillar at the center of the disk, with the whole mounted on a piezoelectric resonator to drive mechanical motion. A laser, used to detect the mechanical motion, was

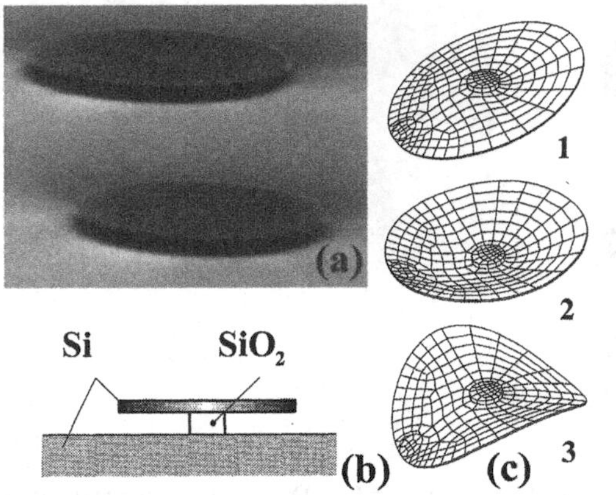

Fig. 9.23. (**a**) SEM image of two of the disks. Measurements were made on a disk with a diameter of 40 μm, 250 nm thick, supported by a 6.7 μm diameter pillar 1 μm tall. (**b**) Cross-section of the resonant structure. (**c**) Lowest three resonant modes, with $f_1 < f_2 < f_3$. Measurements were made for the second resonance at frequency f_2. By permission of the authors [101].

focussed on the edge of the disk, and the reflected signal used to infer the motion. At high enough laser power, the laser generated a large enough change in the effective spring constant of the resonator that parametric gain could be achieved. The structure is displayed in Fig. 9.23, with the three lowest mechanical resonances at frequencies f_1, f_2 and f_3.

The resonator was driven using the external piezoelectric element, with measurements made in vacuum, with the highest quality factor $Q = 1.1 \times 10^4$ observed for the second resonance, at a frequency $f_2 = 0.89$ MHz; all measurements were made for this resonance. As the laser power was increased, thermal stress was generated in the disk, changing the effective spring constant by an amount Δk. This causes a roughly linear shift in resonance frequency f_2 as a function of laser power, with quantitative agreement between the measured frequency shift and that calculated using finite-element simulations.

As the laser power was increased past a certain critical point, an instability was observed in the resonator's response. If the laser is modulated at twice the fundamental frequency, a signal at the fundamental (generated by the piezoelectric element) can be amplified, as in (9.55). In Fig. 9.24, we display the various measurements made on this device, showing gain as a function of frequency and phase.

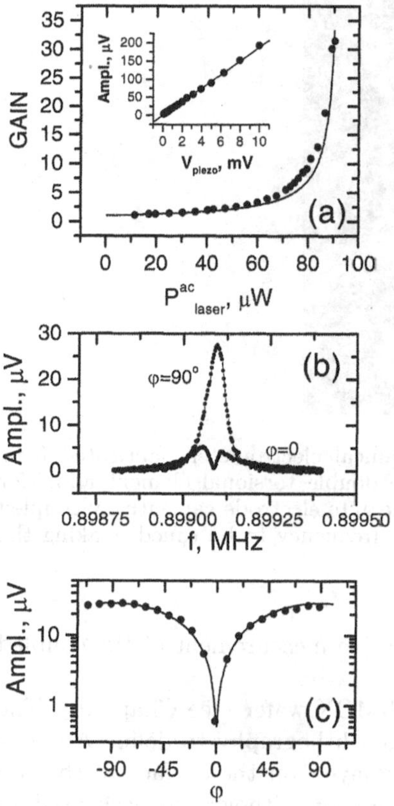

Fig. 9.24. (**a**) Measured gain of the piezoelectric signal at the fundamental frequency, amplified by the laser spring modulation at twice the fundamental frequency, with phase difference $\phi = \pi/2$. The solid line is a fit to the formula (9.55). (**b**) Response measured for the phase fixed at $\phi = 0$ and $\phi = \pi/2$, where the piezoelectric signal was swept through the resonance at 0.89 MHz, and the laser signal modulated at twice the piezoelectric signal. (**c**) Measured response at the peak of the response in (b), as a function of the phase difference between the piezoelectric signal at $\omega_0 = 2\pi f_2$ and the laser signal at $2\omega_0$. The solid line is a fit to the phase dependence in (9.55). By permission of the authors [101].

9.4 Mechanical Electrometer

An example of a measurement instrument based on a nanomechanical structure is the mechanical electrometer, studied by the author and M.L. Roukes [102]. This device, shown in Fig. 9.25, consists of a double torsional resonator that can be driven and detected using the magnetomotive technique. The design includes a gate electrode, 0.5 µm from the torsional element, that allows charge to be coupled to the device. The electrostatic field from the charge modifies the effective torsional spring constant of the resonant element, chan-

Fig. 9.25. Electron micrograph of a mechanical electrometer, fabricated from single- crystal silicon. The electrodes on the double torsional element allow for magnetomotive actuation and sensing, and the gate electrode capacitively coupled to the torsional element allows the resonance frequency to be tuned, making the device a charge detector.

ging its frequency, and allowing a quantitative measurement of the coupled charge.

The structure was fabricated from a SIMOX wafer (see Chap. 10). The substrate was processed using electron-beam lithography to define Au electrodes, 30 nm thick with a 3 nm Cr underlayer, on the surface of the top structural layer of Si. A second layer of aligned lithography was used to define an etch mask, made of Ni, 60 nm thick. The exposed area of the substrate was etched using an anisotropic reactive ion etch, with a gas chemistry of NF_3 mixed with CH_2Cl_2; this etch went through the top Si layer, and partially through the underlying SiO_2 layer. Finally, the structure was freed using dilute hydrofluoric acid to remove the underlying, exposed SiO_2 layer, timing the etch so as to completely free the torsional element while sufficient material beneath the support elements. The smallest lateral feature in the structure is 0.2 μm; hence much latitude exists for scaling these devices to even smaller dimensions.

The torsional resonator has a torsional spring constant κ_0 and torsional moment I. Its fundamental torsional resonance frequency is $\omega_0 = \sqrt{\kappa_0/I} = 2\pi \times 2.61$ MHz, with quality factor $Q = 6500$. The device includes two electrodes, one for sensing motion and the second for driving the mechanical response of the structure. One electrode is in the form of a metal loop tracing the paddle's outer boundary, while the opposing metal gate electrode is fixed to the stationary substrate, at a distance d from the paddle. The mutual capacitance between these electrodes is represented by the parameter C; electrostatic finite-element calculations were used to estimate this capacitance, giving a value of $C = 0.4$ fF. To measure a small charge, the gate electrode is biased by a charge q_0, which yields an electrostatic force between the electrodes of $F_E = q_0^2/Cd$. Small changes in the charge, δq, about the bias point q_0, alter the force by an amount $f = 2\delta q E_0$, where E_0 is the equilibrium electric field. This results in an effective torsional spring constant $\kappa_{\text{eff}} = \kappa_0 + g$, where $g = -\mathrm{d}f_\theta/D\theta = -2\delta q\mathrm{d}E_\theta/\mathrm{d}\theta$ (θ is the paddle torsion angle, E_θ the field component along the unit vector $\hat{\theta}$). This gives rise to a shift from the unperturbed resonance frequency, $\delta\omega/\omega_0 = g/\kappa_0$. Displacement detection is achieved using magnetomotive sensing, with the magnetic field lines in the plane but perpendicular to the torsion axis. An electromotive force is induced when the paddle rotates, due to the change in the magnetic flux linked through metal loop on the paddle.

To perform electrometry, the charge-induced shift in the resonance frequency due to the charge modulation of κ_{eff} is monitored, with the amplitude of vibration maintained in the linear response regime, with peak displacements of about 30 mrad.

Figure 9.26 shows the amplitude of the EMF measured as a function of drive current frequency, with fixed current amplitude. Each trace was taken with a different amount of dc voltage applied to the gate electrode. The range of voltages applied corresponds to a total change in the coupled charge of about 4×10^4 electrons.

A more sensitive detection technique is to fix the frequency of the drive current at the mechanical resonance frequency, and monitor the phase of the induced EMF relative to the phase of the drive current. The inset to Fig. 9.27 shows the frequency-dependent phase responsivity $R = \mathrm{d}\phi/\mathrm{d}q$, which is the measured variation of the EMF phase as a function of dc coupled charge. By measuring the spectral density of the phase noise in this mode of operation,

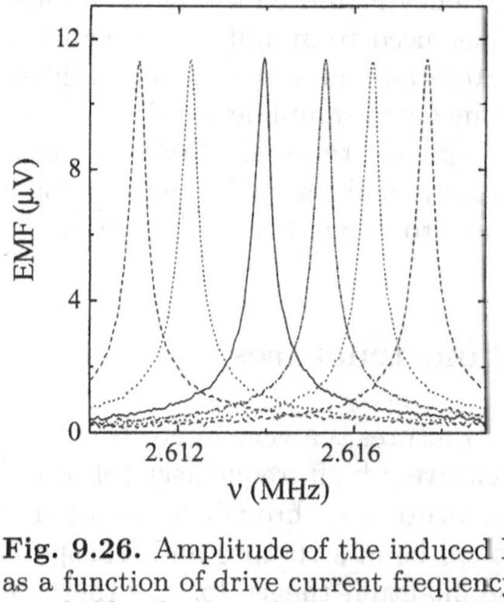

Fig. 9.26. Amplitude of the induced EMF across the small loop of the resonator, as a function of drive current frequency passed through the large loop. The charge coupled by the gate electrode was changed in steps for each trace, with a total range of 4×10^4 e in the coupled charge.

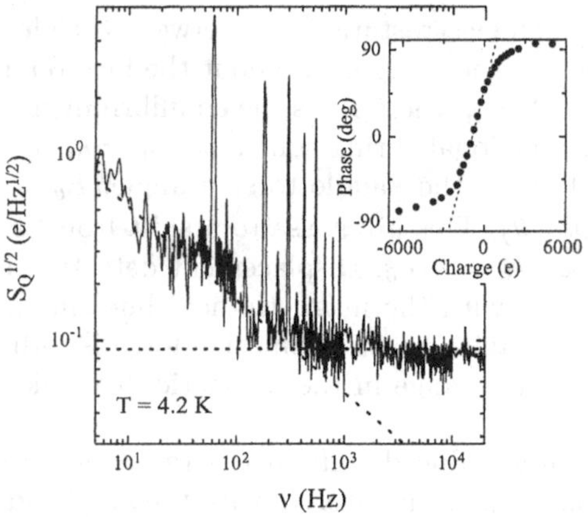

Fig. 9.27. Measured charge noise as a function of frequency. The noise, measured at the output of a mixer, is scaled to the single-sideband value. *Inset*: Measured phase variation as a function of dc coupled charge to the gate.

and using the measured responsivity, the equivalent spectral density of the charge noise is obtained (Fig. 9.27). The charge noise is seen to be dominated by $1/f$ noise at low frequencies, with a noise level of $S_Q^{1/2} = 0.6\ e/\sqrt{\mathrm{Hz}}$ at $\nu = 10$ Hz. Above a frequency of about 500 Hz the noise levels out to a white noise floor of $S_Q^{1/2} = 0.12\ e/\sqrt{\mathrm{Hz}}$. This level of charge noise is competitive with state-of-the-art electrometers based on cryogenically-cooled field-effect transistors. Measurements made in the absence of a magnetic field indicate that the $1/f$ noise emanates from the resonator or from the charge coupled to it, while the white noise floor originates from the room-temperature amplifier.

The measurement of the resonance frequency is limited by thermal noise in the resonator and the first-stage amplifier used to amplify the magnetomotive signal. With a drive amplitude of 30 mrad and a first-stage amplifier noise temperature of 300 K (roughly the value for the amplifier used in the experiment), a voltage bias of 10 V on the gate gives a thermally limited charge noise of $7 \times 10^{-4}\ e/\sqrt{\mathrm{Hz}}$. A state-of-the-art cryogenic amplifier, with a noise temperature of 10 K, would reduce this value to about $1 \times 10^{-4}\ e/\sqrt{\mathrm{Hz}}$.

9.5 Thermal Conductance in Nanostructures

The topic of thermal conductance in nanostructures is a very interesting one, as it probes the behavior of phonons at relatively high frequencies (of order $k_\mathrm{B}T/\hbar$ for thermal conductance at temperature T), through a nanostructure that can have a highly modified phonon band structure and transport properties. We have already discussed the quantum theory for the thermal conductance in such structures (Sect. 3.6.2).

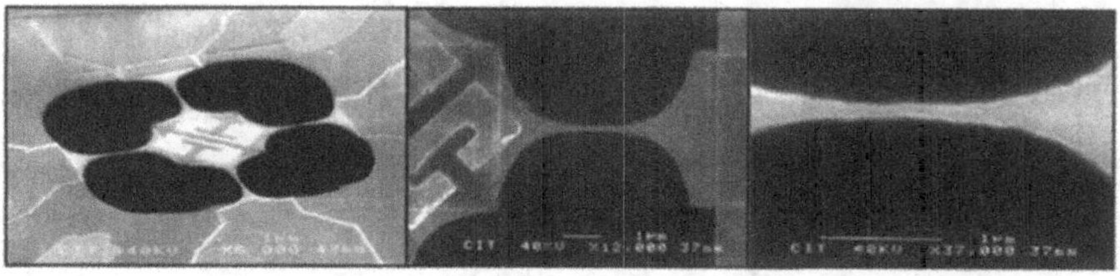

Fig. 9.28. Device used by Schwab et al. to measure the thermal conductance of nanostructure flexural beams. (**a**) Overall view of the device, showing the 12 wirebond pads that converge via thin-film niobium leads into the center of the device. (**b**) View of the suspended device, consisting of a 4×4 μm square. The "c"-shaped objects on the device are gold. (**c**) Close-up of one of the catenoidal beams, displaying the narrowest region which necks down to ≈ 200 nm width. Figure by permission of the authors [37].

The theory for the quantum of thermal conductance was verified in a beautiful experiment by Schwab and Roukes [37]; these authors fabricated a suspended nanoscale structure that consisted of a 4×4 μm^2 block of silicon nitride, 60 nm thick, suspended by four 60 nm thick legs with catenoidal waveguide shapes, with a minimum diameter of 200 nm. The catenoidal shape was used in order to achieve unit transmissivity from the suspended block to the bulk reservoirs [39]. Two metal resistors were patterned on the surface of the suspended block, one of which was used as a heater to generate a temperature gradient along the thin wires, and the second as a thermometer, where a very low noise amplifier, known as a dc SQUID, was used to amplify and measure the electrical Johnson noise in the resistor. From this measurement, the temperature of the metal strip be inferred. A set of SEM micrographs of the device is shown in Fig. 9.28.

Using this device, these authors were able to measure the thermal conductance of the four legs in parallel, and showed that at very low temperatures, below a few hundred mK, this conductance approached the quantized value (3.175), multiplied by sixteen (four modes in each of four suspending wires). In Fig. 9.29 we display the measured thermal conductance measured from 0.1 to 5 K.

At high temperatures, the normalized conductance G/G_Q was observed to have the expected T^2 dependence, while at lower temperatures the value approached the predicted value of $G/G_Q = 16$ (four legs, each with four modes, yielding 16 quanta of conductance). At moderately low temperatures, these authors found that the conductance fell slightly below the quantum limit, calculated assuming unit transmittance. This small discrepancy has since been explained [41] as being due to weak diffusive scattering in the wires, reducing the transmittance, an effect that gets smaller at lower temperatures, as observed in the experiment.

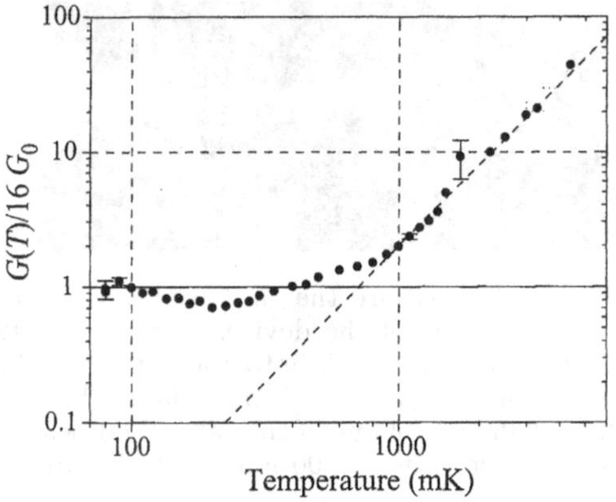

Fig. 9.29. Thermal conductance measured from 0.1 to 1 K. The conductance is plotted in units of the quantum of conductance $G_Q = \pi^2 k_B^2 T/3h$. At high temperatures the expected T^2 dependence is observed, while at lower temperatures the conductance is limited to the predicted quantum dependence. Figure by permission of the authors [37].

9.6 Coupling of Electron Transport and Mechanical Motion

Another interesting area of research involves coupling the mechanical motion of a nanostructure to electron transport through it. One experiment, by H. Park et al. in McEuen's group [103], measured the transport of electrons through a single molecule of C_{60} (fullerene). Detailed measurements of the voltage dependence of the transport rate indicated that the electrons, in passing through the C_{60} molecule, could gain or lose energy in quanta that corresponded to the lowest internal vibration mode of the carbon sphere, as well as to the center-of-mass motion of the molecule, which was trapped by van der Waals' forces at the surface of the gold electrodes used to make electrical contact.

The device was fabricated by first depositing thin gold electrodes on an oxidized Si wafer; the electrodes were patterned using electron beam lithography to a width of 100 nm. A dilute solution of C_{60} molecules suspended in toluene was then deposited on the surface of the sample and dried. The gold electrodes were then broken using a novel electromigration technique [104], creating ~1 nm gaps between the electrodes. Some of the samples prepared in this way had much larger electrical conductance than samples without the C_{60} deposition, which was interpreted as indicating that a C_{60} molecule was trapped in the gap between the two gold electrodes.

The samples with higher electrical conductance were then cooled to $T = 1.5$ K and careful measurements were made of the current I and the differential conductance $\mathrm{d}I/\mathrm{d}V$ as a function of bias voltage V between the

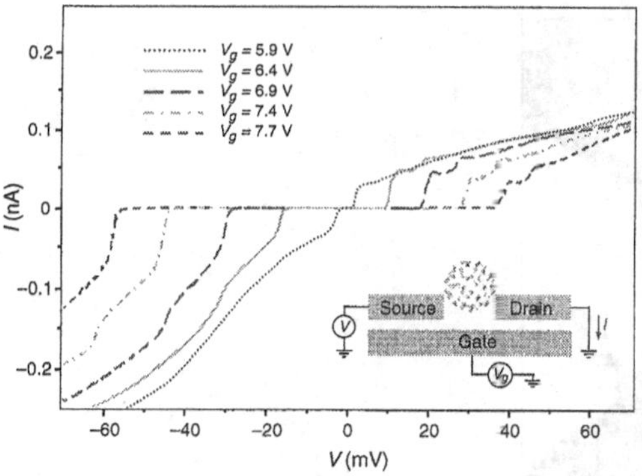

Fig. 9.30. Measured $I - V$ characteristics for a C_{60} molecule trapped between two gold electrodes. Different traces are for different gate voltages. Inset shows a schematic of the device. Figure by permission of the authors [103].

electrodes. In addition, a gate voltage was applied through the heavily doped underlying Si substrate, and its effect monitored as well. The samples displayed strongly non-linear current-voltage $(I - V)$ characteristics, which could be modulated with the gate voltage. The $I - V$ characteristic had a prominent band of voltages centered about $V = 0$ in which no current passed, outside of which the current increased with voltage. The size of this region was strongly dependent on the gate voltage, which could be used to reduce the zero-current region nearly to zero; this is shown in Fig. 9.30. The shape of the $I - V$ characteristic, and its dependence on gate voltage, is very similar to that seen in Coulomb blockade devices such as the single electron transistor [105]. The overall shape of the characteristics were therefore interpreted as being due to charging effects associated with the energy to put a single electron on the C_{60} molecule.

The differential conductance traces in addition displayed striking features at certain bias voltages, which were interpreted as due to special energy-loss processes, that become energetically available to the electrons when their injection energy, from one gold electrode into the molecule, is sufficiently above the extraction energy of the exit electrode. One set of features matched the estimate for the energy associated with a single quantized excitation of the center-of-mass motion of a C_{60} molecule in the van der Waals' interaction potential between the molecule and the adjacent electrode. An estimate of this interaction yields a spring constant of order $k = 70$ N/m, and coupled with the mass M of the molecule gives a resonant frequency $\omega/2\pi = \sqrt{k/M} \approx$ 1.2 THz (1.2×10^{12} Hz), corresponding to an energy $\Delta E = \hbar\omega = 5$ meV. Several of the devices measured displayed resonances on this energy scale.

In addition, resonances were seen at an energy of about 35 meV. These match a calculation for the fundamental vibrational mode of the C_{60} molecule [106], a breathing mode with a frequency of 8 THz. The feature in the

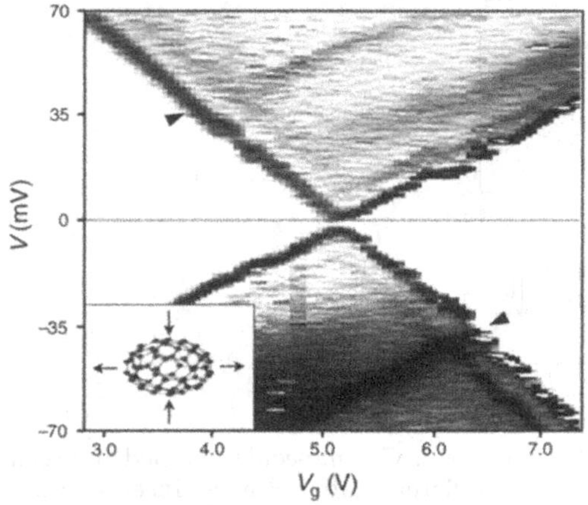

Fig. 9.31. Measured differential conductance characteristic, in greyscale, as function of bias voltage (*vertical axis*) and gate voltage (*horizontal axis*); white indicates zero conductance, black maximum conductance. The arrows indicate the feature associated with the excitation of a normal vibrational mode of the molecule, shown schematically in the inset. Figure by permission of the authors [103].

differential conductance map is shown in Fig. 9.31, along with a schematic of the form of the breathing mode.

These experiments provide one of the first demonstrations of the ability to couple electron transport to the mechanical motion of a molecule. The sensitivity of electron transport, and tunnelling in particular, should lead to other possible probes of the mechanical degrees of freedom.

A second experiment that has explored this aspect of mechanical structures has been the experiment of Erbe and co-workers [107, 108]. This involved the fabrication of a metallized cantilever, suspended between two gold electrodes; the cantilever was electrostatically driven so as to nearly contact each gold electrode in turn. An applied dc voltage then induced electron transport from one gold electrode to the other, via the metallized cantilever. A strong increase in electron current was observed when the electrostatic driving frequency matched a mechanical resonance frequency of the cantilever, due to the larger mechanical motion and thus closer approach to each electrode of the cantilever during each cycle of the motion. The structure was fabricated from a SIMOX substrate, on top of which gold electrodes were defined using electron beam lithography. A second layer of e-beam lithography was then used to define a masking electrode for etching the substrate; the masking material was aluminum. A sequence of dry reactive ion etch steps and wet etches to release the structure were then performed to complete the fabrication. A SEM image of their device is shown in Fig. 9.32. The processing was carefully designed to allow clean tunnel contacts between the gold on the drain and source electrodes as well as on the metal island on the end of the cantilever.

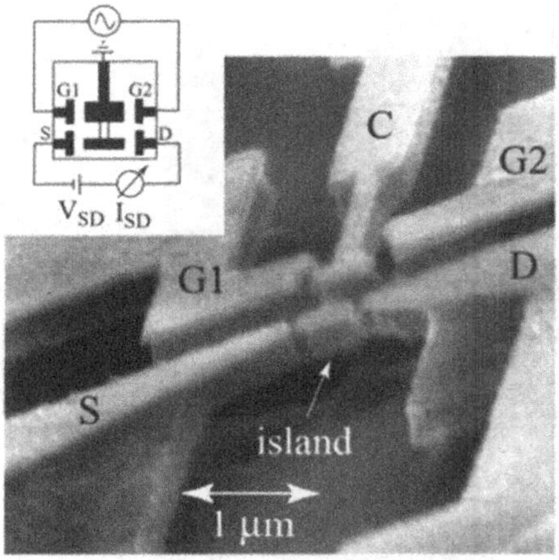

Fig. 9.32. Electron micrograph of the electron shuttle. The cantilevered beam is electrostatically driven by voltages applied to the gates to the gates G_1 and G_2. Electron transport is observed from source (S) to drain (D) through the island at the end of the cantilever. The island is electrically isolated. The inset shows a simplified circuit diagram indicating the source-drain voltage and the driving voltage. Figure by permission of the authors [108].

Electrostatic voltages applied to the electrodes G_1 and G_2 drive the cantilever mechanical motion, and the metal island on the end of the cantilever then approaches each of the source (S) and drain (D) electrodes successively. A dc voltage applied from source to drain then generates a dc transport current via the metal island; for very close approach, the effective tunnel resistance of either electrode to the island will be sufficiently small that one or more electrons is transferred, per mechanical cycle, through the device.

In Fig. 9.33 we display the measured dc current through the device as a function of the electrostatic drive frequency; several peaks are seen in the electron current, corresponding to various mechanical modes of the structure, for which the cantilever-electrode approach is sufficiently large to allow current to flow. The spectrum is complex due to the large number of possible modes for this complex structure, including a number of torsional modes, flexural modes, as well as modes involving the nominally fixed counterelectrodes.

In the absence of the driving gate voltages, it is thought that transport through the device would be dominated by the Coulomb blockade of current onto and off of the center island; electron transport should therefore occur one electron at a time. The large driving voltages needed to initiate the motion of the cantilever however effectively removes the blockade through the gate-island intercapacitance, allowing tens or hundreds of electrons to be transported at a time. Detailed simulations including both the source-drain voltage as well as the effect of the capacitance appear to support this hypothesis [108].

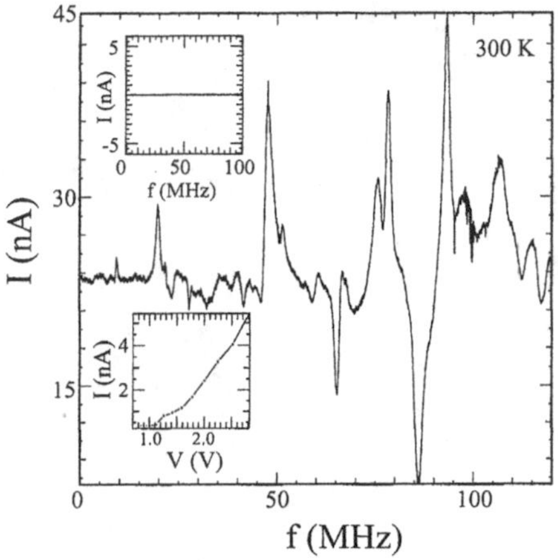

Fig. 9.33. Measurement of the tunnel current from source (S) to drain (D) at room temperature. Finite element simulations show a similarly complex spectrum of mechanical resonances. *Upper inset*: Measurement on a sample with identical layout, where the cantilever was prevented from moving by small silicon connections to the bulk material. No capacitive cross talk or conduction through the substrate could be observed. *Lower inset*: the current in one of the peaks is shown as a function of the driving voltage V. Figure by permission of the authors [108].

9.7 Magnetic Resonance Force Microscopy

One of the most exciting and visionary projects that is underway involving the use of nanomechanical structures is the attempt to develop a microscopic imaging technique known as "magnetic resonance force microscopy". The ultimate goal of this development project is to enable imaging of atoms on a surface, complete with identification of the nuclear species: in other words, single atom identification. The potential applications of such a tour-de-force are enormous, including on-the-spot sequencing of biological proteins and nucleic acids.

A large fraction of the various isotopes of the atomic elements have a non-zero nuclear magnetic moment $\boldsymbol{\mu}_N$; in addition, the electrons in an atom each have an electronic magnetic moment $\boldsymbol{\mu}_B$ (the Bohr magneton). The nucleus has a dimensionless spin $\boldsymbol{I}$, with spin angular momentum $\hbar\boldsymbol{I}$. The nuclear magnetic moment is related to the spin by $\boldsymbol{\mu}_N = \hbar\gamma_N\boldsymbol{I}$, where γ_N is the nuclear gyromagnetic ratio.

The magnetic moment is free to re-orient in space. If a uniform magnetic field $\boldsymbol{B} = B_0\hat{z}$ along the z-axis is applied to such an atom, the spin state with the lowest energy is that where the spin is aligned opposite $\boldsymbol{B}$, i.e. $\boldsymbol{I} = -I\hat{z}$. The other allowed states for the spin are those for which the projection I_z of the spin along the z-axis is quantized, only taking on values from the set of values $I_z \in \{I, I-1, \ldots, -I+1, -I\}$. The energies E of these states are evenly spaced, given by $E = \hbar\,\gamma_N\,I_z\,B_0$, and there is correspondingly a single excitation, or precession, frequency, $\omega_P = \Delta E/\hbar = \gamma_N B_0$. This is also known as the Larmor frequency.

Electrons behave in the way as nuclei, with the gyromagnetic ratio γ_e and magnetic moment μ_B. The spin $\boldsymbol{S}$ of the electron is $\pm\frac{1}{2}$. Electrons bound in orbits about an atom however can also have a significant *angular momentum* $\boldsymbol{L}$ in addition to their spin $\boldsymbol{S}$, so a description of the magnetic response is somewhat more complicated.

For atoms in isolation, the value of the nuclear gyromagnetic ratio γ_N, and that for the nuclear moment μ_N, are precisely known. For example, the proton has $\gamma_N = 2.675 \times 10^8$ Hz/T and $\mu_N = 1.411 \times 10^{-26}$ J/T. Similarly, for the electron, $\mu_B = 9.274 \times 10^{-24}$ J/T, known as the Bohr magneton. Note that the electron has a magnetic moment roughly 1000 times that of the proton; the electron spin Larmor frequencies are correspondingly a thousand times higher (in the same magnetic field), as is the magnetic signal.

A measurement of the nuclear Larmor frequency in a known magnetic field allows identification of the atomic species: The precession frequencies of a nucleus are only very weakly affected by the environment in which the nucleus finds itself, so even in a liquid or solid the frequencies are very close to those for the isolated atom. The weak interaction with the environment means that the nuclear spin can remain unperturbed for a very long time; the Larmor frequencies are therefore very well defined, with very small linewidths, of order of $10^{-3} - 10^{-6}$ Hz, the latter occurring at low temperatures. This very precise frequency selectivity forms the basis of nuclear magnetic resonance (NMR), a very well developed experimental technique whose use has elucidated much about the structure of different molecules and solids. An excellent description of NMR, both theoretical and experimental, can be found in the monograph by Slichter [109].

Nuclear magnetic resonance forms the basis for magnetic resonance imaging (MRI), in which the use of controlled magnetic field gradients allows the extraction of e.g. the hydrogen concentrations at different locations in an object. From such a measurement, the internal composition and chemical makeup of the object can be reconstructed non-invasively. This technique can achieve roughly 1 μm spatial resolution, although as the apparatus used to perform MRI is developed for microscopic techniques, this spatial resolution will doubtless improve.

We briefly describe a one-dimensional version of MRI. A magnetic field, with a gradient $\boldsymbol{B} = (B_0 + bz)\hat{z}$, is applied to the object to be imaged. Here b is the field gradient, in units of T/m. At each point z, a particular type of atom (typically hydrogen) will have a *local* precession frequency $\omega_P(z) = \gamma_N(B_0 + bz)$. A receiver for the NMR signal, tuned to a particular frequency ω_0, will only pick up the signals from atoms with precession frequencies near ω_0, and therefore will only detect atoms with z coordinate near $z_0 = (\omega_0/\gamma - B_0)/b$. If the receiver has a input frequency acceptance window, or bandwidth, $\Delta\omega$, it will detect the atoms centered at this height, in a slice of thickness $\Delta z =$

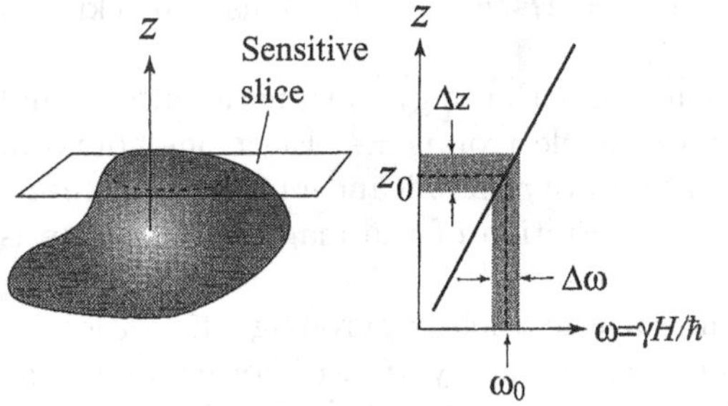

Fig. 9.34. Schematic operation of a one-dimensional magnetic resonance imaging system. The magnetic field gradient generates a precession frequency gradient. A receiver picks out one frequency ω_0 in a band of width $\Delta\omega$, allowing detection of the atoms in a slice at z_0 of width Δz.

$\Delta\omega/\gamma_N b$. The strength of the detected signal, as it is proportional to the number of atoms that generate a signal in the receiver's acceptance window, is then proportional to the number of atoms in this slice (see Fig. 9.34). The uniform field B_0 can then be changed slightly, allowing detection in a slice at a somewhat different location z_0', and by measuring the signal as a function of B_0 the composition of an extended solid along the z axis can be mapped out.

The basic idea for magnetic resonance force microscopy is to extend by several orders of magnitude the spatial resolution of magnetic resonance imaging, and ultimately to achieve single atom resolution with single atom detectability. This is most likely to be accomplished first using electron spin resonance (ESR) rather than nuclear magnetic resonance, due to the much larger electron magnetic moment. However, for element identification, single-atom nuclear imaging is much preferred.

The first proposal for achieving single-atom sensitivity, was published by Sidles [110, 111]. Sidles' approach is to use the very strong local magnetic field gradient that can be generated using a very sharp nanoscale magnet, and couple this with the extremely high force sensitivity achievable using micro- or nanomachined cantilevers, to create a scanned surface probe that can image and detect the local magnetic signal. There are two approaches that can then be used: In one approach ("sample on cantilever"), the sample to be measured is placed on the end of a highly flexible cantilever, and the large field gradient is provided by a separate magnetic tip, which is then scanned over the sample. In the second approach ("magnet on cantilever"), the magnetic tip is fixed to the cantilever, and scanned over a rigidly-held sample. In either case, the sharply pointed magnet and the sample each feel an interaction force, due to the precessing magnetic moments in the sample, and cause the cantilever to flex. The magnet's local magnetic field gradient selects a small volume from the sample that is in resonance with the cantilever, analogous to that in magnetic resonance imaging (see Fig. 9.35). The deflection of the cantilever is

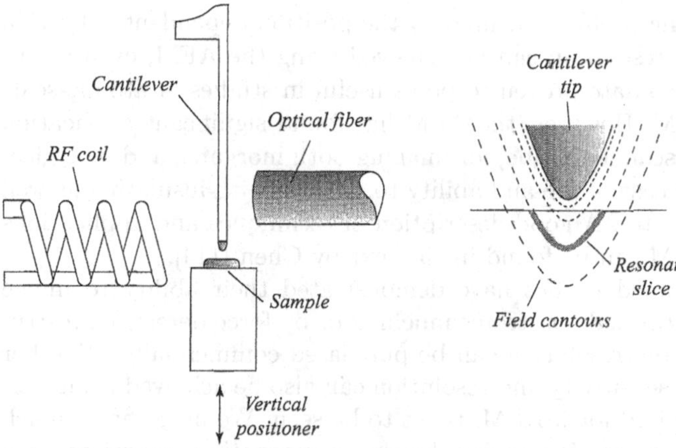

Fig. 9.35. In the magnet-on-cantilever approach, a sharp magnetic tip on a cantilever generates a local magnetic field; in a small volume slice a certain distance from the tip, the total field is such that atoms within that volume have the correct resonance frequency to couple to the cantilever, generating a locally-detected field. On the left we show a system design, adapted from Rugar [113]; this shown the vertical cantilever geometry favored by Rugar's group. On the right we show a detail of the magnetic tip and corresponding resonant slice. The cantilever tip is then scanned over the surface, and the signal strength within the resonant volume is mapped out.

monitored and its response mapped out as a function of the relative position of the sample and the magnetic tip, as one is scanned over the other. An excellent review of the topic can be found in Sidles' paper [112].

Achieving the dual goals of single atom detection and single-atom spatial resolution is very challenging. Single atom imaging has been achieved using both the scanning tunnelling microscope (STM) and the atomic force microscope (AFM). In the STM, a very sharp conducting tip is brought in close proximity (1 nm) to the surface of a conducting substrate. A voltage applied between the tip and substrate then induces an electrical tunnel current, which is exponentially sensitive to the tip-surface spacing. Scanning the tip over the surface then allows measurements of the local surface height, known as profilometry; in very clean samples, typically in ultra-high vacuum applications, single atoms can be imaged and the surface reconstruction studied. More recently, the STM has been used to assemble multi-atom structures. One disadvantage of the STM is that it can only study conducting substrates.

The atomic force microscope provides an alternative approach to single-atom imaging. The AFM comprises a very sharp tip, mounted on the end of a flexible cantilever. The cantilever is (in one mode of operation) in physical contact with the sample, and its deflection monitored by reflecting a laser signal off the back of the cantilever. The tip is scanned over the sample surface, and the resulting profile measured by the position-dependent deflection of the laser. Atomic resolution can be achieved using the AFM, even in air; however, it has not to date proven to be as useful in studies of atomic-scale structure as the STM. However, the AFM has found significant application as a high resolution scanned probe, for imaging both inorganic and biological materials, due to its ease of use and ability to operate with insulating as well as conducting substrates. A good description of techniques and applications of the STM and AFM can be found in the text by Chen [114].

We see that scanned probes have demonstrated their ability to image single atoms, either through electron tunnelling or by force detection; instruments with single-atom resolution can be purchased commercially. Whether this level of imaging sensitivity and resolution can also be achieved using magnetic signals, as desired for MRFM, is yet to be seen. We note that a much simpler approach to magnetic imaging, known as magnetic force microscopy (MFM), has been successfully developed; in this imaging technique, an AFM tip is coated with a thin film of magnetic material and the tip then scanned just above the surface of a magnetic substrate. The AFM tip is deflected due to the tip's interaction with the substrate, with a deflection proportional to the strength of the magnetic force (primarily due to magnetic field gradients). The strength of this deflection is mapped out as a function of position, allowing an image to be built up that relates to the microscopic magnetic composition of the substrate. This is primarily a non-resonant technique: The magnetization of the AFM tip and the substrate remains nominally constant while the image is being scanned, in contrast to magnetic *resonance* microscopy, where the magnetization oscillates at the Larmor frequency, and the strength of the oscillation (and its frequency) are mapped out. It is very unlikely that MFM will achieve single-atom sensitivity.

In order for MRFM to achieve single-atom sensitivity, there are a number of significant challenges that must be met. For example, achieving sufficient force resolution that the magnetic signal from a single atom can be detected is not trivial. The highest force resolution achieved to date is that by Mamin and Rugar [115], at a temperature of 100 mK. This was achieved using a cantilevered beam of silicon with length of 260 μm, width of 3.9 μm and thickness of 0.29 μm. The effective spring constant was 0.26 mN/m and the cantilever resonance frequency was measured to be 4.98 kHz, with a quality factor at the lowest temperature of about 150,000. The corresponding mechanically-determined frequency bandwidth was approximately $\Delta\omega = \omega_0/Q \approx 3 \times 10^{-2}$ Hz. The cantilever motion was detected using a laser interferometer, in which the cantilever formed one side of a Fabry–Perot resonant cavity. The thermal conductivity of the silicon is so poor at low temperatures that significant heating was observed with laser powers of only a few nanowatts. The thermally-induced mechanical motion could be detected down to the lowest operating temperatures; the lowest effective temperature, inferred from the measured thermal noise, corresponded to a resonator temperature of 220 mK, with the physical temperature measured to be 46 mK. The effective integrated displacement noise was measured to be 0.6 Å, corresponding to a force noise of $S_f^{1/2} = 8.2 \times 10^{-19}$ N/$\sqrt{\text{Hz}}$. Averaging the response of the cantilever over one second therefore allows a force sensitivity of better than 10^{-18} N. This level of force sensitivity should allow the detection of single-atom NMR signals, if a sufficiently large magnetic field gradient can be achieved.

The magnetic field gradient obtained at the tip of the cantilever is critical because it both determines the spatial resolution obtained, and the force of the interaction between a precessing nuclear magnetic moment and the cantilever. Given a force noise S_f and a magnetic field gradient b, the minimum number of detectable spins is given by [112]

$$n = \frac{(S_f \Delta\omega)^{1/2}}{b\,\mu} \frac{k_B T}{\mu B}, \tag{9.57}$$

where $\Delta\omega$ is the bandwidth (frequency acceptance window) of the cantilever, and μ the magnetic moment to be detected. The second term gives the fractional spin alignment: At the relatively elevated temperatures and weak fields for these experiments, only a small fraction of the spins will be aligned in the magnetic field B; the spin-aligned fraction is given by this factor. Detecting a single proton with the cantilever described above, with a force noise $S_f^{1/2} = 10^{-18}$ N/$\sqrt{\text{Hz}}$, and a bandwidth of $\Delta\omega/2\pi = 0.03$ Hz, requires a magnetic field gradient $b \approx 3 \times 10^7$ T/m, assuming complete spin alignment. This is quite a large field gradient; to achieve single electron detection, the corresponding gradient is $b \approx 5 \times 10^4$ T/m. The latter figure has recently been achieved by experiments in Sidles' group [116], as well as by Rugar [117], so these two extremely challenging issues appear resolvable, at least for single-electron detection.

Other issues of importance, that are presently under investigation, include the effects of magnetically-induced fluctuations and dissipation [118], as well as the effect of electric field fluctuations between two conducting surface in close proximity to one another [119]. These effects may play a role in determining the ultimate noise limits and place an upper limit on the value of Q in MRFM experiments; as these tend to increase the magnetic noise as well as reduce the effective quality factor, thereby impacting MRFM measurements, it remains to be seen whether the complete experiment, as shown in Fig. 9.35, can succeed in detecting the signal from a single electron or (more challenging still) a single proton.

The results from a number of complete experiments, as opposed to the investigative work described so far, have been published. These experiments all use cantilevers with mechanical resonance frequencies in the few-kHz range, while the ESR and NMR precession frequencies to be detected are typically in the range of $10^7 - 10^{11}$ Hz. A major design choice is how to couple the much higher frequency Larmor signal to the low frequency cantilever resonance. The first published experiment [113] focussed on ESR, also known as electron paramagnetic resonance, and used a sample mounted on a audio-frequency cantilever, with a polarizing permanent magnet placed near the sample. The sample was a material known as DPPH (diphenylpicrylhydrazil), a strong paramagnetic organic insulator. In addition to the constant field B_0 from the permanent magnet, an externally-applied rf field B_1 was used to excite the electron spins in the sample. The magnetization could be modulated by varying the field B_0, or by varying the amplitude and frequency of the rf field B_1. The magnetization has a nonlinear dependence on magnetic field, so by modulating one of these fields at, for example, twice the resonance frequency of the cantilever, the cantilever motion could be excited. This eliminates the possibility of the cantilever being excited by the field itself (as the modulation frequency will not couple directly to the cantilever), and therefore selects out the subharmonic variation of the magnetization. The authors reported that the best results were found by modulating B_0 at twice the cantilever resonant frequency. Using this approach, the minimum detectable signal corresponded to about 1.6×10^6 electron spins, at a Larmor frequency of 800 MHz.

This technique has been extended to allow scanned images to be formed, instead of just a demonstration of signal detection. A second version of the experiment [120, 121] involved a geometry where a small magnet was mounted on the end of the cantilever, and the sample placed on a piezoelectric tube scanner to allow precise sample-tip position control. The magnetic tip was made by gluing a 55 μm sphere of nickel on the end of the cantilever, as in Fig. 9.35. These experiments used a technique known as "cyclic saturation", in which the electrons were periodically polarized by slowly modulating the frequency of the microwave frequency field B_1, which excites the electron spins near the Larmor frequency, as determined by the local field. The periodic modulation in B_1 was chosen to match the cantilever resonance frequency, thereby exciting the cantilever resonance. Imaging was successfully demonstrated, scanning the sample under the fixed cantilever, using both electron paramagnetic resonance in DPPH, as in the earlier experiment, as well as using ferromagnetic resonance (FMR) in yttrium iron garnet (YIG). A spatial resolution of about 5 μm was demonstrated, with spin number sensitivity comparable to the "sample on cantilever" geometry.

A more recent result was recently published [117], which used a roughly 1 μm diameter PrFeB particle glued to the end of a cantilever, which was then milled down using a focussed ion beam to maximize the field gradient b. The cantilever, with dimensions of $190 \times 3 \times 0.85$ μm^3, had a resonance frequency of 21.4 kHz and a quality factor that ranged from 2000 to 200,000, depending on the tip-sample distance. Its displacement was measured using optical interferometry. The sample studied was silica glass, irradiated to form silicon dangling bonds, which have a strong paramagnetic response. The careful treatment of the permanent magnet allowed a field gradient of 10^5 T/m to be achieved, and the authors employed a number of special noise-reducing techniques to maximize the signal-to-noise in the measurements, including a careful timing of the cantilever resonance with the microwave tipping pulse (used to excite the spins). The authors report that they can detect the ESR signal from as few as 100 Bohr magnetons, and were able in this experiment to extract the electron spin relaxation time, as well as investigate the effects of the permanent magnet on spin relaxation.

A number of other groups are developing MRFM technology as well. Sidles' group has made a number of advances, as has a collaboration between Hammel and Roukes. For example, the latter collaboration has made a beautiful experiment imaging ferromagnetic resonance in a single crystal of YIG [122], achieving 15 μm spatial resolution, and were able to detect the magnetostatic resonance modes in the YIG crystal [26].

The significant progress achieved over the past ten years, as evidenced by comparing the results of [113] with those of [117], holds great promise for the potentials of this technique. This very challenging effort is being pursued by a number of excellent groups, and the culmination of the effort may prove that atomic level imaging, of either electron spin or nuclear spin resonance, will prove possible. If this is achieved, the potential impacts of such a powerful tool are quite significant, and would prove an enormous advance in microscopy with direct applications in biomolecular as well as the physical sciences.

10. Nanostructure Fabrication I

The techniques used to fabricate nanoscale structures are continually evolving, and comprise methods that have been developed for over three decades, as well as very new and promising approaches whose potential has not yet been fully realized. Any discussion of nanofabrication techniques will therefore soon be outdated, and can only hope to illustrate the range of possibility. In this and the next chapter, we will focus on a description of the more conventional techniques, outlining their use and their limitations, and in addition we will briefly mention some of the other techniques that are under development. These two chapter therefore cover a significantly different topic from the first nine chapters in this book; here the information discussed is primarily technical, specific to the problem of fabrication and not directed towards developing a theoretical understanding of mechanical structures. It is provided here for the use of researchers interested in developing an understanding of what is involved in fabricating nanostructures, and possibly employing some of these techniques in their own research.

Nanostructures are created in a sequence of fabrication steps, each step typically starting with a definition of a pattern in a temporary material known as a *resist*. The patterning of the resist is followed by either removal of existing material, under the resist, or addition of new material, on top of the resist, in either case masked by the pattern in the resist. A series of these steps are combined to produce a completed device. Much effort is devoted to developing new processes that might offer improved execution speed, repeatability, and resolution, with the hope of achieving a level of control that approaches the molecular scale. At present, however, the conventional approaches continue to offer more flexibility and control, and due to the cost and complexity of semiconductor processing equipment and recipes, revolutionary advances are rare.

We have divided this part of the text into two chapters; this chapter deals with general techniques, both conventional and more development-oriented, and the next chapter deals with the applications of these techniques for structure fabrication. The first section of this chapter outlines the steps required to fabricate a simple beam resonator, using most of the techniques described later; this is followed by a section on lithographic principles, describing the use of photolithography and electron-beam lithography for pattern definition.

We also describe the use of focussed ion beams, neutral atomic beams, selective patterning using scanned probes such as the atomic force and scanned tunnelling microscopes, and soft lithography techniques. The next section discusses material deposition using the conventional thin-film technologies of evaporation, sputtering and chemical vapor deposition. In the final section we discuss etching through the use of wet and dry chemistries.

Nanostructure fabrication is primarily a planar process, and relies on techniques developed by the semiconductor industry for the fabrication of integrated circuits, both from silicon and GaAs substrates. These techniques are fully described in texts on integrated circuit processing, so we only provide outline descriptions of the more generic approaches. Excellent references in this regard are the text by Brodie and Muray [85], and the monograph by Ghandhi [84].

10.1 Materials

The materials used for nanofabrication include insulators, semiconductors and metals. Because of the prevalence of techniques for depositing and etching silicon and gallium arsenide, we focus primarily on these materials. Silicon can be used either in single crystal form, bulk or in a layered substrate (a *heterostructure*), or in polycrystalline films known as *polysilicon*. Gallium arsenide is primarily used in layered structures, grown epitaxially by molecular beam epitaxy or chemical vapor deposition. The silicates, silicon oxide and silicon nitride, are very useful both for their electrical and their mechanical properties; these are almost exclusively used in polycrystalline thin films. Metals range from the noble metals such as gold and copper, to refractory metals such as W and Nb, and include ferromagnets such as Ni, Fe and Co; polycrystalline metal films of many different types can be used, almost without restriction.

The techniques we describe here apply either generically to these materials, such as the lithographic techniques discussed in the next section, or are very specific to a particular material, such as wet etching or dry etching recipes.

10.2 Lithographic Principles

As mentioned above, most processing steps use a temporary layer of a material known as a resist (due its ability to withstand certain acids as well as plasma-based etching processes). The most popular method for patterning a layer of resist, known as *photolithograhy*, employs ultraviolet light, with a wavelength less than 500 nm. This allows patterning to size scales of order 0.5 μm using fairly simple equipment (the present state of the art can achieve under 0.1 μm wide lines, using however quite sophisticated equipment). In addition, a relatively low-cost method for achieving sizes below 0.1 μm employs low to moderate energy electrons tightly focussed in a charged beam, in a technique known as electron beam lithography. In this section we outline the basic principles behind these standard lithographic techniques, and touch on some newer approaches.

10.2.1 Photolithography

Photolithography is at present the principle tool for patterning structures with minimum dimensions above 1 μm, and in advanced applications has been demonstrated at size scales as small as 0.05 μm. It is enormously popular in industrial applications because of its high throughput in comparison to other technologies, and the lack of suitable alternatives has driven the semiconductor industry to continued improvements in minimum feature size and maximum die size (roughly, the field-of-view for the optical system used for lithography)

Photolithography relies on the use of a photosensitive organic film know as *photoresist.* The photoresist, dissolved in a solvent, is puddled onto the substrate to be patterned, and the substrate then spun at a few thousand revolutions per minute to spread the photoresist evenly over its surface, a process known as spin-coating. The solvent is then driven off by baking the coated substrate for a few minutes.

The coated substrate is then selectively exposed, typically by passing ultraviolet light through a *photomask.* A photomask can be made from an optically flat plate of soda lime glass or quartz, on which a chrome or other UV-opaque film is deposited. This film is then patterned using a separate photolithographic process, resulting in a plate with UV-clear and opaque regions. These photolithographic masks may be reused many times.

The photomask and the substrate are placed in registry on a mask aligner, using the substrate boundaries for the first lithographic step, or registry marks from previous lithographic layers for the later steps. The two main variants in mask aligner design are *contact* and *projection* aligners. In contact lithography systems the photomask and the substrate are placed in direct physical contact, so that a 1:1 copy of the photomask pattern is transferred to the photoresist. The entire substrate is typically exposed at one time, so contact lithography is also known as *whole-wafer* lithography. In Fig. 10.1 we illustrate the principle.

In projection lithography, the UV illumination is passed through the photomask, and the optical image of the photomask is subsequently passed through a reducing optical system before exposing the substrate; typically projection mask aligners use 4:1 or 10:1 image reductions. Photomasks used for projection lithography are somewhat less demanding to manufacture than for contact lithography, as the features are proportionally larger. Projection systems are also known as *step-and-repeat* tools, as typically the substrate is

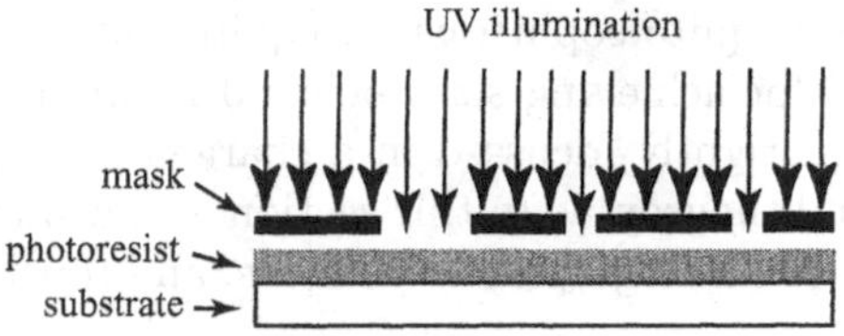

Fig. 10.1. Schematic drawing illustrating contact lithography.

exposed in an array of distinct areas known as *dies*, typically of order 1 cm in size, and each die is exposed separately, and the substrate then moved, or stepped, to the next die location. This process allows separate focus and alignment on each die, significantly increasing the yield for each process step. The projection method is shown schematically in Fig. 10.2.

Mask aligners based on contact lithography are much lower cost than projection aligners. Contact aligners typically use mercury lamps as the UV source, using either the 465 nm or 365 nm emission wavelengths. Feature sizes are limited to of order 1–2 μm and above, although with some care features as small as 0.5 μm can be achieved over small areas. Alignment registry is typically of order 1 μm.

Projection aligners range in sophistication, based on the quality of the optical system, the wavelength of the UV source, and the precision of the registry system. At present, the most common industrial projection aligner is the I-line aligner, so named because it uses the $\lambda = 365$ nm "I" wavelength of a mercury lamp. Projection aligners are primarily limited by diffraction effects; in the far field of the photomask, diffraction limits the linewidth to approximately $\lambda/2NA$, where NA is the numerical aperture of the optical

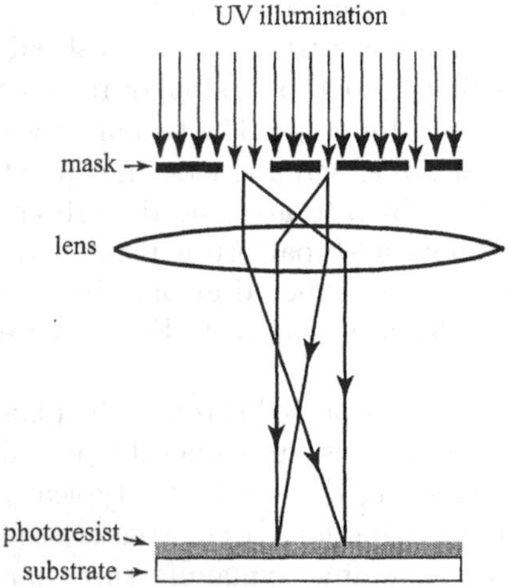

Fig. 10.2. Schematic drawing illustrating projection mask aligner lithography.

system and λ the wavelength of the illumination source. I-line aligners are therefore limited to approximately 0.35 μm linewidths, as most aligner optical systems have NA $\approx$ 1. Aligners using KrF (243 nm) and ArF (193 nm) lasers have been developed, and are now commercially available. Extreme UV systems with even shorter wavelengths are under development; a 10:1 projection system using 13.6 nm radiation has been built capable of exposing a grating with 100 nm features [123]. Researchers have also explored the use of x-rays, although sources with sufficiently large illumination aperture and collimation are extremely expensive to build; the best such source is an electron synchrotron, although plasma sources are being pursued as well. A resolution of 30 nm has been achieved using contact masks and x-rays from a synchrotron [124]. Other approaches, especially non-optical ones, are also being pursued. One such example is the SCALPEL system, which uses projected electron beams; see below.

The minimum linewidth that can be achieved using projection mask aligners is continually being reduced; industry predictions call for 0.10 μm by 2007, and 0.07 μm by 2010; whether these resolutions can be achieved using production-level optical systems is yet to be seen.

The final step in the resist process is resist development, where the exposed (or unexposed) resist is removed by a wet solvent. The solvents used are typically weak basic solutions, and vary from resist to resist. Developing is done by submerging the resist-coated substrate in the developer, possibly diluted with water to slow the dissolution rate, and once developing is complete, the substrate is rinsed with water to remove all traces of the developing solution. After blow-drying the substrate with dry nitrogen, the process is complete.

Photoresists. Photoresists are generally classed as *positive* or *negative* resists. A positive resist is one whose solubility in the developer increases when it is exposed to UV light, while the solubility of a negative resist decreases. Some resists may be transformed from positive to negative (or vice-versa) by exposure to ammonia gas in a baking process, or by pre-exposure to a uniform source of UV light followed by baking prior to the patterned exposure; the specific processes may be obtained from the photoresist manufacturers.

Resists are used in two primary applications: as *etch masks* and as *liftoff masks*. In an etch mask process, the resist is patterned and the substrate is selectively etched through the openings in the resist. An etch mask process is shown in Fig. 10.3.

In a liftoff process, material is deposited (by evaporation or other deposition process) on top of the patterned resist. Openings in the resist allow the material to attach itself to the substrate, while elsewhere it remains on the resist surface. Removal of the resist then takes the deposited material on its surface with it, leaving material where the openings in the resist were. This process, using a *resist bilayer*, is shown schematically in Fig. 10.4. This can also be accomplished by using a chemical soak to harden the top surface

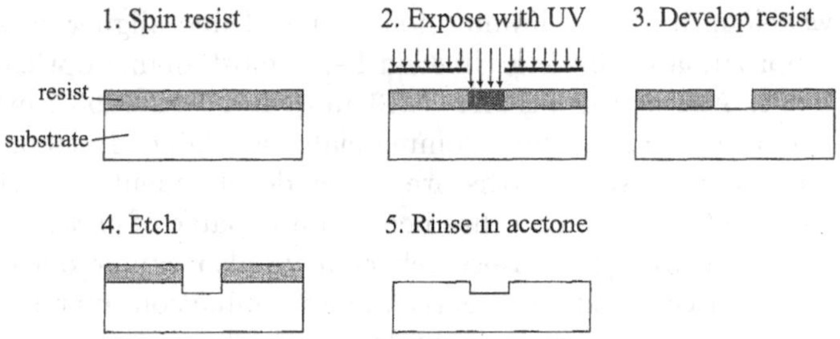

Fig. 10.3. Schematic of a photoresist etch process step.

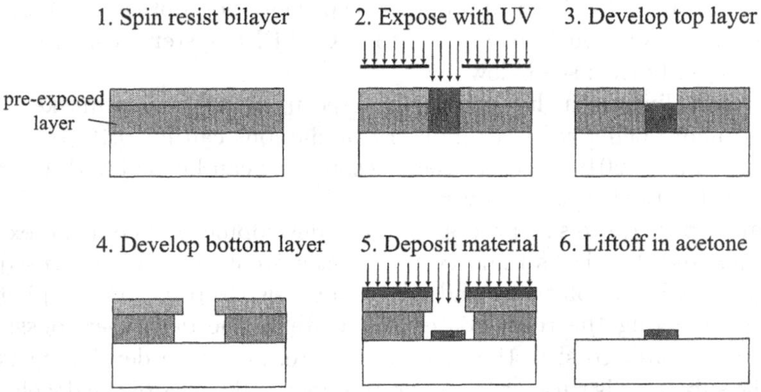

Fig. 10.4. Schematic drawing of liftoff process using a resist bilayer. The bottom resist layer is pre-exposed with a blanket exposure in order to achieve the undercut desired.

of a single resist layer before developing; this can be done with toluene or chlorobenzene (a 5 minute soak is usually sufficient), and generates a fairly well-defined undercut.

Chemical resistance is often an important criterion in selecting a resist, especially when using the resist as an etch mask. Most resists can withstand fairly strong acids, but are less resistant to basic solutions. Many resists are more resistant to buffered solutions, so that when etching SiO_2, for example, the resist will last longer when using buffered hydrofluoric acid (a mixture of hydrofluoric acid and ammonium fluoride) than when using hydrofluoric acid diluted to yield the same etch rate. In situations where a basic etchant must be used, such as when etching Si with potassium hydroxide (KOH), an intermediate mask is often required. Deep wet etching of Si is typically done by first depositing a layer of SiO_2 or silicon nitride (Si_3N_4), patterning this layer using photoresist and either a dry etch (see below) or an acidic wet etch, and then using this base-resistant mask to etch the underlying Si.

Photoresists can also withstand fairly extended dry etch processes, such as reactive ion etching (RIE) and ion beam etching (IBE). These processes are described later in this chapter; they are used to remove substrate material using a combination of chemical and physical processes, and photoresist can be used to restrict their action to defined areas on the substrate.

10.2.2 Electron Beam Lithography

Electron beam lithography provides the lowest cost approach to linewidths significantly less than 1 μm. Systems range from scanning-electron microscopes converted for direct write under the control of a computer, to full lithographic systems with automatic pattern alignment and stitching capability to write over large areas. Electron-beam lithography remains primarily a research and development tool rather than a production tool because it is a *serial-write* process rather than a parallel process, as the electron beam exposes only a single feature at a time rather than exposing an area. This significantly reduces the throughput in comparison to optically-based approaches, and therefore increases the cost of use in a production setting. Certain limited applications, such as direct-write of high density disk drive read heads, do however use electron-beam lithography.

The basic principles are very similar to those of photolithography; the electron beam replaces the UV illumination as the exposure source, and different types of resists are used than for optical processes, but otherwise the concepts remain the same.

The beam of electrons can be created using a number of different sources, ranging from cold Schottky emitters for field-emission systems to tungsten (W) or lanthanum hexaboride (LaB_6) for a thermionic emitter. Schottky emitters use a highly sharpened metal tip as the electron emitter source, which operates near room temperature; very good vacuum is required to achieve reasonable lifetimes for the source. Thermionic emitters heat the electron source to a temperature of order 2000-2500 K, and extract the hot electrons using a high voltage electrode system. Thermionic systems are lower cost than field emission systems, and have less stringent vacuum requirements. Tungsten and LaB_6 sources are the most popular thermionic sources, tungsten because of the lower vacuum requirements and low replacement cost, and LaB_6 because of the stability and brightness. A LaB_6 emitter is typically 1-2 orders of magnitude brighter than a tungsten filament; as the resists used for electron beam lithography have a typical minimum dose measured in electrons per unit area, the time to expose a pattern using a LaB_6 emitter is correspondingly shorter than with a W emitter.

Thermionic emitter systems generate electrons with a fairly large spread in velocities, which limits the viewing and writing resolution; a typical thermionic microscope can resolve 10-20 nm features in viewing, and write features of order 20-50 nm in linewidth. The cold Schottky emitters used by

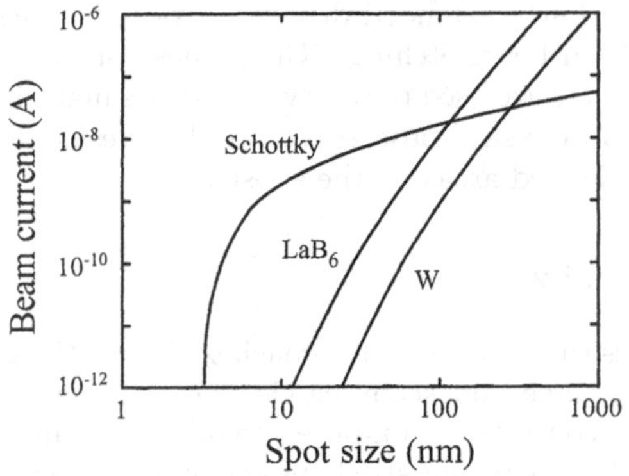

Fig. 10.5. Approximate beam spot size as a function of beam current, for different electron sources. Adapted from Reimer [125].

field-emission microscopes generate a very narrow range of electron velocities. This allows very precise focus and control of the electron beam, and yields extremely high resolution even when a field-emitter is operated at low voltages; in viewing mode resolutions of 1 nm are quoted for a number of commercial systems, and in lithography linewidths below 10 nm have been demonstrated. The electron beam resolution is ultimately limited by electron diffraction effects. In Fig. 10.5 we display the approximate minimum spot size achievable with different emission sources, as a function of total beam current. An excellent review of electron microscopy systems and techniques may be found in the text by Reimer [125].

Once the electrons are extracted from the source, they are accelerated by electrostatic fields to energies ranging from a few keV up to 100-200 keV, and a system of magnetic lenses and beam stop apertures is used to focus and narrow the beam. Magnetic scan coils are used to direct the beam to different points on the sample. A probe used to measure electron current in the beam may also be included, as well as a beam blanker, used to turn the beam on and off rapidly; these last two features are convenient, but not absolutely necessary, in a system used for lithography.

The sample resides in a chamber at the base of the column, on a stage allowing translation and rotation; motorized stages with computer control are convenient for large-area lithography. The electron beam hits the sample, with the electrons passing through the top surface and then scattering within the volume of the sample. Radiation in the form of x-rays and secondary electrons are emitted from the sample, the latter forming the primary imaging mechanism for viewing.

Electron beam lithography is based on the use of electron-beam resists, used in a very similar manner to optical photoresist. The resist is coated on the sample, which is placed in the sample chamber, and the electron beam then writes directly on the resist surface. In the standard positive process,

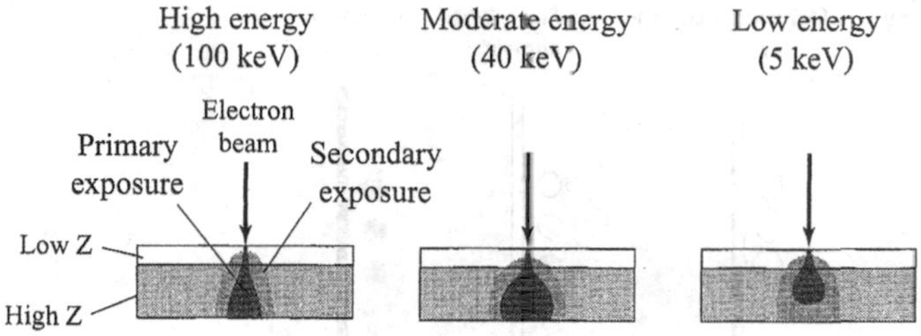

Fig. 10.6. Schematic drawing of backscattering effects at three electron beam energies; at the highest energies, the electrons are transmitted in a narrow cone deep into the substrate, yielding potentially very high resolution; at moderate energies the cone becomes broader with more backscattering exposure, and at very low energies the electrons stop primarily in the top layer and near the surface of the second layer, but the scattering volume is smaller, also leading to potentially very high resolution.

the electrons cause scission of the polymer chains in the resist, increasing its dissolution rate in a developing solution. Techniques have been developed for precision lithography using very low (few kV) to very high voltages (100 keV).

An intrinsic limit to writing resolution, using any electron-beam resist system, comes from electrons scattering within the resist and from electrons backscattering into the resist after passing through it into the substrate. In Fig. 10.6 we display schematic electron scattering profiles for a low atomic weight (small Z) resist on a higher Z substrate. At low energies electrons scatter readily, but travel only small distances after scattering; as the energy increases, the scattering rate decreases, but the range after scattering increases. Two approaches to very high resolution are therefore to work with very narrowly focussed low-energy electrons, with a small exposure volume, or to use very high energy electrons which project forward, leaving only a small exposure area on the top surface of the material. In either case, of course, a very tightly focussed beam must be used to obtain small feature sizes.

The lateral scattering of electrons, in addition to increasing the size of the minimum resolvable pattern, can wreak havoc with exposure patterns that include a large number of closely spaced fine features, or small features placed near larger ones; the scattered electrons from each feature will help to expose the resist of the adjacent features, and can cause overexposure of the resist. This is known as the *proximity effect.* Some examples of this are shown in Fig. 10.7. Pattern design must take this effect into account, by adjusting the pattern dosages in each area to minimize these effects.

A common low-cost electron beam resist is high molecular weight polymethylmethacrylate (PMMA) in a chlorobenzene or anisole solvent; the molecular weights range from 100 to 1000 kD. The thickness of the baked resist can be controlled by changing the coating spin speed, and by diluting the PMMA with solvent; concentrations of 2-5% of 1000 kD PMMA, and

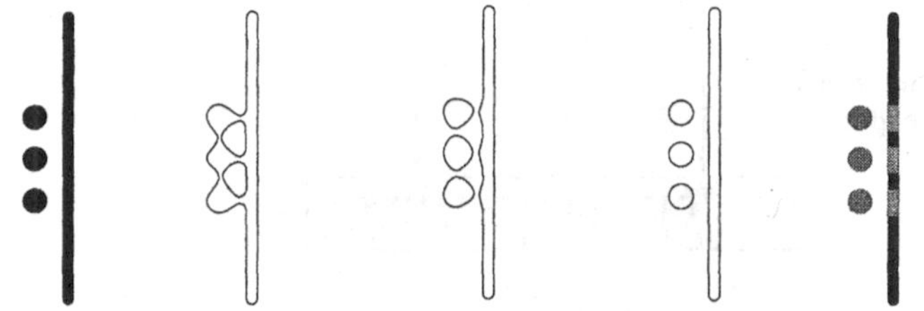

Fig. 10.7. Proximity effect from three circles exposed next to a line; the exposure pattern on the left, with uniform dose, can yield severe proximity effects as shown in the second image. Adjusting the dosages as shown on the right can correct or eliminate these effects, as shown on the third and fourth images.

spin speeds of around 3 krpm, yield thicknesses in the range of 0.1-0.3 μm. The exposure dose for a typical electron acceleration voltage of 40 kV falls in the range of 100-300 $\mu C/cm^2$ for both Si or GaAs substrates. A common developer is a mixture of 1 part methyl isobutyl ketone (MIBK) in 3 parts isopropyl alcohol; developing is complete in 10-20 seconds, although for undercut patterns longer develop times are needed, of order 1 minute.

From the teardrop shapes shown in Fig. 10.6, the exposure process generates a natural undercut profile, yielding a good liftoff geometry: Metal that is deposited on the surface of the PMMA will be disconnected from metal on the substrate surface, so the liftoff will work quite well. More pronounced liftoff geometries may be achieved by using PMMA bilayers, with an underlying layer of lower molecular weight PMMA, which requires a lower electron dosage for dissolution; note that lower molecular weights usually result in lower resolution. Even more extensive undercut may be achieved by replacing the lower PMMA layer with a layer of the copolymer methyl methacrylate, P(MMA-MAA), which is more sensitive to electron dosage (but again yields lower resolution); writing a pattern sufficient to develop out the less sensitive top layer will be more than enough to overdevelop the underlying layer. If a powerful UV source is available, this underlayer can also be pre-exposed using a long UV exposure, giving even more extensive undercut, although the very low sensitivity to UV makes this process inefficient. The underlayer can also be pre-exposed by blanket electron-beam exposure, but this is rather time consuming.

Very high electron dosages cross-link PMMA and make it insoluble in acetone, yielding a negative exposure process with acetone as the developer; a description of this process appears in [126].

Patterned PMMA can be used as an etch mask for dry or wet etch processes. It is however not nearly as robust as photoresist, and can therefore only be used with very weak wet solutions, or for very short times. In dry etch processes PMMA is typically removed quite easily and therefore does not provide an ideal masking material. The standard approach when PMMA is not sufficiently robust to use as a mask is to use it to pattern an intermediate thin film, either by lift-off or by etch, and use the patterned film as the actual etch mask. This of course may create problems later in the process if the patterned film cannot easily be removed once its need has ended. The cross-linked negative image PMMA is very robust and can be used to mask many aqueous and acidic wet etches as well as dry etching techniques such as reactive ion etching.

There are a number of other organic e-beam resists available, many of which are significantly more sensitive than PMMA, requiring lower dose and therefore shorter exposure times; most of these higher sensitivity resists do not achieve the linewidths possible with PMMA, although some of these have better chemical resistance than PMMA, making them useful in e.g. masking of wet etches. One recent example of a resist that appears competitive with PMMA is hydrogen silsesquioxane (HSQ). This novel silicon-bearing organic material has been demonstrated to have a sensitivity comparable to PMMA (in terms of electron dosage), and furthermore has been demonstrated to achieve 10-20 nm linewidths [127]; it is furthermore a negative tone resist, in contrast to PMMA which is positive. The resist is sold dissolved in methyl isobutyl ketone (MIBK) as a solvent, by Dow–Corning marketed under the tradename "FOx-12". Writing at voltages of 40 kV, dosage levels to achieve 10-20 nm linewdiths are in the range of 500-700 $\mu C/cm^2$, quite comparable to PMMA. The resist can also be used as the high-resolution top layer in a bilayer with an optical resist as the underlayer [128].

Some examples of other, somewhat less successful alternatives include P(GMA-co-EA) (poly(glycidyl methacrylate-co-ethyl acrylate), PBS (poly-(butene-1-sulfone)) and PGMA (poly(glycidyl methacrylate)), and as mentioned above, P(MMA-MAA) (copolymer methyl methacrylate). The linewidth resolutions of these range from 0.5 to 1 μm, with dosage requirements of $3 - 8 \times 10^{-7}$ C/cm^2, compared to a resolution of 20 nm for PMMA with a dosage requirement of roughly $\times 10^{-4}$ C/cm^2. P(GMA-co-EA) and PGMA are both negative resists, compared to PMMA and P(MMA-MAA) which are positive (as mentioned above, at high dosages PMMA can be used as a negative resist).

A second example of a very high resolution organic resist, which has been used to pattern sub-10 nm lines, is a calixarene derivative, hexaacetate *p*-methylcalix(6)arene, which is a negative resist similar to HSQ, and is very tolerant to fluorine and chlorine reactive etch chemistries. Unfortunately this resist requires a very high dosage of about 7 mC/cm^2, twenty to seventy times that of PMMA. This resist was first investigated by Fujita et al. [129], who describe a complete process for very small feature fabrication; developing after exposure is done in xylene, and the developer rinsed away in isopropyl alcohol. This resist can be used at energies as low as 5 kV without loss of resolution.

In addition to the organic resists, there are a number of metal halide inorganic resists with which extremely high resolution has been achieved. These include LiF, AlF_2, MgF_2, KCl and NaCl; using very thin films of these materials has allowed lines as small to 1.0-1.5 nm to be written, albeit using high voltages (100 kV) and quite high dosages (ranging from roughly 1 C/cm^2 for AlF_2 to 100 C/cm^2 for NaCl). Some inorganic resists such as SrF_2 and BaF_2 are significantly more sensitive, with dosages comparable to PMMA, but do not achieve the linewidths of the other inorganics; resolutions of order 100 nm have been reported.

10.2.3 Focused Ion Beams

Charged ion beams present a lithographic alternative to electron beams. Ion beams are used in a number of processing techniques, including reactive ion etching, ion beam etching, and as *focussed ion beams*, in which a beam of charged ions is directed through a focusing and scanning system similar to that used for electron beam lithography. The heavy mass and chemical reactivity of charged ions enables focussed ion beams to either pattern an intermediate lithographic layer (PMMA is frequently used) or to directly etch the substrate.

The most common type of focussed ion beam system uses a liquid metal source for the ions, in which a tungsten capillary (possibly with a fine tungsten needle through its center) has liquid metal fed to it from a heated reservoir. A large electrostatic voltage is applied between the capillary and an extraction electrode, which causes the liquid to form a sharply peaked cone known as the *Taylor cone*. The ion emission from this cone is very stable, due to space-charge effects, and can be focussed to a very small diameter. The need for liquid metal sources limits this technology to the use of low-melting point metals such as Ga, Bi, Ge and Hg. At higher extraction voltages, field ionization also allows the use of gaseous ion sources such as Ar, He and Xe. Beam diameters as small as 10 nm with currents of 10^{-11} A are possible [85].

Once ions have been extracted from the source, they pass through a focusing system, and the ion beam is then directed, in a serial-write process, to either expose or ablate an intermediate layer, or to directly etch a substrate. Liquid metal ion sources allow direct etching of substrates such as Si and GaAs, as well as dielectric and metal overlayers. Commercial systems can include a scanning electron microscope for viewing the etch process, allowing real-time etch control. Etch resolutions below 100 nm are possible from such systems; however, the ion source necessarily results in the implantation or coating of the substrate with the metal used in the source, so the electronic and possibly mechanical properties of the etched material are likely to change during the etch process.

Light ions can be used to change the chemical characteristics of an intermediate layer. For example, silicon dioxide (SiO_2), when exposed to a focussed beam of light ions such as H or He at high energies, suffers sufficient radiation damage to significantly increase the etch rate in wet etching; the resulting patterned layer can be used as an etch or implantation mask for an underlying Si layer [85]. Similar radiation damage can be used to pattern intermediate metal layers of Ni or Mo.

PMMA works very well as a ion beam resist; molecular scission occurs when PMMA is exposed to high energy ions, resulting in higher dissolution rates when the exposed PMMA is placed in a suitable developer. The dosage requirements are very similar to those required in electron-beam lithography; however, the much smaller ion spread in the resist allows, in principle, very small feature sizes.

10.2.4 Scanned-Probe Techniques: AFM and STM

The very high viewing resolution achievable with surface-probe instruments such as the atomic force and scanned tunnelling microscopes has fostered much effort to develop these as lithographic tools. Functional electronic devices have been successfully built using local oxidation, scratching and resist-based lithography using the AFM and STM; minimum linewidths achieved using these techniques are below 20 nm. The lack of a robust resist-based technology, allowing both etch and liftoff processes after the patterning step, has however greatly limited the range of application possible with scanned probes. There are efforts underway to remedy this situation, using e.g. self-assembled monolayers as resists (see below), but these have not yet been demonstrated to be widely usable.

The most successful application of scanned probe lithography has been in local oxidation. A natural film of water is formed on substrates kept in air; when a scanned probe is brought in close proximity with the substrate, a small water drop naturally forms between the tip and the substrate. This water is used as part of an electrochemical circuit to generate anodic, local oxidation. A conducting tip, typically of W or Ti, is brought into electrical tunnel contact with the substrate. A small negative voltage (typically a few volts) is applied to the probe tip, and the current of electrons that flows from the tip to the substrate oxidizes the substrate immediately under the tip. The tip is used to trace out a pattern on the surface, and the oxidation pattern follows the tip. This approach has been very successful in drawing oxide lines in Si and in thin films of Ti and Al. It has also been used to pattern GaAs by first depositing a very thin film of Si on GaAs, patterning the Si, and etching to transfer the Si pattern into the GaAs. Features 100 nm wide drawn in Si have been demonstrated by Dagata and coworkers [130], and features 10 nm wide and 30 nm deep have been etched into Si using local oxidation as a mask [131], [132], [133].

Useful applications of local oxidation have involved the fabrication of room temperature single-electron transistors (SETs), as demonstrated by C.F. Quate in Al [134] and by E. Snow [135] and K. Matsumoto [136] in Ti. A very thin metal film of Ti or Al, typically 3-5 nm thick, is deposited on an insulating substrate. The local oxidation process can then oxidize through the entire film thickness, allowing insulating lines drawn with the scanned probe to completely isolate areas of the film from one another. A conducting wire can then be formed by drawing two closely spaced, parallel lines, leaving the conducting wire between them.

Tunnel currents from scanned probe tips can also be used to expose conventional electron beam resists. PMMA has been patterned using an STM tip, allowing the definition of 20 nm wide lines, which were subsequently transferred to AuPd metal lines by evaporation and liftoff [137]. Several groups have successfully patterned the commercial resist SAL-601; STM was used to pattern sub-30 nm lines in this negative resist [138], and a conductive AFM tip was used to pattern 40 nm lines [139]. The AFM can also be used to image the resist immediately after exposure.

Nanoscratching, using a very sharp AFM tip to scrape narrow lines in soft materials such as PMMA or soft metals such as Al, has also been used successfully as a lithographic approach. Microbridges as thin as 10-20 nm have been successfully fabricated using this technique [140].

10.2.5 Imprint Lithography

A recent development in lithographic patterning that shows great promise is the technique of *imprint lithography*, in which a template is patterned and then etched to create a three dimensional relief. The template is physically pressed into a soft plastic or polymer coating on a substrate, transferring the relief into the polymer. The template and substrate are separated, and the relief etched through the remaining polymer to the substrate surface, typically using a dry etch technique. This approach to nanofabrication was proposed and successfully implemented by S.Y. Chou [141] as a method for parallel processing of nanoscale structures. A schematic sequence is illustrated in Fig. 10.8.

The templates used to emboss the pattern are commonly made from Si and SiO_2, patterned using electron-beam lithography and etched using a dry reactive ion etch (see below); the pattern on the template is raised where the embossing is to occur. Typical pattern depths range from 0.2-0.5 μm.

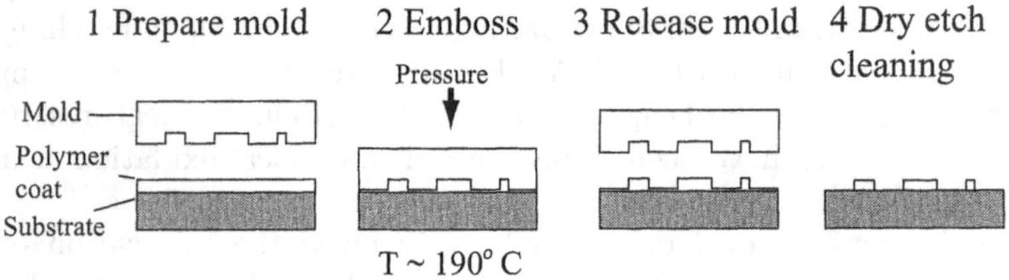

Fig. 10.8. Sequence for imprinting a nanoscale pattern from a mold into a soft polymer.

The molds are used to pattern a number of different soft coating materials; PMMA is very popular, but suffers due to its lack of resistance to dry etching. Very high resolution has been achieved with this material, with linewidths in the 10-20 nm range possible [142], [143]. The mold is pressed into the PMMA at fairly high pressure and temperature, well above the glass temperature of PMMA (in the range of 170-200° C); pressures in the range of 50-100 bar are standard.

A number of polymers other than PMMA have also been used; these include thermoplastic polymers similar to PMMA and thermoset polymers, both of which allow 50 nm linewidths to be realized [144]. Some of these materials have much higher etch resistance than PMMA, so may prove to be more useful.

After embossing, a short dry oxygen reactive ion etch may be necessary to remove material in the embossed regions. The resulting patterned polymer may then be used as an etch mask to pattern the underlying substrate, or as a mask for a metal deposition and liftoff, although the profiles obtained from embossing are not ideal for a liftoff process.

In an interesting variant on these techniques, the use of elastomers such as silicone rubber (RTV) and polydimethylsiloxane (PDMS), stamped out using either hard templates or templates made from photoresist, has allowed the fabrication of microfluidic devices directly from the molded material [145], [146].

In summary, very high resolution patterning of soft polymers has clearly been demonstrated using these techniques, in an implementation that allows parallel processing of large area devices. Functional devices have been successfully fabricated using this technique, such as quantum-point contacts in a two-dimensional electron gas [6], among others.

10.2.6 Self-Assembled Monolayers

There are significant efforts to develop better resist systems for use with all of the patterning technologies. We have mentioned some aspects related to photolithographic and electron beam patterning. Other areas of interest are the use of resists consisting of single, organized monolayers of organic molecules (SAMs, self-assembled monolayers). The molecules used to form SAMs typically consist of a foot, which binds to the substrate (silicon and gold are popular substrates), a body, and a head, which is typically very hydrophobic. Binding to silicon can be achieved by using silane-based feet, containing $SiCl_3$, which react strongly with silicon, while binding to gold surfaces is achieved by using a thiol-containing foot. Silane-based SAMs also bind to SiO_2, and thiol-based SAMs can be used to bind to GaAs. Alkane-based bodies and heads, containing CH_3, are very popular, as they are easily assembled and highly hydrophobic. SAMs that have been successfully demonstrated include OTS (octadecyltrichlorosilane), and OTD (octadecanethiol), the former for use on Si and the latter on GaAs. These have been used as electron-beam resists, and lines 25 nm wide have been exposed using a commercial electron beam lithography system [147].

The extreme hydrophobicity achievable with SAM molecules makes them ideal for masking wet chemical etches, which can transfer patterns into the substrate to a depth of some tens of nm. The most common approach is therefore to coat a substrate with a SAM, expose it, and then transfer the pattern with a wet etch that attacks the substrate. Resolutions as small as 7 nm have been achieved in writing with a high-resolution electron beam lithography system on an SiO_2 substrate, with the pattern subsequently transferred into the substrate using buffered hydrofluoric acid [148].

10.3 Dry Etching Techniques

We now turn from a discussion of lithographic techniques to a description of dry etch processes, which form a central part of semiconductor processing. Plasma-based etching and ion-beam etching are the two main dry etch techniques. In plasma-based etching, a plasma discharge is struck in a low-pressure gas, which generates free ions that then react with the substrate, either in the same space as the plasma, or in a separate chamber into which the ionized gas species diffuse. Ion-beam etching uses ions accelerated to voltages of a few hundred to a thousand volts, which ablate or possibly react with the substrate. In both of these techniques, one can use either reactive gas species, such as chlorine (for GaAs or Si) or fluorine-based gases (primarily for Si, SiO_2 and silicon nitride), or non-reactive species such as Ar and He. Frequently the best results are obtained by mixtures of reactive and non-reactive gases.

10.3.1 Plasma Etching

A plasma is formed when a sufficiently large dc or rf voltage is applied to electrodes in a gas atmosphere, typically at moderately low pressure. The plasma consists of neutral atoms and ions and electrons in a dynamic equilibrium; current in the form of charged ions and electrons flows between the plasma and the electrodes. In most cases the plasma forms a low resistance electrical element, and the voltage applied to the electrodes is dropped between thin neutral sheaths, with very low ion concentration, that form in the gas next to each electrode. For the plasmas commonly used for etching in semiconductor processing, the pressures ($10^{-3} - 1$ torr) and current densities ($10^{-4} - 10^{-1}$ A/cm^2) give bright glow discharges, and the neutral sheaths form dark spaces next to the anode and cathode. The neutral sheath ranges in thickness from 1-10 mm. A thorough discussion of the physics of plasmas in the various limits may be found in the text by Brodie and Muray [85].

In a dc-coupled plasma, the electrodes are directly connected to a high voltage dc power supply, delivering typically a few hundred to a thousand volts. It is somewhat more difficult to ignite plasmas using direct voltage sources, and also to maintain the plasma in a stable state; also, when etching insulators, the charge buildup on the non-conducting samples reduces the effectiveness and directionality of the plasma etching. Most of these problems do not occur, or can be minimized, through the use of rf-coupled plasmas, which are resultingly much more popular.

There are a number of different geometries available for coupling rf power to a plasma. The electrodes can be driven directly by the rf source; in this case, the oscillating rf voltage is distributed symmetrically across the two electrodes, and the plasma typically takes on a positive dc voltage V_p nearly equal to the amplitude of the rf voltage V_{rf}; this occurs because the much lighter electrons leave the plasma through the electrodes, leaving the positively charged ions behind. During each half-cycle of the applied rf voltage, the positive electrode on that part of the cycle approaches the voltage of the plasma, and the negative electrode approaches $-V_{\mathrm{rf}} - V_{\mathrm{p}} \approx -2V_{\mathrm{rf}}$ with respect to the plasma [85]. In the next half cycle the roles of the electrodes are reversed.

A common variant on this design involves placing a capacitor between one of the electrodes and the source, creating a asymmetric voltage distribution. If the discharge time of the capacitor is longer than the rf period, which is usually the case, the capacitively coupled electrode develops a negative dc voltage $-V_{\mathrm{rf}}$, and the electrode connected to the capacitor is therefore at approximately $-2V_{\mathrm{rf}}$ with respect to the plasma. Ions are then always accelerated towards this electrode, while electrons are accelerated towards the directly coupled electrode, resulting in a average dc current through the plasma. The electron current limits both the brightness of the plasma as well as the ion current density. A common innovation that greatly reduces this electron current is to place magnets behind the directly-coupled electrode, so that the electrons are deflected away due to the Lorentz force generated by the magnetic field. This is known as an rf magnetron, and is commonly used in rf-coupled etchers. In Fig. 10.9 we show a schematic diagram for a capacitively-coupled plasma system.

10.3.2 Reactive Ion Etching

Reactive ion etching (RIE) is the most common form for plasma etching, allowing the use of both reactive and non-reactive gases. The plasmas are usually driven by rf voltage sources applied between a pair of parallel plate electrodes, with either symmetric or asymmetric (capacitive) coupling; the sample is placed on one electrode, exposing it to the ions (and electrons, in the symmetric case) that are accelerated through the dark space adjacent to that electrode, when voltage of that electrode is below the plasma voltage. Both reactive and non-reactive species can cause sputter damage, ablating material from the sample; in reactive ion etching, the plasma voltage is usually kept low enough that the ablation process is minimized. The ionized reactive species

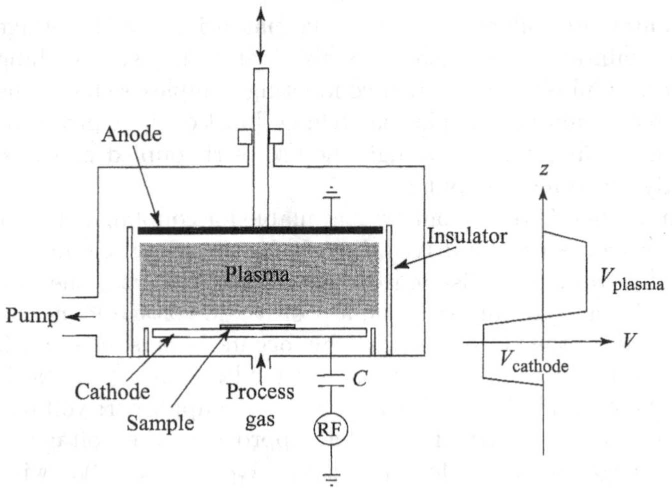

Fig. 10.9. Capacitively-coupled plasma reactive ion etching system; the voltage distribution for the capacitively-coupled RF design is shown on the right. Adapted from [84].

can chemically react with the sample surface, creating volatile intermediate species that either desorb from the surface or are ablated by the low voltage impact ions. At low pressures, this type of reactive etching can yield highly anisotropic results, with very vertical sidewalls; the anisotropy occurs due to the higher rate of ion impact on exposed horizontal surfaces compared to vertical ones, as well as more efficient momentum transfer due to these impacts.

A variant on the use of dc or rf electrode-coupled voltages to ignite and maintain the plasma is to inject power through an microwave-frequency electron cyclotron resonator, or ECR source. In these systems, an electron discharge is confined by magnetic fields so that the discharge electrons execute orbits, perpendicular to the magnetic field, at a well-defined cyclotron frequency. This frequency is typically designed to be 2.45 GHz, matching the frequency of the microwave source that pumps rf power during each cycle of the orbit. The electrons are continually accelerated until they either interact with an ion or drift out of the magnetically confined region. The process gas is fed into the ECR resonator, where it is ionized by collisions with the electrons, and then interacts with the substrate either within the ECR chamber, or in a separate chamber that allows separate dc and rf biasing of the substrate relative to the ECR chamber. This design is very appealing, as the issues related to maintaining a plasma, with the resulting dc and rf voltages between the sample and the plasma, may be completely sidestepped.

10.3.3 Reactive Gas Chemistries

The chemistry that occurs in reactive ion etching is complex, with a number of short-lived intermediate species created when the process gas reacts with the substrate; a full understanding of the various reaction sequences remains a subject of research. However, a number of gas chemistries have been developed that yield selective, anisotropic and isotropic etching (depending on conditions), of different substrates and of intermediate materials. Here we briefly discuss chemistries appropriate to etching GaAs, Si, SiO_2 and silicon nitride, as well as some metals.

Etching rates and specific etching recipes are highly dependent on the specific etching system, geometry and parameters used. The operating pressure, power, and plasma voltage in an RIE system are inter-related, and are very important in determining etch rate and etch anisotropy. Transferring such parameters from one system to another is a matter of trial and error. Absolute etch rates given in the literature, unless applied to more-or-less identical equipment and operating conditions, should therefore be used as guides rather than rules. Relative etching rates are however somewhat more reliable, as these do not vary tremendously given similar etch conditions.

In general, lower pressures and higher plasma voltages yield more anisotropic results, while higher pressures usually give more isotropic etching. Very reactive gases, such as NF_3 and XeF_2, do not need ion bombardment to react with Si, and therefore highly anisotropic etching requires other gases in the process mixture. Some gas mixtures, such as the carbon-based CF_4 and CHF_3, deposit inert films on the substrate sidewall during etching, and this film protects the sidewall, yielding more anisotropic results.

10.3.4 Silicon and Silicate Reactive Ion Etching

Silicon and silicon nitride are etched by fluorinated gases such as SF_6 and CF_4. Free fluorine ions are generated in the plasma through the reduction of the process gas in the plasma. The free fluorine ions then react with the silicon in the substrate, forming the volatile compound SiF_4. The specific chemical reaction are given by the formulas

$$\left.\begin{aligned} \mathrm{Si} + 4\,\mathrm{F} &\rightarrow \mathrm{SiF_4}, \\ \mathrm{SiO_2} + 4\,\mathrm{F} &\rightarrow \mathrm{SiF_4} + \mathrm{O_2}, \\ \mathrm{Si_3N_4} + 12\,\mathrm{F} &\rightarrow 3\,\mathrm{SiF_4} + 2\,\mathrm{N_2}. \end{aligned}\right\} \tag{10.1}$$

GaAs does not react readily with fluorine, so GaAs etch rates in these types of gases are negligible.

Many dry etch recipes based on CF_4 and SF_6 appear in the literature. Pure gas etches are not so commonly used, as adding other gases can improve etch rates and selectivity. For example, when using CF_4, adding small quantities of oxygen to the gas mixture increases the Si etch rate significantly.

The etch rate for Si reaches a maximum at 10% oxygen, making this a favored mixture; the peak occurs because the oxygen reacts with the C in the gas, forming CO and COF_2, and releases larger quantities of atomic fluorine. These reactions are given by

$$\left.\begin{aligned} CF_4 &\rightarrow CF_2 + 2\,F, \\ 2\,CF_4 + O_2 &\rightarrow 2\,COF_2 + 4\,F, \\ 2\,CF_4 + O_2 &\rightarrow 2\,CO + 8\,F. \end{aligned}\right\} \tag{10.2}$$

SiO_2 is also etched by this gas mixture, although the peak etch rate occurs at 20% oxygen. A 4% mixture of oxygen in CF_4 etches silicon at 5-6 times that of SiO_2. The relative etch rates for a CF_4 + 4% O_2 mixture are given by Si:Si_3N_4:SiO_2:AZ 1350 photoresist = 17:3:2.5:1 [84].

Adding small amounts of hydrogen instead of oxygen to a CF_4 plasma can significantly increase the etch rate for SiO_2, and improves the selectivity of oxide over silicon; this is very useful when etching SiO_2 on top of Si, as then significant etching of the underlying Si can be avoided. The hydrogen reduces the amount of atomic fluorine in the plasma, and also reacts with CF_4 to make CHF_3, which etches SiO_2 more rapidly than Si. This can be done either in the form of hydrogen gas, H_2, or by adding CHF_3 to replace some of the CF_4. A 3:1:4 mixture of CF_4:CHF_3:He etches thermally grown SiO_2 at 2-3 times the rate of both single-crystal and polycrystalline Si; LTO and PSG oxides etch about two times faster than thermal SiO_2 in this mixture [149]. A 3:2 mixture of CF_4:H_2 gives an etch ratio of SiO_2:Si of 35:1 [84]. These gas mixtures also etch silicon nitride at a rate comparable to SiO_2; increasing the hydrogen concentration with respect to the CF_4 or CHF_3 increases the nitride etch rate compared to that of oxide; a 2 CF_4:1 CHF_3:2 H_2 mixture etches Si:SiO_2:Si_3N_4 at relative rates of about 1.5:1:2 [149].

Sulfur hexafluoride (SF_6) is another popular silicon etchant. A pure SF_6 gas will etches Si:SiO_2:Si_3N_4 at relative rates of about 1.5:1:3. Adding He to the gas can significantly increase the etch rate of Si compared to oxide or nitride, with relative rates of Si:SiO_2:Si_3N_4 reported to be 10:1:3 [149].

One interesting fluorine-based process gas that does not require a plasma discharge is XeF_2; the fluorine in the gas dissociates easily and reacts quite readily with silicon, so a simple vacuum chamber with a few mbar of XeF_2 will etch silicon at rates of 200-400 nm/min. This gas does not attack SiO_2 noticeably, and etches nitride at quite low rates (10-20 nm/min) [150].

Silicon-based substrates can also be etched with chlorine chemistries, where the chemical reaction generates $SiCl_4$ rather than SiF_4. There are numerous published recipes using Cl_2, $SiCl_4$, BCl_3 and CCl_2F_2, all of which dissociate in the plasma and liberate atomic chlorine, which then reacts with the Si substrate in the process $Si + 4\,Cl \rightarrow SiCl_4$. The lower vapor pressure of $SiCl_4$ compared to SiF_4 makes chlorine-based etching more anisotropic, because the ion bombardment in the plasma is required to remove the chlorides. The etch rate of silicon in chlorine plasmas is in general much higher than that of SiO_2 or silicon nitride, making this an excellent selective etch for Si. A mixture of 2 Cl_2:5 He will etch Si 200-500 times faster than SiO_2, and at 10-15 times the rate for stoichiometric silicon nitride [149].

10.3.5 GaAs and GaAlAs Reactive Ion Etching

Chlorine gases form the basis for etching GaAs and GaAlAs substrates. The vapor pressures of the chlorine-reacted substrate, $GaCl_3$, $AsCl_3$ and $AlCl_3$, are not very high (15 torr, 200 torr and 0.12 torr, respectively, at 75° [84]), so that high substrate temperatures (above 100°) are preferable to eliminate these reaction products. The gases typically used include Cl_2, $SiCl_4$ and BCl_3, the latter two being useful because they eliminate trace water contamination found on the substrate surface.

As discussed in Ghandhi [84], the selective etching of GaAs over AlGaAs is of technological interest. Pure chlorine chemistries do not give a significant differential etch, but the very low reactivity of aluminum fluoride can be taken advantage of by mixing the chlorine-based gases with fluorine-based ones. Mixtures of SF_6 and $SiCl_4$, as well as CCl_2F_2 and He have provided a variable, and controllable, differential etch. A 10% mixture of SF_6 in $SiCl_4$ gives an etch ratio for GaAs:AlGaAs of 100:1, and further increasing the SF_6 content can give relative rates of up to 300.

10.3.6 RIE Etching of Metals

Two types of metal are commonly etched using reactive ion etching: aluminum, because of its utility as an interconnect metallization, and the refractory metals such as Nb and W; Ti is also often etched using RIE.

Aluminum is best etched using chlorine-based chemistry; it is difficult to etch because the native oxide of aluminum is impervious to Cl_2. Gas mixtures therefore usually include BCl_3 to scavenge the oxide; CCl_4 serves the same purpose. A common process therefore uses Cl_2, BCl_3 and CCl_4 or $CHCl_3$ to etch aluminum films. Water vapor can significantly change the etch rate, so aluminum RIE etches usually include a loadlock to maintain a controllable rate. A mixture of 3 Cl_2:5 BCl_3:2 $CHCl_3$:5 N_2 etches aluminum at roughly about 25% higher rate than Si, and at about 5-7 times the rate for SiO_2 and silicon nitride [149].

Tungsten and niobium, as with the other refractory metals, are best etched with fluorine-based gases. SF_6 is a popular choice, which etches W and Nb at about half the rate for Si, while at about twice the rate for SiO_2 [149]. Titanium can be etched with both fluorine and chlorine chemistries.

10.3.7 Masking of RIE Etching

Masking of RIE etching is accomplished by using a patterned layer on top of the substrate to be etched, preferably using a material that etches more slowly than the substrate. If such a material cannot be found, then the mask needs to be sufficiently thick to survive the etch process; the etch profile and resolution will however suffer.

Photoresist can be used in certain gas chemistries (such as $CF_4 + O_2$); however, when used in a parallel-plate rf- or dc-discharge plasma, the ion impacts will harden the resist in a prolonged etch, and makes it very difficult to completely remove; O_2 plasma ashing if often required. ECR-based plasmas, where the plasma can be separately generated and biased with respect to the substrate, are more forgiving to photoresist masks. PMMA, the most common electron-beam resist, is unfortunately not a good RIE mask in general; however, other e-beam resists, such as HSQ (see above) and SAL-601, are more robust and can survive fairly prolonged etching.

Another option is to use photoresist or PMMA to pattern an intermediate masking film, such as SiO_2, Si_3N_4, or a metal film; this of course requires that the film either be compatible with subsequent process steps, or that it can be removed without damaging the remaining material. Silicon-based process masks offer a good choice, as fairly selective wet etches exist for Si (concentrated KOH at 90° C), SiO_2 (hydrofluoric acid, either diluted or buffered) and Si_3N_4 (phosphoric acid).

10.3.8 Ion Beam Etching

Ion beams provide a method for sputter etching of a substrate, where high-energy ions bombard a substrate and ablate it away. When this is done using non-reactive gases such as Ar, the process is known as ion milling; when a chemically reactive gas replaces Ar, this is known as reactive ion beam etching, or RIBE. An intermediate technique is to use Ar as the ionic source, but also use a chemically reactive gas in the process chamber, which is known as chemically assisted ion beam etching, or CAIBE. The most commonly used source for the high-energy ions is the Kaufman ion source [151], shown schematically in Fig. 10.10. A thermionic cathode biased by an extraction voltage provides a source of electrons which serve to ionize the process gas that flows by it, igniting a plasma. The plasma is laterally confined by an axial magnetic field, and the plasma ions are accelerated through a series of grids into the process space. The ions impact the substrate with the acceleration energy, which can range from a few tens of volts to over a kilovolt.

In Ar-based ion milling, the impact energy of the ion, typically a few hundred volts, sputters off one or more atoms from the substrate. Ion milling yields, and therefore the etch rate, vary with the substrate material; some representative values are shown in Table 10.1. The sputter yield is a strong function of the angle of impact; maximum yield occurs not at normal incidence, but at grazing incidence, for angles of about 80° from normal. This is primarily due to the angular dependence of the efficiency of momentum transfer. At angles near normal incidence, the sputter yield is a weak function of angle. This combination yields quite uniform etching for initially planar

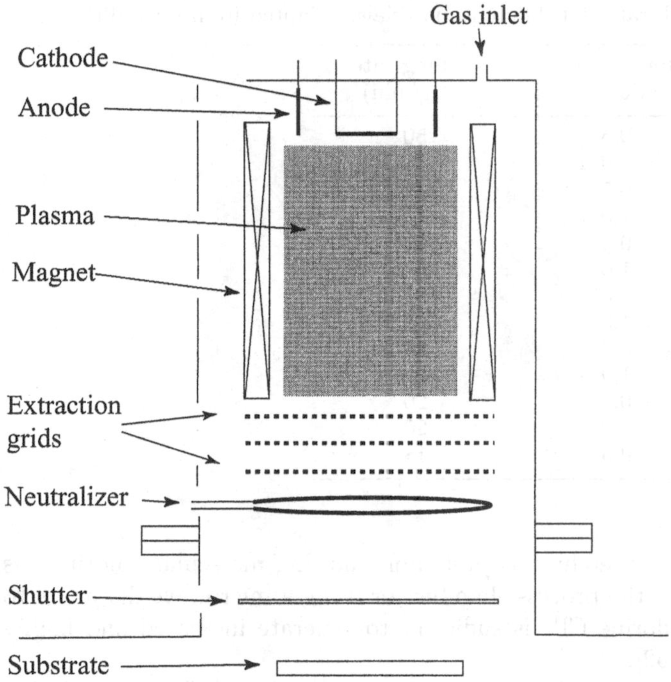

Fig. 10.10. Schematic of a Kaufman ion source, used for ion beam etching. Adapted from [85].

surfaces. Masking can be achieved by using a material that etches more slowly than the substrate. Etch profiles will approximate the mask geometry, but the angular dependence of the etch yield, secondary scattering and redeposition will all affect the etched profiles; for deep etches, trenching of the substrate occurs at the mask edges, due to the high efficiency of oblique incident ions. The parallel removal of the masking material during the etch process reduces the linewidth of masked lines, increases the linewidth of etched trenches, and eventually the masked features will be ablated away.

The addition of reactive gases to the ion mill process gas can significantly enhance the selectively and etch rate of an ion mill process. The reactive ion or chemically assisted etching of GaAs and its related compounds, using molecular or ionized atomic chlorine as the reactive species, can yield very smooth, highly anisotropic etch profiles suitable for laser mirror facets. The process involves the reaction of Cl_2 with the Ga and As atoms in the substrate, which are either volatile or can be made to sublime at elevated temperatures, or are much more easily sputtered away by Ar or other Cl_2 molecules. Masking can be achieved using photoresist, silicon dioxide, silicon nitride, or layers of refractory metals such as tungsten and nickel. Most Kaufman ion sources are not suitable for generating a Cl_2-based plasma within the gun, so that

Table 10.1. Ion mill rates for different materials. Adapted from Ghandhi.

Material	Ion Energy (keV)	Milling rate (nm/min)
Al	0.5	50
	1.0	65
Au	0.5	120
	1.0	175
GaAs	0.5	65
	1.0	260
Si	0.5	35
	1.0	50
SiO_2	0.5	35
	1.0	50
Ti	0.5	20
	1.0	20
W	0.5	18

chemically-assisted ion etching is preferable; flowing molecular chlorine gas across the sample in the process chamber, or using a microwave discharge to generate atomic chlorine Cl*, is sufficient to generate increased and highly selective etching [152].

Metals are frequently etched using ion milling, and some figures for Ar ion milling of metals are given in Table 10.1. Reactive gases such as Cl_2 are often used for ion-mill etching of refractory metals; fluorinated gases can also be used for metals such as W. Si, SiO_2 and silicon nitride can all be ion milled, but the use of fluorinated gases for these materials does not significantly improve etch profiles or etch rates; reactive ion etching (RIE) in a plasma is more effective for these materials.

10.3.9 End-Point Detection

One of the major difficulties in the vertical etching of a structure is in determining the etch depth. Etch rates are usually not sufficiently controllable that timing gives better than 10-20% depth control, and sometimes achieving even this degree of control is very challenging. There are a number of techniques that allow much more precise control of the etch depth. In a situation in which the etch is meant to terminate at the bottom of one layer in a heterostructure substrate (such as the AlGaAs layer in a GaAs/AlGaAs/GaAs heterostructure,or the oxide layer in a SIMOX wafer), end-point can be determined by monitoring either the downstream effluent from the process chamber, or by monitoring the spectroscopic response of a portion of the substrate during the etch. Such techniques are quite precise, but often require quite sophisticated equipment (a rare-gas analyzer in the first case, and an optical spectrometer in the latter).

In situations in which these approaches are not suitable, or where a large number of different processes with different reaction products are expected, another technique that can be used with a high degree of success is to monitor the reflectance of the sample with a simple laser-photodiode combination. The reflected signal can be monitored on an oscilloscope or a strip-chart recorder, and will either yield the reflectivity change (when for instance a highly reflective layer is etched from a less-reflective substrate), or one can count the interference fringes generated when a dielectric layer is removed from above a metal or dielectric with a different index of refraction; one interference fringe will pass for each half-wavelength thickness of upper layer removed. This provides a simple and fairly reliable means of process control.

10.4 Wet Etching Techniques

The dry etching techniques outlined above tend to have higher lateral resolution, and in some cases better depth control, than the wet etching techniques described here. Wet etching is suitable for structures with lateral dimensions above a micron or so, although with care structures with lateral dimensions down to 0.1 μm can be defined. One of the major disadvantages of dry etching is the substrate damage incurred when high-velocity ions impact the substrate surface. Substrate damage can however be minimized using, for example, ECR-generated plasmas, where the plasma voltage can be kept to within a few volts of the substrate, although this frequently implies much lower anisotropy in the etch direction. This is especially relevant when processing organic materials, or high-mobility two-dimensional electron gases that are confined close to the surface of the substrate.

Wet etching conveys two primary advantages: the very high chemical selectivity that can be achieved with certain combinations of mask and substrate, and the very low crystalline and electronic damage induced by the relatively gentle chemical process. Some chemical etches also etch preferentially along certain crystal directions, or etch certain crystal planes very slowly, a feature which can yield extremely smooth etch surfaces with very precise lateral control. The best-known example of this is the KOH-alcohol mixture frequently used to etch crystalline Si, in which the $\langle 111 \rangle$ planes of Si are etched at a rate several hundred times more slowly than the other crystal planes.

Vertical control is typically much more difficult to achieve with wet etching than with dry etching. The reactivities of many wet chemicals change with exposure to air or with time, so that timed etches are notoriously unreliable. When etching layered substrates, in some cases a chemical solution that is sufficiently selective to the etch layer may be found so that the underlayer is used as a stop-etch; this is the case when etching Si, SiO_2 or Si_3N_4, where etches for each of these materials that does not etch the other two exist. There are also etches that can be used to selectively etch GaAs over AlGaAs

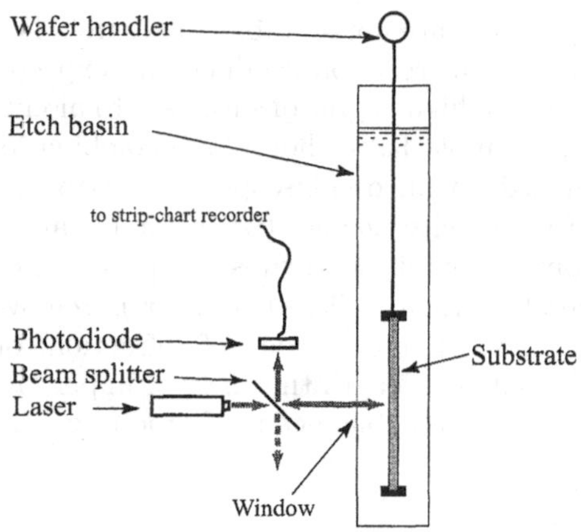

Fig. 10.11. System for laser end-point detection in a wet-etch process station.

and vice-versa, although the selectivity is not as high as for the silicon-based materials. One solution to detecting end-point in a wet etch is to use the same type of laser-photodiode end-point detection system as was described for the dry-etch process, modified for use in wet etch processes. An example of a wet-process end-point detection system is shown in Fig. 10.11.

Wet etching involves a number of general principles; the etchant in solution reacts at the substrate surface, through a complex interaction which frequently depends on the electrical properties of the substrate. Reaction products are typically soluble in the etchant, and must be transported away; fresh etchant must also be transported from the bulk solution to the substrate surface. Etch rates are therefore determined by transport rates to and from the substrate surface, as well as by the reaction rate at the surface. The latter is a strong function of the solvent temperature; an increase of about 10° C in the solvent temperature typically increases the reaction rate by a factor of two. Transport rates may be increased by stirring the solution, or by using ultrasonic agitation during the etch; spray etching, where the substrate is placed above the solvent surface and a pressurized spray directed at the substrate, can lead to significantly higher and in some cases more regular etch rates.

Etchants are in general characterized as either isotropic or anisotropic, and in the latter case most are crystalline-specific, reacting at different rates when etching different crystal planes. Isotropic etches are sometimes termed *polishing etches* as they tend to remove defects and protrusions from the surface. Some chemicals share both characteristics, with moderate selectivity to different crystal directions.

10.4.1 Silicon Etchants

Isotropic Etching of Silicon. The most popular isotropic silicon etch is a mixture of nitric acid (HNO_3) and hydrofluoric acid (HF), dissolved in water or acetic acid (CH_3COOH). Acetic acid yields higher etch rates than water for a given ratio of HNO_3 to HF. Both HNO_3 and HF are centrally involved in the etch process, the former reacting with Si to generate SiO_2, and the latter used to dissolve the SiO_2 into solution. A wide range of different mixtures of the two acids may be used; at high HF concentrations, the rate is determined by the concentration of HNO_3, and is roughly independent of HF concentration. At high HNO_3 concentration, the rate is determined by the HF concentration. A by-weight mixture of 60% HF (using a 50% concentration source bottle) and 40% HNO_3 (from a 70% concentration source bottle) yields an isotropic etch rate of about 1 mm/min; diluting this to 50% (by weight) in acetic acid reduces the rate to about 0.2 mm/min, while the same dilution in water yields a rate of about 10 μm/min. Isoetch curves for the HF:HNO_3:acetic acid and water system may be found in the literature (see e.g. [84]).

A similar etch replaces the HF with ammonium fluoride (NH_4F), which is more neutral and can be used with photoresist [149].

A few popular silicon etch mixtures are given in Table 10.2.

Table 10.2.

Mixture	Etch rate	Comments
30 ml HF 50 ml HNO_3 30 ml CH_3COOH	80 μm/min	CP-4A etch, very high rates
120 ml HNO_3 5 ml NH_4F 60 ml H_2O	0.15 μm/min	Isotropic silicon etch. Photoresist-compatible [149].
20 ml HF 150 ml HNO_3 50 ml CH_3COOH	5 μm/min	Planarizing etch
100 ml H_2O_2 37 g NH_4F	0.7 μm/min	pH $\approx$ 7.0, nearly neutral
20 ml HF 1 ml $KMnO_4$	0.3 μm/min	

Crystallographic Silicon Etches. A very popular anisotropic, crystallographic etch for Si is potassium hydroxide (KOH) diluted in water, with the possible addition of ethyl alcohol. This etchant is highly selective, etching the ⟨100⟩ and ⟨110⟩ planes at roughly the same rate but hardly etching ⟨111⟩

faces at all; this appears to be due to the high bond density in the ⟨111⟩ planes. Etching rates at room temperature are very low; when heated to about 80-90°, however, etch rates increase significantly. The solution must also be stirred during the etch, as otherwise the etch stratifies and yields non-uniform etching. A mixture of 1 part KOH to 2 parts water (by weight) etches ⟨100⟩ planes of Si at about 1.0-1.4 μm/min [153], [149]. The relative etch rates for ⟨100⟩:⟨110⟩:⟨111⟩ planes is about 300:600:1, although the rate for ⟨111⟩ planes varies quite a bit in the literature. Note that the KOH pellets used to make this mixture contain about 10% water, so the actual mixture is approximately 30% KOH.

The KOH etch can be masked by both SiO_2 and silicon nitride; SiO_2 etches at between 50-100 nm/min, so that long etches will not reproduce the mask pattern exactly. Silicon nitride however is not known to be etched by KOH, so it makes an excellent mask.

The very high etch rates for ⟨100⟩ and ⟨110⟩ planes means that mask openings on a ⟨100⟩-oriented wafer will result in triangular pits, whose shape depends on the shape of the mask opening. In Fig. 10.12 we display the results of etching through a rectangular and a circular opening in a mask, with the rectangular openings aligned with the ⟨110⟩ flats on the wafer edges. The ⟨111⟩ planes are tilted at 54.75° to the ⟨100⟩ planes.

The use of other substrate orientations yields other etch pit geometries; ⟨110⟩-oriented Si, when masked with a rectangular mask aligned with the ⟨111⟩ flats on the wafer edges, will yield vertical sidewalls that can extend through the wafer on two opposing sides of the etch pit; the other two sides expose the ⟨111⟩ planes inclined at 35.26° to the wafer surface.

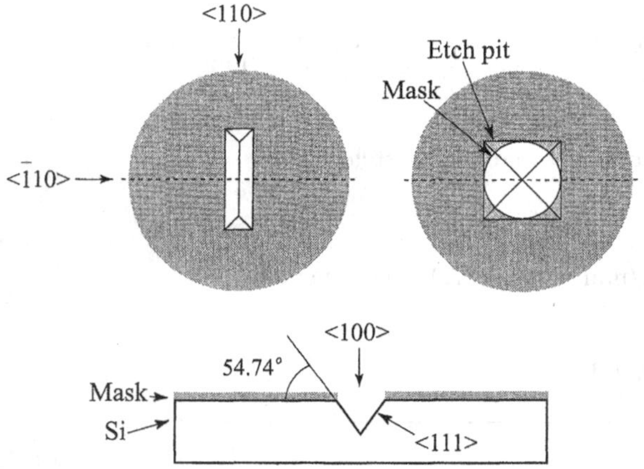

Fig. 10.12. A view of a KOH-etched ⟨100⟩ Si substrate, etched through a rectangular (*top left*) and circular mask opening (*top right*). A cross-section of the substrate taken along the dashed line in either case has the appearance shown in the lower part of the figure.

A frequent variant on the KOH-H_2O mixture is to add isopropanol, in e.g. a 1 KOH:1 isopropanol:3 water mixture (proportions by weight). When heated to 80-90° C, this etches in a slower but more uniform manner, reducing the need for stirring and somewhat reducing the etch rate of the ⟨110⟩ planes in comparison to the ⟨111⟩ planes, while leaving the ⟨100⟩ rates relatively unchanged.

There are two other common crystallographic etches for Si. Ethylenediamine pyrocathecol (EDP) is a mixture typically made of 45 g pyrocathecol, 250 cm^3 of ethylene diamine and 120 cm^3 of water; when heated to 100° C, the etch rate is about 1 μm/min, and increases to about 2 μm/min at 120° C [153]. The etch rate for ⟨110⟩ planes is roughly that of ⟨100⟩ planes, while the etch rate for ⟨111⟩ is negligible. EDP etches SiO_2 at about 0.5 nm/min, so that silicon dioxide masks are quite suitable for this etch. Silicon nitride is not noticeably etched by EDP, so that nitride masks may be used as well. A useful feature of EDP is that it does not etch heavily-doped *p*-type Si, so that boron-implanted or diffused Si may be used as an effective stop layer.

Another popular etchant is tetramethyl ammonium hydroxide (TMAH) [154], which has the advantage of not containing Na and not attacking aluminum thin films that contain trace amounts of Si, both important advantages when this etch is used with silicon MOS circuits; this etchant also has very low toxicity compared to EDP. Solutions are made by mixing the water-complexed solid $(CH_3)_4$-NOH · 5 H_2O with water; etch rates for ⟨100⟩ and ⟨110⟩ planes in Si increase with decreasing TMAH concentration, down to a few percent by weight TMAH. A 25% solution (1 part TMAH:3 parts DI water by weight), when heated to 80° C, etches ⟨100⟩ Si at 0.4 μm/min, ⟨110⟩ planes etch at 0.6 μm/min, and ⟨111⟩ planes etch at 0.04 μm/min, with the rates roughly doubling every 10° C [155]. A 5% by weight solution etches at roughly double these rates.

Silicate Etchants. We now briefly discuss the various wet etches that are available for etching the silicates SiO_2 and Si_3N_4. Values for the rates are taken from Ghandhi [84] and from the review article by Williams and Muller [149].

Silicon dioxide is best etched using hydrofluoric and buffered hydrofluoric acid; these also etch many metals (Al and Ti are etched very quickly by HF; Au however is not). Other glass films such as phosphosilicate glass (PSG), or low-temperature oxides (LTO) grown by PECVD or LPCVD, are also etched by these acids. The rates given here are for SiO_2 grown on single-crystal silicon in a oxygen furnace, exposed to water-saturated oxygen; dry-oxygen oxidation gives very similar results, but other glasses etch at rates that are significantly higher; some values are given for these as well.

Concentrated hydrofluoric acid (50% from a source bottle) etches SiO_2 at 1.8-2.3 μm/min; when diluted with water, the etch rate drops, so that 1 part 50% HF:10 parts water by volume etches SiO_2 at 25 nm/min, and diluted 1 part 50% HF:25 parts H_20 etches at 10 nm/min [149]. Buffered hydrofluoric acid, also known as buffered oxide etch (BOE), a commercially available mixture of hydrofluoric acid and ammonium fluoride, is available in a number of mixture ratios. A popular mixture, 5:1 BHF, is a mixture of 5 parts 40% ammonium fluoride to 1 part 49% hydrofluoric acid (parts by weight), which can be masked with photoresist for fairly prolonged etches, and for brief etches with PMMA. This mixture etches furnace-grown SiO_2 at 100 nm/min, annealed PSG at about 400 nm/min, and unannealed PSG at about 700 nm/min.

Silicon nitride films are wet etched using concentrated, heated phosphoric acid. Phosphoric acid (85% by weight source bottle) is heated to 120-160° C, and etches both stoichiometric and low-stress nitride films at 2-4 nm/min [149]. This etch is difficult to mask, as it etches photoresist fairly rapidly (40-100 nm/mn) as well as PSG glass, both annealed (1-2.5 nm/min) and unannealed (4 nm/min). However, annealed PSG glass is commonly used, because of its ease of deposition and subsequent removal. Dry etching of silicon nitride, using fluorine-based RIE, is often the preferred technique.

Metal Film Etching. The most commonly used metal films include Al, Ti, Cr, and Au. To a lesser extent the other noble metals (Ag and Pt) are also used, and metals such as Nb, Cu, W, and Ni are also used in some applications. These films are deposited using a range of deposition techniques, primarily by thermal or electron beam evaporation, and by sputtering.

Aluminum is easily removed using a commercial aluminum etchant (Aluminum Etchant Type A; Transene Corp., Rowley MA), which consists of 80% phosphoric acid, 5% nitric acid, 5% acetic acid and 10% water (all quantities by volume). When heated to 50° C, this etches Al at about 600 nm/min, and does not attack silicon, the various silicon oxides or the nitrides. This etch does however attack GaAs. An alternative etch is a simple dilution of hydrochloric acid, 1 part 37% by weight HCl:2 parts water, which etches Al much more slowly than aluminum etchant, but does not attack GaAs. Photoresist may be used to mask both the Type A etchant and the HCl-based etchant.

Titanium is etched by dilute hydrofluoric acid; adding hydrogen peroxide increases the etch rate substantially. A by-volume mixture of 20 parts water:1 part 49% HF:1 part 30% H_2O_2, at room temperature, will etch Ti at about 900 nm/min, and also etches PSG at 150-200 nm/min, and wet or dry oxide at about 10-15 nm/min [149].

Chromium is etched by hydrochloric-acid based etchants. One Cr etch is a mixture of 37% HCl:1 part glycerine (by volume), which etches Cr at about 80 nm/min; a second Cr etch is 1 part (by volume) 37% HCl:9 parts saturated $CeSO_4$ solution, with about the same etch rate [84].

Gold and platinum are etched by *aqua regia*, a mixture of 3 parts 37% HCl acid with 1 part 70% HNO_3, by volume. This etches Au at 25-50 μm/min, and Pt at 20 μm/min, if the Pt is pre-soaked for a few seconds in concentrated HF to remove a native oxide. Pt may also be etched without the pre-soak by doubling the amount of HCl and heating the solution to 85° C, where it etches at about 40-50 nm/min. An alternative etchant for Au consists of an iodine-based solution of 4 g KI:1 g I_2:40 ml H_2O, which etches Au at 0.5-1 µm/min, and may be masked with photoresist [84].

Ag is etched by a mixture of 1 part (by volume) 29% NH_4OH:1 part 30% H_2O_2:4 parts methanol, which etches Ag at 350 nm/min [84].

Tungsten (W) is easily removed using hydrogen peroxide at bottle strength (30%) at room temperature. An etch rate of about 20 nm/min is obtained [149], and the peroxide will not noticeably etch other silicon-based or metal films.

10.5 Thin Film Deposition

10.5.1 Evaporation and Sputtering Deposition of Thin Films

Thin films of many metals and insulators are most easily deposited through the use of thermal evaporation or by sputtering in either a dc or an rf plasma.

Thermal Evaporation. Thermal evaporation is performed in a vacuum system with a base pressure of 10^{-6} Torr or lower, in which the substrate and the material to deposit (the source) are placed. The source material is heated to the point where its vapor pressure becomes sufficiently high to generate a substantial flow of material from the source, and the atoms flow ballistically, without colliding with residual gas atoms, to the substrate surface, where typically a high fraction (~30%) of the atoms adhere and form a film. Metals typically must be heated to temperatures of at least 400-500° C, higher for refractory metals such as Ti or Ni. This can done by Joule heating, where the source is placed in a boat or on a wire filament made of a high temperature material (typically W or Mo), through which a dc current of order 100 A is run. Alternatively, in an electron gun evaporator, a high voltage (10 kV) beam of electrons is magnetically steered to hit the source, which is placed in a W or carbon crucible, and the crucible in turn placed in a water-cooled copper hearth. The electron gun either heats a small area of the evaporant, or heats the entire crucible, until the vapor pressure of the evaporant is sufficiently high. In Fig. 10.13 we display the vapor pressure curves for a number of common metals. A vapor pressure of order 10^{-2} torr is required to achieve a reasonable evaporation rate. The pressure curves are for a vapor in equilibrium with a liquid at temperature T, which is not achieved in a thermal evaporator; however, the figure is indicative of the range of temperatures required for common metals.

The ballistic transport from the evaporation source to the substrate means that material will only be deposited only at points on the substrate that are on a line of sight to the evaporation source; points blocked by intervening material, such as a mechanical mask in contact with the substrate, or a photoresist or other mask patterned on the substrate, will not receive any

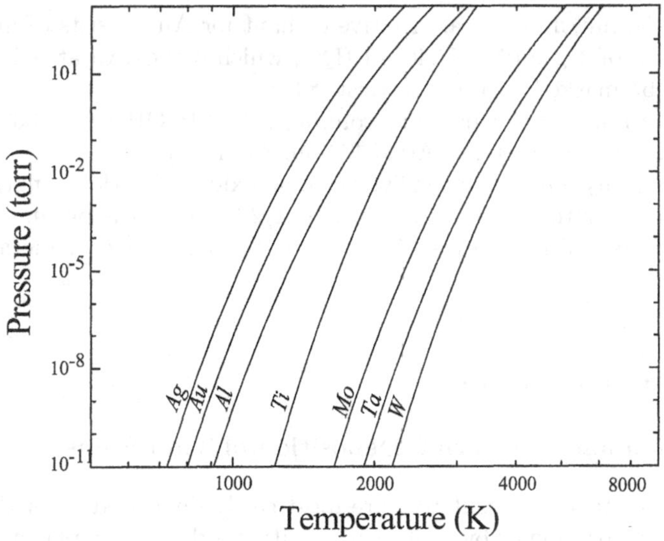

Fig. 10.13. Vapor pressure as a function of liquid temperature for a number of different metals.

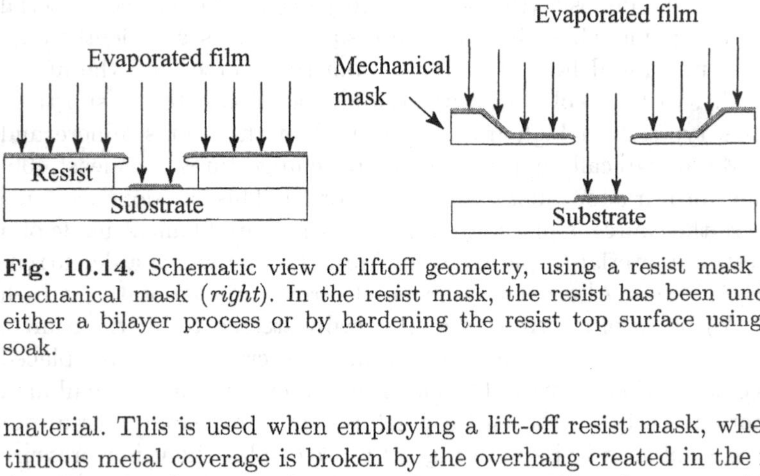

Fig. 10.14. Schematic view of liftoff geometry, using a resist mask (*left*) and a mechanical mask (*right*). In the resist mask, the resist has been undercut using either a bilayer process or by hardening the resist top surface using a chemical soak.

material. This is used when employing a lift-off resist mask, where the continuous metal coverage is broken by the overhang created in the resist mask (see Fig. 10.14); it can also be used to pattern films by removable, reusable mechanical masks, which has been successfully demonstrated using silicon nitride masks down to size scales of order 100 nm [156].

The materials in Fig. 10.13 are all metals; insulators may also be evaporated in both Joule-heated and electron beam evaporators, but these tend to sublime from the solid rather than evaporate from the liquid. The poor thermal and electrical conductivity of insulating materials make evaporation somewhat difficult; in Joule-heated evaporators, it is often necessary to use small furnace-type sources in order to achieve reasonable rates, such as when evaporating silicon monoxide (SiO). In electron beam evaporators, insulators will charge up from the electron current, and either deflect the beam in a random manner, or will spit material out of the crucible, contaminating the substrate and the evaporator. In general, it is somewhat easier to deposit insulators through rf sputtering rather than evaporation.

Sputter Deposition. In a sputtering system, a plasma is struck in an inert gas such as Ar, using either dc or rf high-voltage electric fields. Ions from the plasma are accelerated by static electric fields, generated either by a constant voltage bias in dc sputtering, or by using a blocking capacitor in rf sputtering (see Sect. 10.3.1). The high energy ions are directed to strike a target made from the source material to be deposited. Atoms are ablated from the source, and are then transported to the substrate by residual kinetic energy. The sputtered atoms pass through the plasma and the residual gas in the process chamber, and typically undergo a number of scattering events with the gas molecules; sputtering is therefore not a ballistic process, and therefore allows better step coverage on pre-patterned substrates. Suspended liftoff masks however do not work quite as well.

Metals can be deposited using either rf or dc sputtering, with deposition rates of a few tens of nm/sec possible. Insulators must typically be deposited using rf sputtering, as the voltage drop across the insulator in dc sputtering strongly attenuates the ionic acceleration from the plasma and gives much smaller deposition rates. In rf sputtering the insulator acts as part of a series capacitance, changing the tuning of the rf network but not impacting the deposition rates.

Sputter guns typically include magnets to confine the motion of the electrons used to ionize the sputter gas, which is then accelerated to the sputter target to generate the neutral sputter atoms. One common type of sputter gun is the *planar magnetron*, in which magnets are placed behind a planar cathode, made of the sputter material, and the anode faces the cathode, trapping the plasma between the two plates; the substrate is placed on the anode face. The other common type of sputter gun is the *circular magnetron*, where the anode is in a circular ring surrounding a planar cathode, behind which magnets are placed; the plasma is localized in the space between the anode and cathode, and the substrate is placed facing the sputter gun. This geometry is illustrated in Fig. 10.15.

Liftoff of sputtered films is however typically not as clean as with evaporated films, although the use of resist masks with very large undercuts (see Fig. 10.14) can be used. If very cleanly defined edges are required, it is sometimes preferable to use a subsequent etch process to remove unwanted material after a uniform, unpatterned sputter deposition.

Chemically reactive sputtering can also be performed, by mixing a reactive gas with the neutral gas used to generate the plasma. Silicon nitride films can be deposited by rf sputtering from a Si target, with a gas mixture of Ar and N_2; the partial pressure of N_2 is adjusted to control the stoichiometry of the deposited film. Silicon dioxide may similarly be deposited by using O_2 in

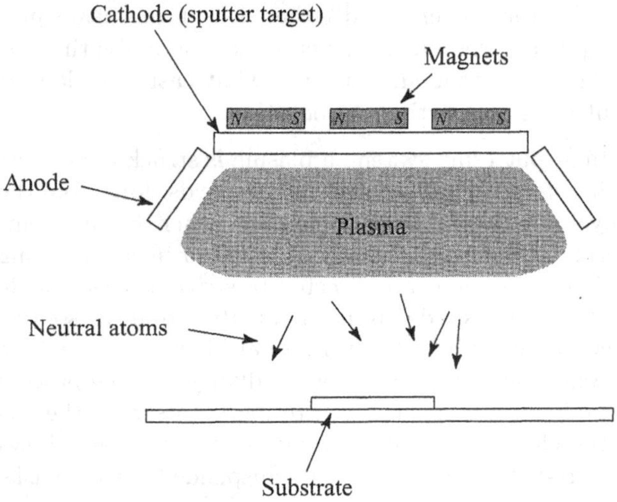

Fig. 10.15. Geometry for a rf circular magnetron, in which the cathode is a metal disk with magnets placed it, and the anode is a tilted cylinder surrounding the circumference of the cathode. The substrate is placed in line-of-sight to the cathode, and the neutral sputtered material arrives at its surface with a distribution of incoming angles.

the place of N_2. The quality, uniformity, and void fraction of these types of films is rarely as good as in CVD-deposited films, but they may nonetheless serve as insulating films or masks for later process steps.

Film Thickness Monitoring. The thickness of the deposited material is typically monitored using a quartz crystal microbalance. These rely on measuring the mechanical resonance frequency of a single crystal of quartz, which moves to lower frequencies as the mass of the evaporant deposited on the crystal surface increases. The resonance frequency is monitored using a phase-locked loop, coupled to readout circuitry via electrodes placed on the quartz surface. Commercial quartz microbalance systems include calibrations for a wide range of different evaporants, so that the thickness can be read out directly, and allow for feedback control of the evaporation process, to maintain constant deposition rates, as well as full programming features to control the entire deposition process.

10.5.2 Chemical Vapor Deposition

In chemical vapor deposition, reactive gases flow over the substrate, and are made to react at the substrate surface by thermally-driven chemical reactions, or by plasma excitation. Thermal oxidation of silicon, although normally not thought of in this way, is a form of chemical vapor deposition. Both low-pressure chemical vapor deposition (LPCVD) and atmospheric-pressure

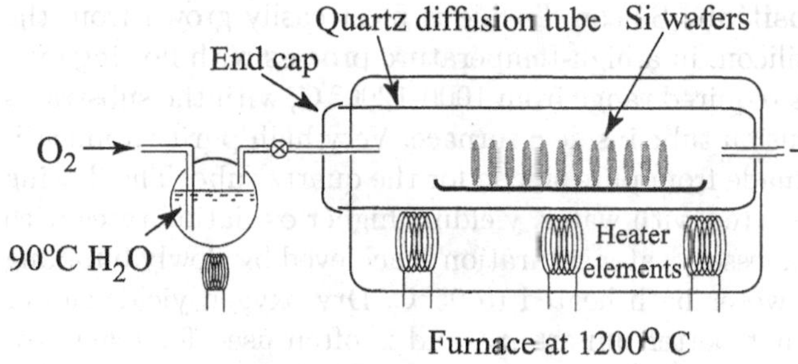

Fig. 10.16. Apparatus for the wet oxidation of silicon. Dry oxides may be grown by bypassing the heated water bath. Chemical vapor deposition (APCVD and LP-CVD) systems use a very similar arrangement, although without the use of a water bath.

CVD (APCVD) use reactors very similar to those for thermal oxidation (see Fig. 10.16); they differ in the materials used to construct them, the sophistication of the seals on the reaction chamber, and possibly the use of hazardous gas treatment systems at the exhaust.

Plasma-enhanced chemical vapor deposition differs from LPCVD and APCVD in that an RF discharge is used to generate the chemically reactive species, and typically involves lower substrate temperatures. Film quality is however somewhat poorer, with more defects such as pinholes, and more residual strain in the deposited films than in LPCVD-grown films. A typical PECVD system is shown in Fig. 10.17.

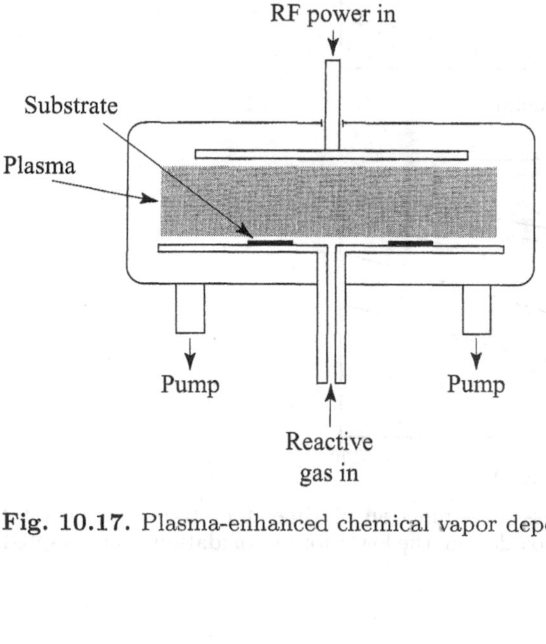

Fig. 10.17. Plasma-enhanced chemical vapor deposition (PECVD) system.

Silicon Oxide Deposition. Silicon dioxide is most easily grown from the thermal oxidation of silicon, in a high-temperature process with flowing oxygen. The temperatures required range from 1000-1200° C, with the substrates placed in a quartz diffusion tube inside a furnace. Very high purity applications substitute a tube made from pure silicon for the quartz tube. The flowing oxygen gas can be saturated with water, yielding higher oxidation rates with thicker final oxide thicknesses; water saturation is achieved by slowly bubbling the oxygen through a water bath heated to 95°C. Dry oxygen yields slower oxidation with somewhat better uniformity, and is often used for transistor gate oxides. In Fig. 10.16 we display a simple assembly for furnace oxidation of silicon; this system assembly is very similar to that used for LPCVD and APCVD (see above).

Oxidation rates are high at the beginning of the oxidation cycle, but decrease as the oxide thickens. Use of wet oxidation allows much thicker final oxides than with dry oxidation; using oxygen overpressures of several atmospheres allows thicknesses as high as 2 μm to be achieved in a one hour oxidation [157]. In Fig. 10.18 we display typical oxidation curves for wet and dry oxidation.

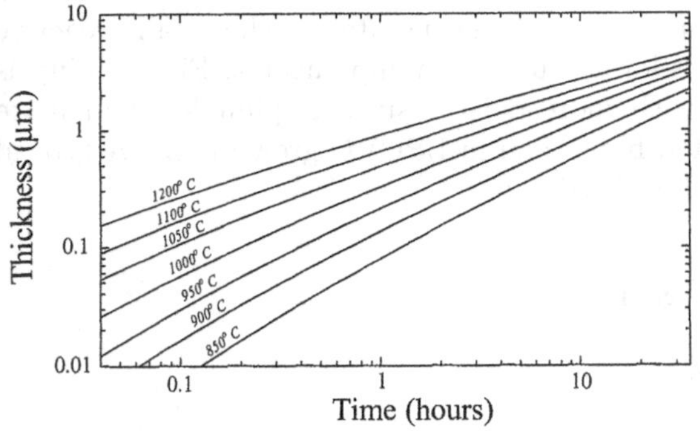

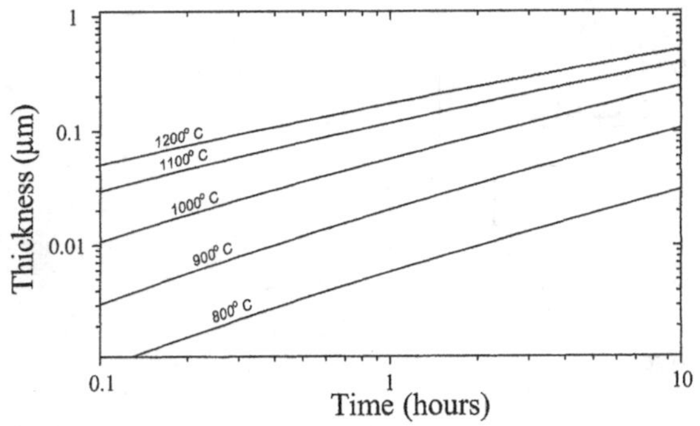

Fig. 10.18. Oxidation thickness as a function of time, for ⟨100⟩ single-crystal silicon. The upper plot is for wet oxidation, the lower for dry oxidation. Plot adapted from Ghandhi [84].

There are a number of other techniques for depositing silicon oxide. Using low-pressure chemical vapor deposition (LPCVD) to deposit SiO_2 generates an oxide that is not as robust or dense as thermally-grown oxide, but can be grown at much higher thicknesses and at lower temperatures. The process involves a gas mixture of silane (SiH_4) and oxygen, with the chemical reaction

$$SiH_4 + O_2 \rightarrow SiO_2 + 2\,H_2. \tag{10.3}$$

This reaction will proceed at temperatures as low as 450° C. The growth rate in an LPCVD process is about 10–15 nm/min [84], with low pinhole density and low stress. Plasma-enhanced CVD, using a system of the form shown in Fig. 10.17, can also be used using the same reaction (10.3), as well as with gas mixtures where the O_2 is replaced by NO_2 or CO_2. PECVD growth using silane and N_2O can be performed at low temperatures (300° C) with growth rates of 60 nm/min [84].

Oxides other than SiO_2 may also be grown using CVD systems. Phosphosilicate glass (PSG) may be grown through the reaction of silane, phosphine (PH_3) and oxygen at temperatures around 350-450° C. The phosphine reacts with oxygen to form P_2O_5, which is then incorporated in the silicate SiO_2 glass network. PSG is sometimes preferable to SiO_2 because it has higher density and fewer voids than amorphous SiO_2, and therefore forms a better mask for Zn or Sn diffusion in GaAs, and traps alkali ions such as Na more efficiently. Borophosphosilicate glass (BSPG), where both B_2O_5 and P_2O_5 are incorporated into the glass, is even more effective at blocking diffusion of alkali ions; this glass may also be grown by PECVD or LPCVD through the oxidation of silane, phosphine and B_2H_6, in the former case at 350–400° C, and in the latter case at 400–450° C. Both PSG and BSPG have lower reflow temperatures than pure glass, allowing better step coverage and allowing better metal film coverage in later process steps.

Typically LTO, PSG and BSPG are annealed to relieve stress and further densify the glass. Typical anneal processes use an atmosphere of flowing N_2 gas, and heat the material to around 1000° C for an hour [149].

Silicon Nitride Deposition. Silicon nitride, in stoichiometric form Si_3N_4, can be deposited by sputtering silicon in an RF discharge while flowing nitrogen through the process chamber. A much more reliable and popular approach is to deposit nitride by plasma-enhanced chemical vapor deposition (PECVD), or for low-stress applications, using low-pressure chemical vapor deposition (LPCVD). Reaction gases include both pure silane (SiH_4) mixed with ammonia, in the reaction

$$3\,SiH_4 + 4\,NH_3 \rightarrow Si_3N_4 + 12\,H_2, \tag{10.4}$$

or using SiH_2Cl_2 with ammonia in the reaction

$$3\,SiH_2Cl_2 + 4\,NH_3 \rightarrow Si_3N_4 + 6\,H_2 + 6\,HCl. \tag{10.5}$$

LPCVD parameters for low-stress nitride, which is silicon-rich, use an overpressure of SiH_2Cl_2 with respect to NH_3, at a chamber pressure of 200 mT and a temperature of about 800° C [149]. Stoichiometric silicon nitride can be grown using LPCVD with an overpressure of ammonia, with similar growth parameters. The same reaction may be performed in a PECVD reactor, with much lower temperatures (300° C) with growth rates of about 20 nm/min; the remnant stress is however usually quite high, although both compressive and tensile strain can result, and with careful tuning a near zero-stress film can be achieved.

Polysilicon Deposition. Polycrystalline silicon, or polysilicon, can be deposited by evaporation or sputtering. The best film qualities are achieved by CVD, typically by decomposition of silane (SiH_4). In an LPCVD reactor, growth is at 600-650°C, with a pressure of 300 mT. The polycrystalline films, especially those grown at higher temperatures, are very fine grained and yield a mirror finish. The undoped resistivity is typically of order 500 Ω-cm. This material is usually quite low stress, and does require an anneal step after deposition.

Doping of Silicon and Gallium Arsenide. The control of the electrical resistance and carrier type is of critical importance in fabricating many active semiconductor devices. Both single-crystal Si and GaAs, and polycrystalline Si layers, may be doped during the growth phase by addition of dopant atoms to the growth process. Some dopant atoms create shallow levels, with bound state energies close to the conduction band (n-type) or valence band (p-type) in the pure crystal; these dopants are used to control the carrier density and resistivity of the semiconductor. An n-type dopant greatly increases the number of electron-like carriers, moving the Fermi energy in the semiconductor from mid-gap towards the conduction band. P-type dopants increase the number of holes, and move the Fermi energy towards the valence band. Other dopants create states with energies nearer the mid-gap region of the semiconductor, and are called *deep level donors*;these are frequently used to reduce the minority carrier lifetime in high-speed electronic applications. In silicon, the preferred shallow-level dopants are As and P for n-type doping, and B for p-type doping. Au and Ni are commonly used deep-level dopants in Si. In GaAs, Si is a favored n-type dopant because it is easy to incorporate in an MBE growth process chamber. Zn is frequently used as a dopant as well, and acts as an n-type dopant when it is located at interstitial sites in the GaAs lattice, while acting as a p-type dopant when it substitutes Ga in the lattice. At lower concentrations, Zn acts primarily as a donor.

Doped polycrystalline films may be grown by adding phosphine (PH_3) to the gas flow, with otherwise the same growth parameters as for undoped polysilicon. This yields n-type polysilicon by incorporation of the P atoms in the film. p-type polysilicon may be grown by adding B_2H_6 to the gas flow in place of PH_3, which results in an increase in the growth rate for small B_2H_6/SiH_4 ratios. Resistivities of order 0.005-0.01 Ω-cm result from this type of doping.

Diffusion Doping. Two primary methods used to dope a semiconductor after it has been deposited are through high-temperature diffusion of dopants in a carrier gas, and through implantation of dopant ions in a high-energy process. Furnace-based diffusion systems are very similar to those used for oxidation of silicon, and are shown schematically in Fig. 10.19. Liquid diffusion sources are very similar to that shown in Fig. 10.16, with the water bath replaced by the liquid source, through which a carrier gas is bubbled.

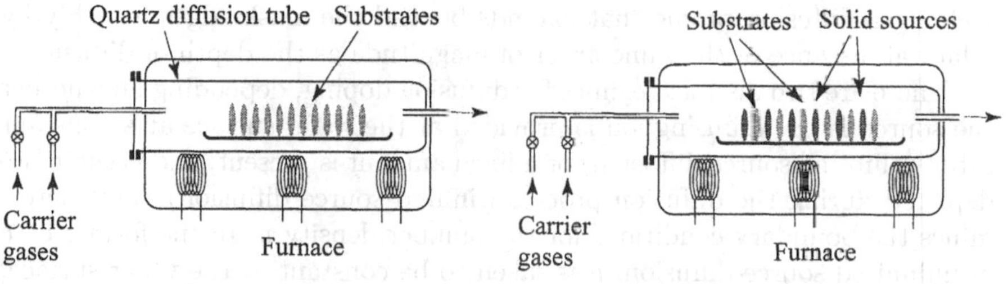

Fig. 10.19. Diffusion furnace using both gas sources (*left*) and solid sources (*right*). Figure adapted from Ghandhi [84].

Another technique for doping is to use spun-on organosilanes doped with the appropriate dopant atom. A layer of the doped organosilane is spun onto the sample, and the substrate then baked at 200° C, forming a doped SiO_2 layer on the substrate. A high-temperature bake then allows diffusion from the oxide into the substrate.

Diffusion is controlled by *Fick's law*, in which the number density $n(x,t)$ of the diffusant satisfies a differential equation. In a one-dimensional situation, which describes the diffusion of a uniformly distributed sheet of material along the direction perpendicular to the sheet, the form of the diffusion equation is given by

$$\frac{\partial n(x,t)}{\partial t} = \frac{\partial}{\partial x}\left(D\frac{\partial n(x,t)}{\partial x}\right). \tag{10.6}$$

The constant D is the diffusivity constant, with SI units of m^2/sec. The value the diffusion constant depends on the physical situation and is strongly temperature-dependent. Ions diffusing in a crystal lattice have their diffusivity controlled by the energy it takes to move the ion from one point in the lattice to the next; if the total energy barrier is ΔE, the diffusivity will have the form

$$D = D_0 e^{-\Delta E/k_B T}, \tag{10.7}$$

where D_0 is the constant of proportionality. From this expression, the diffusivity can clearly be greatly increased by raising the temperature; typical values for either substitutional or interstitial defects are of order ΔE =1-3 eV, corresponding to temperatures $T = \Delta E/K_B \approx$1100-3500 K. The constant D_0 is weakly temperature dependent.

In a situation where diffusion occurs in more than one direction, such as when diffusing material through an aperture in an otherwise impervious mask, the three-dimensional form of the diffusion equation is given by

$$\frac{\partial n(\boldsymbol{r},t)}{\partial t} = \nabla\left(D\nabla n(\boldsymbol{r},t)\right). \tag{10.8}$$

Atoms diffuse both into the substrate and around the edges of the mask, yielding a diffusion profile that extends beyond the mask edges, roughly by a lateral distance of the same order of magnitude as the depth of diffusion.

There are two distinct regimes for diffusion doping, depending on whether the source of the diffusing ion is provided at the wafer surface at a constant rate (unlimited source diffusion) or a fixed amount is present, and is therefore depleted during the diffusion process (limited source diffusion). This determines the boundary conditions for the number density n. In the former case of unlimited source diffusion, n is taken to be constant at the wafer surface, $n(0,t) = n_0$. Solving Fick's equation in one dimension, (10.6), the number density has the functional dependence

$$n(x,t) = n_0 \operatorname{erfc}\left(\frac{x}{2\sqrt{Dt}}\right), \tag{10.9}$$

where erfc(x) is the *complementary error function.* In the case of limited source diffusion, the number density has the form

$$n(x,t) = \frac{N_0}{\sqrt{\pi Dt}} \exp\left(-\left(\frac{x}{2\sqrt{Dt}}\right)^2\right), \tag{10.10}$$

where N_0 is the initial surface density of the dopant source, in atoms/m^2. In Fig. 10.20 we display the profiles for both cases.

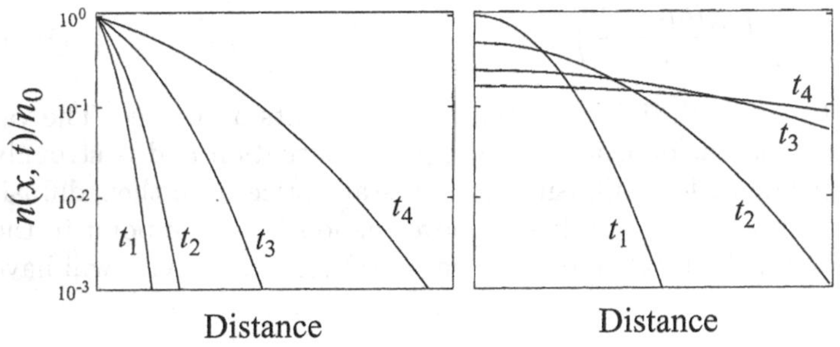

Fig. 10.20. Diffusion profiles for an unlimited source (*left*) and a limited source (*right*). The plots show the concentration as a function of position for different geometrically-spaced times, with $t_4 = 4t_3 = 4t_2 = 4t_1$.

10.5.3 Silicon Doping

***P*-type Silicon Doping.** Silicon is typically *p*-type doped by diffusion of boron. A common method for doing this is to use solid boron nitride (BN) disks in a solid source diffusion system, as shown in Fig. 10.19. The BN disks must be preoxidized in the furnace, with flowing oxygen at around 1000° C for 30 minutes, so that the disks form a skin of B_2O_5. This skin is then the actual source of B. Doping rates can be increased by adding small amounts of water vapor to the gas flow during the doping cycle, or by adding oxides such as BaO or SiO_2 to the nitride disks.

Another technique is to bubble oxygen through a liquid boron source, such as trimethyl borate ($2(CH_3O)_3B$), boron tribromide (BBr_3) or boron trichloride (BCl_3). In all of these the boron reacts with the oxygen and forms B_2O_5 on the silicon surface during the diffusion. One gaseous source that may be used is diborane (B_2H_6), but this gas is explosive and must be highly diluted with argon; oxygen is required in the gas flow to form B_2O_5 on the silicon surface.

***N*-type Silicon Doping.** The primary dopants for creating *n*-type silicon are phosphorous and arsenic. Phosphorous in general is delivered in the form of P_2O_5 to the silicon surface, where it reacts with Si to form free phosphorous atoms, through the process

$$2\,P_2O_5 + 5\,Si \rightarrow 4\,P + 5\,SiO_2. \tag{10.11}$$

Solid materials used as sources of phosphorous include ammonium dihydrogen phosphate (NH_4HPO_4) and ammonium monohydrogen phosphate ($(NH_4)_2HPO_4$). Diffusion occurs with furnace temperatures of order 500-900° C. Disc-shaped sources include silicon pyrophosphate in an inert binder, and aluminum phosphate $Al(PO_3)_3$. Typically oxygen is included in the carrier gas mix to form P_2O_5, which is transported in vapor form to the silicon surface.

A commonly used liquid source is phosphorous oxychloride ($POCl_3$), through which a carrier gas including oxygen is bubbled; alternatively, phosphorous tribromide (PBr_3) may be used.

Phosphine, a highly toxic and explosive gas, is the most commonly used direct phosphorous gas-source. Phosphine (PH_3) is highly diluted with Ar or N_2, and in the presence of oxygen the phosphine reacts above 400° C to form P_2O_5. Great care must be taken with the source gas as well as the exhaust.

Another commonly used *n*-type dopant for silicon is arsenic, which has a very low diffusivity and therefore allows better control when subsequent process steps involve high temperatures. Solid arsenic sources include arsenic trioxide (As_2O_3), which is heated separately to 200–250° C and transported in a N_2 carrier gas to the diffusion furnace. Alternatively aluminum arsenate ($AlAsO_4$) in an inert binder may be placed in the diffusion tube with the substrates. Carrier gases must not include oxygen, as silicon oxide layers will mask the arsenic diffusion. Gaseous arsine (AsH_3) may also be used. The most popular method for As doping of silicon is however by ion implantation (see below).

GaAs Doping. The diffusion doping of gallium arsenide is rather difficult, as at the elevated temperatures necessary for driving in the dopant, the arsenic in the GaAs substrate dissociates, and the surface of the substrate begins to deteriorate. Successful approaches have therefore used sealed ampoules with an overpressure of arsenic, or using doped oxide sources which form caps over the substrate. Ion implantation is significantly more popular and useful than these techniques.

10.5.4 Ion Implantation

Ion implantation is a very popular method for doping semiconductors. An ion implanter can be used for a wide variety of dopant types, and the processing is done primarily at low temperatures. The depth and range of depths (the straggle) of the ions can be fairly well controlled. Ion implantation does require significant capital costs, and the damage to the substrate crystal is significant, requiring a fairly high-temperature anneal after the implantation.

An ion implanter uses a high energy source (100-500 keV) of ions, that can be selected for their energy in order to control implant depth. The ions are directed to the substrate, which may be masked using a number of masking materials, such as SiO_2 or silicon nitride. The ions hit the substrate and penetrate to an average depth R_p, determined by the ion type and energy. Scattering within the crystal gives a range of depths ΔR_p, typically smaller than or of the order of the average depth R_p, and also a lateral spread ΔR_t, defining a volume of order $\Delta R_p(\Delta R_t)^2$ in which the ions come to rest.

In Fig. 10.21 we display the ion implantation depth R_p for Si and GaAs. The range of ion depths is roughly proportional to the depth, with a constant of proportionality given by [85]

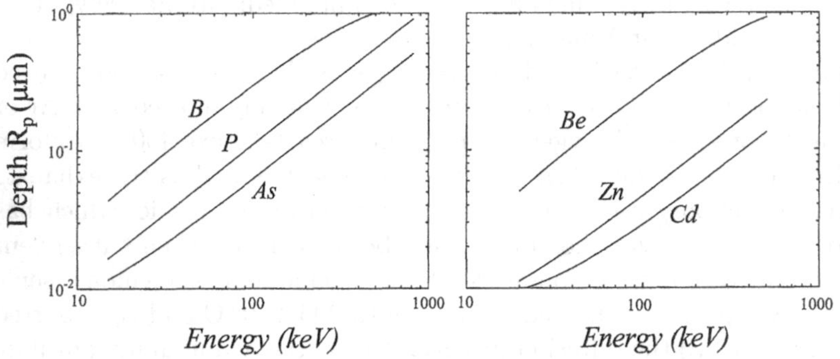

Fig. 10.21. Ion implantation depth R_p as a function of energy, for different dopant ions. The left figure is for implantation in silicon, the right for implantation in GaAs. Figure adapted from Brodie and Muray [85].

$$\Delta R_p \approx \frac{2\sqrt{M_{\rm ion}M_{\rm lattice}}}{3(M_{\rm ion}+M_{\rm lattice})}R_p, \tag{10.12}$$

where $M_{\rm ion}$ is the ion mass, and $M_{\rm lattice}$ is the substrate atomic mass. The lateral spread ΔR_t is typically of the same order of magnitude as the depth spread ΔR_p.

The significant crystal damage caused by the high energy ion implants can be annealed by either high-temperature bakes or by rapid thermal annealing (RTA). Silicon typically requires annealing for 15-30 minutes at 800-1000° C; this depends on the type and energy of the implant. Gallium arsenide requires roughly the same temperatures, but due to the problem of dissociation from arsenic loss, the anneal must either be carried out in an overpressure of arsenic gas, or using an inert cap on the substrate surface, such as silicon nitride or aluminum oxide. More details on annealing cycles and ion implantation and activation may be found in the text by Ghandhi [84].

10.6 Tunnel Junction Fabrication: Suspended Resist Mask

We briefly describe a method for fabricating metal-insulator-metal tunnel junctions, most applicable for the fabrication of small size (0.01-10 μm^2 area) junctions. These devices may be made using either normal metals or superconductors (when operated at low temperatures). They are useful because the physics of electron tunnelling through metal oxide barriers is relatively simple, and can be used for a number of applications: As thermometers, as high-frequency mixers, and, for very small size scales and low temperatures, form the basic element of a nearly quantum-mechanically limited amplifier known as the single-electron transistor. An excellent review of single-electron charging effects in metal tunnel junctions may be found in the text by Devoret and Grabert [105]; a more recent review includes a discussion of ultrasmall single-electron transistors fabricated using Ti or Si thin films patterned using atomic-force microscopy or scanning tunnelling microscopy [136], [158].

The technique we describe uses a suspended PMMA resist mask that allows multiple evaporations at different angles, so that the tunnel junctions can be completed in a single vacuum cycle of a thermal or electron-beam evaporator. There are a number of techniques for fabricating such masks, typically using double layers of electron-beam resist.

In one approach, a thick bottom layer of copolymer, P(MMA-MAA), is first spun on the substrate and then baked, with a target thickness of 0.4-1.0 μm (see Sect. 10.2.2 for more details on fabrication recipes and materials). Next a thin top layer of high molecular weight, high resolution PMMA is spun on and baked; 960 KD molecular weight PMMA, dissolved 3-5% by weight in a solvent, is a popular choice, yielding a top layer thickness of about 0.1 μm. The double resist layer is then exposed with an electron beam lithography system, with careful dose control to achieve the smallest linewidths, so that both layers are exposed and developed simultaneously. The much higher sensitivity and lower resolution of the bottom layer allows a significant undercut to be created.

In a variant on this approach, the first PMMA layer (possibly using the copolymer P(MMA-MAA)) is spun on and baked, and a thin layer of Ge is then thermally evaporated on the surface of this layer, with a thickness in the range of 40-100 nm. A second layer of PMMA is then spun and baked on, and after exposure in the electron-beam lithography system, the top layer is developed to yield the exposed pattern. An anisotropic RIE etch (see Sect. 10.3.2), typically using fluorine chemistry, is then used to transfer the pattern into the underlying Ge layer. The bottom PMMA layer is then removed using a combination of anisotropic followed by isotropic oxygen plasma RIE, allowing a deep undercut of the patterned Ge layer. The final structure, with a suspended "bridge" of PMMA or Ge, in either case then has the appearance shown in Fig. 10.22.

The suspended bridge structure is placed in a thermal or e-beam evaporator to fabricate the actual tunnel junction. The evaporator must include the ability to remotely tilt the sample with respect to the evaporation source, without breaking vacuum; good angle control, of order 1° or better, is required. An evaporation is made at one angle, casting a shadow of the mask pattern on the substrate. The metal most commonly used is Al, as the oxide of Al is of very high quality, with a large barrier and good mechanical and chemical stability. After the first evaporation, the Al film is oxidized in a slight pressure of O_2 (0.1-1 Torr for a few minutes), the oxygen pumped out, the sample tilted to a new angle, and a second metal evaporation performed. A correct combination of angles will yield a final pattern as shown in Fig. 10.22. The second

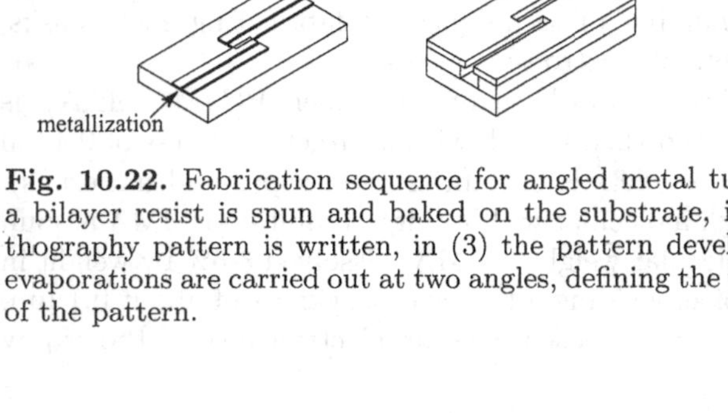

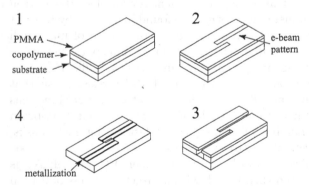

Fig. 10.22. Fabrication sequence for angled metal tunnel junctions; in step (1), a bilayer resist is spun and baked on the substrate, in (2) the electron beam lithography pattern is written, in (3) the pattern developed out, and in (4) metal evaporations are carried out at two angles, defining the tunnel junction in the center of the pattern.

metal can be Al, or if a normal metal-insulator-superconducting tunnel junction is desired, the second metal can be Cu, Ag, or Pt, all of which will not poison or short out the Al oxide barrier. A single-electron transistor (SET) is formed by joining two overlap junctions of the form shown in Fig. 10.22; the smaller the overlap area, the smaller the tunnel junction area and therefore its capacitance, which typically dominates the capacitance of the SET. Very small tunnel junction areas, used to fabricate SETs that can operate at liquid nitrogen temperatures, have been achieved by careful control of this process; see [103].

11. Nanostructure Fabrication II

In this final chapter, we outline a few of the fabrication sequences developed for fabricating nanoscale mechanical structures, using a range of substrates: Bulk, single crystal silicon; buried SiO_2 heterostructures fabricated by SIMOX (Separation by IMplantation of OXygen) or by a wafer bonding technique; silicon single-crystal regrowth over patterned bulk silicon; and single-crystal GaAs heterostructures. We conclude with a brief discussion of critical point drying.

11.1 Bulk Single Crystal Nanomachining

Silicon micromachining was first proposed and advanced by researchers who had easy access to single-crystal silicon, and who primarily used photoresist and wet etching techniques to pattern and etch structures from this material. Papers outlining the techniques, with useful references, include those of Bean [159] and Petersen [160], describing both techniques and applications of silicon MEMS (micro-electromechanical systems).

The development of methods to deposit low-stress polycrystalline silicon and silicon nitride films since that time has driven much of MEMS applications and research; most MEMS structures today are fabricated from polycrystalline materials, due to the flexibility given by the ability to deposit material on top of pre-patterned substrates, and the relative ease with which quite high quality films can be fabricated. The advantages associated with having several different material types (polysilicon, silicon nitride and silicon dioxide) are not to be underestimated. Problems of reproducibility of materials properties have been partially dealt with through engineering approaches. Size scales for lateral device features in polycrystalline-based MEMS are mostly well above 1 µm, and complex structures are usually hundreds of microns across. Indeed, as many interesting applications involve active MEMS devices on the 1 mm size scale, much effort has been devoted to developing technology for the difficult window in the 100 µm-10 mm size scale, rather than attempting to develop very small scale approaches.

Fabrication of working structures at the sub-1 µm size scale is challenging because it involves the use of advanced lithographic techniques, more sophisticated pattern transfer techniques than are offered by wet etch processes, and is not completely compatible with the polycrystalline materials on which much attention has been focussed. Issues of residual strain and control of grain size become important when the size scales are of the order of, or smaller than, the grain size in the base material. The planarity of structures with small minimum dimensions but large overall areas are difficult to maintain when the material retains residual stress, which remains a challenging aspect for polycrystalline micromachining. Efforts by a number of groups have therefore been directed towards the development of techniques for fabricating active structures from single-crystal materials, in which the lack of intrinsic strain can prove a very useful feature; the reproducible nature of the substrate and lack of grain structure are also quite appealing. The materials of choice to date have been single-crystal silicon and single-crystal GaAs.

Much of the development of techniques for nanofabrication of single-crystal silicon and GaAs has been done by the group of N.C. MacDonald; in this section we outline the techniques developed by his group as an illustration of one approach to fabricating single-crystal mechanical nanostructures. For a review, see [161] and [162].

MacDonald's group has developed a number of techniques for fabricating mechanically suspended structures with single-crystal silicon, with acronyms such as SCREAM (Single-Crystal Reactive Etch and Metallization) [163] and COMBAT (Cantilevers by Oxidation for Mechanical Beams And Tips) [164]. The themes underlying these fabrication techniques are similar; here we described the SCREAM process. This process, in which polycrystalline materials are used to protect the single-crystal substrate, but are subsequently removed, applies equally well to GaAs as silicon [165]. We first outline the procedure for Si, and then recap the approach as applied to GaAs fabrication.

The approach is based on the use of non-reactive coatings on the silicon to protect the base layer during anisotropic and isotropic process steps. The SCREAM process uses thermally grown and PECVD deposited oxide as well as nitride coatings. In Fig. 11.1 we show one process flow sequence; a variant on this can be used to fabricate higher aspect-ratio structures.

The first step in the process is to grow a thick thermal oxide, of order 1-1.5 μm, on the surface of a silicon wafer. Photoresist is coated on the oxidized surface, patterned, and used as an etch mask for a fluorine-based reactive-ion etch through the oxide, stopping on the silicon surface; this can be done using a timed etch with a CF_4-H_2 gas mixture, which etches SiO_2 much faster than Si (see Chap. 10). The photoresist is then stripped off, and the patterned oxide used as a mask for a deep (2-4 μm) etch of the underlying silicon.

Following the anisotropic etch into the silicon, the exposed Si sidewalls must be protected from the next step. This can be done by growing a second, thinner oxide layer, using either thermal oxidation or PECVD deposited oxide, with a thickness of 0.2-0.3 μm sufficient to provide the necessary protection. The entire structure can then be metallized with aluminum, as is done in the SCREAM process, or the metallization step can be postponed,

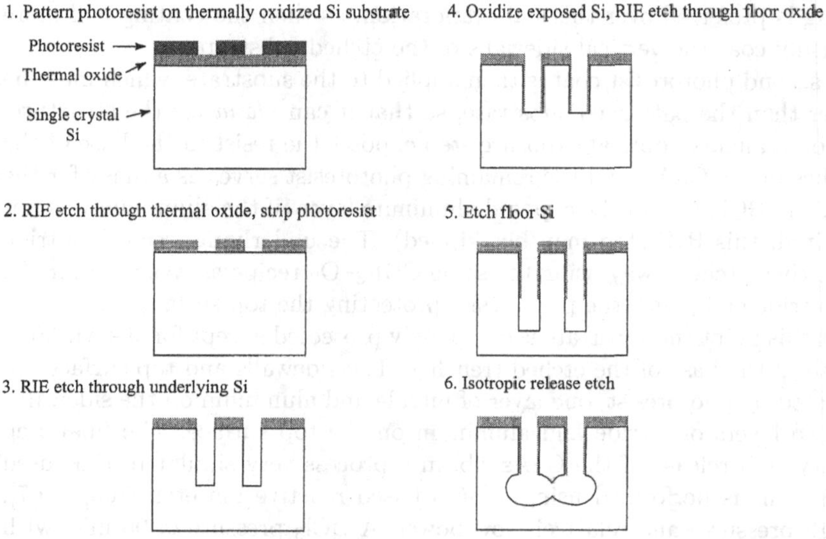

Fig. 11.1. Process for patterning suspended beams from single-crystal Si. Details are given in the text. Adapted from [163].

as is shown in Fig. 11.1. The oxide deposited at the base of the etched Si structure is removed by a short RIE, which stops before the oxide on the top surface of the structures is removed. A second Si RIE is then performed, generating a deeper structure with unprotected sidewalls at the structure base; this etch is 0.4-0.6 μm deep.

The next step in the process is a isotropic SF_6 reactive ion etch, that attacks all the exposed silicon and releases the entire structure. The structure is finally metallized by sputter deposition, allowing capacitively-coupled drive and sense of the suspended structures.

The GaAs process is very similar to the one described here [165]. Starting with a bulk GaAs wafer, a 350 nm thick layer of PECVD silicon nitride is deposited on the substrate. Optical lithography is used to pattern a photoresist film spun on top of the nitride film, and the patterned film serves as a mask for a CHF_3-O_2 reactive ion etch through the nitride film, stopping on the GaAs top surface. The patterned nitride then serves as a mask for the chemically-assisted ion beam etch (CAIBE) of the GaAs substrate, using Cl_2 as the reactive gas in a Kaufman ion source. This etch is allowed to proceed to a depth of about 10 μm. Any remaining photoresist is then stripped off using a wet photoresist remover, and possibly a high-pressure isotropic oxygen plasma etch.

A second PECVD nitride layer, 300 nm thick, is then deposited uniformly over the entire wafer. Metallization of the structure can then be done at this point; in the published process, the authors sputtered 250 nm of Al. Sputter coating is preferred over thermal evaporation, as then the coating will more uniformly coat the vertical sidewalls of the etched substrate.

A second photoresist coat is then applied to the substrate, which must be thicker than the patterned substrate, so that it can *planarize* the structure. A second, aligned, optical exposure step exposes the resist in the base of the trenches in the GaAs, and the remaining photoresist serves as a mask for the timed Cl_2-BCl_3 RIE of the exposed aluminum layer (if the aluminum was not deposited, this RIE step may be skipped). The underlying, second, nitride film is then etched away using the same CHF_3–O_2 recipe as was used for the first nitride etch, with the photoresist protecting the top surface.

At this point the substrate is completely protected except for the windows opened in the base of the etched trenches. The sidewalls and top surface are protected by photoresist, one layer of nitride and aluminum on the sidewalls, and two layers of nitride and aluminum on the top surface. The final step is a dry etch release of the GaAs rib, in a process very similar to that used for Si. This is performed using a BCl_3-based reactive ion etch [166], [167], at high pressures and relatively low power. A BCl_3 pressure of 90 mT, with a dc bias of 75 V, yielded a GaAs etch rate of 180 nm/min, compared to a nitride etch rate of 7 nm/min. This highly selective and isotropic etch can completely undercut the GaAs ribs; a 30 minute etch was sufficient to release a 0.4 μm wide beam from the substrate.

In the final step, any remaining photoresist is removed using a dry oxygen isotropic plasma etch. The aluminum metallization is then exposed, allowing capacitive drive and detect of the completed structures.

This process sequence has a minimum linewidth determined primarily by the resolution of the first photolithographic process step. A high-resolution mask aligner can yield minimum linewidths of 0.1-0.3 μm. Alternatively, electron-beam lithography can be applied to these processes, with a slight modification, allowing linewidths below 0.1 μm to be achieved; in the case of the silicon-based SCREAM process, a single self-aligned e-beam write is needed, as the remaining process steps are self-aligned. Such a process has been published by the author [86].

Note that the process as described can be completed using all low-temperature processing, with the exception of the first thermal oxidation step. Furthermore only a single lithographic step is required to complete the structures; the remaining steps are all self-aligned.

The fundamental structure formed by this process is a straight, cantilevered or doubly-clamped beam. The process steps are however independent of crystal plane orientation, so that arbitrary orientations and circular structures are easily fabricated. More complex structures, such as torsional oscillators, interdigitated capacitive displacement actuators and sensors, and a number of other structures may be created by linking beams in e.g. right-angle arrays that are suspended as a whole.

11.1.1 Silicon Regrowth Structures

A remarkable approach to fabricating suspended beams and membranes from a wafer of single-crystal silicon has been developed by Sato et al. [168], [169]. This technique is based on etching closely spaced rectangular holes vertically through the top surface of a single-crystal Si wafer, with hole dimensions of 0.25 × 0.55 μm, etched to a depth of 2.3 μm. The holes were etched by reactive-ion etch through a silicon oxide mask consisting of holes arranged either singly, in a line with center-to-center spacing of about 1 μm, or in a regular array with the same spacing. After RIE etching, the oxide mask was removed by wet etch, and the patterned wafer was placed for 10 minutes in a hydrogen annealing furnace at 1100° C, with flowing hydrogen gas at 10 Torr.

During the anneal, Si diffuses along the surface of the wafer, and the upper half of the etched trench slowly fills, consuming silicon from the adjacent top surface and the bottom half of the trench, until the top of the trench is filled in, leaving behind a roughly spherical void buried below a single-crystal surface. The radius of the void is roughly twice that of the original trench. Two adjacent trenches, spaced by less than the final void diameter, will result in the voids joining together, forming a buried cylindrical tube. A line of trenches can form a tube of indefinite length, and an array of trenches can join to form a membrane of silicon above a plate-shaped void.

Such structures can clearly provide a novel way to create suspended beams or membranes, where a final step of etch processing would free the beam or membrane from the adjoining top surface silicon. This technique may prove a highly useful approach to suspended nanostructure fabrication.

11.2 Epitaxial Heterostructure Nanomachining

11.2.1 SIMOX and SLICE Wafers

A great simplification of the bulk silicon process can be achieved through the use of heterostructure substrates. One of the major hurdles in bulk processing, and the need for of more than one of the process steps outlined above, is undercutting the suspended structures. Using a substrate that includes a sacrificial underlayer, which can be removed through a selective wet or dry etch that does not attack the top, structural layer, nor the bottom substrate "handle", significantly simplifies the process flow, without incurring an inordinate increase in the substrate cost.

The SIMOX (Separation by IMplantation of OXygen) process is used to create wafers from 4 to 6 inches in diameter that include a bulk substrate, an amorphous, insulating SiO_2 layer in the range of 0.05-1 μm thick, and a top Si layer typically 0.1-0.2 μm thick. These are created by ion implanting a large dose of oxygen ions through the top Si surface, forming a buried layer of amorphous SiO_2 mixed with Si. The implanted wafer is then annealed for several hours at quite high temperatures. The anneal allows the separation of the SiO_2 from the Si into a smooth buried layer, and the recrystallization of the Si on the top surface into a single-crystal layer above the buried SiO_2 layer. Descriptions of the process appear in the literature [170, 171].

A standard SIMOX wafer has a 0.4 μm thick buried silicon dioxide layer, and a 0.2 μm thick top Si layer. The top silicon layer is single crystal but does contain a relatively large number of defects, which are mostly threading dislocations running from the oxide underlayer through the top Si layer. Defect densities in the range of 10^3-$10^5/\mathrm{cm}^2$ are typical. The top Si layer has a thickness uniformity of better than 15 nm, and the buried oxide thickness is uniform to ±3%. A sketch of the structure appears in Fig. 11.2.

For nanomachining applications, the top Si layer is patterned, with holes cut through it into the underlying oxide layer, and the final suspension step is completed using a selective SiO_2 etch, such as hydrofluoric acid, to remove the oxide while leaving the top structural layer intact. An alternative to SIMOX wafers are wafers known as "SLICE" wafers, in which a highly thinned single crystal Si wafer, acting as the structural layer, is bonded to another wafer that has an oxide layer grown on it. The bonded wafer has the same structural form as a SIMOX wafer; this approach however gives more flexibility in choosing the oxide and structural layer thicknesses, and prices per wafer are comparable.

A suspended nanostructure is fabricated from a SIMOX substrate through a process similar to, but simpler than, that used for bulk Si. A description of

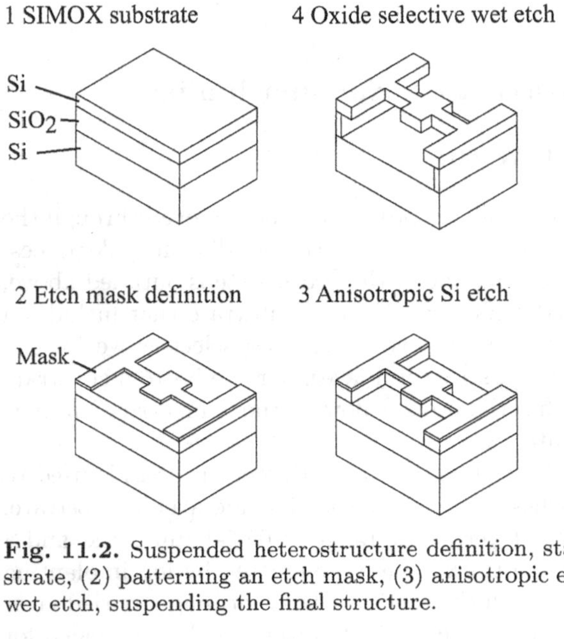

Fig. 11.2. Suspended heterostructure definition, starting (1) with a SIMOX substrate, (2) patterning an etch mask, (3) anisotropic etch of the Si and (4) sacrificial wet etch, suspending the final structure.

such a process appears in [102], and is sketched out in Fig. 11.2. In the first step, a SIMOX substrate is coated with PMMA for electron beam lithography, with the PMMA patterned for use as a metal liftoff mask. A RIE-resistant mask, using for example a refractory metal such as Ni, is deposited on the patterned PMMA and then lifted off; a Ni thickness of 50–100 nm is more than sufficient. Other metals or insulators may be used as well; aluminum is another common choice. An alternative approach is to use a RIE-resistant electron beam resist such as SAL-601.

The patterned film serves as a mask for anisotropic RIE of the top Si layer, which can be performed using a fluorine- or chlorine-based chemistry, such as CF_4 or Cl_2. The RIE etch is timed to completely cut through the top Si layer, terminating on or below the surface of the buried oxide. The undercut etch is then performed by submerging the substrate in hydrofluoric acid or buffered HF, for a time sufficient for the oxide beneath the suspended structure to be fully removed.

Metal electrodes for driving and detecting the mechanical motion of the suspended structure may be easily integrated using a lithography step prior to the step that defines the RIE mask. A patterned PMMA film can be used as a liftoff mask for the electrode metallization, and the lithography step that defines the RIE mask is aligned to the completed metal electrode pattern. The remainder of the process remains unchanged. Alternatively, the metallization layer may be defined after the structure has been suspended; we have found that the process steps involved in coating and patterning PMMA on a suspended substrate, as well as lifting off the PMMA after metal evaporation, does not necessarily destroy the suspended structures. Larger suspended areas are more likely to cause problems than small, but we have completed such processes on structure with suspended areas up to 10–20 μm^2, supported by 0.2 μm diameter legs several microns long.

Structures fabricated using the SIMOX substrates are shown in Fig. 11.3 and Fig. 11.4.

Fig. 11.3. Set of 0.2×0.2 μm^2 cross-sections single-crystal beams, with lengths ranging from 3 to 5 μm.

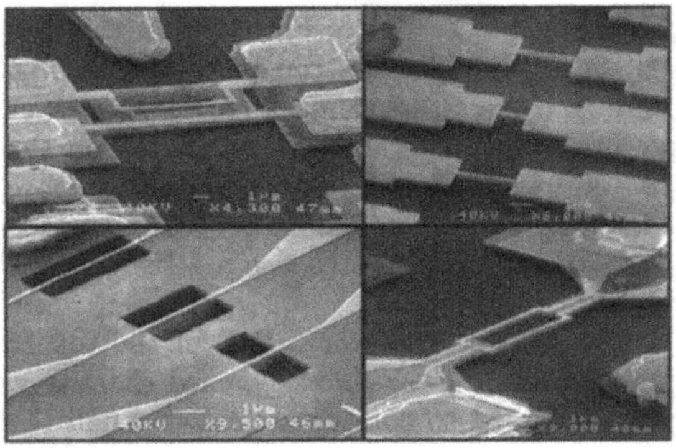

Fig. 11.4. Examples of structures that can be fabricated using the techniques outlined here.

11.3 Polycrystalline Nanomachining

There is a vast literature on fabricating increasingly intricate devices and structures from polycrystalline materials. These materials form the basis of the MEMS industry, and the ability to fabricate moving structures with size scales of a few tens of microns has led to much innovative research and development. Covering even a fraction of the range of effort in this area is beyond the reach of this section, so we shall instead outline two approaches to fabricating simple resonator structures, one based on the use of polycrystalline silicon and other on polycrystalline silicon nitride.

Silicon nitride can be grown using LPCVD techniques to yield films with very low residual stress, a desirable feature when suspending long beams or membranes of thin nitride. There are several commercial ventures that will provide low-stress nitride films grown using these techniques.

There are a number of simple fabrication techniques for creating suspended nitride films. One very straightforward approach is to begin with a single-crystal silicon wafer, typically with a ⟨100⟩ orientation, and deposit a thin film of low-stress nitride on its surface (most growth facilities will deposit this film on both sides of the wafer; either side of a double-side polished wafer can then be used). Thicknesses from 0.1 to 1 μm are typical. The back side of the wafer is then patterned using optical lithography. First a window is opened in the backside nitride, either by wet etching with phosphoric acid or with fluorine-based RIE using gases such as CF_4, SF_6 or CHF_3 (see Chap. 10). Next the Si wafer is etched (from the back), which can be done very simply using a prolonged KOH-based etch, which will stop when it reaches the front side nitride, leaving a pyramidal opening through the wafer. The faces of the pyramid are tilted at about 54.75° to the ⟨100⟩ silicon surface, so that the front side opening is smaller than that on the back side. Alternatively the silicon can be etched using a RIE process; deep vertical etches through the thickness of a silicon wafer (350-500 μm thick) can be performed using the patented Bosch process, typically implemented in a dedicated RIE etcher using an ICP- or ECR-generated plasma. If the wafer with the suspended nitride film is handled delicately, further front side lithographic processing followed by dry RIE etching can be used to pattern the front-side nitride. Very delicate structures, with suspended lengths of a few millimeters or even centimeters, may be fabricated in this manner.

In Fig. 11.5 we outline the steps of this fabrication process.

A somewhat simpler approach does not remove the entire thickness of the backside silicon. The front side nitride can first be patterned, and the holes opened in the nitride then used as a wet etch or a dry etch mask. The nitride film pattern, masked by photoresist or some other RIE-etch mask, is then transferred into the underlying silicon through an anisotropic silicon etch, which can be done using CF_4 at a pressure of a few mTorr. The undercut can then be completed using a isotropic (typically high-pressure) plasma etch, such as SF_6 at 100 mTorr.

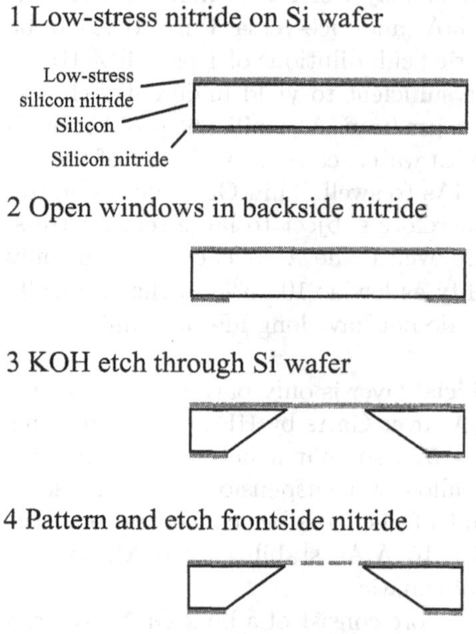

Fig. 11.5. Silicon nitride suspended structure fabrication. A wafer with a low-stress nitride film on both sides is backside patterned and the Si wafer etched through using the crystallographic KOH etch. The front side nitride film is then patterned. Details are in the text.

11.4 GaAs Heterostructure-Based Nanofabrication

A smaller community of researchers has developed techniques to fabricate mechanical structures using single-crystal heterostructures based on gallium arsenide. Structures have been developed by Hjort and coworkers [172], [173], [174], and Uenishi et al. [175]. These authors took advantage of the different chemical reactivity of aluminum-doped GaAs ($Al_xGa_{1-x}As$) to that of pure GaAs, and the fact that $Al_xGa_{1-x}As$ can be grown epitaxially on GaAs substrates, for any concentration $0 \leq x \leq 1$, and vice versa. $Al_xGa_{1-x}As$ may be etched using both dilute hydrofluoric acid (HF) and concentrated hydrochloric acid (HCl), neither of which will noticeably etch GaAs. A layer of GaAs supported by AlGaAs may therefore be suspended by a timed etch in either of these acids. By growing a heterostructure consisting of a GaAs base wafer, a sacrificial AlGaAs layer, whose thickness can range from a few monolayers to several microns, and a top structural GaAs layer, substrates very similar in function to the SIMOX wafers described above may be generated.

The concentration of Al in the sacrificial layer can be as much as 100%, as single-crystal AlAs may be grown on GaAs and vice-versa. Pure AlAs can be cleanly removed using dilute hydrofluoric acid; dilutions of 1 part 49% HF:10 parts water (measures by volume) are sufficient to yield undercut etches of AlAs, with etch rates in the range 0.1-1 μm/min. A significant problem with pure AlAs, however, is that it reacts with water, causing oxidation of the Al, and the oxidizing process causes the AlAs to swell. Thin GaAs layers on top of layers of AlAs exposed to air are therefore subject to large tensile stress, and tend to crack and fail. This occurs even if the AlAs is exposed for only a few hours to air with relative humidity as low as 10-20%, so that typically substrates with this type of underlayer do not have long lifetimes unless kept in very dry or vacuum conditions.

More typically, therefore, the sacrificial layer is only partially doped with Al. The selective etching of $Al_xGa_{1-x}As$ over GaAs by HF and HCl remains even for concentrations as low as $x \sim 5\%$, so that a heterostructure with a $Al_{0.05}Ga_{0.95}As$ sacrificial layer will allow wet suspension using the same chemicals as above. On the other end of the possible concentrations, the addition of even small amounts of Ga to AlAs stabilizes the Al, so that exposure to humid air is no longer problematic.

A typical heterostructure might therefore consist of a bulk GaAs wafer, a 1 μm thick AlGaAs sacrificial layer, and a 0.1 μm thick GaAs layer, all grown by molecular beam epitaxy. Fabrication of suspended beams using this type of structure then follows a fabrication procedure very similar to that for SIMOX, where electron-beam lithography is used to define a RIE mask for etching through the top structural layer. Alternatively, the top layer may be etched using chlorine-based chemically-assisted ion-beam etching (CAIBE), with a protective mask made from Ni or Ti [36]. The RIE or CAIBE etch can either be stopped after etching through the GaAs top layer, or continued through the sacrificial layer. These processes have a somewhat slower etch rate for AlGaAs than for GaAs, but are not selective enough that the etch will stop at the interface. The underlying sacrificial layer is then removed by a timed wet etch in dilute (10-20%) hydrofluoric acid, or in concentrated hydrochloric acid (bottle strength), as described above.

More complex structures may be fabricated by adding process steps. Metallization can be completed by a lithographic step completed before the sacrificial etch; Au with either a Cr or Ti adhesion layer can be pre-patterned, and if protected by the etch mask during the RIE or CAIBE etch, will not be affected by a HF- or HCl-based wet etch. Capacitive or current-bearing leads can thereby be integrated in the structure fabrication.

GaAs may be doped n-type through the introduction of Si dopants during the growth phase. n-doped GaAs has some useful properties; it forms a temperature- and strain-sensitive piezoresistor, allowing its use as either an integrated thermometer or a strain sensor. A beautiful demonstration of integrated thermal conductance measurements was described by Tighe, Worlock and Roukes [36]. These authors began with a substrate consisting of GaAs, a sacrificial AlAs layer (1 µm thick), a structural layer (300 nm thick) and a n^+-doped layer (150 nm thick). The n^+ layer was patterned into two sub-micron meander patterns using electron beam lithography, and etched by Cl_2 CAIBE, etching through the n^+ layer but stopping on the structural layer. Ohmic contacts to the n^+ layer were made, and leads contacted the n^+ layer using patterned Nb (a superconductor below about 9 K). The Nb leads and n^+ layer were then protected by a etch mask patterned using a third e-beam defined layer, and Cl_2 CAIBE used to etch through the structural layer, forming a block of structural GaAs suspended by GaAs legs. The n^+ meanders were centered on the suspended block, each of which could be used as a heater or as a thermometer. The superconducting Nb leads ran along the suspended legs to electrical contacts on the bulk substrate. This structure was then used to measure the thermal conductance of the legs at temperatures from 0.5-6 K, from which the thermal phonons in this temperature range could be showed to have mean free paths determined by the diameter of the legs, with roughly 30% specular scattering at the boundaries.

11.5 Critical Point Drying

A common problem in the completion of a mechanically suspended structure is that the typical method for releasing a structure uses a wet etch (for instance, hydrofluoric acid for etching the buried oxide layer in a SIMOX wafer, and hydrochloric and hydrofluoric acid for etching the buried AlGaAs layer in GaAs-based heterostructures). The wet etch is rinsed in deionized water, or possibly isopropyl or methyl alcohol, and the sample must then be removed from the solution and dried. The drying process can destroy delicate structures because of the large surface tension forces exerted on the structure as

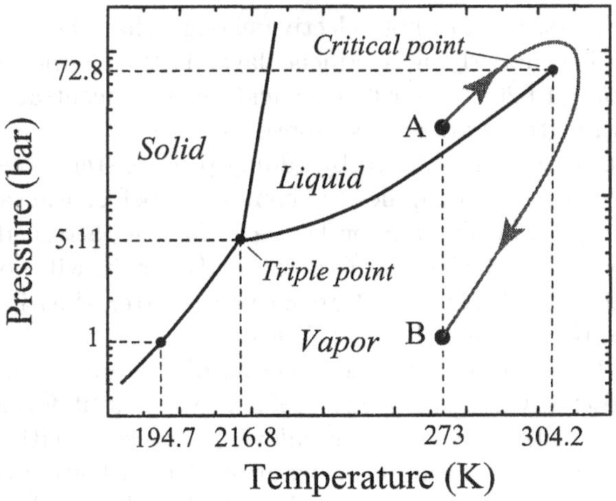

Fig. 11.6. Phase diagram for CO_2, showing the triple and critical points. A sample is dried by submerging it in pressurized CO_2 liquid at room temperature (point *A*), heating and pressurizing the liquid so that it passes around the critical point (*gray line*), and cooling and releasing the pressure to return the sample to room temperature in the vapor phase (point *B*).

the solvent evaporates. One technique, that was primarily developed by biologists studying delicate structures that also needed to be dried, is to use the fact that carbon dioxide (CO_2), in liquid form, is a good solvent for ethanol as well as acetone, and furthermore can be relatively easily pressurized and heated so as to pass around the critical point of the liquid-vapor phase diagram. This allows drying a sample without exposing the delicate parts to the vapor-liquid interface at which the surface tension is generated. In Fig. 11.6 we show the pressure-temperature phase diagram for CO_2, showing the triple point at which the solid, liquid and vapor phases coexist, as well as the critical point where the interface between the vapor and liquid phases vanishes. A path followed during the critical point drying process is also shown.

There are a number of commercial systems built for critical point drying. These consist of a "bomb" into which the sample, submerged in ethyl alcohol, acetone or amyl acetate, is placed. The bomb is then filled with CO_2 liquid, at room temperature, and sealed, at a pressure of 5 bar. The bomb is then heated gently, raising the temperature and pressure of the CO_2, until the critical point has been passed, and the pressure in the bomb then gradually released while slowly allowing the volume to cool. The commercial systems typically include a temperature-controlled water source, with a pump for circulating the heated water through a jacket built into the bomb, and a pressure sensor and thermometer in the volume of the bomb. These systems are quite simple and reliable, and allow extremely delicate samples to be dried without collapse.

A. Mathematical Tools

A.1 Scalars, Vectors, Tensors

We present a quick summary of the mathematics of scalar numbers, vectors, and nth rank tensors, an understanding of which is necessary for the mathematical theory of deformable solids.

Quantitative descriptions of physical systems involve the use of numbers (mass, volume, height) and vectors (position, acceleration, force, magnetic field). On occasion matrices, or tensors, also prove necessary: rotations of coordinate systems or of vectors are most easily performed using the 3×3 rotation matrix, the vector relation between current density $\boldsymbol{J}$ and electric field $\boldsymbol{E}$ in a conductor involves the 3×3 conductivity tensor σ, and angular momentum $\boldsymbol{L}$ and angular velocity $\boldsymbol{\omega}$ are related through the 3×3 moment of inertia tensor I. The description of the behavior of deformable solids, which is a significant part of the discussion of this text, relies heavily on the use of tensors to describe what is going on. We therefore begin with a short discussion of scalars, vectors and tensors, and how they transform under changes of coordinates. A more thorough discussion of these topics can also be found in Arfken [3], or in Morse and Feshbach [4].

The term tensor tends to be somewhat forbidding, as it is not commonly used, and few undergraduate or graduate level physics and engineering courses introduce them or describe their properties. Numbers are actually zero-rank tensors, and vectors are first-rank tensors. Two-dimensional tensors (i.e. ones with two indices) are second-rank tensors, and so on. Tensors are distinguished from the more general class of matrices by the rules that govern their transformations under changes of coordinate systems.

Scalars, or numbers, in general do not change when one changes one's coordinate system, or frame of reference (unless the change is a relativistic one, which we do not discuss here). The mass or volume of an object remains the same if it is turned upside down, or if the laboratory in which it is measured is turned on its side. Scalars are therefore known as *invariant* quantities.

A.1.1 Vectors

Vectors in three-dimensional space are represented by a set of three numbers, so a vector $\boldsymbol{A}$ would be written as (a_1, a_2, a_3) in one coordinate system and

(a'_1, a'_2, a'_3) in another. Vectors and tensors that describe physical quantities usually do not change when the reference frame, or coordinate system, changes. However, the *representation* of the vector or tensor, in other words the set of numbers (a_1, a_2, a_3) that defines its value in a particular coordinate system, does change under a change of reference frame.

We note that some care must be taken with this statement of invariance. Often one locates the position of a point in space with a vector referenced to the origin of the coordinate system. If the origin is changed under a coordinate transformation, then the vector that references that point will be different due to the change in reference. This is not a violation of the statement of invariance: The vector that now defines the location is a different one. Vectors between points that are themselves invariant under the transformation are invariant; in other words, *difference vectors*, or relative position vectors, are invariant. We want to find the mathematical relation between the coordinate representations of a vector $\boldsymbol{A}$, expressed in two different coordinate systems.

To proceed, we first specify the original coordinate system by using a set of three mutually orthogonal, unit length vectors $\hat{\boldsymbol{x}}_i$ $(i = 1 \text{ to } 3)$, and define a new, primed, coordinate system with a similar set of vectors $\hat{\boldsymbol{x}}'_i$ $(i = 1 \text{ to } 3)$. Each of these forms a Cartesian coordinate system; we will not discuss the transformations for cylindrical, spherical or more general curvilinear systems, which can be found in the references [4] and [3].

The relation between the primed and unprimed coordinate axes will in general involve two distinct transformations: A translation of the origin, described by the translation vector $\boldsymbol{X}$ of the origin of the primed axes with respect to the unprimed axes, and a reorientation of the axes, represented by the rotation tensor R. See Fig. A.1.

The rotation tensor R is a 3×3 matrix of numbers, where R_{ij} is the value in the ith row and jth column of the matrix. The values R_{ij} are defined by the angles between the axes in the two reference frames. We form the ij^{th} component R_{ij} of the rotation tensor from the relative angle θ_{ij} between the

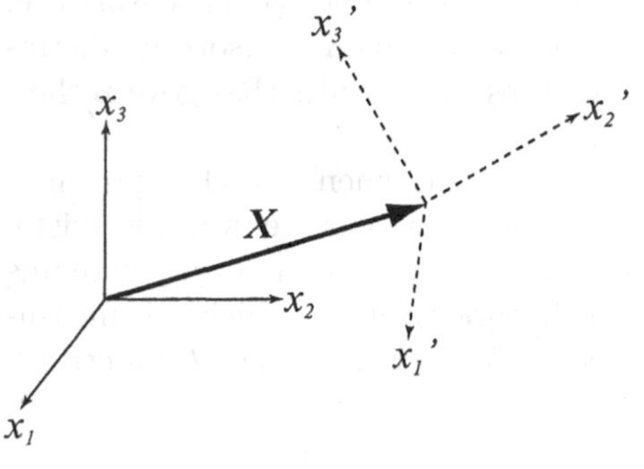

Fig. A.1. Relative displacement and orientation of primed and unprimed coordinate axes.

direction of $\hat{\boldsymbol{x}}_i$ and $\hat{\boldsymbol{x}}'_j$:

$$\begin{aligned} R_{ij} &= \cos(\theta_{ij}) \\ &= \hat{\boldsymbol{x}}_i \cdot \hat{\boldsymbol{x}}'_j, \qquad (i, j = 1 \text{ to } 3). \end{aligned} \tag{A.1}$$

The elements $\cos(\theta_{ij})$ are known as the *direction cosines* of the x'_j axes with respect to the x_i axes. The transpose is formed by interchanging rows for columns, so that $(\mathsf{R}^T)_{ij} = R_{ji}$. An important property of the rotation tensor is that its transpose is its own inverse, i.e. $\mathsf{R} \cdot \mathsf{R}^T = \mathsf{R}^T \cdot \mathsf{R} = \mathbf{1}$, where 1 is the identity tensor, with one's in the diagonal entries and zeros in all the of-diagonal entries. Another way to state this is that to undo a coordinate transformation described by R, one must multiply by the inverse of R, which is R^T.

Any vector $\boldsymbol{A}$, giving for example the displacement between two points fixed in space, can be expressed by its representation, or coordinate expansion, in the two coordinate systems connected by R. This is given by (A.2):

$$\left.\begin{aligned} \boldsymbol{A} &= a_1\hat{\boldsymbol{x}}_1 + a_2\hat{\boldsymbol{x}}_2 + a_3\hat{\boldsymbol{x}}_3 \\ &= a'_1\hat{\boldsymbol{x}}'_1 + a'_2\hat{\boldsymbol{x}}'_2 + a'_3\hat{\boldsymbol{x}}'_3. \end{aligned}\right\} \tag{A.2}$$

Vectors are frequently tabulated in a column format, where one must take care to specify the reference frame in which the components are calculated:

$$\left.\begin{aligned} \boldsymbol{A} &= \begin{bmatrix} a_1 \\ a_2 \\ a_3 \end{bmatrix} \qquad (\textit{unprimed axes}) \\ &= \begin{bmatrix} a'_1 \\ a'_2 \\ a'_3 \end{bmatrix} \qquad (\textit{primed axes}). \end{aligned}\right\} \tag{A.3}$$

The relation between the two coordinate representations of $\boldsymbol{A}$ is given by the following relations:

$$a'_i = \sum_{j=1}^{3} R_{ij}\, a_j, \qquad (i = 1 \text{ to } 3), \tag{A.4}$$

$$a_i = \sum_{j=1}^{3} R_{ji} a'_j. \qquad (i = 1 \text{ to } 3). \tag{A.5}$$

If we abuse the notation somewhat, and write $\boldsymbol{A}$ as the coordinate representation of the vector in the unprimed system, while $\boldsymbol{A}'$ is its representation in the primed system, then we can write the coordinate transformation as

$$\boldsymbol{A}' = \mathsf{R} \cdot \boldsymbol{A}, \tag{A.6}$$

$$\boldsymbol{A} = \mathsf{R}^T \cdot \boldsymbol{A}'. \tag{A.7}$$

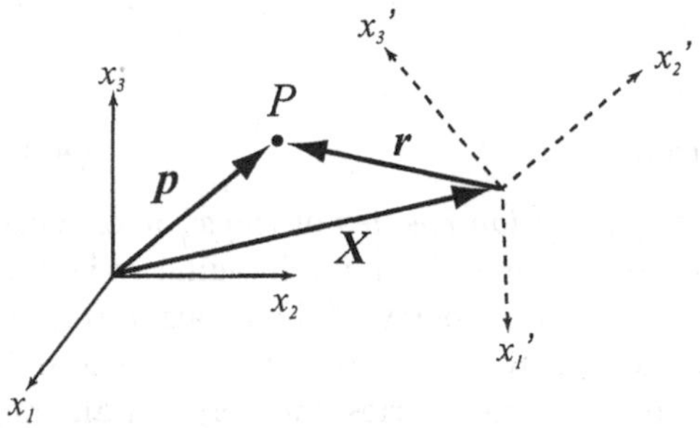

Fig. A.2. The point P defined by two absolute position vectors $\boldsymbol{p}$ and $\boldsymbol{r}$, from two coordinate systems whose origins are displaced by $\boldsymbol{X}$.

A vector will always transform under these relations; the distinction between covariant and contravariant vectors, which some readers may be familiar with, does not appear in the Cartesian coordinate system. We note as a special case that the components a'_i ($i = 1$ to 3) of the representation of the unit vector $\hat{\boldsymbol{x}}_j$ in the primed coordinate system are just the elements of the column R_{ij} ($i = 1$ to 3) of the rotation tensor. Similarly, the components a_i ($i = 1$ to 3) of the representation of the unit vector $\hat{\boldsymbol{x}}'_j$ in the unprimed system are just the elements of the row R_{ji} ($i = 1$ to 3).

As noted above, there are certain instances where the transformation between two coordinate systems is more complex. Vectors that give the absolute position of a point P fixed in space, that is, the position referenced to the origin, are a good example. In the unprimed system this is the vector $\boldsymbol{p}$; in the primed coordinate system, the position is given by the vector $\boldsymbol{r}$, as is shown in Fig. A.2. The vectors $\boldsymbol{p}$ and $\boldsymbol{r}$ are different, so these vectors do not transform according to the relations (A.4) and (A.5). The vectors themselves are related through the change in origin between the primed and unprimed reference frames, $\boldsymbol{p} = \boldsymbol{r} + \boldsymbol{X}$. The transformation of the coordinate representation of these two vectors is then given by

$$p_i = X_i + \sum_{j=1}^{3} R_{ji} r'_j \tag{A.8}$$

$$r'_i = \sum_{j=1}^{3} R_{ij}(p_j - X_j). \tag{A.9}$$

A.1.2 Tensors

We now outline the transformation relations for a two dimensional, second rank tensor. In general, a matrix, which is a tabular arrangement of numbers, is not a tensor; tensors are by definition square, with equal numbers of rows and columns, and must transform according to the relations that we give below. We will almost exclusively deal with 3×3 tensors. By an extension of the relations given by (A.4) and (A.5), we can write the transformation for the coordinate representation of a second-rank, 3×3 tensor T, represented in the original reference frame by the components T_{ij}, and in the new reference frame by the components T'_{ij}:

$$T'_{ij} = \sum_{m=1}^{3}\sum_{n=1}^{3} R_{im}R_{jn}T_{mn} \qquad (i,j = 1 \text{ to } 3), \tag{A.10}$$

$$T_{ij} = \sum_{m=1}^{3}\sum_{n=1}^{3} R_{mi}R_{nj}T'_{mn}. \tag{A.11}$$

Another way to write these transformations is using tensor notation, where by abuse of our notation T is the coordinate representation of the tensor in the unprimed coordinate system, and T' its representation in the primed coordinate system:

$$\left.\begin{aligned} \mathsf{T}' &= \mathsf{R}\cdot\mathsf{T}\cdot\mathsf{R}^T \\ \mathsf{T} &= \mathsf{R}^T\cdot\mathsf{T}\cdot\mathsf{R}. \end{aligned}\right\} \tag{A.12}$$

Note that T and T' in this notation represent the same tensor, but have different numerical values.

Extensions to third-rank tensors, with three indices, and higher-rank tensors are simple extensions of these rules. We emphasize that, just as for vectors, the tensor as an object is invariant under a coordinate transformation, but its *coordinate representation*, or the values of the numbers in the indexed, tabulated form, change according to these rules. The primed and unprimed notation used in (A.12) distinguishes the coordinate representations of what is a single tensor T.

Example A.1: Rotation Tensor. As an example we consider the situation shown in Fig. A.3, where the new (primed) coordinate system has the same origin as the original (unprimed) system, but is rotated by 45° about the original z-axis, and again by 45° about the new x-axis. These correspond to the Euler angle rotations $\theta = \pi/4$ and $\phi = \pi/4$, respectively; see e.g., Marion and Thornton [45]. The rotation tensor for this arrangement can be constructed from the rotation tensors for the two successive rotations. The first rotation tensor has the form

$$\mathsf{R}_\phi = \begin{bmatrix} 1/\sqrt{2} & 1/\sqrt{2} & 0 \\ -1/\sqrt{2} & 1/\sqrt{2} & 0 \\ 0 & 0 & 1 \end{bmatrix}, \tag{A.13}$$

and the second tensor the form

$$\mathsf{R}_\theta = \begin{bmatrix} 1 & 0 & 0 \\ 0 & 1/\sqrt{2} & 1/\sqrt{2} \\ 0 & -1/\sqrt{2} & 1/\sqrt{2} \end{bmatrix}. \tag{A.14}$$

The full rotation is then $\mathsf{R} = \mathsf{R}_\theta \cdot \mathsf{R}_\phi$, and is given by

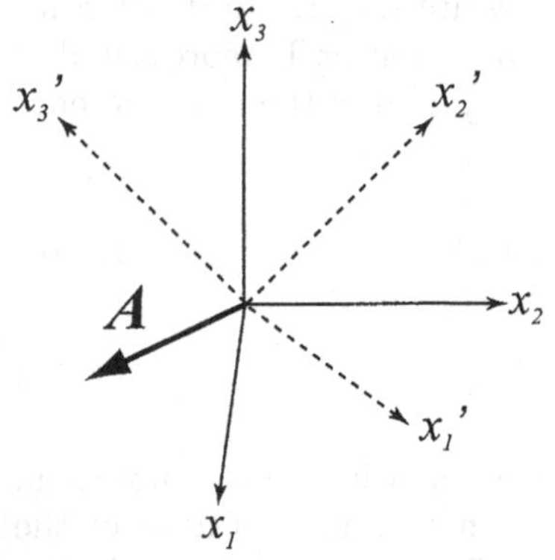

Fig. A.3. Example A.1

$$\mathsf{R} = \begin{bmatrix} 1/\sqrt{2} & 1/\sqrt{2} & 0 \\ -1/2 & 1/2 & 1/\sqrt{2} \\ 1/2 & -1/2 & 1/\sqrt{2} \end{bmatrix} . \tag{A.15}$$

A vector $\boldsymbol{A}$ given by the column vector $[1, -1, 0]^T$ in the unprimed reference frame is then represented by the components $\mathsf{R} \cdot [1, -1, 0]^T = [0, -1, 1]^T$ in the primed frame.

A tensor T, represented in the unprimed system by the values

$$\mathsf{T} = \begin{bmatrix} 1/\sqrt{2} & -1 & 3/2 \\ 1/\sqrt{2} & 1 & -3/2 \\ 0 & \sqrt{2} & 3/\sqrt{2} \end{bmatrix} , \tag{A.16}$$

has the form in the primed system given by

$$\mathsf{T} = \begin{bmatrix} 1 & 0 & 0 \\ 0 & 2 & 0 \\ 0 & 0 & 3 \end{bmatrix} , \tag{A.17}$$

A.2 Eigenvectors and Eigenvalues

A fundamental theorem of linear algebra states that any symmetric tensor may be brought into diagonal form by an orthogonal transformation. Hence, for any symmetric matrix T, with $T_{ij} = T_{ji}$, there exists a rotation R such that

$$\mathsf{T}' = \mathsf{R} \cdot \mathsf{T} \cdot \mathsf{R}^T = \begin{bmatrix} T_1' & 0 & 0 \\ 0 & T_2' & 0 \\ 0 & 0 & T_3' \end{bmatrix} , \tag{A.18}$$

where T_1', T_2' and T_3' are known as the *eigenvalues* or *principal values* of the tensor T. If the elements of T are all real, then the eigenvalues T_i' are also all real.

The rotation tensor R has a geometric meaning: applied to the unprimed coordinate system, it rotates the set of unit vectors $\hat{\boldsymbol{x}}_i$ to a new set of axes $\hat{\boldsymbol{x}}_i'$ which are known as the *principal axes* of the tensor T. For a coordinate system oriented along the tensor's principal axes, the tensor is diagonal. These axes are also known as the *eigenvectors* of the tensor T, with eigenvalues T_i', and satisfy the relations

$$\mathsf{T} \cdot \hat{\boldsymbol{x}}_i' = T_i' \hat{\boldsymbol{x}}_i'. \tag{A.19}$$

The eigenvectors and eigenvalues are found by solving for the eigenvalues T_i' by way of (A.19). We replace T_i' by the variable λ, subtract $\lambda \hat{\boldsymbol{x}}_i'$ from both sides of (A.19), and find the *secular equation*

$$(\mathsf{T} - \lambda \mathsf{1}) \cdot \hat{\boldsymbol{x}}_i' = 0. \tag{A.20}$$

This has a non-trivial solution only if the determinant of the tensor quantity is zero:

$$\det(\mathsf{T} - \lambda \mathsf{1}) = 0. \tag{A.21}$$

This determinant yields an equation known as the *secular equation*, which has the form for a general tensor given by

$$\begin{aligned}\lambda^3 - & (T_{11} + T_{22} + T_{33})\lambda^2 + \\ & (T_{11}T_{22} + T_{22}T_{33} + T_{11}T_{33} - T_{12}^2 - T_{13}^2 - T_{23}^2)\lambda - \\ & (T_{11}T_{22}T_{33} - T_{11}T_{23}^2 - T_{22}T_{13}^2 - T_{33}T_{12}^2 + 2T_{12}T_{13}T_{23}) = 0\end{aligned} \tag{A.22}$$

The three solutions to this cubic equation in λ are guaranteed to be real, and give the three eigenvalues $\lambda = T_1'$, T_2' and T_3'.

The eigenvalues T_i' are independent of the coordinate system, and the form of the cubic equation in λ is also independent: The coefficients in particular do not change if the tensor T is expressed in a different coordinate system. Hence these coefficients are said to be *invariant*, and we define these as Σ_1, Σ_2 and Σ_3, with

$$\left.\begin{aligned}\Sigma_1 &= T_{11} + T_{22} + T_{33}, \\ \Sigma_2 &= T_{11}T_{22} + T_{11}T_{33} + T_{22}T_{33} - T_{12}^2 - T_{13}^2 - T_{23}^2, \\ \Sigma_3 &= T_{11}T_{22}T_{33} - T_{11}T_{23}^2 - T_{22}T_{13}^2 - T_{33}T_{12}^2 + 2T_{12}T_{13}T_{23}.\end{aligned}\right\} \tag{A.23}$$

The eigenvectors $\hat{\boldsymbol{x}}_i'$ that correspond to the three eigenvalues T_i' are found by substituting the eigenvalues into the equation (A.20) and solving the resulting set of *two* independent equations. These equations determine the direction of the eigenvectors, and the lengths are then set to unity. The rotation tensor R, relating the unprimed (original) coordinates x_i to the eigenvector axes x_i', is then found from direction cosines, $\cos\theta_{ij} = \hat{\boldsymbol{x}}_i' \cdot \hat{\boldsymbol{x}}_i$.

In some cases the three eigenvalues T_1', T_2' and T_3' are not all different. If two of the eigenvalues are equal, say $T_1 = T_2$, this reflects a cylindrical symmetry of the tensor. In such a case, the choice of eigenvectors $\hat{\boldsymbol{x}}_1'$ and $\hat{\boldsymbol{x}}_2'$ is not unique: any two unit vectors perpendicular to the third eigenvector $\hat{\boldsymbol{x}}_3$ (which is unique), and chosen to be perpendicular to one another, will work, reflecting the underlying cylindrical symmetry. If all three of the eigenvectors are the same, the tensor is spherically symmetric, and will be diagonal in any coordinate system. In such a case all three eigenvectors may be chosen arbitrarily, as long as they are mutually perpendicular.

A.3 The Dirac Delta Function

The Dirac δ-function is a useful tool, belonging to a class of mathematical functions known as distribution functions. There are equivalent definitions for this function in one, two and three dimensions. Consider, in one dimension, a function $d_w(x)$ defined so that it takes on the value w for the coordinate x in the range $-w/2 < x < w/2$, and zero elsewhere; the integral of $d_w(x)$ over all x is unity. Now take the limit where $w \to 0$, so the width of the function gets narrower, but its amplitude larger in order to maintain an area of unity. We define the delta function $\delta(x)$ as the function in this limit:

$$\delta(x) = \lim_{w \to 0} d_w(x). \tag{A.24}$$

This is sketched in Fig. A.4.

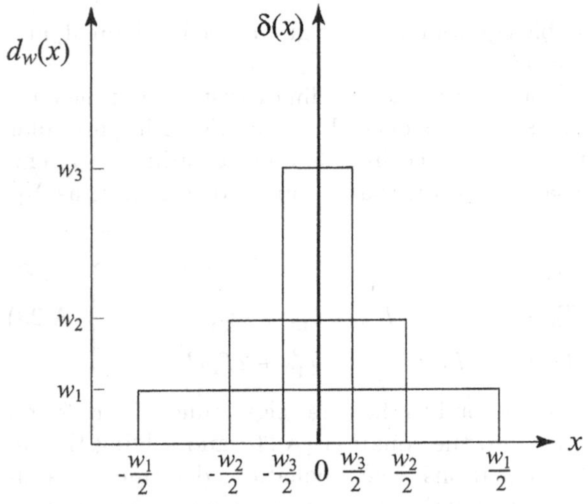

Fig. A.4. Limiting construction $d_w(x)$ for the Dirac δ-function.

The function $\delta(x)$ is thus a function with zero value for $x \neq 0$, and a value at $x = 0$ such that its integral is unity:

$$\left.\begin{array}{rcl} \delta(x) & = & 0, \qquad (x \neq 0) \\ \int_{-\infty}^{\infty} \delta(x)\,\mathrm{d}x & = & 1. \end{array}\right\} \tag{A.25}$$

Some properties of this function are given below:

$$\int_{-\infty}^{\infty} f(x)\,\delta(x-a)\,\mathrm{d}x = f(a), \tag{A.26}$$

$$\delta(\frac{x}{a} - b) = a\delta(x - ab) \tag{A.27}$$

$$\int_{-\infty}^{\infty} f(x)\,\delta'(x-a)\,\mathrm{d}x = -f'(a), \tag{A.28}$$

$$\delta(f(x)) = \sum_i \frac{1}{\left|\frac{\mathrm{d}f}{\mathrm{d}x}(x_i)\right|}\,\delta(x - x_i), \tag{A.29}$$

where in the last relation the x_i are the zeroes of the function $f(x)$.

An integral relation, of use when working with Fourier transforms, is given by

$$\frac{1}{2\pi}\int_{-\infty}^{\infty} \mathrm{e}^{\mathrm{i}k(x-x')}\mathrm{d}k = \delta(x - x'). \tag{A.30}$$

Another useful relation when dealing with Fourier series is

$$\sum_{n=-\infty}^{\infty} \delta(x - na) = \frac{1}{a}\sum_{m=-\infty}^{\infty} \mathrm{e}^{2\pi i m x/a}, \tag{A.31}$$

with the equivalent relation

$$\sum_{m=-\infty}^{\infty} \mathrm{e}^{\mathrm{i}mgx} = \frac{2\pi}{g}\sum_{n=-\infty}^{\infty} \delta(x - n\frac{2\pi}{g}). \tag{A.32}$$

B. Compatibility Relations for Stress and Strain

The strain tensor $\mathsf{S}(\boldsymbol{r})$ is defined through various derivatives of the relative displacement vector $\boldsymbol{u}(\boldsymbol{r})$. In component form, the defining relations are

$$S_{ij} = \frac{1}{2}\left(\frac{\partial u_i}{\partial x_j} + \frac{\partial u_j}{\partial x_i}\right). \tag{B.1}$$

As the nine terms in the strain tensor are defined in terms of only three displacement terms, these cannot be chosen arbitrarily, but must satisfy certain constraints. The constraints are stated through relations involving the second derivatives of the strain tensor, and are usually termed the *compatibility relations*, first derived by Saint–Venant in 1860. These are fully worked out in the text by Sokolnikoff [44]. The relations are given in compact form by (B.2),

$$\frac{\partial^2 S_{ij}}{\partial x_k \partial x_l} + \frac{\partial^2 S_{kl}}{\partial x_i \partial x_j} - \frac{\partial^2 S_{ik}}{\partial x_j \partial x_l} - \frac{\partial^2 S_{jl}}{\partial x_i \partial x_k} = 0. \tag{B.2}$$

The indices i, j, k, and l all run from 1 to 3, generating 81 separate equations. These 81 equations are not all linearly independent, and it turns out that only 6 independent relations are needed, which are given by

$$\left.\begin{aligned}
\frac{\partial^2 S_{11}}{\partial x_2 \partial x_3} &= \frac{\partial}{\partial x_1}\left(-\frac{\partial S_{23}}{\partial x_1} + \frac{\partial S_{31}}{\partial x_2} + \frac{\partial S_{12}}{\partial x_3}\right), \\
\frac{\partial^2 S_{22}}{\partial x_1 \partial x_3} &= \frac{\partial}{\partial x_2}\left(-\frac{\partial S_{13}}{\partial x_2} + \frac{\partial S_{12}}{\partial x_3} + \frac{\partial S_{23}}{\partial x_1}\right), \\
\frac{\partial^2 S_{33}}{\partial x_1 \partial x_2} &= \frac{\partial}{\partial x_3}\left(-\frac{\partial S_{12}}{\partial x_3} + \frac{\partial S_{23}}{\partial x_1} + \frac{\partial S_{13}}{\partial x_2}\right), \\
2\frac{\partial^2 S_{12}}{\partial x_1 \partial x_2} &= \frac{\partial^2 S_{11}}{\partial x_2^2} + \frac{\partial^2 S_{22}}{\partial x_1^2}, \\
2\frac{\partial^2 S_{23}}{\partial x_2 \partial x_3} &= \frac{\partial^2 S_{22}}{\partial x_3^2} + \frac{\partial^2 S_{33}}{\partial x_3^2}, \\
2\frac{\partial^2 S_{13}}{\partial x_1 \partial x_3} &= \frac{\partial^2 S_{33}}{\partial x_1^2} + \frac{\partial^2 S_{11}}{\partial x_3^2}.
\end{aligned}\right\} \tag{B.3}$$

All valid strain tensors S must satisfy these relations.

An equivalent set of relations can be found for the stress tensor T in a linear, elastic material, using the linear stress–strain relations discussed in Chap. 5. For a material that requires the full 6×6 compliance matrix c, the resulting relations are quite complex. For a isotropic material, however, the relations are relatively simple. Taking an isotropic material with Young's modulus E and Poisson's ratio ν, the relation between stress T and strain S is given by (see Chap. 5)

$$S_{ij} = \frac{1+\nu}{E} T_{ij} - \frac{\nu}{E} \left(T_{11} + T_{22} + T_{33}\right) \delta_{ij}, \tag{B.4}$$

where δ_{ij} is the Kronecker delta, $\delta_{ij} = 1$ if $i = j$ and $\delta_{ij} = 0$ if $i \neq j$. We can also write this using the invariant $\Sigma_1 = T_{11} + T_{22} + T_{33}$ (see Sect. 4.5.3).

Inserting the relations (B.4) in the compatibility relations (B.2), we obtain, after some manipulations, the set of equations

$$\begin{aligned} &\frac{\partial^2 T_{ij}}{\partial x_k \partial x_l} + \frac{\partial^2 T_{kl}}{\partial x_i \partial x_j} - \frac{\partial^2 T_{ik}}{\partial x_j \partial x_l} - \frac{\partial^2 T_{jl}}{\partial x_i \partial x_k} \\ &= \frac{\nu}{1+\nu} \left(\delta_{ij} \frac{\partial^2 \Sigma_1}{\partial x_k \partial x_l} + \delta_{kl} \frac{\partial^2 \Sigma_1}{\partial x_i \partial x_j} - \delta_{ik} \frac{\partial^2 \Sigma_1}{\partial x_j \partial x_l} - \delta_{jl} \frac{\partial^2 \Sigma_1}{\partial x_i \partial x_k} \right), \end{aligned} \tag{B.5}$$

with i, j, k and l again running each from 1 to 3. Just as for the strain compatibility relations, most of these are redundant, and can be simplified. In addition, by using the equations for force equilibrium,

$$\frac{\partial T_{ij}}{\partial x_j} + f_i = 0, \tag{B.6}$$

where $\boldsymbol{f}$ is the local body force density (see Chap. 4), the resulting system of equations can be simplified to a set of six relations, the *Beltrami–Michell compatibility equations.* To express these, we use the Laplace operator ∇^2 and the divergence operator ∇:

$$\left.\begin{aligned} \nabla^2 T_{11} + \frac{1}{1+\nu}\frac{\partial^2 \Sigma_1}{\partial x_1^2} &= -\frac{\nu}{1-\nu}\nabla\cdot\boldsymbol{f} - 2\frac{\partial f_1}{\partial x_1}, \\ \nabla^2 T_{22} + \frac{1}{1+\nu}\frac{\partial^2 \Sigma_1}{\partial x_2^2} &= -\frac{\nu}{1-\nu}\nabla\cdot\boldsymbol{f} - 2\frac{\partial f_2}{\partial x_2}, \\ \nabla^2 T_{33} + \frac{1}{1+\nu}\frac{\partial^2 \Sigma_1}{\partial x_3^2} &= -\frac{\nu}{1-\nu}\nabla\cdot\boldsymbol{f} - 2\frac{\partial f_3}{\partial x_3}, \\ \nabla^2 T_{12} + \frac{1}{1+\nu}\frac{\partial^2 \Sigma_1}{\partial x_1 \partial x_2} &= -\left(\frac{\partial f_1}{\partial x_2} + \frac{\partial f_2}{\partial x_1}\right), \\ \nabla^2 T_{13} + \frac{1}{1+\nu}\frac{\partial^2 \Sigma_1}{\partial x_1 \partial x_3} &= -\left(\frac{\partial f_1}{\partial x_3} + \frac{\partial f_3}{\partial x_1}\right), \\ \nabla^2 T_{23} + \frac{1}{1+\nu}\frac{\partial^2 \Sigma_1}{\partial x_2 \partial x_3} &= -\left(\frac{\partial f_2}{\partial x_3} + \frac{\partial f_3}{\partial x_2}\right). \end{aligned}\right\} \tag{B.7}$$

These equations were first derived by Michell in 1900; the equations are simpler in the absence of body forces, and were also derived by Beltrami in 1892. We note that these relations only hold for a isotropic material, but will be approximately correct for a nearly isotropic one.

C. Notation

The various symbols used in the text are listed alphabetically along with their metric units in *Systéme International* (SI), and the conversion and units in CGS. To convert from SI to CGS units, multiply the quantity in SI units by the conversion factor.

Table C.1. Mechanical Quantities.

Description	Symbol	SI units	Conversion factor	CGS units
Angular frequency	ω	Hz	1	Hz
Body force	$\boldsymbol{f}$	N/m^3	0.1	$dyne/cm^3$
Displacement	$\boldsymbol{u}$	m	10^2	cm
Displacement tensor	D	1	1	1
Dilatation	Δ	1	1	1
Dilatational wave velocity	c_ℓ	m/s	10^2	cm/s
Elastic stiffness	c_{ij}	N/m^2	10	$dyne/cm^2$
Elastic compliance	s_{ij}	m^2/N	0.1	$cm^2/dyne$
Energy	E	J	10^7	erg
Force	$\boldsymbol{F}$	N	10^5	dyne
Lagrangian	$\mathcal{L}$	J	10^7	erg
Lamé constants	λ	N/m^2	10	$dyne/cm^2$
	μ	N/m^2	10	$dyne/cm^2$
Mass	m	kg	10^3	g
Mass density	ρ	kg/m^3	10^3	g/cm^3
Poisson's ratio	ν	1	1	1
Polar moment of inertia	I_i	m^4	10^4	cm^4
Position	$\boldsymbol{r}$	m	10^2	cm
Power	P	W	10^7	statW (erg/s)
Rotation tensor	Ω, R	1	1	1
Scalar potential	Φ	m^2	10^4	cm^2
Stress tensor (matrix form)	T	N/m^2	10	$dyne/cm^2$
Stress tensor (six-vector form)	τ	N/m^2	10	$dyne/cm^2$
Strain tensor (matrix form)	S	1	1	1
Strain tensor (six-vector form)	ϵ	1	1	1
Surface force	$\boldsymbol{t}$	N/m^2	10	$dyne/cm^2$
Thermal conductivity	κ	W/m-K	10^5	erg/s-cm-K
Time	t	sec	1	sec
Torque	$\boldsymbol{N}$	N-m	10^7	dyne-cm
Vector potential	$\boldsymbol{H}$	m^2	10^4	cm^2
Wave velocity (transverse and longitudinal)	c_t and C_ℓ	m/s	10^2	cm/s
Wave number	$\boldsymbol{q}$	1/m	10^{-2}	1/cm
Young's modulus	E or Y	N/m^2	10	$dyne/cm^2$

Table C.2. Electromagnetic Quantities.

Description	Symbol	SI units	Conversion factor	CGS units
Electric charge	Q	C (Coulomb)	3×10^9	statcoulomb
Electric displacement	$\boldsymbol{D}$	C/m^2	$12\pi \times 10^5$	statvolt/cm
Electric field	$\boldsymbol{E}$	V/m	10^{-4}	statvolt/cm
Electric potential	V	V (Volt)	1/300	statvolt
Magnetic induction	$\boldsymbol{B}$	T (Tesla)	10^4	gauss
Polarization	$\boldsymbol{P}$	C/m^2	3×10^5	cm^{-3}

References

1. Neil W. Ashcroft and N. David Mermin. *Solid State Physics.* Saunders College, Philadelphia PA, 1976.
2. H. Goldstein, C. Poole, and J. Safko. *Classical Mechanics.* Addison–Wesley, San Francisco, 3rd edition, 2002.
3. Arfken. *Mathematical Methods for Physicists (3rd edition).* Academic Press, Orlando, FL, 1985.
4. P.M. Morse and H. Feshbach. *Methods of Theoretical Physics.* McGraw–Hill, New York, 1953.
5. L.D. Landau, E.M. Lifshitz, and L.P. Pitaevskii. *Statistical Physics (4th edition).* Pergamon Press, Oxford, 1980.
6. I. Martini, D. Eisert, M. Kamp, L. Worschech, Alfred Forchel, and Johannes Koeth. Quantum point contacts fabricated by nanoimprint lithography. *Appl. Phys. Lett.*, 77:2237–2239, 2000.
7. C. Cohen-Tannoudji, B. Diu, and F. Laloë. *Quantum Mechanics.* John Wiley and Sons, New York, 1977.
8. Robert W.G. Wyckoff. *Crystal Structures.* John Wiley and Sons, New York, 2nd edition, 1971.
9. D.C. Champeney. *Fourier Transforms and Their Physical Applications.* Academic Press, New York, 1973.
10. J.M Ziman. *Principles of the Theory of Solids.* Cambridge University Press, Cambridge, 2nd edition, 1972.
11. J.M Ziman. *Electrons and Phonons.* Oxford University Press, Oxford, 1962.
12. M. Brockhouse et al. *Physical Review*, 128:1099, 1962.
13. S. Tamura. Isotope scattering of large-wave-vector phonons in GaAs and InSb: Deformation-dipole and overlap-shell models. *Physical Review B*, 30:849–854, 1984.
14. Karl F. Graff. *Wave Motion in Elastic Solids.* Dover Publications, New York, 1975.
15. B.A. Auld. *Acoustic Fields and Waves in Solids.* Wiley and Sons, New York, 2nd edition, 1990.
16. Leonard I. Schiff. *Quantum Mechanics.* McGraw–Hill, New York, 3rd ed. edition, 1968.
17. S. Tamura, J.A. Shields, and J.P. Wolfe. Lattice dynamics and elastic phonon scattering in silicon. *Physical Review B*, 44:3001–3011, 19891.
18. P.G. Klemens. Decay of high-frequency longitudinal phonons. *J. Appl. Phys.*, 38:4573–4576, 1967.
19. M. Lax, P. Hu, and V. Narayanamurti. Spontaneous phonon decay selection rule: N and U processes. *Physical Review B*, 23:3095–3097, 1981.
20. K. Okubo and S. Tamura. Two-phonon density of states and anharmonic decay of large-wave-vector LA phonons. *Physical Review B*, 28:4847–4850, 1983.

21. S. Tamura. Spontaneous decay rates of LA phonons in quasi-isotropic solids. *Physical Review B*, 31:2574–2577, 1985.
22. S. Tamura. Spontaneous decay of TA phonons. *Physical Review B*, 31:2595–2598, 1985.
23. A.G. Every, W. Sachse, K.Y. Kim, and M.O. Thompson. Phonon focusing and mode-conversion effects in silicon at ultrasonic frequencies. *Physical Review Letters*, 65:1446–1449, 1990.
24. H. Maris and S. Tamura. Anharmonic decay and the propagation of phonons in an isotopically pure crystal at low temperatures: Applications to dark matter detection. *Physical Review B*, 47:727–739, 1993.
25. A.S. Nowick and B.S. Berry. *Anelastic Relaxation in Crystalline Solids*. Academic Press, New York, 1972.
26. Charles Kittel. *Introduction to Solid State Physics (7th edition)*. John Wiley and Sons, New York, 1996.
27. Warren P. Mason. Effect of impurity and phonon processes on ultrasonic attenuation. In Warren P. Mason, editor, *Lattice Dynamics*, number IIIB in Physical Acoustics: Principles and Methods, chapter 6, pages 235–286. Academic Press, New York, 1965.
28. P.G. Klemens. Effect of thermal and phonon processes on ultrasonic attenuation. In Warren P. Mason, editor, *Lattice Dynamics*, number IIIB in Physical Acoustics: Principles and Methods, chapter 5, pages 201–234. Academic Press, New York, 1965.
29. W.P. Mason and T.B. Bateman. *J. Acoust. Soc. Am.*, 36:644, 1964.
30. M.L. Roukes, M.R. Freeman, R.S. Germain, R.C. Richardson, and M.B. Ketchen. *Phys. Rev. Lett.*, 55:422–425, 1985.
31. F.C. Wellstood, C. Urbina, and J. Clarke. Hot-electron effects in metals. *Phys. Rev. B*, 49:5942–5955, 1994.
32. F.C. Wellstood. *Excess Noise in the dc SQUID: 4.2 K to 20 mK*. PhD thesis, University of California, Berkeley, October 1988.
33. T. Klitsner and R.O. Pohl. *Physical Review B*, 34:6045–, 1986.
34. T. Klitsner and R.O. Pohl. *Physical Review B*, 36:6551–, 1987.
35. T. Klitsner, J.E. van Cleve, H.E. Fischer, and R.O. Pohl. Phonon radiative heat transfer and surface scattering. *Physical Review B*, 38:7576–7594, 1988.
36. T.S. Tighe, J.M. Worlock, and M.L. Roukes. Direct thermal conductance measurements on suspended monocrystalline nanostructures. *Applied Physics Letters*, 70:2687–2689, 1997.
37. K.C. Schwab, E.A. Henriksen, J.M. Worlock, and M.L. Roukes. Measurement of the quantum of thermal conductance. *Nature*, 404:974–977, 2000.
38. C.S. Yung, D. R. Schmidt, and A.N. Cleland. Thermal conductance and electron-phonon coupling in mechanically suspended nanostructures. *Appl. Phys. Lett. (*in press*)*, 2002.
39. L.G.C. Rego and G. Kirczenow. Quantized thermal conductance of dielectric quantum wires. *Phys. Rev. Lett.*, 81:232–235, 1998.
40. M.C. Cross and R. Lifshitz. Elastic wave transmission at an abrupt junction in a thin plate, with application to heat transport and vibrations in mesoscopic systems. *cond-mat/0011501 (unpublished)*, 2001.
41. D.H. Santamore and M.C. Cross. Effect of surface roughness on the universal thermal conductance. *Phys. Rev. B*, 63:184306, 2001.
42. R.L. Powell and W.A. Blanpied. *Thermal conductivity of metals and alloys at low temperatures, Circular 556*. National Bureau of Standards, Washington D.C., 1954.
43. Irving H. Shames. *Mechanics of Deformable Solids*. Prentice–Hall, Englewood Cliffs, N.J., 1964.

44. I.S. Sokolnikoff. *Mathematical Theory of Elasticity.* McGraw-Hill, New York, 1956.
45. Jerry B. Marion and Stephen T. Thornton. *Classical Dynamics of Particles and Systems.* Harcourt Brace Publishing, Fort Worth, TX, 1995.
46. R.T. Gunther. *Early Science in Oxford*, volume 8. Oxford University Press, Oxford, 1931.
47. M.B. Viani, T.E. Schaffer, A. Chand, M. Rief, H.E. Gaub, and P.K. Hansma. Small cantilevers for force spectroscopy of single molecules. *Journal of Applied Physics*, 86:2258–2262, 1999.
48. M.B. Viani, T.E. Schaffer, G.T. Paloczi, I. Pietrasanta, B.L. Smith, J.B. Thompson, M. Richter, M. Rief, H.E. Gaub, K.W. Plaxco, A.N. Cleland, H.G. Hansma, and P.K. Hansma. Fast imaging and fast force spectroscopy of single biopolymers with a new atomic force microscope designed for small cantilevers. *Review of Scientific Instruments*, 70:4300–4303, 1999.
49. Warren P. Mason. *Piezoelectric Crystals and their Application to Ultrasonics.* Van Nostrand Co., New York, NY, 1950.
50. K.-H. Hellwege and A.H. Hellwege, editors. *Landolt–Börnstein: Numerical data and functional relationships in science and technology*, volume 11. Springer–Verlag, Berlin, 1979.
51. S. Nakamura and G. Fasol. *The blue laser diode : GaN based light emitters and lasers.* Springer–Verlag, Berlin, 1997.
52. O. Ambacher. Growth and applications of Group-III Nitrides. *J. Phys. D: Applied Physics*, 31:2653–2710, 1998.
53. K.-H. Hellwege and O. Madelung, editors. *Landolt–Börnstein: Numerical data and functional relationships in science and technology*, volume 18. Springer–Verlag, Berlin, 1984.
54. W.N. Sharpe, B. Yuan, R. Vaidyanathan, and R.L. Edwards. New test structures and techniques for measurement of mechanical properties of MEMS materials. *Proceedings of the SPIE - Microlithography and Metrology in Micromachining II*, 2880:78–91, 1996.
55. M.T. Kim. Influence of substrates on the elastic reaction of films for the microindentation tests. *Thin Solid Films*, 283:12–16, 1996.
56. O. Tabata, S. Sugiyama, and M. Takigawa. Control of internal stress and Young's modulus of Si_3N_4 and polycrystalline Si using the ion implantation technique. *Applied Physics Letters*, 56:1314–1316, 1990.
57. V. Ziebart, O. Paul, U. Munch, and H. Baltes. A novel method to measure Poisson's ratio of thin films. *Thin-Films - Stresses and Mechanical Properties VII - Symposium - Boston MA*, pages 27–32, 1998.
58. R. von Mises. On Saint–Venant's principle. *Bulletin of the American Mathematical Society*, 51:555–562, 1945.
59. W.C. Young. *Roarke's Formulas for Stress and Strain, 6th ed.* McGraw–Hill, New York, 1989.
60. J.D. Jackson. *Classical Electrodynamics (3rd edition).* John Wiley and Sons, New York, 1975.
61. A.E.H. Love. *A treatise on the mathematical theory of elasticity.* Dover Publications, New York, 1944.
62. S. Timoshenko, D.H. Young, and Jr. W. Weaver. *Vibration Problems in Engineering.* Wiley and Sons, New York, 1974.
63. J. Wolfe. *Acoustic Phonons.* Van Nostrand, 1998.
64. G.A. Northrop and J.P. Wolfe. Ballistic phonon imaging in germanium. *Comp. Phys. Comm.*, 28:103, 1980.
65. D.C. Hurley and J.P. Wolfe. Phonon focusing in cubic crystals. *Physical Review B*, 32:2568, 1985.

66. J.W.S. Rayleigh. On waves propagated along the plane surface of an elastic solid. *Proc. Lond. Math. Soc.*, 17:4–11, 1887.
67. Abraham I. Beltzer. *Acoustics of Solids.* Springer–Verlag, Berlin, Germany, 1988.
68. F. Reif. *Fundamentals of Statistical and Thermal Physics.* McGraw–Hill, New York, 1965.
69. C. Zener. *Elasticity and Anelasticity of Metals.* University of Chicago Press, Chicago, 1948.
70. T. Suzuki, K. Tsubono, and H. Hirakawa. *Physics Letters A*, 67:10–, 1978.
71. R. Lifshitz and M.L. Roukes. Thermoelastic damping in micro- and nanomechanical systems. *Physical Review B*, 61:5600–5609, 2000.
72. L.D. Landau and E.M. Lifshitz. *Theory of Elasticity (3rd edition).* Pergamon Press, Oxford, 1995.
73. I.N. Sneddon and R. Hill. *Progress in Solid Mechanics*, volume 1. North–Holland, Amsterdam, 1960.
74. A. Akhiezer. *J. Phys. (Moscow)*, 1:277, 1939.
75. H.E. Bömmel and R. Dransfeld. *Physical Review*, 117:1245, 1960.
76. R.O. Woodruff and H. Ehrenreich. *Phys. Rev.*, 123:1553, 1961.
77. A.B. Pippard. *Phil. Mag.*, 46:1104–1114, 1955.
78. J.A. Rayne and C.K. Jones. Ultrasonic attenuation in normal metals and superconductors: Fermi–surface effects. In Warren P. Mason, editor, *Lattice Dynamics*, number VII in Physical Acoustics: Principles and Methods, chapter 3, pages 149–218. Academic Press, New York, 1970.
79. J. Roger Peverey. Ultrasonics and the fermi surfaces of monovalent metals. In Warren P. Mason, editor, *Lattice Dynamics*, number VII in Physical Acoustics: Principles and Methods, chapter 9, pages 353–378. Academic Press, New York, 1970.
80. W.F. Egan. *IEEE Trans. Instr. Meas.*, 37:240–244, 1988.
81. W.P. Robins. *Phase Noise in Signal Sources.* Peter Peregrinus Ltd., London, 1982.
82. D.W. Allan. Statistics of atomic frequency standards. *Proc. IEEE*, 54:221–230, 1966.
83. W.F. Egan. *Frequency Synthesis by Phase Lock.* Wiley and Sons, New York, 1981.
84. Sorab K. Ghandhi. *VLSI Fabrication Principles.* Wiley and Sons, New York, 1994.
85. V. B. Braginsky and F. Y. Khalili. *Quantum Measurement.* Cambridge University Press, Cambridge, 1992.
86. A. N. Cleland and M. L. Roukes. Fabrication of high frequency nanometer scale mechanical resonators from bulk Si crystals. *Appl. Phys. Lett.*, 69:2653–2656, 1996.
87. S. C. Arney and N. C. MacDonald. *J. Vac. Sci. Tech. B*, 6:341, 1988.
88. C. A. Zorman, A. J. Fleischman, A. S. Dewa, M. Mehregany, C. Jacob, and P. Pirouz. *J. Appl. Phys.*, 78:5136, 1995.
89. Y. T. Yang, K. L. Ekinci, X. M. H. Huang, L. M. Schiavone, M. L. Roukes, C. A. Zorman, and M. Mehregany. Monocrystalline silicon carbide nanoelectromechanical systems. *Appl. Phys. Lett.*, 78:162–164, 2001.
90. A. N. Cleland, M. Pophristic, and I. Ferguson. Single-crystal aluminum nitride nanomechanical resonators. *Applied Physics Letters*, 79:2070–2072, 2001.
91. B. Yurke, D.S. Greywall, A.N. Pargellis, and P.A. Busch. Theory of amplifier noise evasion in a oscillator employing a nonlinear resonator. *Phys. Rev. A*, 51:4211–4229, 1995.

92. D.S. Greywall, B. Yurke, P.A. Busch, A.N. Pargellis, and R.A. Willett. Evading amplifier noise in nonlinear oscillators. *Phys. Rev. Lett.*, 72:2992–2995, 1994.
93. K. Wang, A.C. Wong, and C.T. Nguyen. VHF free-free beam high-Q micromechanical resonators. *J. Microelectromechanical Systems*, 9:347–360, 2000.
94. S. Timoshenko and D.H. Young. *Elements of Strength of Materials.* Van Nostrand, Princeton, N.J., 1968.
95. J. R. Clark, W. T. Hsu, and C. T.-C. Nguyen. High-Q VHF micromechanical contour-mode disk resonators. *Technical Digest, IEEE Int. Electron Devices Meeting, San Francisco CA*, pages 399–402, 2000.
96. A.H. Nayfeh and D.T. Mook. *Nonlinear Oscillations.* Wiley Interscience, New York, 1979.
97. D. Rugar and P. Grütter. Mechanical parametric amplification and thermomechanical noise squeezing. *Physical Review Letters*, 67:699–702, 1991.
98. K.L. Turner, S. A. Miller, P. G. Hartwell, N.C. MacDonald, S.H. Strogatz, and S.G. Adams. Five parametric resonances in a microelectromechanical system. *Nature*, 396:149–151, 1998.
99. W.C. Tang, M.G. Lim, and R.T. Howe. Electrostatic comb drive levitation and control method. *J. MicroElectroMechan. Syst.*, 1:170–178, 1992.
100. D. W. Carr, S. Evoy, L. Sekaric, H. G. Craighead, and J. M. Parpia. Parametric amplification in a torsional microresonator. *Applied Physics Letters*, 77:1545–1547, 2000.
101. M. Zalatudinov, A. Olkhovets, A. Zehnder, B. Ilic, D. Czaplewski, H. G. Craighead, and J. M. Parpia. Optically pumped parametric amplification for micromechanical oscillators. *Appl. Phys. Lett.*, 78:3142–3144, 2001.
102. A. N. Cleland and M. L. Roukes. Nanometer scale mechanical electrometry. *Nature*, 320:160–161, 1998.
103. Yu. A. Pashkin, Y. Nakamura, and J. S. Tsai. Room-temperature al single-electron transistor made by electron-beam lithography. *Appl. Phys. Lett.*, 76:2256–2258, 2000.
104. H. Park, A.K.L. Lim, J. Park, A.P. Alivisatos, and P.L. McEuen. Fabrication of metallic electrodes with nanometer separation by electromigration. *Appl. Phys. Lett.*, 75:301–303, 1999.
105. Hermann Grabert and Michel H. Devoret, editors. *Single Charge Tunneling*, volume 294 of *NATO ASI Series.* Plenum Press, New York, 1992.
106. R. Heid, L. Pintschovius, and J.M. Godard. Eigenvectors of internal vibrations of C60: Theory and experiment. *Phys. Rev. B*, 56:5925–5936, 1997.
107. A. Erbe, R.H. Blick, A. Tilke, A. Kriele, and J.P. Kotthaus. *Appl. Phys. Lett.*, 73:3751, 1998.
108. A. Erbe, C. Weiss, W. Zwerger, and R.H. Blick. Nanomechanical resonator shuttling electrons at radio frequencies. *Phys. Rev. Lett.*, 87:096106, 2001.
109. C.P. Slichter. *Principles of Magnetic Resonance.* Springer, New York, 3rd edition, 1989.
110. J.A. Sidles. *Appl. Phys. Lett.*, 58:2854–2856, 1991.
111. J.A. Sidles. *Rev. Sci. Inst.*, 63:3881–3899, 1992.
112. J.A. Sidles, J.L. Garbini, K.J. Bruland, D. Rugar, O. Züger, S. Hoen, and C.S. Yannoni. *Rev. Mod. Phys.*, 67:249–265, 1995.
113. D. Rugar, C.S. Yannoni, and J.A. Sidles. Mechanical detection of magnetic resonance. *Nature*, 360:563–566, 1992.
114. C. Julian Chen. *Introduction to Scanning Tunneling Microscopy.* Oxford University Press, New York, 1993.
115. H.J. Mamin and D Rugar. Sub-attonewton force detection at millikelvin temperatures. *Appl. Phys. Lett.*, 79:3358–3360, 2001.

116. K.J. Bruland, W.M. Dougherty, J.L. Garbini, J.A. Sidles, and S.H. Chao. Force-detected magnetic resonance in a field gradient of 250000 Tesla per meter. *Appl. Phys. Lett.*, 73:3159–3161, 1998.
117. B.C. Stipe, H.J. Mamin, T.D. Stowe, T.W. Kenny, and D. Rugar. Electron spin relaxation near a micron-size ferromagnet. *Phys. Rev. Lett.*, 87:277602–1–4, 2001.
118. B.C. Stipe, H.J. Mamin, T.D. Stowe, T.W. Kenny, and D. Rugar. Magnetic dissipation and fluctuations in individual nanomagnets measured by ultrasensitive cantilever magnetometry. *Phys. Rev. Lett.*, 86:2874–2877, 2001.
119. B.C. Stipe, H.J. Mamin, T.D. Stowe, T.W. Kenny, and D. Rugar. Noncontact friction and force fluctuations between closely spaced bodies. *Phys. Rev. Lett.*, 87:096801–096804, 2001.
120. K. Wago, D. Botkin, C. S. Yannoni, and D. Rugar. Paramagnetic and ferromagnetic resonance imaging with a tip-on-cantilever magnetic resonance force microscope. *Appl. Phys. Lett.*, 72:2757–2759, 1998.
121. K. Wago, D. Botkin, C. S. Yannoni, and D. Rugar. Force-detected electron-spin resonance:adiabatic inversion, nutation, and spin echo. *Phys. Rev. B*, 57:1108–1114, 1998.
122. M.M. Midzor, P.E. Wigen, D. Pelekhov, W. Chen, P.C. Hammel, and M.L. Roukes. Imaging mechanisms of force detected FMR microscopy. *J. Appl. Phys.*, 87:6493–6495, 2000.
123. G.D. Kubiak et al. *J. Vac. Sci Technol. B*, 12:3820–, 1994.
124. F. Cerrina. *Proc. IEEE*, 84:644, 1997.
125. Ludwig Reimer. *Scanning Electron Microscopy.* Springer–Verlag, New York, 1985.
126. I. Zailer, J.E.F. Frost, V. Chabasseur-Molyneux, C.J.B. Ford, and M. Pepper. Cross-linked PMMA as a high resolution negative resist for electron beam lithography and applications for physics of low-dimensional structures. *Semicond. Sci. Technol.*, 11:1235–1238, 1996.
127. H. Namatsu, T. Yamaguchi, M. Nagase, K. Yamasaki, and K. Kurihara. *Microelectron. Eng.*, 41/42:331, 1998.
128. F. C. M. J. M. van Delft, J. P. Weterings, A. K. van Langen-Suurling, and H. Romijn. Hydrogen silsesquioxane/novolak bilayer resist for high aspect ratio nanoscale electron-beam lithography. *J. Vac. Sci. Tech. B*, 18:3419–3423, 2000.
129. J. Fujita, Y. Ohnishi, Y. Ochiai, and S. Matsui. Ultrahigh resolution of calixarene negative resist in electron beam lithography. *Appl. Phys. Lett.*, 68:1297–1299, 1996.
130. J.A. Dagata, J. Schneir, H.H. Harary, C.J. Evans, M.T. Postek, and J. Bennet. *Appl. Phys. Lett.*, 56:2001, 1990.
131. E.S. Snow, P.M. Campbell, and P.J. McArr. *Appl. Phys. Lett.*, 63:749, 1993.
132. E.S. Snow, W.H. Juan, S.W. Pang, and P.M. Campbell. *Appl. Phys. Lett.*, 66:1729, 1995.
133. E.S. Snow, P.M. Campbell, and F.K. Perkins. *Proc. IEEE*, 85:601, 1997.
134. S.W. Park, H.T. Soh, C.F. Quate, and S.I. Park. *Appl. Phys. Lett.*, 67:2415, 1995.
135. E.S. Snow, D. Park, and P.M. Campbell. *Appl. Phys. Lett.*, 69:269, 1996.
136. K. Matsumoto, M. Ishii, K. Segawa, Y. Oka, B.J. Vartanian, and J.S. Harris. *Appl. Phys. Lett.*, 68:34, 1996.
137. M.A. McCord and R.F.W. Pease. *J. Vac. Sc. Technol. B*, 6:293, 1988.
138. E.A. Dobisz and C.R.K. Marian. *Appl. Phys. Lett.*, 58:2526, 1991.
139. K. Wilder, H.T. Soh, A. Atalar, and C.F. Quate. *J. Vac. Sci. Technol. B*, 15:1811, 1997.

140. T.A. Jung, A. Moser, H.J. Hug, D. Brodbeck, R. Hofer, H.R. Hidber, and U.D. Schwarz. *Ultramicroscopy*, 42-44:1446, 1992.
141. S.Y. Chou, P.R. Krauss, and P.J. Renstrom. *Appl. Phys. Lett.*, 67:3114, 1995.
142. S.Y. Chou and P.R. Krauss. *Microelect. Eng.*, 35:237, 1997.
143. Mingtao Li, Jian Wang, Lei Zhuang, and S. Y. Chou. Fabrication of circular optical structures with a 20 nm minimum feature size using nanoimprint lithography. *Appl. Phys. Lett.*, 76:673, 1999.
144. H. Schulz, H.-C. Scheer, and T. Hoffmann et al. New polymer materials for nanoimprinting. *J. Vac. Sci. Tech. B*, 18:1861, 2000.
145. S.R. Quake and A. Scherer. From micro- to nanofabriction with soft materials. *Science*, 290:1536–1540, 2000.
146. M.A. Unger, H.P. Chou, T. Thorsen, A. Scherer, and S.R. Quake. Monolithic microfabricated valves and pumps by multilayer soft lithography. *Science*, 288:113–116, 2000.
147. M.J. Lercel, R.C. Tiberio, P.F. Chapman, H.G. Craighead, C.W. Sheen, A.N. Parikh, and D.L. Allara. *J. Vac. Sci. Technol. B*, 11:2823, 1993.
148. M.M. Leivo, J.P. Pekola, and D.V. Averin. Efficient Peltier refrigeration by a pair of normal metal/insulator/superconductor tunnel junctions. *Appl. Phys. Lett.*, 68:1996–1998, 1996.
149. E.R. Williams and R.S. Muller. Etch rates for micromachining processing. *J. Microelectromech. Sys.*, 5:256–262, 1996.
150. E. Hoffman, B. Warneke, E. Kruglick, J. Weigold, and K.S. Pister. 3D structures with piezoresistive sensors in standard CMOS. *Proc. IEEE Microelectromechanical Systems 1995*, pages 288–293, 1995.
151. H.R. Kaufman. Technology of electron-bombardment ion thrusters. *Adv. Electron. Electron Phys.*, 36:265–273, 1974.
152. P. Tikanyi, D.K. Wagner, A.J. Roza, J.J. Vollmer, C.M. Harding, R.J. Davis, and E.D. Wolf. High-power AlGaAs/GaAs single quantum well lasers with chemically-assisted ion beam etch mirrors. *Applied Physics Letters*, 50:1640, 1987.
153. G. Kaminsky. Micromachining of silicon mechanical structures. *J. Vac. Sci. Technol. B*, 3:1015–1024, 1985.
154. O. Tabata, R. Asahi, H. Funabashi, K. Shimaoka, and S. Sugiyama. Anisotropic etching os silicon in TMAH solutions. *Sensors and Actuators A*, 34:51–57, 1992.
155. A. Hanneborg E. Steinsland, T. Finstad. Etch rates of single-crystal silicon in TMAH. *Sensors and Actuators*, 86:73–80, 2000.
156. M.M. Deshmukh, D.C. Ralph, M. Thomas, and J. Silcox. Nanofabrication using a stencil mask. *Appl. Phys. Lett.*, 75:1631–1634, 1999.
157. S.C. Su. Low-temperature silicon processing techniques for vlsic fabrication. *Solid State Technol.*, page 72, March 1981.
158. T.P. Pearsall, editor. *Quantum Semiconductor Devices and Technologies*. Kluwer Academic Pulishers, Boston MA, 2000.
159. K.E. Bean. Anisotropic etching of silicon. *IEEE Trans. Electron. Dev.*, ED 25:1185–1192, 1978.
160. K.E. Petersen. Silicon as a mechanical material. *Proc. IEEE*, 70:420–457, 1982.
161. N.C. MacDonald. Nanostructure in motion: Microinstruments for moving nanometer scale objects. In G. Timp, editor, *Nanotechnology*, chapter 3, pages 89–159. AIP Press, New York, 1999.
162. N.C. MacDonald. SCREAM microelectromechanical systems. *Microelectron. Eng.*, 32:49–73, 1996.

163. N.C. MacDonald J.J. Yao, S. Arney. Fabrication of high-frequency two-dimensional nanoactuators for scanned probe devices. *J. Microelectromech. Systems*, 1:14–22, 1992.
164. Z.L. Zhang and N.C. MacDonald. Integrated silicon process for microdynamic vacuum field-emission cathodes. *J. Vac. Sci. Tech. B*, 4:2538–2543, 1993.
165. Z.L. Zhang and N.C. MacDonald. Fabrication of submicron high-aspect ratio GaAs resonators. *J. Microelectromech. Sys.*, 2:66–73, 1993.
166. S.S. Cooperman, H.K. Choi, H.H. Sawin, and D.F. Kolesar. Reactive ion etching of GaAs and AlGaAs in a BCl_3-Ar discharge. *J. Vac. Sci. Tech. B*, 7:41–46, 1989.
167. G.J. Sonek and J.M. Ballantyne. Reactive ion etching of GaAs using BCl_3. *J. Vac. Sci. Tech. B*, 2:653–657, 1984.
168. T. Sato, K. Mitsutake, I. Mizushima, and Y. Tsunashima. *Jap. J. Appl. Phys.*, 39:5033, 2000.
169. I. Mizushima, T. Sato, S. Taniguchi, and Y. Tsunashima. Empty space in silicon technique for fabricating a silicon-on-nothing structure. *Appl. Phys. Lett.*, 77:3290–3293, 2000.
170. M.A. Guerra. The status of SIMOX technology. *Solid State Tech.*, November 1990.
171. M. Guerra, B. Cordts, and T. Smick et al. Manfacturing technology for 200 mm SIMOX wafers. *Microelect. Eng.*, 22:351–354, 1993.
172. J. Söderkvist and K. Hjort. The piezoelectric effect of GaAs used for resonators and resonant sensors. *J. Micromech. Microeng.*, 4:28–34, 1994.
173. K. Hjort, J. Söderkvist, and J.A. Schweitz. Gallium Arsenide as a mechanical material. *J. Micromech. Microeng.*, 4:1–13, 1994.
174. B. Hök and K. Hjort. Micromachining in GaAs and related compounds. In H. Seidel, editor, *Advances in Sensors and Actuators*, chapter 9. Elsevier, Amsterdam, 1994.
175. Y. Uenishi, H. Tanaka, and H. Ukita. *IEEE Trans. Electron Dev.*, 41:1778, 1994.

Index